Nanomaterials in Bionanotechnology

Emerging Materials and Technologies

Series Editor:
Boris I. Kharissov

Nanomaterials in Bionanotechnology
Fundamentals and Applications

Edited by
Ravindra Pratap Singh and Kshitij RB Singh

CRC Press
Taylor & Francis Group
Boca Raton London New York

CRC Press is an imprint of the
Taylor & Francis Group, an **informa** business

First edition published 2022
by CRC Press
6000 Broken Sound Parkway NW, Suite 300, Boca Raton, FL 33487-2742

and by CRC Press
2 Park Square, Milton Park, Abingdon, Oxon, OX14 4RN

© 2022 Taylor & Francis Group, LLC

CRC Press is an imprint of Taylor & Francis Group, LLC

Library of Congress Cataloging-in-Publication Data
Names: Pratap Singh, Ravindra, editor. | Singh, Kshitij R. B., editor.
Title: Nanomaterials in bionanotechnology : fundamentals and applications /
edited by Ravindra Pratap Singh and Kshitij RB Singh.
Description: First edition. | Boca Raton, FL : CRC Press, 2021. |
Series: Emerging materials and technologies |
Includes bibliographical references and index. |
Summary: "This book offers a comprehensive view of nanomaterials in biotechnology,
from fundamentals to applications. It explains the basics of nanomaterial properties,
synthesis, biological synthesis, and chemistry and demonstrates how to use
nanomaterials to overcome problems in agricultural, environmental, and
biomedical applications. This work will serve as a reference for industry
professionals, advanced students, and researchers working in the discipline
of bionanotechnology"— Provided by publisher.
Identifiers: LCCN 2021009049 (print) | LCCN 2021009050 (ebook) |
ISBN 9780367689445 (hbk) | ISBN 9780367689469 (pbk) | ISBN 9781003139744 (ebk)
Subjects: LCSH: Nanostructured materials. | Biomedical materials.
Classification: LCC TA418.9.N35 N32885 2021 (print) |
LCC TA418.9.N35 (ebook) | DDC 660.6028/4—dc23
LC record available at https://lccn.loc.gov/2021009049
LC ebook record available at https://lccn.loc.gov/2021009050

ISBN: 978-0-367-68944-5 (hbk)
ISBN: 978-0-367-68946-9 (pbk)
ISBN: 978-1-003-13974-4 (ebk)

Typeset in Times
by codeMantra

Contents

Acknowledgments

It gives us immense pleasure to acknowledge Prof. Shri Prakash Mani Tripathi, Honorable Vice-Chancellor, Indira Gandhi National Tribal University, Amarkantak, India, and Professor Ajaya Kumar Singh, Department of Chemistry, Govt. V. Y. T. PG. Autonomous College, Durg, Chhattisgarh, India, for providing constant assistance in all the possible ways. It is also our great pleasure to acknowledge and express our enormous debt to all the contributors who have provided their quality material to prepare this book. We are grateful to our beloved family members, who joyfully supported and stood with us in many hours of our absence to finish this book project. Thanks are also due to Gabrielle Vernachio, Boris I. Kharissov, Allison Shatkin, and the entire publishing team for their patience and extra care in publishing this book.

Ravindra Pratap Singh
Kshitij RB Singh

Preface

The current need of the time is development in the biomedical, environmental, and agricultural domains; nanotechnology has played a very crucial role in the development of these fields, as nanomaterial-based devices, nanomedicines, nanopesticides, etc. have revolutionized the various domains of science by operating at the nanometer scale (1–100 nm). Nanomaterials are new cutting-edge materials for developing applications in communications, sensing, biosensing, energy storage, data storage, optics, transmission, environmental protection, cosmetics, biology, and medicine due to their unique optical, mechanical, electrical, and magnetic properties. According to the World Health Organization (WHO), 12.6 million people die every year as a result of polluted working conditions/environment such as air, soil, water, exposure to chemicals, change in climate, and UV radiation. Thus, these changes in the environment lead to 100 different types of medical implications, and this also shows that biomedical, agricultural, and environmental domains are interconnected fields, as bad environmental conditions and contaminated agricultural produces can cause health issues in humans, and if the level of toxins causing these issues is high, it can also lead to death. Nanomaterials of the 21st century are advancing the scientific domains and have proved to be novel materials for developing futuristic tools and techniques for therapeutics, diagnostics, bioremediation, pollutant detection, food packaging and processing, etc. Hence, these nanomaterials can be the solution to the problems pertaining to these fields. Nanomaterials and their nanocomposites are also gaining much attention in these fields. The demand is to tolerate high stress, and these nanocomposite materials, due to their properties such as long durability and strength, have the capacity to tolerate high-stress levels. Hence, nanocomposites can also play a crucial role in solving problems faced by humanity.

This book, *Nanomaterials in Bionanotechnology: Fundamentals and Applications*, will be an outstanding collection of current research on nanomaterials and their applications in biomedical, environmental, and agricultural fields, along with their prospects. This book will provide a nanomaterial property, synthesis, and chemistry concept to help students and professionals better understand the nanomaterials and their vast applications. This will also help them to understand and address day-to-day problems related to agriculture, environment, and biomedical and how they can overcome these challenges using nanomaterials. This book will serve as a reference book for professionals, students, scientists, researchers, and academicians in this subject area.

AIMS AND SCOPE

This book, *Nanomaterials in Bionanotechnology: Fundamentals and Applications*, will attract a wide range of readers from all fields. This book in the first few chapters will introduce the nanomaterial concept in detail, and then other remaining chapters will highlight applications of nanomaterials in different domains.

Below we have listed few facets covering major points that will be addressed in this book:

- Nanomaterials and their properties, synthesis, biological synthesis, chemistry, and applications including biomedical, environmental, and agricultural.
- Different types of nanomaterials and their properties.
- Applications of nanomaterials and their composites.
- Nanomaterials for environmental analysis, detection, and monitoring of heavy metals, chemical toxins, and water pollutants.
- Nanomaterials-based biosensors and instrumentation.
- Nanomaterials for diagnosis and therapeutics of diseases.
- Nanomaterials for food processing and packaging.
- Nanomaterials for agricultural waste management.

Editors

Dr. Ravindra Pratap Singh did his B.Sc. from Allahabad University, India and his M.Sc. and Ph.D. in Biochemistry from Lucknow University, India. He is currently working as an Assistant Professor in the Department of Biotechnology, Indira Gandhi National Tribal University, Amarkantak, Madhya Pradesh, India. He has previously worked as a scientist at various esteemed laboratories globally, such as Sogang University, South Korea and IGR, Paris. His work and research interests include biochemistry, biosensors, nanobiotechnology, electrochemistry, material sciences, and applications of biosensors in biomedical, environment, agriculture, and forensics. He has to his credit several reputed national and international honors and awards. Dr. Singh authored over 30 articles in international peer-reviewed journals and more than 20 book chapters of international repute, and he serves as a reviewer of many reputed international journals and is also a member of many international societies. He is currently also involved in editing various books that will be published in internationally reputed publication houses, namely IOP Publishing, CRC Press, Elsevier, and Springer Nature. Moreover, he is book series editor of "Emerging advances in Bionanotechnology," CRC Press, Taylor and Francis Group and Guest managing editor of Materials Letters, Elsevier, to edit a special issue on "Special Issue on Smart and Intelligent Nanobiosensors: Multidimensional applications".

Mr. Kshitij RB Singh is a postgraduate in Biotechnology from Indira Gandhi National Tribal University, Amarkantak, Madhya Pradesh, India. He is currently working in the laboratory of Dr. Ajaya Singh, Department of Chemistry, Government V.Y.T. PG. Autonomous College Durg, Chhattisgarh, India. He has many publications to his credit and has authored more than ten book chapters published in the internationally reputed presses, namely Elsevier, Springer Nature, and CRC Press. He is currently also involved in the editing of books with international publishing houses, including CRC Press, IOP Publishing, Elsevier, and Springer Nature. His research interest is in biotechnology, biochemistry, epidemiology, nanotechnology, nanobiotechnology, biosensors, and materials sciences.

Contributors

Juliana Bunmi Adetunji
Department of Biochemistry Sciences
Osun State University
Osogbo, Nigeria

Charles Oluwaseun Adetunji
Department of Microbiology
Edo State University Uzairue
Edo State, Nigeria

Gözde Koşarsoy Ağçeli
Department of Biology
Hacettepe University
Ankara, Turkey

Osikemekha Anthony Anani
Department of Biological Science
Edo State University Uzairue
Edo State, Nigeria

A. Avila
Microelectronics Center (CMUA),
    Department of Electrical and
    Electronics Engineering
Los Andes University
Bogotá, Colombia

Najla Bentrad
Department of Biology and Physiology
    of Organisms
University of Sciences and Technology
    Houari Boumediene (USTHB)
Bab Ezzouar, Algeria

Ruth Ebunoluwa Bodunrinde
Department of Microbiology
Federal University of Technology Akure
Akure, Nigeria

Pankaj Kumar Chauhan
Faculty of Applied Sciences and
    Biotechnology
Shoolini University
Solan, India

Parveen Chauhan
Faculty of Applied Sciences and
    Biotechnology
Shoolini University
Solan, India

Hye Kyu Choi
Department of Chemical and
    Biomolecular Engineering
Sogang University
Seoul, South Korea

Jeong-Woo Choi
Department of Chemical and
    Biomolecular Engineering
Sogang University
Seoul, South Korea

Jin-Ha Choi
Department of Chemical and
    Biomolecular Engineering
Sogang University
Seoul, South Korea

V. Dhinakaran
Department of Mechanical Engineering
Chennai Institute of Technology
Kundrathur, India

Kanika Dulta
Faculty of Applied Sciences and
    Biotechnology
Shoolini University
Solan, India

Asma Hamida-Ferhat
Department of Pharmacy
University of Algiers
Bab Ezzouar, Algeria

Abel Inobeme
Department of Chemistry
Edo State University Uzairue
Edo State, Nigeria

Karthik Kannan
Center for Advanced Materials
Qatar University
Qatar

Yokraj Katre
Department of Chemistry
Kalyan PG College
Bhilai, India

Rameshroo Kenwat
Department of Pharmacy
Indira Gandhi National Tribal
    University
Amarkantak, India

L. Sivarama Krishna
Faculty of Ocean Engineering
    Technology and Informatics
Universiti Malaysia Terengganu
Kuala Nerus, Malaysia

Pramod Kumar
Institute of Lung Biology, Helmholtz
    Zentrum Munich, Germany

Sejon Lee
Department of Semiconductor Science
Dongguk University
Seoul, South Korea

G.A. Lanza
Microelectronics Center (CMUA),
    Department of Electrical and
    Electronics Engineering
Los Andes University
Bogotá, Colombia

B.D. Malhotra
Department of Biotechnology
Delhi Technological University
Rohini, India

Eneyew Tadesse Melaku
Food Science and Applied Nutrition
Addis Ababa Science and Technology
    University
Addis Ababa, Ethiopia

Elaine Gabutin Mission
Instituto Bioeconomia
University of Valladolid
Valladolid, Spain

Syed Muzammil Munawar
Department of Chemistry &
    Biochemistry
C. Abdul Hakeem College
Melvisharam, India

E. Kayalvizhi Nangai
Department of Physics
K. Ramakrishnan College of
    Technology
Samayapuram, India

Vanya Nayak
Department of Biotechnology
Indira Gandhi National Tribal
    University
Amarkantak, India

Ankur Ojha
Food Science and Technology
NIFTEM
Sonipat, India

Olugbemi T. Olaniyan
Department of Physiology
Edo State University Uzairue
Edo State, Nigeria

Frances. N. Olisaka
Department of Biological Sciences
  Faculty of Science
Benson Idahosa University
Benin City, Nigeria

Avinash C. Pandey
Nanotechnology Application Centre
University of Allahabad
Prayagraj, India
  & Inter University Accelerator Centre
New Delhi, India

Wadzani Dauda Palnam
Department of Agronomy
Federal University
Gashua, Nigeria

Rishi Paliwal
Department of Pharmacy
Indira Gandhi National Tribal
  University
Amarkantak, India

Shivani Rai Paliwal
SLT Institute of Pharmaceutical
  Sciences\
Guru Ghasidas Vishwavidyalaya
Bilaspur, India

J.A. Perez-Taborda
Microelectronics Center (CMUA),
  Department of Electrical and
  Electronics Engineering
Los Andes University
Bogotá, Colombia

M. Velayutham Pillai
Kalasalingam Academy of Research
  and Education
Srivilliputhur, India

Devi Radhika
Department of Chemistry
Jain-Deemed to be University
Bengaluru, India

Shweta Rathee
Food Science and Technology
NIFTEM
Sonipat, India

Khaleel Basha Sabjan
Department of Chemistry &
  Biochemistry
C. Abdul Hakeem College
Melvisharam, India

S. Sankar
Department of Semiconductor Science
Dongguk University
Seoul, South Korea

S. Saravanan
Department of Mechanical Engineering
K. Ramakrishnan College of
  Technology
Samayapuram, India

Sreeja K. Satheesh
Atonarp Microsystems
Halasuru, India

Otmar Schmid
Institute of Lung Biology, Helmholtz
  Zentrum Munich, Germany

Minkyu Shin
Department of Chemical and
  Biomolecular Engineering
Sogang University
Seoul, South Korea

Ajaya Kumar Singh
Department of Chemistry
Govt. V.Y.T. PG. Autonomous College
Durg, India

Anurag Singh
Food Science and Technology
NIFTEM
Sonipat, India

Kshitij RB Singh
Department of Chemistry
Govt. V. Y. T. PG. Autonomous College
Durg, India

Ravindra Pratap Singh
Department of Biotechnology
Indira Gandhi National Tribal
    University
Amarkantak, India

Srishti Singh
School of Biotechnology and
    Bioinformatics
D Y Patil Deemed to be University
Navi Mumbai, India

Pratima R. Solanki
Special Centre for Nanoscience (SCNS)
Jawaharlal Nehru University
New Delhi, India

S. Sreevidya
Department of Chemistry
Kalyan PG College
Bhilai, India

Kirtana Sankara Subramanian
Department of Food Science
University of Melbourne
Melbourne, Australia

C. M. Naga Sudha
Department of Computer Technology
Anna University MIT Campus
Chennai, India

R. Suriyaprabha
School of Nanoscience
Central University of Gujarat
Gandhinagar, India

Jinho Yoon
Department of Chemical and
    Biomolecular Engineering
Sogang University
Seoul, South Korea & Department of
    Chemistry and Chemical Biology
Rutgers, The State University of New
    Jersey
Piscataway, New Jersey

1 Introduction to Nanomaterials
An Overview toward Broad-Spectrum Applications

Kshitij RB Singh
Govt. V. Y. T. PG. Autonomous College

Pratima R. Solanki
Jawaharlal Nehru University

B.D. Malhotra
Delhi Technological University

Avinash C. Pandey
University of Allahabad and Inter
University Accelerator Centre

Ravindra Pratap Singh
Indira Gandhi National Tribal University

CONTENTS

1.1   INTRODUCTION

Nanomaterials are utilized to develop optoelectronic devices, electronic devices, bio-sensors, nanodevices, and solar cells, due to their unique properties compared to their bulk forms such as miniaturized size, insulating nature, elasticity, electrical conductivity, mechanical strength, and high reactivity. They are classified based on geometry, morphology, composition, uniformity, and agglomeration. Geometrically, they are 0D, 1D, 2D, or 3D; morphologically, they are spherical, flat, needle, or random orientations (e.g., nanotubes and nanowires) with various shapes such as helices, belts, zigzags, with high aspect ratio and spherical, oval, cubic, and pillar, with low aspect ratio, and occurs in powder, colloidal, and suspension forms. Based on composition, nanomaterials are either a single-constituent material or a composite of several materials such as polymers, metals, ceramics, and alloys. Engineered nanomaterials can be synthesized by gas-phase processes, mechanical processes, vapor deposition synthesis, coprecipitation, etc., in an agglomerated state or dispersed uniformly in a matrix to their chemistry and electrostatic properties. Clusters or agglomerates are possible due to their surface energy, and it can be avoided with the proper chemical treatment to become uniform. Based on nanomaterial composition, they are classified as monometallic, bimetallic, trimetallic, metal oxide, magnetic, hybrid, semiconductor, composite, etc. Further, many nanomaterials can be classified based on their orientations, morphologies, and characteristics, including quantum dots (QDs), nanowires, nanotubes, nanofibers, nanofluids, nanobelts, nanoribbons, nanocapsules, nanosprings, nanosheets, and nanocomposites [1–3].

Nanomaterials are widely used nowadays for various applications in various domains such as biomedical, environmental, and agricultural. It was recently reported that gold nanomaterials that exhibit unique optical properties could be utilized in colorimetric sensors for a wide range of applications [4]. Further, Zhou et al. (2020) reported carbon nanofibers utilizing biomedicine, environmental science, energy storage, and materials science. These functional nanomaterials are being used as environmental adsorbents, supercapacitors, batteries, fuel cells, solar cells, sensors, biosensors, antibacterial materials, tissue engineering, and sharp memory materials [5]. Moreover, recently bioactive proteins and nanomaterial-based advanced version of biosensors were fabricated for the point-of-care diagnosis, environmental monitoring, and food safety [6]. Further, it is well established that nanomaterials (nano-ferric oxide, carbon nanotubes (CNTs), graphene oxide (GO), fly ash, and steel fibers), when utilized in the concrete, result in improved durability and sustainability by enhancing mechanical features, which offer them a variety of applications in the field of mechanical engineering [7].

This chapter of the book entitled "Nanomaterials in Bionanotechnology: Fundamentals and Applications" highlights the basic utility of the nanomaterials in the biomedical, environmental, and agricultural domains along with the current trends and prospects. Further, this book will elaborate on nanomaterial classification, synthesis, and properties in **Chapter 2**, followed by **Chapter 3** about biomaterials' biological synthesis. Furthermore, the chemical aspects in the fabrication of various technologies based on nanomaterial are discussed in **Chapter 4**. **Chapter 5** deals with the emerging nanocomposites and their multifunctional applications, followed by **Chapter 6**, which deals with the current nanomaterial scenario in the environmental, agricultural, and biomedical domains. **Chapter 7** of this book deals with nanomaterials for environmental hazard analysis, monitoring, and removal, and **Chapter 8** deals with the recent agriculture trends based on nanomaterials. Further, **Chapter 9** and **Chapter 10** deal with the role of nanomaterials in the food sector, **Chapter 11** and **Chapter 12** deal with the role of nanomaterials in disease diagnosis, and the last chapter of this book, **Chapter 13**, deals with the potentialities of nanomaterials as nanomedicine for the treatment of various types of diseases.

1.2 BIOMEDICAL APPLICATIONS

Nanomaterials have potential biomedical applications (Figure 1.1) using either inorganic or organic–inorganic materials with unique properties such as physicochemical, optical, magnetic, and stimuli-responsive at the 1- to 100-nm scale for drug delivery, targeted drug delivery, gene delivery, bioimaging, biosensors, cell labeling, and photoablation therapy [8–11]. Dai et al. (2020) reported novel methods based on nanomaterials for the early diagnosis of hepatocellular carcinoma, a liver cancer known worldwide, as its early detection and treatment can improve patients' lives [12]. Koo et al. (2020) reported magnetic nanomaterial-based electrochemical biosensors for selective detection of cancer biomarkers associated with cell surface proteins of tumor cells and their nucleic acids [13]. Nuclear medicine imaging is a diagnostic approach for cancers to detect gamma rays. Recently, Ge et al. (2020) reported nanomaterial-based radioactive tracers for early and accurate diagnosis of cancers [14]. In modern biomedical sciences, exosomal cancer biomarkers are known for the early diagnosis and treatment of cancer. Shao and Xiao (2020) reported nanomaterial-based optical biosensors to detect exosomal cancer biomarkers [15]. Khan et al. (2020) reported the nanozyme, next-generation biomaterials in biomedical and industrial biosensing and therapeutic activities [16]. Furthermore, Zhang et al. (2020) reported the next-generation artificial enzymes used in detection, diagnosis, and therapy [17]. Li et al. (2020) reported nanozyme-based composite materials, an intelligent and multifunctional therapeutic agent for preventing or resisting bacterial biofilms [18]. Yang et al. (2020) reported details about bacterial infection-related diseases that are causing health problems globally as the use of antibiotics is causing antibiotic resistance and for overcoming these problems, nanozymes, inorganic nanostructures with enzymatic activities, have shown great potential owing to its

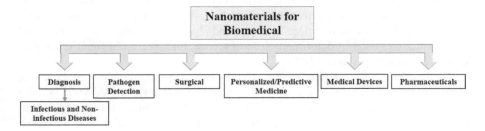

FIGURE 1.1 Broad-spectrum applications of nanomaterials in the biomedical domain.

excellent antimicrobial properties and negligible biotoxicities [19]. Johnson et al. (2020) reported graphene nanoribbons, strips of single-layer graphene, amphiphilic in nature, and having unique properties and high surface area, can be used in biomedical applications such as in gene therapy, drug delivery, antimicrobial therapy, anticancer therapy, photothermal therapy, bioimaging, and sensing [20]. Li et al. (2020) reported selenium-containing nanomaterials, which exhibit good biocompatibility and can be used as redox stimuli, drug delivery carriers, and anticancer drugs for the treatment of cancer, as it is the global health problem and causes economic burden worldwide [21]. Further, Joshi et al. (2020) reported nanomaterial-based electrochemical sensing techniques to detect antibiotic residues in environmental media, human fluids, and food and beverage samples. To some extent, it was an effort to resolve this problem, as antibiotic is an essential medicine, but uncontrolled use and disposal cause serious environmental and health concerns [22]. Apart from the above, Wang et al. (2020) reported titanium-based nanomaterials for photocatalysis, electronics, energy, and biomedicine/biomedical, as previously the Ti-based nanomaterials were used in photothermal, photodynamic, and sonodynamic therapy and drug delivery systems [23].

Furthermore, nanomaterials based on phosphorus have unique physicochemical, optical, and biological properties and are widely used in biology and nanomedicine [24]. Nanomaterials can be used to improve various traditional detection methods such as ELISA (enzyme-linked immunosorbent assay), and improved ELISA-based nanomaterials have immense applications for environmental monitoring, clinical diagnosis, and food quality control. This improved ELISA is based on colorimetric, fluorescent, electrochemical, photothermal, and Raman scattering using nanomaterials as signal reporters, enzyme mimics, and loading carriers [25]. The biologically synthesized nanomaterials from the green route are widely utilized in bio-labeling, bioimaging, pharmaceutical, environmental monitoring, and food packaging [26]. Photothermal therapy is a noninvasive therapeutic technique using nanomaterials for drug-resistant bacteria and bacterial biofilms. Thus, Chen et al. (2020) reported the photothermal therapy-involved multifunctional theranostic nanoplatforms [27]. Ehtesabi (2020) has reported that saliva is a bio-fluid with different biomarkers for the detection by biosensing technology and carbon nanomaterial (CNTs, carbon dots, graphene, GO, reduced GO (rGO), and graphitic carbon nitride)-based biosensors have a potential role to be utilized for diagnosis, treatment, formulating, monitoring, and managing patients with different abnormalities such as glucose, amino acids, hormones, cancer antigens, cancer biomarkers, viruses, bacteria, sialic acid, uric acid, and dopamine [28]. Further, it has been reported that nature has natural nanomaterials with different functionality, size, and shape, for example, metal NPs and QDs, which are used in colorimetric, fluorescence, and electrochemical sensing to detect the analyte of interest [29].

Nanomaterials have numerous electronics, food, and biomedicine applications, resulting in human exposure and biological and toxic effects. Wu et al. (2020) reported mitochondrial toxicity of nanomaterials, which change mitochondrial morphology and cytochrome C release, alter the membrane potential, and overcome these drawbacks, and engineered nanomaterials are being synthesized, such as silica nanoparticles, magnetic nanoparticles, QDs, dendrimers, polymeric nanoparticles, nanofibers, graphene, CNTs, liposomes, and micelles [30,31]. Aggregation of proteins causes health disorders, for example, amyloid fibrils in the brain, floaters in the vitreous of the eye, and clots in the blood. Sauvage et al. (2020) reported nanomaterials to avoid and destroy protein aggregates due to nanomaterials' stimuli-responsive features against harmful protein aggregates [32]. Further, Zeng and Chen (2020) reported surface modifications of black phosphorus-based nanomaterials (2D) to achieve safe therapeutic effects that offers them multifunctional applications in the field of biomedicine [33]. Biomedical engineering is in demand to utilize highly functionalized biomaterials. Thus, Park et al. (2020) reported smart or functional organic/inorganic hybrid biomaterials as nanoparticles and nanocomposites in the fields of medical applications for diagnosing and treating various human diseases [34]. Liu et al. (2020) reported treatment of myocardial infarction using conductive nanomaterials by cardiac tissue engineering [35]. Ehtesabi (2020) reported a carbon nanomaterial-based biosensor that could be utilized to detect various human pathogenic viruses (human immunodeficiency virus (HIV), influenza virus, Zika virus, hepatitis virus, dengue virus, Ebola virus, and adenovirus) as they are the main cause of illness and death globally [36]. Nanomaterials play an immense role in cancer treatment by chemotherapy and radiation to manage cancer. Nanomaterials were recently used as nanocarriers loaded with chemotherapeutic agents for effective therapeutic efficacy, as it is based on nanotechnological applications, and due to this, it has very few off-target effects and does not harm the body. Figure 1.2 depicts the role of nanotechnology in bioimaging and how it is safer than other traditional techniques. Hence, theragnostic therapy and imaging based on nanomaterials have sound pharmacokinetics and pharmacodynamics, which has offered them potentialities as an effective carcinogenic therapy agent [37].

The spread of infectious diseases such as severe acute respiratory syndrome coronavirus 2 (SARS-CoV-2) is a serious threat to human health. Thus, there is an urgent

FIGURE 1.2 Diagrammatic representation of nanomaterials as an agent with fewer side effects in bioimaging of cancer disease diagnosis.

need for a promising technique for detecting these deadly infectious disease pathogens without sophisticated tools/techniques; for this, paper-based analytical device and even colorimetric strategies are promising as the eye can easily interpret them, and it will be a kind of point-of-care testing device. Nguyen and Kim (2020) developed a colorimetric pathogen detection kit using nanomaterial-mediated paper-based biosensors [38]. Further, it was reported that in Alzheimer's disease (a neurodegenerative disorder), nanomaterials as nanocarriers (CNTs, QDs, magnetic nanoparticles, multifunctional liposomes, polymeric nanocapsules, and nanoemulsions) loaded with bioactive molecules are very much helpful in treating this disorder. In addition to this, nanomaterials such as Ag, Au, graphene, CNTs, magnetic nanoparticles, polymers, and QDs are utilized in the fabrication of electrochemical and optical nanobiosensors to identify Alzheimer's disease biomarkers to increase early diagnosis of these types of disorders [39]. Nanomaterials can be very useful in autoimmune disease prevention, diagnosis, and treatment by multi-active probes such as drugs, biomolecules, and target ligands. Chai et al. (2020) reported nanomaterial-based immunotherapeutic strategies and autoimmune pathogenesis [40].

Further, Mao et al. (2020) have reported a new electrochemical aptasensor-based technology to monitor and evaluate community-wide illicit drug use, as these drugs are crime-oriented drugs frequently used in our society and monitoring of such drugs is very important for combating the crimes [41]. Apart from this, Jung et al. (2021) reported nanomaterials used in biomedical applications using surface engineering of nanostructured porous silicon materials in bioimaging, biosensors, drug delivery systems, diagnosis, and therapy [42]. In our contemporary society, infectious diseases cannot be cured properly due to drug-resistant bacterial strains. Ogunsona et al. (2020) reported the effective treatment regimen using nanoparticles as an antimicrobial agent to kill pathogens or limit microbial growth before human infection. The engineered antimicrobial materials are useful in antimicrobial applications and highly effective in water treatment, healthcare, alternative food packaging, and antifouling coatings [43]. Sahu et al. (2020) reported the utilization of nanomaterial-based delivery of O_2 to remove endogenous H_2O_2 at the tumor site for reversing hypoxia. Hypoxia is an imbalance between oxygen supply and consumption for cancer growth and metastasis, an obstacle to cancer therapy [44]. Furthermore, Wang et al. (2020) reported Prussian blue as an antidote to treat thallium poisoning as Prussian blue-based nanomaterials are widely used for cancer treatment, drug delivery, and molecular imaging, and they are future candidates for biomedical applications [45]. Ahmadian et al. (2020) reported nanomaterial-based drug resistance screening techniques, as its need of current global scenario due to the global healthcare crisis, which decreases drug efficacy and develops toxicities [46]. Nowadays, a wide range of nanomaterials are widely used as an immunomodulator for immunomodulation in vaccination or cancer immunotherapy [47]. Sarkar et al. (2020) reported lysozymes, which have potential biomedical applications, and when these lysozymes are incorporated with proteins to form nanocomposite material, they exhibit multifaceted applications [48]. Nanocarriers such as dendrimers, lipids, inorganic nanoparticles, natural polymers, and nanoemulsion-loaded lactoferrin (a glycoprotein) have an enormous role as diagnostic and therapeutic agents [49]. Furthermore, gold nanoparticles are widely used in both the biological and chemical fields for catalysis, biological sensing, diagnosis, therapy, and imaging. AuNPs have shown colloidal stability, which offers them potentialities for cancer imaging/diagnosis to treat cancer properly [50]. Moreover, the biologically synthesized nanoparticles using a plant containing phytochemicals as bio-reductant for nanoparticle synthesis are potent antimicrobial agents and therapeutic agents for various cancer treatments [51].

Bionanotechnology deals with biological systems for biological applications. Shah et al. (2020) reported stimuli-responsive bionanomaterials for expansion/contraction, activation, and self-assembly [52]. Moreover, Zeng et al. (2020) reported gadolinium-based nanoparticles for magnetic resonance imaging and computed tomography imaging. In the same study, gadolinium hybrid-based nanocomposites have also been used in multimodal imaging loaded with anticancer drugs, nucleic acids, and photosensitizers to eliminate tumor cells with reduced toxicity [53]. Hussain et al. (2020) reported nano-scaled materials, which have impacted medicine and health by using various targeted drug delivery devices to diagnose and treat various neurodegenerative disorders [54]. Nanomaterial-based orthopedic implants have also gained a lot of attention by using bioimplant engineering for organ functionality, as nanomaterials help in the growth of different tissues by adhesion, proliferation, differentiation, and migration with stimuli-responsive behavior [55]. Farzin et al. (2020) reported biosensors for early detection of HIV infection using nanomaterials to determine HIV gene, CD^{4+} cells, p24, p17, and HIV-1 [56]. Furthermore, Farzin et al. (2020) reported fluorescent QD-based electrochemical biosensors to determine tumor markers, depression markers, and inflammatory biomarkers [57]. Figueroa et al. (2020) reported viruses as nanomaterials for biomedical applications [58]. Carbon monoxide shows multifaceted cellular function regulation and is useful for cancer treatment. Zhou et al. (2020) reported carbon monoxide-releasing nanoplatforms for cancer therapy [59]. Sharifi et al. (2020) reported nanobiosensor devices for detecting cancer biomarkers and for breast cancer diagnosis [60]. Nanotheranostics is an emerging diagnostic and therapeutic tool. Siafaka et al. (2020) reported gold, silver, polymeric, carbon-based, and liposomal nanoparticles as multifunctional nanotheranostic agents [61]. Dhara and Mahapatra (2020) reported that nanomaterial-based microfluidic nanobiosensors have potent detection sensitivity and selectivity for detecting cardiovascular disease biomarkers. Cardiovascular disease is the root cause of human mortality associated with these risk factors. Early disease diagnosis and management are very important; for diagnosing this disease, C-reactive protein and cardiac troponin I are specific markers of inflammation that are clinically correlated with cardiovascular diseases [62]. Cernat et al. (2020) reported graphene-based electrochemical sensors for the simultaneous detection of neurotransmitters for early diagnosis and therapy of neurodegenerative diseases [63]. Neurodegenerative disorders and brain tumors are affecting the brain. The delivery of therapeutic agents into the brain does not cross the blood–brain barrier, which is a very complicated and challenging aspect. Thus, Henna et al. (2020) reported nanocarriers such as polymeric nanoparticles, carbon nanoparticles, and lipid-based nanoparticles for easy penetration of the blood–brain barrier; they can be used as efficient carriers for drug delivery into the brain. They can also be used as a central nervous system therapeutic agent due to neurodegenerative activity to treat brain disorders [64].

Prostate cancer is a deadly cancer in older men, and early diagnosis of this disease facilitates disease control and treatment. Negahdary et al. (2020) reported immunosensors, aptasensors, and peptide sensors using antibodies, aptamers, and peptides to detect prostate-specific antigens [65]. Xiang et al. (2020) reported aptamer-based biosensors for detecting carcinoembryonic antigen (a tumor marker), which is over-expressed in breast, ovarian, gastric, lung, colorectal, and pancreatic cancer, as the diagnosis and therapeutic evaluation of such diseases are important [66]. Further, Ghitman et al. (2020) reported the PLGA (polylactic-co-glycolic acid)–lipid hybrid nanoparticles for drug delivery systems with minimized side effects and safe clinical applications [67], and recently, Cheng et al. (2020) reported the health monitoring systems based on nanomaterials, which are very important in preventive medicine to improve quality of life [68]. Bone is a tissue of the human body made up of osteoblasts and osteoclasts, and bone regeneration is possible by bone tissue engineering using various biocomposite scaffolds; thus, Adithya et al. (2020) reported nanosheets, 2D metallic and nonmetallic nanosheets, which strengthen the biocomposite scaffolds using gelatin, chitosan, collagen, polymers, bioceramics, etc., and also provide flexibility and durability [69]. Recently, engineered nanomaterials have gained much attention for delivering therapeutic agents, as they result in improved drug bioavailability and bio-stability and minimize the systemic side effects of amyotrophic lateral sclerosis motor neuron disease, i.e., loss of muscle control [70]. The diagnosis of acute myocardial infarction is very important for the survival of patients. Pourali et al. (2020) reported amperometry/voltammetry biosensors to detect biomarkers of acute myocardial infarction such as cardiac troponin (cTnI, cTnT) isoforms using nanomaterials [71]. Chronic kidney disease is a burden to global public health. Its early detection and effective therapy are mandatory for properly curing the disease. Ma et al. (2020) reported nanoparticles with variable particle size, charge, shape, and density of targeting ligands to diagnose and treat disease [72]. Disease diagnosis and treatment are the main challenges in healthcare, and for achieving the same, Horne et al. (2020) reported nanofiber-based immunosensors for detecting diseases and monitoring specific biomarkers of certain diseases and illnesses such as malaria and cancer [73].

Carbon dots are photoluminescent with many potential applications. Miao et al. (2020) reported heteroatom-doped carbon dots to show catalytic, optical, and biological properties. Owing to these properties, they have enormous applicability, as nanoprobes, optoelectronic devices, catalysis, and biomedicine [74]. From monotherapy to combination therapy, this therapy enhances therapeutic efficacies and diminishes undesired side effects for fighting cancer. Cheng et al. (2020) reported a silica-based nanosystem for effective cancer treatments using photo-chemotherapy [75]. Zhu et al. (2020) reported MXene-based optical biosensing such as photoluminescence, electrogenerated chemiluminescence, surface plasmon resonance (SPR) and surface-enhanced Raman scattering (SERS) photoelectrochemical, colorimetric, and photothermal using two kinds of titanium carbide for the applications of diagnostic and also biological researches [76]. Szuplewska et al. (2020) reported 2D carbides, nitrides, and carbonitrides of transition metals used as therapeutics for anti-cancer treatment and also in biosensing and bioimaging. They are also very efficient agents for environmental and antimicrobial treatments [77]. Sivasankarapillaiet al.

(2020) reported that MXenes (Ti_3C_2, Nb_2C, Ti_2C) are well-established nanomaterials with their physical, chemical, optical, and electronic properties and biocompatibility for biomedical applications such as cancer theranostics, drug delivery, biosensing probes, auxiliary agents for hyperthermia, and photothermal therapy [78].

1.3 ENVIRONMENTAL APPLICATIONS

The major threat to the environment is the release of heavy metal ions, pesticides, chemical toxins, etc., into the water bodies, food, and atmosphere, that possess a severe global threat to health. To solve these issues, nanomaterials have enormous applications in environmental monitoring and analysis (Figure 1.3). Nanobiosensors have brought a novel, cost-effective technology for environmental analysis and moni-toring; for example, heavy metal waste (Pb, Cd, Hg, etc.) from metallurgic industries pollutes the biosphere because of the toxic nature of heavy metals. Apart from this, the use of pesticides in large quantities to enhance agricultural produce yields is the main cause of environmental pollution as residues of pesticides are found in water and food products, a direct human interaction route, which directly affects human health. For solving these issues, nanomaterial-based immunosensors and genosensors are very effective tools [79–83]. Nsibande and Forbes (2016) reported QD-based nanobiosen-sors for environmental analysis and monitoring, as QD-based nanobiosensors can be commercialized and used routinely for the detection and monitoring of environmental pollutants such as pesticides, heavy metals, and chemical toxins, and the QD-based nanobiosensors as a metal sensor for toxic heavy metal ion detection in water are also very effective and needed, as contamination of heavy metals is the major cause of water pollution [84]. Duan et al. (2011) fabricated ZnS QD nanomaterial-based sensors to detect Hg^{2+} traces in water samples [85]. Further, Zhao et al. (2013) reported an ultra-sensitive method for the detection of Pb^{2+} by utilizing dithizone-functionalized CdSe/CdS [86]. Further, ZnS QDs and Mn^{2+} QDs were utilized to make nanobiosensors for the detection of different pesticides such as cyphenothrin [87], pentachlorophenol

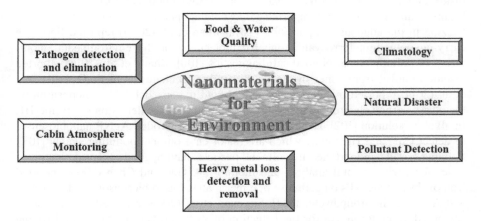

FIGURE 1.3 Representation of potential broad-spectrum applications of nanomaterials in the environment domain.

[88], and acetamiprid [89]. Ramnani et al. (2016) reported carbon-based nanomaterial biosensors, which impart unique electrical and physical properties and offer various applications to detect and monitor environmental pollutants such as heavy metal ions, explosives, pathogens, and pesticides [90]. Aptasensors (DNA-functionalized CNT biosensors) are used for heavy metal ion interaction with DNA. Liu and Wei (2008) reported a DNA–SWCNT (single-walled carbon nanotube)- modified glassy carbon electrode for the detection of arsenic(III) [91]. Gong et al. (2013) reported SWCNT-functionalized duplex polyT:polyA, a label-free chemiresistive biosensor to detect Hg^{2+} [92]. Liu et al. (2013) reported nanobiosensors associated with rGO field-effect transistors for the detection of rotavirus (waterborne virus), which causes diarrhea [93]. Lian et al. (2014) reported a DNA-wrapped metallic SWCNT electrochemical biosensor for Pb^{2+} detection in water [94].

Pesticides are used in agriculture to inhibit the growth of certain plants and animals, which increases the yield of crops. However, the toxic pollutants that are harmful to human health are organophosphates, which cause cholinergic dysfunction in humans and animals. CNT-based enzymatic biosensors are utilized for the detection of pesticides by using enzymes, namely organophosphorus hydrolase, butyrylcholinesterase, tyrosinase, and acetylcholinesterase. Further, Du et al. (2007) reported multiwalled CNT–chitosan composite acetylcholinesterase biosensor for the detection of triazophos, an organophosphorus insecticide [95]. Oliveira et al. (2010) reported dendrimer-based nanomaterials (branched polymers) and starlike structures, and these synthesized nanomaterials have a variety of applications such as in vitro diagnosis, gene therapy, regenerative medicine, and therapeutics [96]. Castillo et al. (2015) reported aptamer biosensor for the detection of aflatoxin B1, a food contaminant [97]. Snejdarkova et al. (2003) reported choline oxidase and acetylcholinesterase (AChE) immobilized with polyamidoamine (PAMAM) fourth-generation-based biosensor to detect dimethyl-2,2-dichlorovinyl- phosphate, carbofuran, and eserine pesticides [98]. Liposomes are artificial spherical vesicles consisting of a phospholipid bilayer surrounded by an aqueous cavity, which can bear different molecules inside, and these liposomes have a variety of environmental monitoring applications. Baumner and Schmid (1998) reported a liposomal amperometric sensor for detecting triazine in the aqueous sample [99]. Viswanathan et al. (2006) reported liposome carrying potassium ferrocyanide-labeled molecules for the environmental monitoring of pollutants and biomedical diagnostics [100]. Zhang et al. (2016) reported arsenic removal from water by strong adsorption mechanisms of Fe_3O_4 [101]. Wu et al. (2014) reported manganese oxide and gold nanoparticle electrodeposition on the glassy carbon electrode with cyclic voltammetry for the detection of arsenic(III) in alkaline solution [102]. Li et al. (2013) reported dye-coded AgNP nanobiosensor to detect Cu^{2+} and Hg^{2+} ions, which are major environment-polluting agents [103]. Polymer-based nanoparticles are colloidal particles having good encapsulation properties for environmental analysis and monitoring. Guo and Chen (2005) reported iron oxyhydroxide NPs on cellulose polymer matrix-based biosensors to detect As^{3+} and As^{5+} ions in groundwater [104]. Sugunan et al. (2005) reported a nanobiosensor based on chitosan–Au for the detection of Zn^{2+} and Cu^{2+} [105]. Su and Huang (2007) reported the synthesis of TiO_2–polypyrrole by in situ photopolymerization for the detection of changes in humidity for proper monitoring and maintenance of environmental pollutants [106]. Inorganic arsenic is a toxic heavy metal and poses a problem to the environment and human health, and for solving these problems, Xu et al. (2020) reported nanomaterial-based sensors to detect inorganic arsenic and also improve the analytical performance when compared with the traditional detection method. Various types of nanomaterial-based principles and strategies have been established to detect inorganic arsenic using inhibitions of natural enzymes, aptamers, and anodic stripping voltammetry [107].

Nanomaterials are used in catalysis, oil processing, sensors, medicine, food, energy storage, building materials, etc. Some nanomaterials may be harmful when leaked into the environment, and to analyze such environmental contaminants is mandatory [108]. Nanomaterials, either natural or synthetic, that are daily produced, transformed, and exported into the environment negatively affect the environment; high usage of synthetic nanomaterials makes them discharged into the atmosphere, water sources, soil, and landfill waste and susceptible to the biota or human beings [109]. Carbon-based nanostructures such as graphite, graphene, fullerene, CNTs, diamond, and carbon black have been used in the oil and gas industry for drilling and enhanced oil recovery [110]. Nanomaterials have become part of our daily lives, but few nanomaterials are not easily degraded by the environment [111]. Synthesis of iron nanoparticles and it's nanocomposite mediated by plant sources, microorganisms, cellulose, biopolymers, hemoglobin, and glucose are useful for environmental applications [112]. Nanomaterials are very popular in every aspect of life or ecosystem in air, soil, and water, but their toxicity could not be ruled out in the case of aquatic life [113]. Molten-salt synthesis-based nanomaterials are used in various electrical, catalytic, magnetic, and optoelectronic applications [114]. Lead poisoning is due to lead (a toxic heavy metal) that poses public health risks in developing countries. Fluorescent nanosensors have been used to detect Pb^{2+} ions in environmental media (soil, water, food, etc.) [115]. Filik and Avan (2020) reported carbonaceous nanomaterial-based electrochemical sensors for simultaneous electrochemical sensing of environmental contaminants (dihydroxybenzene isomers) [116].

Metal–organic framework-based nanomaterials have immense potential for catalysis, sensors, energy storage, and conversions, which offers them potentialities for environmental cleaning and monitoring, especially for wastewater treatment, air purification, and target-specific sensing by utilization of bimetal–organic framework-derived nanomaterials [117]. Nanomaterial-based biosensors are very effective for detecting airborne pathogens, and these pathogens are the main cause of common cold, flu, asthma, tuberculosis, botulism, anthrax, and pneumonia [118]. Rapid industrialization and urbanization have eased people's lives and resulted in pollution to the natural environment, jeopardizing living beings, aquatic life, food and water resources, or planet ecosystem. When released into the aqueous system, the heavy metal ions/radionuclides cause environmental pollution, and to eliminate it, zeolitic imidazolate framework nanomaterials have potential utility [119]. Further, molecularly imprinted polymers (MIPs) based on carbon nanomaterials are utilized to detect various environmental pollutants, as carbon nanomaterials have an important role in pollution control, heavy metal removal, and the detection of toxic gases and antibiotics [120,121]. MXenes are a new type of nanomaterial, for example, 2D titanium carbide, which has photocatalytic degradation of organic pollutants to check environmental pollution [122]. Water is polluted by dyes, heavy metals, and pathogens, and they harm human and animal health. Nano-sized materials, tubes, and composites as adsorbents can remove heavy metals from polluted environmental sources such as rivers, ponds, and lakes [123].

Graphitic carbon nitride acts as a photocatalyst; for example, graphite–C_3N_4-based 2D/2D composites have the potential role in the degradation of pollutants, H_2 generation, CO_2 reduction, and photocatalytic disinfection [124]. Marine oil spills are sources of water pollution that disturbs the ecological framework of the oceans' aquatic biodiversity and by utilization of magnetic nanocomposites, nanofluids, polymers, functionalized superparamagnetic iron oxide nanoparticles, and magnetic nanoadsorbents play a very effective role in removal of oil spills to protect oceanic life. Hence, nanomaterials have enormous utility for enhancing the oil recovery of oil spills in the oceans [125,126]. The MIPs use carbon dots, CNTs, graphene, and carbon electrodes as a substrate to identify and analyze biological, environmental, and food samples [127]. Endocrine-disrupting chemicals have a serious threat to humans and animals' health. Concerning this, a variety of nanomaterial-based fabricated biosensors are used to detect estrogens in food and environmental samples [128]. Mercury is a poisonous heavy metal that possess an adverse effect on human health, and for the detection of Hg^{2+} ions, nanobiosensors have proven potential applications. Recently, nanozymes such as AuNPs and PtNPs are also being used to detect Hg^{2+} ions quantitatively [129]. Toxic gases in the environment cause harmful effects to humans and animals, and it is an important aspect that is needed to be addressed currently, and for doing so, Wang (2020) reported a quartz crystal microbalance platform using various sensing materials such as metal–organic frameworks to detect gases such as formaldehyde, toluene, and acetone [130]; further, Fadillah et al. (2020) reported functional polymers for analyzing toxic gases, heavy metals, and aromatic compounds [131].

Water pollution due to the increase in industrial development, population, and climate change globally is a major concern, and to overcome this issue, the utility of nanomaterials for water remediation of organic pollutants is the best alternative [132,133]. Carbon-based nanomaterials are widely used as unique sorbents for the removal of a variety of contaminants; thus, Ghorbaniet al. (2020) reported the carbon-based nanomaterials/nanocomposites for the detection and removal of lead(II) ions from water and wastewater samples [134]. Chromium and arsenic are toxic metals, and for the removal of these toxic ions, there are many nanomaterials to remove them from contaminated water sources (wastewater). Furthermore, Lal et al. (2020) reported the various carbonaceous nanomaterials for removing Cr and As ions from wastewater [135]. Environmental pollution has increased due to the increase in the number of industries that release hazardous substances into the environment as waste; thus, water treatment technologies based on adsorption using adsorbents to remove inorganic and organic contaminants from water are very much needed. However, Jeon et al. (2020) reported an MXene-based adsorbent nanomaterial that can remove inorganic and organic contaminants from water and improve water quality [136]. Zeidman et al. (2020) reported synthetic silica-based nanocomposite materials to remove antibiotics from the aqueous phase by photocatalytic degradation [137]. Chenab et al. (2020) reported engineered nanomaterials as adsorbents and photocatalysts to remove heavy metal ions and organic dyes by photocatalytic degradation [138]. Basak et al. (2020) reported biofunctionalized nanomaterials to remediate monocyclic and polycyclic aromatic hydrocarbons [139]. Kumar et al. (2020) reported graphene to remove volatile organic compounds due to the adsorptive capability [140]. Lye et al. (2020) reported nanomaterials such as metal- or carbon-based materials, which are emerging adsorbents used in water treatment processes to adsorb aquatic dissolved organic matter and alter the fate, transport, and toxic effects [141]. Bolade et al. (2020) reported the green synthesized iron nanoparticles for environmental remediation [142]. Chloramphenicol is an antibiotic that is potentially harmful to human health. Dong et al. (2020) reported an electrochemical aptamer sensor for chloramphenicol detection [143]. Kegl et al. (2020) reported removal of rare earth elements from wastewater by the adsorption process using efficient nanomaterials [144].

Water quality deterioration is a worldwide problem due to the presence of toxic organic and inorganic pollutants. Gusain et al. (2020) reported adsorption-based water technologies using carbon nanomaterials as adsorbents in water treatment [145]. Wadhawan et al. (2020) reported nanoadsorbents for water purification by removing heavy metal ions from wastewater [146]. Yadav et al. (2020) reported carbon-based nanomaterials to enhance the forward osmosis membrane's mechanical strength at the early stage of laboratory investigation [147]. Nasrollahzadeh et al. (2021) reported that natural biopolymers, polymeric organic molecules derived from living organisms, are greener, sustainable, and eco-friendly materials and utilized as nanocatalyst sorbents, i.e., polysaccharide-supported metal/metal oxide, for eliminating pollutants and contaminants from wastewater [148]. Ihsanullah (2021) reported boron nitride-based materials for water remediation from an aqueous medium [149]. Mmelesi et al. (2020) reported cobalt ferrite nanoparticles and its composites to remove pollutants and they also demonstrated its antimicrobial activities [150]. Nunes et al. (2020) reported membrane technology using renewable nanomaterials for membrane production and water purification [151]. Zhou et al. (2021) reported laccase, a biocatalyst for removing heavy metals, and for water purification when it is effectively immobilized [152].

1.4   AGRICULTURAL APPLICATIONS

Agriculture is a diversified field of self-sustaining economic development of any country. It requires nanotechnical interventions for food processing, food safety, food quality or quality assurance, food security, disaster risk management, diagnosis, and local and global prevention; thus, nanomaterials have enormous applications in the agriculture field (Figure 1.4). Further, agriculture has challenges for enhancing crop productivity and sustainability, food safety of livestock, and natural agricultural resources. Agriculture is proved as the main source of raw materials to food industries and is also the backbone of developing countries for their economic growth and development. Nowadays, this sector faces huge problems due to urbanization, incorrect methodology to cultivate the soil and utilize land and water, and misutilization and management of pesticides and fertilizers. Natural resources are critical factors for economic growth and development phenomena pertaining to the

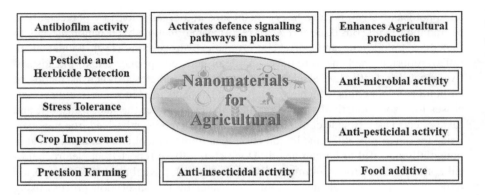

FIGURE 1.4 Illustrations of potential applications of nanomaterials in the agricultural domain.

large population's poverty and hunger globally. Agriculture is the science of soil, crops, and livestock and is an economic boon of any country for the source of livelihood to improve social welfare in both urban and rural areas. However, accessibility of resources such as soil and water quality is declining for agriculture and is creating a big economical loss. This continued stress on agricultural resources increases due to overpopulation, high demand for food, and a constant increase in pesticides, insecticides, herbicides, and heavy metals in soil. Nanomaterial-based nanobiosensors can solve all these challenges or issues to change agricultural-based food systems to improve agricultural products' quality and boost the national economy toward sustainable agriculture with quality products and less cost by detecting target analyte of interest such as pesticides, insecticides, herbicides, and various microbial pathogens in food and drink, food process industries, food safety, food security, healthcare, and environment. Nanomaterials such as metals and metal oxide nanoparticles (Au, Ag, Cu, Co, ZnO, TiO_2, Fe_3O_4, MgO, etc.), magnetic nanoparticles, CNTs, graphene, dendrimers, polymeric, and QD nanoparticles are widely used for the detection of the agricultural analytes of interest. Agriculture using nanomaterials can enhance crops, soil, and livestock productivity, which feed our populations free from any implications. Nanomaterial-based biosensors have potential applications in agriculture to improve crop health by detecting plant pathogens, pesticides, herbicides, and soil testing. Nanoparticles are utilized in the diagnostic tool to detect plant pathogens [153–157].

In developing countries, consumption of foods contaminated with pathogenic foodborne bacteria is posing illness locally and globally. Foodborne pathogens are a serious threat to the animal's production and health via food poisoning, gastroenteritis, etc. The nanomaterial-based device can detect and remove food contaminants, pathogens, banned dyes, adulterants, antibiotics, hormones, and allergens. Brock et al. (2011) reported nanosensors to detect plant pathogens, viruses, and soil nutrients [158]. Viswanathan et al. (2006) reported liposomal poly(3,4-ethylenedioxythiophene)-coated CNT immunosensor for the detection of cholera toxin [159]. Wang and Alocijia (2015) reported a nanobiosensor based on functionalized Fe_3O_4 NP- and AuNP-conjugated monoclonal antibodies to detect *Escherichia coli O157:H7* [160]. Kim et al. (2013) reported immobilized anti-*Salmonella* polyclonal antibodies on streptavidin–biotin onto the QD surface nano-biosensor to detect *Salmonella sp.* in food [161].

Mycotoxins are fungi/mold-derived toxic chemicals that are natural contaminants found in foodstuffs and animal feed products that threaten human health and are most commonly known as hepatotoxic, nephrotoxic, carcinogenic, and mutagenic, for example, aflatoxin, ochratoxin, and zearalenone. Mycotoxins are a threat to human health, which influence the crops, feed, and food products, especially in rainy seasons, and this results in great economic losses globally. Masikini et al. (2015) reported the detection of fumonisins using polyaniline–CNT-doped palladium tel-luride QDs. There are several nanostructured materials such as QDs, nanoparticles (metal/metal oxide, polymer), nanowires, nanotubes or nanorods, and graphene, and they can be functionalized and immobilized by various biomolecules as receptors such as antibodies, enzymes, and DNA/RNA aptamers for the detection of food toxicants, adulterants, and pathogens [162]. Parker and Tothill (2009) reported the detection of aflatoxin M1 contaminants in the milk by an electrochemical immuno-sensor [163]. Xu et al. (2013) reported the detection of aflatoxin B1 in peanuts [164]. Eldin et al. (2014) reported an immunosensor using Au nanorod conjugated with anti-aflatoxin B1 polyclonal antibody to detect aflatoxin B1 in peanuts, chilies, maize, and rice [165]. Bonel et al. (2010) reported an electrochemical immunosensor constructed by AuNPs conjugated polyclonal antibodies to detect ochratoxin A [166].

Pesticides are used to protect plants and animals and as an herbicide to control weeds, a fungicide to control fungi, and insecticide to control insects, but spray-ing heavy dose causes toxicity to human/animals. Zhang et al. (2008) reported the detection of pesticide content in food by an acetylcholinesterase nanobiosensor [167]. Norouzi et al. (2010) reported monocrotophos and organophosphate pesticide detec-tion by an electrochemical biosensor [168]. Zheng et al. (2011) reported the detection of paraoxon and parathion pesticides by an optical nanobiosensor using acetylcho-linesterase and CdTe QDs [169]. Guan et al. (2012) reported the detection of dichlor-vos pesticides by acetylcholinesterase biosensors [170]. Song et al. (2015) reported the detection of carbamate pesticide by an electrochemical biosensor fabricated by AuNPs/(3-mercaptopropyl)-trimethoxysilane/Au sensing surface [171]. Haddaoui and Raouafi (2015) reported the detection of chlortoluron herbicide by a nanobiosen-sor based on tyrosinase inhibition [172].

Veterinary drug residues are biological contaminants; they are mostly antibiot-ics used in farm animals to treat animal diseases and enhance animal growth [173]. Antibiotic use in animals causes serious risks in animals and humans, which leads to an outbreak [174]. Veterinary drugs are available in the black market to use in the poultry industry without consulting a veterinary doctor [175]. Wu et al. (2015) reported detecting chloramphenicol by using an aptamer-based fluorescence bio-sensor [176]. Simmons et al. (2020) reported the detection of ampicillin in milk by utilizing an aptamer biosensor [177]. Furthermore, Hou et al. (2013) reported the detection of oxytetracycline by aptamer-based cantilever array sensors [178].

Food safety and security and climate changes are important issues and need advanced scientific interventions to resolve human health. Nutritious food is important in the diet, which should be economical, safe, and sufficient, but our diet is contaminated, so there is an urgent need for food safety and food security. Sastry et al. (2013) reported a nanobiosensor to detect toxic natural food contaminants for food accessibility and food utilization [179]. Food contamination is causing food poisoning due to foodborne pathogens. However, food preservation using chemicals to increase food life, which causes illness, and to develop a system to identify the fresh or good quality of food is very much needed. This smart system can detect the freshness of food such as dairy items, meat, and fruits, and it is only possible to fabricate such type of system by utilization of biological systems and engineered nanomaterials for making smart nanobiosensors, which can detect vitamins, antibiotics, food spoilage, and microbial contaminants. For achieving these objectives, Inbaraj and Chen (2016) reported a nanobiosensor for detecting a bacterial pathogen in meat [180].

Nanomaterials exhibit unique properties and diverse applications, which can be applied to increase water surface area-to-volume ratio and a source of electroluminescence light in a hydroponic system and act as coatings and surface hardeners in crop-processing machinery [181]. Pesticides are highly used hazardous toxic chemicals in large-scale agricultural practices. Nanomaterial-based electrochemical biosensors are used to detect pesticides by receptors (DNA/RNA, antibodies, enzymes) and bioreceptors such as aptamers, and MIP electrode modified by graphene, CNTs, and metal and conducting polymer nanoparticles [182]. Fluorescence carbon nanomaterial-based biosensors are also used to detect pesticides such as organophosphorus, neonicotinoid, and carbamate pesticides in environmental, food, and biological samples, which help protect the ecosystem, provide food safety, and prevent diseases using carbon nitride nanosheets, carbon dots, graphene, and QDs [183]. Polysaccharides, lipids, and proteins are biomaterials with renewability, are biocompatible, biodegradable, and tunable, and are multi-active binding sites that possess functions for emerging applications; further, a few food-derived nanomaterials and bionanocomposites are also widely used for biosensing and smart packaging of food product applications [184,185].

Nanobiosensing technology for analysis of pesticides by nanobiosensors using nanomaterials with bioelements is currently perhaps the best sensing method [186]. Nanomaterials have promoted sustainable agriculture to improve crop production and protection [187]. The use of nanomaterials on diverse sensing principles was established to detect pesticides/herbicides in food, water, and soil [188]. Antibiotic resistance is a serious challenge in food packaging, and to overcome this problem, nanobiomaterials can play a crucial role as they demonstrate great antifungal, antibacterial, and antioxidant potential that disrupt membrane due to oxidative stress [189]. The nanofertilizers and nanopesticides can enhance crop production by increasing yield and altering crop quality. Engineered CeO_2 nanoparticles, CuO, and ZnO have increased proteins, sugars, starch, and essential metallic elements such as Ca, Mg, and P in several crops. However, future research needs to validate these nanomaterials' toxic effects on human health if consumed as nano-enabled agricultural products [190]. Arsenic is a toxic contaminant, which is a serious carcinogen that threatens environmental and human health. The arsenic-removing technology for the field-testing has been widely worked upon by researchers in this field and is successfully developed by utilizing nanomaterials for fabricating biosensors to have rapid detection of arsenic. However, nanomaterial-based aptamer sensors using optical and electrochemical technologies are also developed for arsenic detection [191]. Heavy metal contamination is a global environmental problem and a threat to aquatic life and humans. Nanomaterials are used as nanosorbents to remove heavy metals such as Cr(VI) ions from wastewater [192]. Nanozymes are nanomaterials that can catalyze biochemical reactions for various biosensing applications to detect contaminants to manage food quality and safety [193]. Food product quality is an important issue and challenges controlling diffusion through nanomaterials in food and pharmaceutical packaging [194]. Nanostructured materials as lipase-based nanobiocatalytic systems using CNTs, nano-silica, graphene/GO, metal nanoparticles, magnetic nanostructures, metal–organic frameworks, and hybrid nanoflowers are useful to support matrices for the immobilization of lipase enzyme for utility in dairy, food, pharmaceutical, and detergent industries [195]. The packaging systems are used to protect products from physical impacts for storage and transportation. Nanofabrication for smart food packaging systems can show gas barrier, antibacterial, and protective effects in the food industry or the smart food packaging industry [196]. Mycotoxins present in foodstuffs have serious health hazards. Nanomaterial-based sensing has a great potential to detect mycotoxins [197]. Enzymes have played important roles in food science. Thus, nanozymes have a scope of application in food sciences for bioprocessing in food analysis, detection, and quantification [198]. The food safety issue has concerned people for the specific detection of food pollutants. The functionalized Fe_3O_4 nanoparticles have the innovation of food quality and safety detection in the agrifood industry [199].

Immunosensors are used to detect pesticide residues and in healthcare and environmental monitoring [200]. Hydroxyapatite nanoparticles are nanomaterials used for plant's phosphorus nutrition. These nanofertilizers are applied to soil bearing soybean, sorghum, pea, and pak choi plants [201]. Edible food packaging uses plasticizers and nanocomposites, but no evidence on biodegradability, toxicological data, less marketing, and no awareness have been reported pertaining to customer acceptance and food safety [202]. Safety and public health issues are a big problem in handling nanomaterials [203]. The bioelectronic tongues are used for food and beverage control and characterization [204]. Biopolymer-based polysaccharides derived from seaweed are used in food packaging [205]. Meat, fish, and their derived products are food that deteriorates very rapidly when stored improperly. Edible films and coatings can preserve the meat, fish, and their derived products. These films and coatings are of biopolymers derived to improve the sensory and quality characteristics of packaged products and extend shelf life [206].

1.5   ENERGY

The energy-based applications of smart nanomaterials (Figure 1.5) are in the process of being established with a highly efficient production method, energy storage, and conversion. The smart nanomaterials, namely metal/oxide nanoparticles, nanowires, QDs, and nanotubes, are highly utilized nanomaterials for photovoltaics, batteries, hydrogen storage, and supercapacitors. These nanomaterials

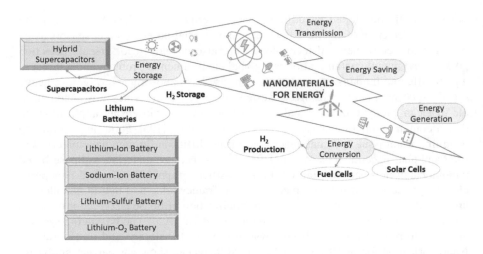

FIGURE 1.5   Broad-spectrum applications of nanomaterials in the energy domain.

have shown hardness/strength, luminescence, electrical conductivity, magnetism, band gap, etc. [207]. Lithium–ion batteries are sensitive to extreme temperature with long durability and cooling using liquid-based, air-based, and phase change material-based nanomaterials [208]. The oxygen reduction reaction is possible in fuel cells containing electrocatalysts to improve the cathodic reaction. The electrospun nanomaterial utilizes important electrocatalysts such as heteroatom-doped carbon nanofibers, transition metal/carbon nanofiber composites, and carbon-free nanofibers, which shows chemical stability and structural diversity to oxygen reduction reaction [209]. Biomass-derived nanomaterials are of great interest in green chemistry and renewable energy as they are eco-friendly with unique properties and functionalities for energy, sensing, catalytic, biomedical, and environmental applications [210]. Boron carbon nitride has varied structures and diverse energy applications such as in batteries, supercapacitors, water purification, and biosensing [211]. Doped TiO_2 nanomaterials mediated photocatalytic property utilized in energy consumption [212]. Immobilized lipase on/in support of nanomaterial acts as a biocatalyst, which is well known for biodiesel production. Zhong et al. (2020) reported immobilized lipase-based biocatalytic systems to produce biodiesel from waste materials [213].

Local and global demand for energy is currently increasing. Hoang et al. (2020) reported metal-based nanomaterials used in electrochemical CO_2 reduction reactions [214]. Black phosphorus, graphene, organic frameworks, boron nitride, and carbon nitride are widely used as photocatalysts in energy [215]. Copper sulfides are nanomaterials with diverse applications in catalysis, photovoltaics, sensors, electronics, energy storage, etc. Majumdar (2020) reported supercapacitive applications using copper sulfide nanosystems [216]. Carbon nanomaterials are used in high-power technology [217]. Vanadium-based nanomaterials have been used in high-capacity lithium batteries [218]. Sodium–ion storage devices are large-scale energy storage systems. Electrospun materials are used in sodium–sulfur batteries, sodium–ion batteries, sodium–air batteries, and sodium–selenium batteries [219]. Carbonaceous nanomaterials are widely used in energy conversion and storage applications due to sp hybridization. For example, graphene QDs, graphdiyne, and carbyne are used in supercapacitors, lithium–ion batteries, and solar cells for future energy-oriented applications [220]. Lithium–sulfur batteries have some demerits. To overcome this issue, porous carbon nanocomposites and graphene nanocomposites are being used in lithium–sulfur batteries for electric energy storage [221]. Nanocellulose and its fibrillated form derived from biomass have the potential applications in energy [222]. The hydrogen economy is the future of long-term energy storage, transportation, and its usage using metal–organic framework hybrid clusters for energy storage [223]. For example, 2D materials, such as graphene-based photocatalysts and graphitic carbon nitride-based photocatalysts, are used in solar fuels [224].

Electrocatalytic water splitting is a renewable energy type, which utilizes electrocatalytic nanomaterials into hydrogen and oxygen evolution [225]. Dye-sensitized solar cells are the third-generation photovoltaic technologies based on photoconversion [226]. Nanocellulose is used in electrochemical energy storage to increase conductivity [227]. Wastewater discharges could be utilized for biodiesel production [228]. In biomass production and bioremediation, algae are exploited in biosensing applications to detect analytes of agro-environmental interest. Various algae's and their photosynthetic subcomponents can be used for continuous environmental monitoring by utilizing algae-based biosensing [229]. Graphene and CNTs have been found to have application in energy storage and energy conversion. However, an agglomeration of these materials causes a problem. The hybridization of CNTs with grapheme composite can prevent the agglomeration behavior and generate a synergistic effect. The graphene/CNT composites have been widely utilized in energy-related applications [230]. Microalgal biohydrogen is a carbon-free source of sustainable renewable energy [231]. $VOx@Graphene$ and $VSx@Graphene$ nanocomposites possess high practical and economic value [232]. Phase change materials are used in thermal energy storage due to their low thermal conductivity and thermophysical properties [233,234]. Ultrathin (2D) materials are displaying a wide range of different extraordinary properties. Borophene is a magical 2D material after graphene, which has displayed a high metallic character and is used in batteries, catalysis, gas storage, and sensors (for sensing various toxic gases) [235]. Solar energy is solving the global energy crisis using various semiconductors. Zhang et al. (2020) reported CdS nanocomposites for solar energy utilization [236]. Zhou et al. (2020) reported high-performance electrochemical energy storage devices using low-strain titanium-based oxide electrodes to replace the carbonous materials, for example, $Li_4Ti_5O_{12}$ and $TiNb_2O_7$ [237].

1.6 MISCELLANEOUS APPLICATIONS

1.6.1 AUTOMOBILE INDUSTRY

Nanomaterials with controlled size, chemical composition, and atomic structures create macroscopic materials with unique characteristics and functionalities in the automotive sector.

In automobile tires, nano-oxide fillers, carbon black, carbon nanofibers, graphene, and nanoclay improve tires' lifetime. Nanostructured tungsten nanospheres, boric acid, graphene, and copper nanoparticles are used as a fluid lubricant in the car, improving mechanical properties. CNTs and Ag nanowires are utilized in the battery for automobile applications, and these materials have a future in the automotive industry [238]. Transformer oil-based nanofluids are useful in air-conditioning, refrigeration systems, heating, solar cell, heat pipes, systems, and many others [239]. Nanomaterial-based cooling/heating working fluids are much more potent and eco-friendly fluids in refrigerants and lubricants as nanorefrigerants/nanolubricants control boiling, condensation, and heat transfer rate and the thermophysical properties. Carbon nanomaterials are useful in tribological research, such as preparing the engine oils, self-lubricating materials, nanolubricants, and coatings to improve anti-friction/wear properties of the worn surfaces in automobiles [240].

1.6.2 DENTISTRY

The role of nanomaterials in dentistry is gaining much attention these days due to nanomaterials' unique structures and properties, which help in preventive dentistry, dental diagnostics, prosthodontics, dental materials, conservative and aesthetic dentistry, endodontics, implantology, periodontics, and regenerative dentistry. The nanomaterials used in dentistry are nanoparticles/nanotubes/nanofibers, nanocomposites, antimicrobial nanomaterials, and bio-mineralized nanoparticles as a coating material [241]. Dental treatment is very difficult because of biofilm presence, and to overcome this, antimicrobial dental materials such as polymeric and inorganic nanoscopic agents are used to inhibit microorganism proliferation. Nanofillers in dental materials could be utilized to enhance microbicidal ability. The antibacterial agents can remove microbial infection [242].

1.7 CONCLUSIONS

Nanomaterials are used in diagnostic and therapeutic applications at the molecular scale. In the biomedical field, accurate and early diagnosis of several diseases is needed for effective treatment and prevention. Several analytical biosensing systems such as electrochemical, electrical, and optical utilize specific biological reactions (antigen–antibody binding, enzymatic reaction, aptamer-based binding, etc.) with metal nanomaterials such as gold, silver, platinum, and other pure metals for the improvement of their diagnostic functions for the detection of biomarker [243], which has been explored in this chapter in detail. Further, fluorescent nanohybrids can be used to detect and destroy pathogenic microbes, and MRI can be improved by using nanomaterials as contrasting agents. In therapeutics, nanomaterials are used in targeted delivery and sustained release of drug molecules and play a key role in tissue engineering and antimicrobial activity.

Environmental pollution such as in water, food, and air is the main cause of health issues in humans and animals. Nanomaterials have a wide range of properties that offer them a wide range of environmental monitoring applications using QDs, carbon, dendrimers, liposome, metal and metal oxides, and polymers, which are elaborated in detail in this chapter. In this chapter, we have also talked about biosensors based on nanomaterials (nanobiosensors), which have various applications in the environmental domain. Further, safe drinking and clean water are important challenges globally. Nasrollahzadeh et al. (2021) reported sustainable nanomaterials as adsorbents for water and wastewater treatment and purification, which are utilized for safe drinking and irrigation. The elimination of contaminants and heavy metals using nanosorbents from wastewater, drinking water, and groundwater is utilized to treat and purify wastewater [244]. We have discussed in detail the role of nanomaterials in wastewater treatment strategies.

The food production and food industry via food processing, preservation, and packaging is beneficial to our farmers. Nanomaterial-based nanobiosensors are being developed to detect plant pathogens and test soil quality to improve plant health and support sustainable agriculture by enhancing crop productivity. The detection of food pathogens, pesticides, antibiotics, and food contaminants has to be done using nanobiosensors toward food safety to check threat to human health [245]. In this chapter, agricultural applications based on nanomaterials for food processing and packaging, agricultural contaminant detection, crop yield enhancement and stress tolerance, antipesticidal activity, precision farming, and antibiofilm activity are discussed in detail.

This chapter has elaborated the broad-spectrum application of nanomaterials in the four domains (biomedical/clinical, environmental, agricultural, and energy), but apart from these domains, the nanomaterials also have potentialities in other important domains such as automobiles, dentistry, military and defense, and bioterrorism. The efforts have also been made to discuss the latest development/recent trends and prospects in these four domains.

ACKNOWLEDGMENTS

KRBS is thankful to Dr. Ajaya Singh and Principal of Govt. V. Y. T. PG. Autonomous College, Durg, India, for providing a working platform for completing this work diligently and smoothly. PRS is thankful to the vice-chancellor of Jawaharlal Nehru University, New Delhi, India, for providing support. BDM thanks the Science and Engineering Board (DST-SERB, Govt. of India) for the award of the Distinguished Fellowship (SB/DF/011/2019). ACP is thankful to his institution University of Allahabad, Prayagraj, UP, India, and Inter-University Accelerator Centre, New Delhi, India. RPS is thankful to Hon'ble Vice-chancellor of Indira Gandhi National Tribal University, Amarkantak, MP, India, for providing financial assistance to work smoothly and diligently.

REFERENCES

1. Singh, R.P., Singh, P., & Singh, K.R.B. 2021. Introduction to Composite Materials: Nanocomposites and their Potential Applications. in *Composite Materials: Properties, Characterisation and Applications*. CRC Press, Taylor & Francis, Boca Raton.
2. Singh R.P. 2019. *Potential of Biogenic Plant-Mediated Iron and Iron Oxide Nanostructured Nanoparticles and Their Utility*. Chap 04, *Nanotechnology in the Life Sciences: Plant Nanobionics, Vol. 2. Approaches in Nanoparticles Biosynthesis and Toxicity*, Chap 04. Editor: Ram Prasad, Springer Nature 2019, Switzerland AG. ISBN 978-3-030-16378-5.

3. Singh R.P. 2019. *Potentialities of Biogenic Plant-Mediated Copper and Copper Oxide Nanostructured Nanoparticles and Their Utility Chap 05, Nanotechnology in the Life Sciences: Plant Nanobionics, Vol. 2. Approaches in Nanoparticles Biosynthesis and Toxicity*, Chap 05. Editor: Ram Prasad, Springer Nature 2019, Switzerland AG. ISBN 978-3-030-16378-5.

4. Yu, L., Song, Z., et al. 2020. Progress of gold nanomaterials for colorimetric sensing based on different strategies. *TrACTrends in Analytical Chemistry* **127**, 115880. Doi: 10.1016/j.trac.2020.115880.

5. Zhou, X., Wang, Y., Gong, C., Liu, B., & Wei, G. 2020. Production, structural design, functional control, and broad applications of carbon nanofiber-based nanomaterials: A comprehensive review. *Chemical Engineering Journal* **402**, 126189. Doi: 10.1016/j.cej.2020.126189.

6. Zhang, L., Ying, Y., Li, Y., & Fu, Y. 2020. Integration and synergy in protein-nanomaterial hybrids for biosensing: Strategies and in-field detection applications. *Biosensors and Bioelectronics* **154**, 112036. Doi: 10.1016/j.bios.2020.112036.

7. Makul. N. 2020. Advanced smart concrete-A review of current progress, benefits and challenges. *Journal of Cleaner Production* **274**, 122899. Doi: 10.1016/j.jclepro.2020.122899.

8. Singh, R.P., Choi, J.W., Tiwari, A., & Pandey, A.C. 2014. Functional Nanomaterials for Multifarious Nanomedicine, in *Biosensors Nanotechnology* (eds Tiwari, A. & Turner, A. P.F.), John Wiley & Sons, Inc., Hoboken, NJ.

9. Singh, K.R.B., Sridevi, P., & Singh, R.P. 2020, Potential applications of peptide nucleic acid in biomedical domain. *Engineering Reports* 2, e12238. Doi: 10.1002/eng2.12238.

10. Singh, K.R.B., Nayak, V., Sarkar, T., & Singh, R.P. 2020. Cerium oxide nanoparticles: Properties, biosynthesis and their biomedical application. *RSC Advances* 10, 27194–27214. Doi: 10.1039/D0RA04736H.

11. Nayak, V., Singh, K.R.B., Singh A.K., & Singh, R.P. 2021. Potentialities of selenium nanoparticles in biomedical sciences. *New Journal of Chemistry*. Doi: 10.1039/D0NJ05884J

12. Dai, Y., Han, B., Dong, L., Zhao, J., & Cao, Y. 2020. Recent advances in nanomaterial-enhanced biosensing methods for hepatocellular carcinoma diagnosis. *TrACTrends in Analytical Chemistry* **130**, 115965. Doi: 10.1016/j.trac.2020.115965.

13. Koo, K.M., Soda, N., & Shiddiky, M.J.A. 2021. Magnetic nanomaterial-based electrochemical biosensors for the detection of diverse circulating cancer biomarkers. *Current Opinion in Electrochemistry* **25**, 100645. Doi: 10.1016/j.coelec.2020.100645.

14. Ge, J., Zhang, Q., Zeng, J., Gu, Z., & Gao, M. 2020. Radiolabeling nanomaterials for multimodality imaging: New insights into nuclear medicine and cancer diagnosis. *Biomaterials* **228**, 119553. Doi: 10.1016/j.biomaterials.2019.119553.

15. Shao, B., & Xiao, Z. 2020. Recent achievements in exosomal biomarkers detection by nanomaterials-based optical biosensors-A review. *Analytica Chimica Acta* **1114**, 74–84. Doi: 10.1016/j.aca.2020.02.041.

16. Khan, S., Sharifi, M., et al. 2020. In vivo guiding inorganic nanozymes for biosensing and therapeutic potential in cancer, inflammation and microbial infections. *Talanta* 121805. Doi: 10.1016/j.talanta.2020.121805.

17. Zhang, X., Lin, S., et al. 2020. Advances in organometallic/organic nanozymes and their applications. *Coordination Chemistry Reviews* 213652. Doi: 10.1016/j.ccr.2020.213652.

18. Li, Y., Zhu, W., Li, J., & Chu, H. 2020. Research progress in nanozyme-based composite materials for fighting against bacteria and biofilms. *Colloids and Surfaces B: Biointerfaces* 111465. Doi: 10.1016/j.colsurfb.2020.111465.

19. Yang, D., Chen, Z., Gao, Z., Tammina, S.K., & Yang, Y. 2020. Nanozymes used for antimicrobials and their applications. *Colloids and Surfaces B: Biointerfaces* **195**, 111252. Doi: 10.1016/j.colsurfb.2020.111252.

20. Johnson, A.P., Gangadharappa, H.V., & Pramod, K. 2020. Graphene nanoribbons: A promising nanomaterial for biomedical applications. *Journal of Controlled Release* **325**, 141–162. Doi: 10.1016/j.jconrel.2020.06.034.
21. Li, T., & Xu, H. 2020. Selenium-containing nanomaterials for cancer treatment. *Cell Reports Physical Science* **1**(7), 100111. Doi: 10.1016/j.xcrp.2020.100111.
22. Joshi, A., & Kim, K.H. 2020. Recent advances in nanomaterial-based electrochemical detection of antibiotics: Challenges and future perspectives. *Biosensors and Bioelectronics* **153**, 112046. Doi: 10.1016/j.bios.2020.112046.
23. Wang, X., Zhong, X., & Cheng, L. 2020. Titanium-based nanomaterials for cancer theranostics. *Coordination Chemistry Reviews* 213662, ISSN 0010-8545. Doi: 10.1016/j.ccr.2020.213662.
24. Tang, Z., Kong, N., et al. 2020. Phosphorus science-oriented design and synthesis of multifunctional nanomaterials for biomedical applications. *Matter* **2**(2), 297–322. Doi: 10.1016/j.matt.2019.12.007.
25. Zhao, Q., Lu, D., Zhang, G., Zhang, D., & Shi, X. 2021. Recent improvements in enzyme-linked immunosorbent assays based on nanomaterials. *Talanta* **223**(1), 121722. Doi: 10.1016/j.talanta.2020.121722.
26. Rawtani, D., Rao, P.K., & Hussain, C.M. 2020. Recent advances in analytical, bioanalytical and miscellaneous applications of green nanomaterial. *TrAC Trends in Analytical Chemistry* **133**, 116109. Doi: 10.1016/j.trac.2020.116109.
27. Chen, Y., Gao, Y., Chen, Y., Liu, L., Mo, A., & Peng, Q. 2020. Nanomaterials-based photothermal therapy and its potentials in antibacterial treatment. *Journal of the Controlled Release Society* **328**, 251–262. Doi: 10.1016/j.jconrel.2020.08.055.
28. Ehtesabi, H. 2020. Carbon nanomaterials for salivary-based biosensors: a review. *Materials Today Chemistry* **17**, 100342. Doi: 10.1016/j.mtchem.2020.100342.
29. Kaviya, S. 2020. Synthesis, self-assembly, sensing methods and mechanism of biosource facilitated nanomaterials: A review with future outlook. *Nano-Structures & Nano-Objects* **23**, 100498. Doi: 10.1016/j.nanoso.2020.100498.
30. Wu, D., Ma, Y., Cao, Y., & Zhang, T. 2020. Mitochondrial toxicity of nanomaterials. *Science of the Total Environment* **702**, 134994. Doi: 10.1016/j.scitotenv.2019.134994.
31. Kumar, R., Aadil, K.R., Ranjan, S., & Kumar, V.B. 2020. Advances in nanotechnology and nanomaterials based strategies for neural tissue engineering. *Journal of Drug Delivery Science Technology* **57**, 101617. Doi: 10.1016/j.jddst.2020.101617.
32. Sauvage, F., Schymkowitz, J., et al. 2020. Nanomaterials to avoid and destroy protein aggregates. *Nano Today* **31**, 100837. Doi: 10.1016/j.nantod.2019.100837.
33. Zeng, G., & Chen, Y. 2020. Surface modification of black phosphorus-based nanomaterials in biomedical applications: Strategies and recent advances. *Acta Biomaterialia* **118**, 1–17. Doi: 10.1016/j.actbio.2020.10.004.
34. Park, W., Shin, H., et al. 2020. Advanced hybrid nanomaterials for biomedical applications. *Progress in Materials Science* **114**, 100686. Doi: 10.1016/j.pmatsci.2020.100686.
35. Liu, W., Zhao, L., Wang, C., & Zhou, J. 2020. Conductive nanomaterials for cardiac tissues engineering. *Engineered Regeneration* **1**, 88–94. Doi: 10.1016/j.engreg.2020.09.001.
36. Ehtesabi, H. 2020. Application of carbon nanomaterials in human virus detection. *Journal of Science: Advanced Materials and Devices*, ISSN 2468-2179. Doi: 10.1016/j.jsamd.2020.09.005.
37. Tan, Y.Y., Yap, P.K., et al. 2020. Perspectives and advancements in the design of nanomaterials for targeted cancer theranostics. *Chemicobiological Interactions* **329**, 109221. Doi: 10.1016/j.cbi.2020.109221.
38. Nguyen, Q.H., & Kim, M.I.l. 2020. Nanomaterial-mediated paper-based biosensors for colorimetric pathogen detection. *TrAC Trends in Analytical Chemistry* **132**, 116038. Doi: 10.1016/j.trac.2020.116038.

39. Bilal, M., Barani, M., Sabir, F., Rahdar, A., & Kyzas, G.Z. 2020. Nanomaterials for the treatment and diagnosis of Alzheimer's disease: An overview. *Nano Impact* **20**, 100251. Doi: 10.1016/j.impact.2020.100251.

40. Chai, L.X., Fan, X.X., et al. 2020. Low-dimensional nanomaterials enabled autoimmune disease treatments: Recent advances, strategies, and future challenges. *Coordination Chemistry Reviews* 213697. Doi: 10.1016/j.ccr.2020.213697.

41. Mao, K., Zhang, H., et al. 2020. Nanomaterial-based aptamer sensors for analysis of illicit drugs and evaluation of drugs consumption for wastewater-based epidemiology. *TrACTrendsin Analytical Chemistry* **130**, 115975. Doi: 10.1016/j.trac.2020.115975.

42. Jung, Y., Huh, Y., & Kim, D. 2021. Recent advances in surface engineering of porous silicon nanomaterials for biomedical applications. *Microporous and Mesoporous Materials* **310**, 110673. Doi: 10.1016/j.micromeso.2020.110673.

43. Ogunsona, E.O., Muthuraj, R., Ojogbo, E., Valerio, O., & Mekonnen, T.H. 2020. Engineered nanomaterials forantimicrobial applications: A review. *Applied Materials Today* **18**, 100473. Doi: 10.1016/j.apmt.2019.100473.

44. Sahu, A., Kwon, I., & Tae, G. 2020. Improving cancer therapy through the nanomaterials-assisted alleviation of hypoxia. *Biomaterials* **228**, 119578. Doi: 10.1016/j.biomaterials.2019.119578.

45. Wang, X., & Cheng, L. 2020. Multifunctional Prussian blue-based nanomaterials: Preparation, modification, and theranostic applications. *Coordination Chemistry Reviews* **419**, 213393. Doi: 10.1016/j.ccr.2020.213393.

46. Ahmadian, E., Samiei, M., et al. 2020. Monitoring of drug resistance towards reducing the toxicity of pharmaceutical compounds: Past, present and future. *Journal of Pharmaceutical Biomedecal Analysis* **186**, 113265. Doi: 10.1016/j.jpba.2020.113265.

47. Kubackova, J., Zbytovska, J., & Holas, O. 2020. Nanomaterials for direct and indirect immunomodulation: A review of applications. *European Journal of Pharmaceutical Sciences* **142**, 105139. Doi: 10.1016/j.ejps.2019.105139.

48. Sarkar, S., Gulati, K., Mishra, A., & Poluri K.M. 2020. Protein nanocomposites: Special inferences to lysozyme based nanomaterials. *International Journal of Biological Macromolecules* **151**, 467–482. Doi: 10.1016/j.ijbiomac.2020.02.179.

49. Agwa, M.M., & Sabra, S. 2020. Lactoferrin coated or conjugated nanomaterials as an active targeting approach in nanomedicine. *International Journal of Biological Macromolecules* ISSN 0141-8130. Doi: 10.1016/j.ijbiomac.2020.11.107.

50. Falahati, M., Attar, F., et al. 2020. Gold nanomaterials as key suppliers in biological and chemical sensing, catalysis, and medicine. *Biochimica et Biophysica Acta-General Subjects* **1864**(1), 129435. Doi: 10.1016/j.bbagen.2019.129435.

51. Bachheti, R.K., Fikadu, A., Bachheti, A., & Husen, A. 2020. Biogenic fabrication of nanomaterials from flower-based chemical compounds, characterization and their various applications: A review. *Saudi Journal of Biological Sciences* **27**(10), 2551–2562. Doi: 10.1016/j.sjbs.2020.05.012.

52. Shah, R.A., Frazar, E.M., & Hilt, J.Z. 2020. Recent developments in stimuli responsive nanomaterials and their bionanotechnology applications. *Current Opinion in Chemical Engineering* **30**, 103–111. Doi: 10.1016/j.coche.2020.08.007.

53. Zeng, Y., Li, H., et al. 2020. Engineered gadolinium-based nanomaterials as cancer imaging agents. *Applied Materials Today* **20**, 100686. Doi: 10.1016/j.apmt.2020.100686.

54. Hussain, Z., Thu, H.E., et al. 2020. Nano-scaled materials may induce severe neurotoxicity upon chronic exposure to brain tissues: A critical appraisal and recent updates on predisposing factors, underlying mechanism, and future prospects. *Journal of Controlled Release* **328**, 873–894. Doi: 10.1016/j.jconrel.2020.10.053.

55. Kumar, S., Nehra, M., et al. 2020. Nanotechnology-based biomaterials for orthopaedic applications: Recent advances and future prospects. *Materials Science and Engineering: C* **106**, 110154. Doi: 10.1016/j.msec.2019.110154.

56. Farzin, L., Shamsipur, M., Samandari, L., & Sheibani, S. 2020. HIV biosensors for early diagnosis of infection: The intertwine of nanotechnology with sensing strategies. *Talanta* **206**, 120201. Doi: 10.1016/j.talanta.2019.120201.
57. Farzin, M.A., & Abdoos, H. 2020. A critical review on quantum dots: From synthesis toward applications in electrochemical biosensors for determination of disease-related biomolecules. *Talanta* 121828. Doi: 10.1016/j.talanta.2020.121828.
58. Figueroa, S.M., Fleischmann, D., & Goepferich, A. 2020. Biomedical nanoparticle design: What we can learn from viruses. *Journal of Controlled Release*, ISSN 0168-3659. Doi: 10.1016/j.jconrel.2020.09.045.
59. Zhou, Y., Yu, W., Cao, J., & Gao, H. 2020. Harnessing carbon monoxide-releasing platforms for cancertherapy. *Biomaterials* **255**, 120193. Doi: 10.1016/j.biomaterials.2020.120193.
60. Sharifi, M., Hasan, A., Attar, F., Taghizadeh, A., & Falahati, M. 2020. Development of point-of-care nanobiosensors for breast cancers diagnosis. *Talanta* **217**, 12109. Doi: 10.1016/j.talanta.2020.121091.
61. Siafaka, P.I., Okur, N.U., Karantas, I.D., Okur, M.E., & Gundogdu, E.A. 2020. Current update on nanoplatforms as therapeutic and diagnostic tools: A review for the materials used as nanotheranostics and imaging modalities. *Asian Journal of Pharmaceutical Sciences* ISSN 1818-0876. Doi: 10.1016/j.ajps.2020.03.003.
62. Dhara, K., & Mahapatra, D.R. 2020. Review on electrochemical sensing strategies for C-reactive protein and cardiac troponin I detection. *Microchemical Journal* **156**, 104857. Doi: 10.1016/j.microc.2020.104857.
63. Cernat, A., Stefan, G., Tertis, M., Cristea, C., & Simon, I. 2020. An overview of the detection of serotonin and dopamine with graphene-based sensors. *Bioelectrochemistry* **136**, 107620. Doi: 10.1016/j.bioelechem.2020.107620.
64. Henna, T.K., Raphey, V.R., et al. 2020. Carbon nanostructures: The drug and the delivery system for brain disorders. *International Journal of Pharmaceutics* **587**, 119701. Doi: 10.1016/j.ijpharm.2020.119701.
65. Negahdary, M., Sattarahmady, N., & Heli, H. 2020. Advances in prostate specific antigen biosensors-impact of nanotechnology. *ClinicaChimica Acta* **504**, 43–55. Doi: 10.1016/j.cca.2020.01.028.
66. Xiang, W., Lv, Q., Shi, H., Xie, B., & Gao, L. 2020. Aptamer-based biosensor for detecting carcinoembryonic antigen. *Talanta* **214**, 120716. Doi: 10.1016/j.talanta.2020.120716.
67. Ghitman, J., Biru, E.I., Stan, R., & Iovu, H. 2020. Review of hybrid PLGA nanoparticles: Future of smart drug delivery and theranostics medicine. *Materials & Design* **193**, 108805. Doi: 10.1016/j.matdes.2020.108805.
68. Cheng, M., Zhu, G., et al. 2020. A review of flexible force sensors for human health monitoring. *Journal of Advanced Research* **26**, 53–68. Doi: 10.1016/j.jare.2020.07.001.
69. Adithya, S.P., Sidharthan, S.D., Abhinandan, R., Balagangadharan, K., & Selvamurugan, N. 2020. Nanosheets-incorporated bio-composites containing natural and synthetic polymers/ceramics for bone tissue engineering. *International Journal of Biological Macromolecules* **164**, 1960–1972. Doi: 10.1016/j.ijbiomac.2020.08.053.
70. Wang, G.Y., Rayner, S.L., Chung, R., Shi, B.Y., & Liang, X.J. 2020. Advances in nanotechnology-based strategies for the treatments of amyotrophic lateral sclerosis. *Materials Today Bio* **6**, 100055. Doi: 10.1016/j.mtbio.2020.100055.
71. Pourali, A., Rashidi, M.R., Barar, J., Djavid, G.P., & Omidi, Y. 2020. Voltammetric biosensors for analytical detection of cardiac troponin biomarkers in acute myocardial infarction. *TrAC Trends in Analytical Chemistry* 116123. Doi: 10.1016/j.trac.2020.116123.
72. Ma, Y., Cai, F., et al. 2020. A review of the application of nanoparticles in the diagnosis and treatment of chronic kidney disease. *Bioactive Materials* **5**(3), 732–743. Doi: 10.1016/j.bioactmat.2020.05.002.

73. Horne, J., McLoughlin, L., Bridgers, B., & Wujcik, E.K. 2020. Recent developments in nanofiber-based sensors for disease detection, immunosensing, and monitoring. *Sensors and Actuators Reports* **2**(1), 100005. Doi: 10.1016/j.snr.2020.100005.

74. Miao, S., Liang, K., et al. 2020. Hetero-atom-doped carbon dots: Doping strategies, properties and applications. *Nano Today* **33**, 100879. Doi: 10.1016/j.nantod.2020.100879.

75. Cheng, Y.J., Hu, J.J., Qin, S.Y., Zhang, A.Q., & Zhang, X.Z. 2020. Recent advances in functional mesoporous silica-based nanoplatforms for combinational photo-chemotherapy of cancer. *Biomaterials* **232**, 119738. Doi: 10.1016/j.biomaterials.2019.119738.

76. Zhu, X., Zhang, Y., Liu, M., & Liu, Y. 2021. 2D titanium carbide MXenes as emerging optical biosensing platforms. *Biosensors and Bioelectronics* **171**, 112730. Doi: 10.1016/j.bios.2020.112730.

77. Szuplewska, A., Kulpinska, D., et al. 2020. Future applications of MXenes in biotechnology, nanomedicine, and sensors. *Trends in Biotechnology* **38**(3), 264–279. Doi: 10.1016/j.tibtech.2019.09.001.

78. Sivasankarapillai, V.S., Somakumar, A.K., et al. 2020. Cancer theranostic applications of MXene nanomaterials: Recent updates. *Nano-Structures & Nano-Objects* **22**, 100457. Doi: 10.1016/j.nanoso.2020.100457.

79. Singh, R.P. Shukla, V.K., Yadav, R.S., Sharma, P.K., Singh, P.K., & Pandey, A.C. (2011). Biological approach of zinc oxide nanoparticles formation and its characterization. *Advanced Materials Letters* **2**(4), 313–317.

80. Singh, R.P., Tiwari, A., & Pandey, A.C. 2011. Silver/polyaniline nanocomposite for the electrocatalytic hydrazine oxidation. *Journal of Inorganic and Organometallic Polymers and Materials* **21**, 788–792.

81. Singh, R.P. 2011. Prospects of nanobiomaterials for biosensing. *International Journal of Electrochemistry*, publisher SAGE-Hindawi journal collection. Vol. **2011**, Review Article ID 125487, 30 p. Doi:10.4061/2011/125487.

82. Singh, R.P. 2012. Prospects of organic conducting polymer modified electrodes: Enzymosensors, Editor: Benjamín R. Scharifker. *International Journal of Electrochemistry*, publisher SAGE-Hindawi journal collection. Academic Vol. **2012**, Article ID 502707, 14 p. Doi: 10.1155/2012/502707.

83. Fernandes, M., Singh, K.R.B., Sarkar, T., Singh, P., & Singh, R.P. 2020. Recent applications of magnesium oxide (MgO) nanoparticles in various domains. *Advanced Materials Letters* **11**, 20081543.

84. Nsibande, S. A., & Forbes, P. B. C. (2016). Fluorescence detection of pesticides using quantum dot materials-A review. *Analytica Chimica Acta* **945**, 9–22. Doi: 10.1016/j.aca.2016.10.002.

85. Duan, J., Jiang, X., Ni, S., Yang, M., & Zhan, J. (2011). Facile synthesis of N-acetyl-l-cysteine capped ZnS quantum dots as an eco-friendly fluorescence sensor for Hg2+. *Talanta* **85**(4), 1738–1743. Doi: 10.1016/j.talanta.2011.06.071.

86. Zhao, Q., Rong, X., Ma, H., & Tao, G. (2013). Dithizone functionalized CdSe/CdS quantum dots as turn-on fluorescent probe for ultrasensitive detection of lead ion. *Journal of Hazardous Materials* **250–251**, 45–52. Doi: 10.1016/j.jhazmat.2013.01.062.

87. Ren, X., & Chen, L. (2015). Quantum dots coated with molecularly imprinted polymer as fluorescence probe for detection of cyphenothrin. *Biosensors and Bioelectronics* **64**, 182–188. Doi: 10.1016/j.bios.2014.08.086.

88. Lin, B., Yu, Y., Li, R., Cao, Y., & Guo, M. (2016). Turn-on sensor for quantification and imaging of acetamiprid residues based on quantum dots functionalized with aptamer. *Sensors and Actuators B: Chemical* **229**, 100–109. Doi: 10.1016/j.snb.2016.01.114.

89. Yang, M., Han, A., Duan, J., Li, Z., Lai, Y., & Zhan, J. (2012). Magnetic nanoparticles and quantum dots co-loaded imprinted matrix for pentachlorophenol. *Journal of Hazardous Materials* **237–238**, 63–70. Doi: 10.1016/j.jhazmat.2012.07.064.

90. Ramnani, P., Saucedo, N. M., & Mulchandani, A. (2016). Carbon nanomaterial-based electrochemical biosensors for label-free sensing of environmental pollutants. *Chemosphere* **143**, 85–98. Doi: 10.1016/j.chemosphere.2015.04.063.
91. Liu, Y., & Wei, W. 2008. Layer-by-layer assembled DNA functionalized single-walled carbon nanotube hybrids for arsenic(III) detection. *Electrochemistry Communications* **10** (6), 872–875. Doi: 10.1016/j.elecom.2008.03.013
92. Gong, J.L., Sarkar, T., Badhulika, S., & Mulchandani, A. 2013. Label-free chemiresistive biosensor for mercury (II) based on single-walled carbon nanotubes and structure-switching DNA. *Applied Physics Letters* **102**(1), 013701. Doi: 10.1063/1.4773569.
93. Liu, F., Kim, Y. H., Cheon, D. S., & Seo, T. S. 2013. Micropatterned reduced graphene oxide based field-effect transistor for real-time virus detection. *Sensors and Actuators B: Chemical* **186**, 252–257. Doi: 10.1016/j.snb.2013.05.097.
94. Lian, Y., Yuan, M., & Zhao, H. 2014. DNA wrapped metallic single-walled carbon nanotube sensor for Pb (II) detection. *Fullerenes, Nanotubes and Carbon Nanostructures* **22**(5), 510–518. Doi: 10.1080/1536383X.2012.690462.
95. Du, D., Huang, X., Cai, J., & Zhang, A. (2007). Amperometric detection of triazophos pesticide using acetylcholinesterase biosensor based on multiwall carbon nanotube–chitosan matrix. *Sensors and Actuators B: Chemical* **127**(2), 531–535. Doi: 10.1016/j.snb.2007.05.006.
96. Oliveira, J. M., Salgado, A. J., Sousa, N., Mano, J. F., & Reis, R. L. 2010. Dendrimers and derivatives as a potential therapeutic tool in regenerative medicine strategies-A review. *Progress in Polymer Science* **35**(9), 1163–1194. Doi: 10.1016/j.progpolymsci.2010.04.006.
97. Castillo, G., Spinella, K., Poturnayová, A., Snejdarkova, M., Mosiello, L., & Hianik, T. 2015. Detection of aflatoxin B1 by aptamer-based biosensor using PAMAM dendrimers as immobilization platform. *Food Control* **52**, 9–18. Doi: 10.1016/j.foodcont.2014.12.008.
98. Snejdarkova, M., Svobodova, L., Nikolelis, D. P., Wang, J., & Hianik, T. 2003. Acetylcholine biosensor based on dendrimer layers for pesticides detection. *Electroanalysis* **15**(14), 1185–1191. Doi: 10.1002/elan.200390145.
99. Baumner, A. J., & Schmid, R. D. 1998. Development of a new immunosensor for pesticide detection: a disposable system with liposome-enhancement and amperometric detection. *Biosensors and Bioelectronics* **13**(5), 519–529. Doi: 10.1016/S0956-5663(97)00131-0
100. Viswanathan, S., Wu, L., Huang, M.R., & Ho, J. A. 2006. Electrochemical immunosensor for cholera toxin using liposomes and Poly(3,4-ethylenedioxythiophene)-coated carbon nanotubes. *Analytical Chemistry* **78**(4), 1115–1121. Doi: 10.1021/ac051435d.
101. Zhang, X., Zeng, T., Hu, C., Hu, S., & Tian, Q. 2016. Studies on fabrication and application of arsenic electrochemical sensors based on titanium dioxide nanoparticle modified gold strip electrodes. *Analytical Methods* **8**(5), 1162–1169. Doi: 10.1039/C5AY02397A.
102. Wu, S., Zhao, Q., Zhou, L., & Zhang, Z. 2014. Stripping analysis of trace arsenic based on the MnO x /AuNPs composite film modified electrode in alkaline media. *Electroanalysis* **26**(8), 1840–1849. Doi: 10.1002/elan.201400219.
103. Li, F., Wang, J., Lai, Y., Wu, C., Sun, S., He, Y., & Ma, H. 2013. Ultrasensitive and selective detection of copper (II) and mercury (II) ions by dye-coded silver nanoparticle-based SERS probes. *Biosensors and Bioelectronics* **39**(1), 82–87. Doi: 10.1016/j.bios.2012.06.050.
104. Guo, X., & Chen, F. 2005. Removal of arsenic by bead cellulose loaded with iron oxyhydroxide from groundwater. *Environmental Science & Technology* **39**(17), 6808–6818. Doi: 10.1021/es048080k.

105. Sugunan, A., Thanachayanont, C., Dutta, J., & Hilborn, J. G. 2005. Heavy-metal ion sensors using chitosan-capped gold nanoparticles. *Science and Technology of Advanced Materials* **6**(3–4), 335–340. Doi: 10.1016/j.stam.2005.03.007.

106. Su, P.G., & Huang, L.N. 2007. Humidity sensors based on TiO_2 nanoparticles/polypyrrole composite thin films. *Sensors and Actuators B: Chemical* **123**(1), 501–507. Doi: 10.1016/j.snb.2006.09.052.

107. Xu, X., Niu, X., et al. 2020. Nanomaterial-based sensors and biosensors for enhanced inorganic arsenic detection: A functional perspective. *Sensors and Actuators B: Chemical* **315**, 128100. Doi: 10.1016/j.snb.2020.128100.

108. Saleh, T.A. 2020. Trends in the sample preparation and analysis of nanomaterials as environmental contaminants. *Trends in Environment and Analytical Chemistry* **28**, e00101. Doi: 10.1016/j.teac.2020.e00101.

109. Malakar, A., Kanel, S.R., Ray, C., Snow, D.D., & Nadagouda, M.N. 2020. Nanomaterials in the environment, human exposure pathway, and health effects: A review. *Science of the Total Environment*, 143470. Doi: 10.1016/j.scitotenv.2020.143470.

110. Hajiabadi, S.H., Aghaei, H., Aghamohammadi, M.K., & Shorgasthi, M. 2020. An overview on the significance of carbon-based nanomaterials in upstream oil and gas industry. *Journal of Petroleum Science Engineering* **186**, 106783. Doi: 10.1016/j.petrol.2019.106783.

111. Saleh, T.A. 2020. Nanomaterials: Classification, properties, and environmental toxicities. *Environment Technology Innovations* **20**, 101067. Doi: 10.1016/j.eti.2020.101067.

112. Mondal, P., Anweshan, A., & Purkait, M.K. 2020. Green synthesis and environmental application of iron-based nanomaterials and nanocomposite: A review. *Chemosphere* **259**, 127509. Doi: 10.1016/j.chemosphere.2020.127509.

113. Bakshi, M.S. 2020. Impact of nanomaterials on ecosystems: Mechanistic aspects in vivo. *Environmental Research* **182**, 109099. Doi: 10.1016/j.envres.2019.109099.

114. Gupta, S.K., & Mao, Y. 2020. A review on molten salt synthesis of metal oxide nanomaterials: Status, opportunity, and challenge. *Progress in Materials Science* 100734. Doi: 10.1016/j.pmatsci.2020.100734.

115. Singh, H., Bamrah, A., et al. 2020. Nanomaterial-based fluorescent sensors for the detection of lead ions. *Journal of Hazardous Materials* 124379. Doi: 10.1016/j.jhazmat.2020.124379.

116. Filik, H., & Avan, A.A. 2020. Review on applications of carbon nanomaterials for simultaneous electrochemical sensing of environmental contaminant dihydroxybenzene isomers. *Arabian Journal of Chemistry* **13**(7), 6092–6105. Doi: 10.1016/j.arabjc.2020.05.009.

117. He, Y., Wang, Z., et al. 2020. Metal-organic framework-derived nanomaterials in environment related fields: Fundamentals, properties and applications. *Coordination Chemistry Review*, 213618. Doi: 10.1016/j.ccr.2020.213618.

118. Bhardwaj, S.K., Bhardwaj, N., et al. 2021. Recent progress in nanomaterial-based sensing of airborne viral and bacterial pathogens. *Environment International* **146**, 106183. Doi: 10.1016/j.envint.2020.106183.

119. Liu, Y., Pang, H., et al. 2021. Zeolitic imidazolate framework-based nanomaterials for the capture of heavy metal ions and radionuclides:Areview. *Chemical Engineering Journal* **406**, 127139. Doi: 10.1016/j.cej.2020.127139.

120. Singh, M., Singh, S., Singh, S.P., & Patel, S.S. 2020. Recent advancement of carbon nanomaterials engrained molecular imprinted polymer for environmental matrix. *Trends in Environmental Analytical Chemistry* **27**, e00092. Doi: 10.1016/j.teac.2020.e00092.

121. Peng, Z., Liu, X., et al. 2020. Advances in the application, toxicity and degradation of carbon nanomaterials in environment: A review. *Environment International* **134**, 105298. Doi: 10.1016/j.envint.2019.105298.

122. Feng, X., Yu, Z., et al. 2020. Review MXenes as a new type of nanomaterial for environmental applications in the photocatalytic degradation of water pollutants. *Ceramics International*, ISSN 0272-8842. Doi: 10.1016/j.ceramint.2020.11.151.

123. Jawed, A., Saxena, V., & Pandey, L.M. 2020. Engineered nanomaterials and their surface functionalization for the removal of heavy metals: A review. *Journal of Water Process Engineering* **33**, 101009. Doi: 10.1016/j.jwpe.2019.101009.

124. Zhang, X., Yuan, X., et al. 2020. Powerful combination of 2D g-C3N4 and 2D nanomaterials for photocatalysis: Recent advances. *Chemical Engineering Journal* **390**, 124475. Doi: 10.1016/j.cej.2020.124475.

125. Lashari, N., & Ganat, T. 2020. Emerging applications of nanomaterials in chemical enhanced oil recovery: Progress and perspective. *Chinese Journal of Chemical Engineering* **28**(8), 1995–2009. Doi: 10.1016/j.cjche.2020.05.019.

126. Singh, H., Bhardwaj, N., Arya, S.K., & Khatri, M. 2020. Environmental impacts of oil spills and their remediation by magnetic nanomaterials. *Environmental Nanotechnology, Monitoring & Management* **14**, 100305. Doi: 10.1016/j.enmm.2020.100305.

127. Pandey, H., Khare, P., Singh, S., & Singh, S.P. 2020. Carbon nanomaterials integrated molecularly imprinted polymers for biological sample analysis: A critical review. *Materials Chemistry and Physics* **239**, 121966. Doi: 10.1016/j.matchemphys.2019.121966.

128. Lu, X., Sun, J., & Sun, X. 2020. Recent advances in biosensors for the detection of estrogens in the environment and food. *TrAC Trends in Analytical Chemistry* **127**, 115882. Doi: 10.1016/j.trac.2020.115882.

129. Hasan, A., Nanakali, N.M.Q., et al. 2020. Nanozyme-based sensing platforms for detection of toxic mercury ions: An alternative approach to conventional methods. *Talanta* **215**, 120939. Doi: 10.1016/j.talanta.2020.120939.

130. Wang, L. 2020. Metal-organic frameworks for QCM-based gas sensors: A review. *Sensors and Actuators A: Physical* **307**, 111984. Doi: 10.1016/j.sna.2020.111984.

131. Fadillah, G., Saputra, O.A., & Saleh, T.A. 2020. Trends in polymers functionalized nanostructures for analysis of environmental pollutants. *Trends in Environmental Analytical Chemistry* **26**, e00084. Doi: 10.1016/j.teac.2020.e00084.

132. Lu, F., & Astruc, D. 2020. Nanocatalysts and other nanomaterials for water remediation from organic pollutants. *Coordination Chemistry Review* **408**, 213180. Doi: 10.1016/j.ccr.2020.213180.

133. Singh, R.P. 2019. *Nanocomposites: Recent Trends, Developments and Applications*. CRC Press, November 15, 2018 Forthcoming, Reference-552, CRC Press, 2018. pp 552 Chap 2, Advances in Nanostructured Composites: Volume 1: Carbon Nanotube and Graphene Composites. 1st Edition, Mahmood Aliofkhazraei.

134. Ghorbani, M., Seyedin, O., & Aghamohammadhassan, M. 2020. Adsorptive removal of lead (II) ion from water and wastewater media using carbon-based nanomaterials as unique sorbents: A review. *Journal of Environmental Management* **254**, 109814. Doi: 10.1016/j.jenvman.2019.109814.

135. Lal, S., Singhal, A., & Kumari, P. 2020. Exploring carbonaceous nanomaterials for arsenic and chromium removal from wastewater. *Journal of Water Process Engineering* **36**, 101276. Doi: 10.1016/j.jwpe.2020.101276.

136. Jeon, M., Jun, B.M., et al. 2020. A review on MXene-based nanomaterials as adsorbents in aqueous solution. *Chemosphere* **261**, 127781. Doi: 10.1016/j.chemosphere.2020.127781.

137. Zeidman, A.B., Narvaez, O.M.R., Moon, J., & Bandala, E.R. 2020. Removal of antibiotics in aqueous phase using silica-based immobilized nanomaterials: A review. *Environment Technology Innovation* **20**, 101030. Doi: 10.1016/j.eti.2020.101030.

138. Chenab, K.K., Sohrabi, B., Jafari, A., & Ramakrishna, S. 2020. Water treatment: functional nanomaterials and applications from adsorption to photodegradation. *Materials Today Chemistry* **16**, 100262. Doi: 10.1016/j.mtchem.2020.100262.

139. Basak, G., Hazra, C., & Sen, R. 2020. Biofunctionalized nanomaterials for *in situ* clean-up of hydrocarbon contamination: A quantum jump in global bioremediation research. *Journal of Environment Management* **256**, 109913. Doi: 10.1016/j.jenvman.2019.109913.

140. Kumar, V., Lee, Y.S., et al. 2020. Potential applications of graphene-based nanomaterials as adsorbent for removal of volatile organic compounds. *Environment International* **135**, 105356. Doi: 10.1016/j.envint.2019.105356.

141. Lye, Q.V., Maqbool, T., et al. 2020. Characterization of dissolved organic matter for understanding the adsorption on nanomaterials in aquatic environment: A review. *Chemosphere*, 128690. Doi: 10.1016/j.chemosphere.2020.128690.

142. Bolade, O.P., Williams, A.B., & Benson, N.U. 2020. Green synthesis of iron-based nanomaterials for environmental remediation: A review. *Environmental Nanotechnology, Monitoring & Management* **13**, 100279. Doi: 10.1016/j.enmm.2019.100279.

143. Dong, X., Yan, X., et al. 2020. Ultrasensitive detection of chloramphenicol using electrochemical aptamer sensor: A mini review. *Electrochemistry Communication* **120**, 106835. Doi: 10.1016/j.elecom.2020.106835.

144. Kegl, T., Kosak, A., et al. 2020. Adsorption of rare earth metals from wastewater by nanomaterials: A review. *Journal of Hazardous Materials* **386**, 121632. Doi: 10.1016/j.jhazmat.2019.121632.

145. Gusain, R., Kumar, N., & Ray, S.S. 2020. Recent advances in carbon nanomaterial-based adsorbents for water purification. *Coordination Chemistry Reviews* **405**, 213111. Doi: 10.1016/j.ccr.2019.213111.

146. Wadhawan, S., Jain, A., Nayyar, J., & Mehta, S.K. 2020. Role of nanomaterials as adsorbents in heavy metal ion removal from waste water: A review. *Journal of Water Process Engineering* **33**, 101038. Doi: 10.1016/j.jwpe.2019.101038.

147. Yadav, S., Saleem, H., et al. 2020. Recent developments in forward osmosis membranes using carbon-based nanomaterials. *Desalination* **482**, 114375. Doi: 10.1016/j.desal.2020.114375.

148. Nasrollahzadeh, M., Sajjadi, M., Iravani, S., & Varma, R.S. 2021. Starch, cellulose, pectin, gum, alginate, chitin and chitosan derived (nano)materials for sustainable water treatment: A review. *Carbohydrate Polymers* **251**, 116986. Doi: 10.1016/j.carbpol.2020.116986.

149. Ihsanullah, I. 2021. Boron nitride-based materials for water purification: Progress and outlook. *Chemosphere* **263**, 127970. Doi: 10.1016/j.chemosphere.2020.127970.

150. Mmelesi, O.K., Masunga, N., et al. 2020. Cobalt ferrite nanoparticles and nanocomposites: Photocatalytic, antimicrobial activity and toxicity in water treatment. *Materials Science in Semiconductor Processing* 105523. Doi: 10.1016/j.mssp.2020.105523.

151. Nunes, S.P., Emecen, P.Z.C., et al. 2020. Thinking the future of membranes: Perspectives for advanced and new membrane materials and manufacturing processes. *Journal of Membrane Science* **598**, 117761. Doi: 10.1016/j.memsci.2019.117761.

152. Zhou, W., Zhang, W., & Cai, Y. 2021. Laccase immobilization for water purification: A comprehensive review. *Chemical Engineering Journal* **403**, 126272. Doi: 10.1016/j.cej.2020.126272.

153. Singh, R.P. 2016. Nanobiosensors: Potentiality towards Bioanalysis. *Journal of Bioanalysis & Biomedicine* **8**, e143. Doi: 10.4172/1948-593X.1000e143.

154. Singh, R.P. 2017. Application of nanomaterials towards development of nanobiosensors and their utility in agriculture, in *Nanotechnology: An Agricultural Paradigm* (eds Prasad, R., Kumar, M., & Kumar, V.), Springer Publisher, New York, USA, Ch 14, 293–303.

155. Singh R.P. 2019. *Utility of Nanomaterials in Food Safety*. Chap 11. Food Safety and Human Health Chap 11. (eds Singh, R.L. and Mondal, S.K.), Elsevier Inc ISSN: 2523-8027.

156. Singh, R.P. 2021. Recent trends, prospects and challenges of nanobiosensors in agriculture. in *Biosensors in Agriculture: Recent Trends and Future Perspectives*. Springer Nature.

157. Singh, R.P., & Singh, K.R. 2021. Nanobiotechnology in animal production and health. in *Advances in Animal Genomics*, (eds Mondal, S., & Singh, R.L.), Elsevier, Cambridge, MA.
158. Brock, D.A., Douglas, T.E., Queller, D.C., & Strassmann, J.E. 2011. Primitive agriculture in a social amoeba. *Nature* **469**(7330), 393–396.
159. Viswanathan, S., Wu, L., Huang, M., & Ho, J. 2006. Electrochemical immunosensor for cholera toxin using liposomes and poly(3,4-ethylenedioxythiophene)-coated carbon nanotubes. *Analytical Chemistry* **78**(4), 1115–1121.
160. Wang, Y., & Alocijia, E.C. 2015. Gold nanoparticle-labeled biosensor for rapid and sensitive detection of bacterial pathogens. *Journal of Biological Engineering* **9**, 16.
161. Kim, G., Park, S.B., Moon, J., & Lee, S. 2013. Detection of pathogenic Salmonella with nanobiosensors. *Analytical Methods* **5**, 5717–5723.
162. Masikini, M., Mailu, S.N., et al. 2015. A fumonisins immunosensor based on polyanilino-carbon nanotubes doped with palladium telluride quantum dots. *Sensors* **15**, 529–546.
163. Parker, C.O., & Tothill, I.E. 2009. Development of an electrochemical immunosensor for aflatoxin M (1) in milk with focus on matrix interference. *Biosensors and Bioelectronics* **24**(8), 2452–2457.
164. Xu, X., Liu, X., Li, Y., & Ying, Y. 2013. A simple and rapid optical biosensor for detection of aflatoxin B1 based on competitive dispersion of gold nanorods. *Biosensors and Bioelectronics* **47**, 361–367.
165. Eldin, T.A.S., Elshoky, H.A., & Ali, M.A. 2014. Nanobiosensor based on gold nanoparticles probe for aflatoxin B1 detection in food. *International Journal of Current Microbiology and Applied Sciences* **3**(8), 219–230.
166. Bonel, L., Vidal, J., Duato, P., & Castillo, J. 2010. Ochratoxin A nanostructured electrochemical immunosensors based on polyclonal antibodies and gold nanoparticles coupled to the antigen. *Analytical Methods* **2**, 335–341.
167. Zhang, S., Shan, L., et al. 2008. Study of enzyme biosensor based on carbon nanotubes modified electrode for detection of pesticides residue. *Chinese Chemical Letters* **19**, 592–594.
168. Norouzi, P., Pirali-Hamedani, M., Ganjal, M.R., & Faridbod, F. 2010. A novel acetylcholinesterase biosensor for determination of monocrotophos using FFT continuous cyclic voltammetry. *International Journal of Electrochemical Sciences* **5**, 1434–1446.
169. Zheng, Z., Zhoub, Y., Li, X., Liua, S., & Tang, Z. 2011. Highly-sensitive organophosphorous pesticide biosensors based on nanostructured films of acetylcholinesterase and CdTe quantum dots. *Biosensors and Bioelectronic* **26**, 3081–3085.
170. Guan H., Zhang F., Yu J., & Chi D. (2012). The novel acetylcholinesterase biosensors based on liposome bioreactors-chitosan nanocomposites film for detection of organophosphates pesticides. *Food Research International* **49**(1), 15–21.
171. Song, Y., Chen, J., & Wang, L.A. 2015. Simple electrochemical biosensor based on AuNPs/MPS/Au electrode sensing layer for monitoring carbamate pesticides in real samples. *Journal of hazardous* **304**, 103–109.
172. Haddaoui, M., & Raouafi, N. 2015. Chlortoluron-induced enzymatic activity inhibition in tyrosinase/ZnO NPs/ SPCE biosensor for the detection of ppb levels of herbicide. *Sensors Actuators B Chemistry* **219**, 171–178.
173. McEwen, S.A., & Fedorka-Cray, P.J. 2002. Antimicrobial use and resistance in animals. *Clinical Infectious Diseases* **34**(3), 93–106.
174. Courvalin, P. 2008. Predictable and unpredictable evolution of antibiotic resistance. *Journal of International Medicine* **264**, 4–16.
175. Idowu, F., Junaid, K., Paul, A., Gabriel, O., Paul, A., Sati, N., Maryam, M., & Jarlath, U. 2010. Antimicrobial screening of commercial eggs and determination of tetracycline residue using two microbiological methods. *International Journal of Poultry Sciences* **9**(10), 959–962.

176. Wu, S., Zhang, H., et al. 2015. Aptamer-based fluorescence biosensor for chloramphenicol determination using upconversionnanoparticles. *Food Control* **50**, 597–604. Doi: 10.1016/j.foodcont.2014.10.003
177. Simmons, M.D., Miller, L.M., Sundstrom, M.O., & Johnson, S. 2020. Aptamer-based detection of ampicillin in urine samples. *Antibiotics* **9**(10), 655. Doi: 10.3390/antibiotics9100655.
178. Hou, H., Bai, X., et al. 2013. Aptamer-based cantilever array sensors for oxytetracycline detection. *Analytical Chemistry* **85**(4), 2010–2014. Doi: 10.1021/ac3037574.
179. Sastry, R.K., Anshul, S., & Rao, N.H. 2013. Nanotechnology in food processing sector-An assessment of emerging trends. *Journal of Food Science and Technology.* **50**(5), 831–41.
180. Inbaraj, B.S., & Chen, B.H. 2016. Nanomaterial-based sensors for detection of food-borne bacterial pathogens and toxins as well as pork adulteration in meat products. *Journal of Food and Drug Analysis* **24**(1), 15–28.
181. Ndukwu, M.C., Ikechukwu-Edeh, C.E., et al. 2020. Nanomaterials application in greenhouse structures, crop processing machinery, packaging materials and agro-biomass conversion. *Materials Science for Energy Technologies* **3**, 690–699. Doi: 10.1016/j.mset.2020.07.006.
182. Wang, W., Wang, X., et al. 2020. Recent advances in nanomaterials-based electrochemical (bio)sensors for pesticides detection. *TrAC Trends in Analytical Chemistry* **132**, 116041. Doi: 10.1016/j.trac.2020.116041.
183. Su, D., Li, H., Yan, X., Lin, Y., & Lu, V. 2020. Biosensors based on fluorescence carbon nanomaterials for detection of pesticides. *TrAC* Trends in *Analytical Chemistry* 116126. Doi: 10.1016/j.trac.2020.116126.
184. Zhou, R., Zhao, L., et al. 2020. Recent advances in food-derived nanomaterials applied to biosensing. *TrAC Trends in Analytical Chemistry* **127**, 115884. Doi: 10.1016/j.trac.2020.115884.
185. Mohammadi, Z., & Jafari, S.M. 2020. Detection of food spoilage and adulteration by novel nanomaterial-based sensors. *Advance Colloidal Interface Science* **286**, 102297. Doi: 10.1016/j.cis.2020.102297.
186. Christopher, F.C., Kumar, P.S., Christopher, F.J., Joshiba, G.J., & Madhesh, P. 2020. Recent advancements in rapid analysis of pesticides using nano biosensors: A present and future perspective. *Journal of Cleaner Production* **269**, 122356. Doi: 10.1016/j.jclepro.2020.122356.
187. Usman, M., Farooq, M., et al. 2020. Nanotechnology in agriculture: Current status, challenges and future opportunities. *Science of the Total Environment* **721**, 137778. Doi: 10.1016/j.scitotenv.2020.137778.
188. Kumar, V., Vaid, K., Bansal, S.A., & Kim, K.H. 2020. Nanomaterial-based immunosensors for ultrasensitive detection of pesticides/herbicides: Current status and perspectives. *Biosensors and Bioelectronics* **165**, 112382. Doi: 10.1016/j.bios.2020.112382.
189. Carvalho, A.P.A.D., & Conte Jr, C.A. 2020. Green strategies for active food packagings: A systematic review on active properties of graphene-based nanomaterials and biodegradable polymers. *Trends in Food Science & Technology* **103**, 130–143. Doi: 10.1016/j.tifs.2020.07.012.
190. Gomez, A., Narayan, M., et al. 2021. Effects of nano-enabled agricultural strategies on food quality: Current knowledge and future research needs. *Journal of Hazardous Materials* **401**, 123385. Doi: 10.1016/j.jhazmat.2020.123385.
191. Mao, K., Zhang, H., et al. 2020. Nanomaterial-based aptamer sensors for arsenic detection. *Biosensors and Bioelectronics* **148**, 111785. Doi: 10.1016/j.bios.2019.111785.
192. Aigbe, U.O., & Osibote, O.A. 2020. A review of hexavalent chromium removal from aqueous solutions by sorption technique using nanomaterials. *Journal of Environmental Chemical Engineering* **8**(6), 104503. Doi: 10.1016/j.jece.2020.104503.

193. Wang, W., & Gunasekaran, S. 2020. Nanozymes-based biosensors for food quality and safety. *TrAC Trends in Analytical Chemistry* **126**, 115841. Doi: 10.1016/j.trac.2020.115841.
194. Jildeh, N.B., & Matouq, M. 2020. Nanotechnology in packing materials for food and drug stuff opportunities. *Journal of Environmental Chemical Engineering* **8**(5), 104338. Doi: 10.1016/j.jece.2020.104338.
195. Bilal, M., & Iqbal, H.M.N. 2020. Armoring bio-catalysis via structural and functional coordination between nanostructured materials and lipases for tailored applications. *International J. Biological Macromolecules* ISSN 0141-8130. Doi: 10.1016/j.ijbiomac.2020.10.239.
196. Park, S., Jeon, Y., et al. 2020. Nanoscale manufacturing as an enabling strategy for the design of smart food packaging systems. *Food Packaging and Shelf Life* **26**, 100570. Doi: 10.1016/j.fpsl.2020.100570.
197. Yang, Y., Li, G., et al. 2020. Recent advances on toxicity and determination methods of mycotoxins in foodstuffs. *Trends in Food Science &Technology* **96**, 233–252. Doi: 10.1016/j.tifs.2019.12.021.
198. Zhang, Y., Rui, X., & Simpson, B.K. 2021. Trends in nanozymes development versus traditional enzymes in food science. *Current Opinion in Food Science* **37**, 10–16. Doi: 10.1016/j.cofs.2020.08.001.
199. Wang, L., Huang, X., et al. 2020. Applications of surface functionalized Fe_3O_4 NPs-based detection methods in food safety. *Food Chemistry* 128343. Doi: 10.1016/j.foodchem.2020.128343.
200. Fang, L., Liao, X., & et al. 2020. Recent progress in immunosensors for pesticides. *Biosensors and Bioelectronics* **164**, 112255. Doi: 10.1016/j.bios.2020.112255.
201. Maghsoodi, M.R., Ghodszad, L., & Lajayer, B.A. 2020. Dilemma of hydroxyapatite nanoparticles as phosphorus fertilizer: Potentials, challenges and effects on plants. *Environmental Technology &Innovation* **19**, 100869. Doi: 10.1016/j.eti.2020.100869.
202. Jeevahan, J.J., Chandrasekaran, M., et al. 2020. Scaling up difficulties and commercial aspects of edible films for food packaging: A review. *Trends in Food Science &Technology* **100**, 210–222. Doi: 10.1016/j.tifs.2020.04.014.
203. Deschenes, L., & Ells, T. 2020. Bacteria-nanoparticle interactions in the context of nanofouling. *Advances in Colloid and Interface Science* **277**, 102106. Doi: 10.1016/j.cis.2020.102106.
204. Skladal, P. 2020. Smart bioelectronic tongues for food and drinks control. *TrAC Trends in Analytical Chemistry* **127**, 115887. Doi: 10.1016/j.trac.2020.115887.
205. Junior, L.M., Vieira, R.P., Jamroz, E., & Anjos, C.A.R. 2021. Furcellaran: An innovative biopolymer in the production of films and coatings. *Carbohydrate Polymers* **252**, 117221. Doi: 10.1016/j.carbpol.2020.117221.
206. Umaraw, P., Munekata, P.E.S., et al. 2020. Edible films/coating with tailored properties for active packaging of meat, fish and derived products. *Trends in Food Science &Technology* **98**, 10–24. Doi: 10.1016/j.tifs.2020.01.
207. Yadav, R.S., Singh, R.P., Verma, P., Tiwari, A., & Pandey, A.C. 2012. in *Intelligent Nanomaterials: Processes, Properties, and Applications. Smart Nanomaterials for Space and Energy Applications*, 213–249. Chap 6. Doi: 10.1002/9781118311974.ch6.
208. Kumar, P., Chaudhary, D., et al. 2020. Critical review on battery thermal management and role of nanomaterial in heat transfer enhancement for electrical vehicle application. *Journal of Energy Storage* 32, 102003. Doi: 10.1016/j.est.2020.102003.
209. Xu, Z., Zhao, H., et al. 2020. Noble-metal-free electrospun nanomaterials as electrocatalysts for oxygen reduction reaction. *Materials Today Physics* **15**, 100280. Doi: 10.1016/j.mtphys.2020.100280.
210. Huang, J., Liu, J., & Wang, J. 2020. Optical properties of biomass-derived nanomaterials for sensing, catalytic, biomedical and environmental applications. *TrAC Trends in Analytical Chemistry* **124**, 115800. Doi: 10.1016/j.trac.2019.115800.

211. Nehate, S.D., Saikumar, A.K., Prakash, A., Sundaram, K.B. 2020. A review of boron carbon nitride thin films and progress in nanomaterials. *Materials Today Advances* **8**, 100106. Doi: 10.1016/j.mtadv.2020.100106.
212. Varma, K.S., Tayade, R.J., et al. 2020. Photocatalytic degradation of pharmaceutical and pesticide compounds (PPCs) using doped TiO_2 nanomaterials: A review. *Water-Energy Nexus* **3**, 46–61. Doi: 10.1016/j.wen.2020.03.008.
213. Zhong, L., Feng, Y., et al. 2020. Production and use of immobilized lipases in/on nanomaterials: A review from the waste to biodiesel production. *International Journal of Biological Macromolecules* **152**, 207–222. Doi: 10.1016/j.ijbiomac.2020.02.258.
214. Hoang, V.C., Gomes, V.G., & Kornienko, N. 2020. Metal-based nanomaterials for efficient CO_2 electroreduction: Recent advances in mechanism, material design and selectivity. *Nano Energy* **78**, 105311. Doi: 10.1016/j.nanoen.2020.105311.
215. Lai, N., An, B., et al. 2021. Future roadmap on nonmetal-based 2D ultrathin nanomaterials for photocatalysis. *Chemical Engineering Journal* **406**, 126780. Doi: 10.1016/j.cej.2020.126780.
216. Majumdar, D. 2020. Recent progress in copper sulfide based nanomaterials for high energy supercapacitor applications. *Journal of Electroanalytical Chemistry* 114825. Doi: 10.1016/j.jelechem.2020.114825.
217. Liu, Y., Ge, Z., Li, Z., & Chen, Y. 2021. High-power instant-synthesis technology of carbon nanomaterials and nanocomposites. *Nano Energy* **80**, 105500. Doi: 10.1016/j.nanoen.2020.105500.
218. Xu, X., Xiong, F., Meng, J., An, Q., & Mai, L. 2020. Multi-electron reactions of vanadium-based nanomaterials for high-capacity lithium batteries: challenges and opportunities. *Materials Today Nano* **10**, 100073. Doi: 10.1016/j.mtnano.2020.100073.
219. Wang, L., Yang, G., et al. 2020. One-dimensional nanomaterials toward electrochemical sodium-ion storage applications via electrospinning. *Energy Storage Materials* **25**, 443–476. Doi: 10.1016/j.ensm.2019.09.036.
220. Wang, Y., Yang, P., Zheng, L., Shi, X., Zheng, H. 2020. Carbon nanomaterials with sp2 or/and sp hybridization in energy conversion and storage applications: A review. *Energy Storage Materials* **26**, 349–370. Doi: 10.1016/j.ensm.2019.11.006.
221. Knoop, J.E., & Ahn, S. 2020. Recent advances in nanomaterials for high-performance Li-S batteries. *Journal of Energy Chemistry* **47**, 86–106. Doi: 10.1016/j.jechem.2019.11.018.
222. Teo, H.L., & Wahab, R.A. 2020. Towards an eco-friendly deconstruction of agro-industrial biomass and preparation of renewable cellulose nanomaterials: A review. *International Journal of Biological Macromolecules* **161**, 1414–1430. Doi: 10.1016/j.ijbiomac.2020.08.076
223. Singh, R., Altaee, A., & Gautam, S. 2020. Nanomaterials in the advancement of hydrogen energy storage. *Heliyon* **6**(7), e04487. Doi: 10.1016/j.heliyon.2020.e04487.
224. Fung, C.M., Tang, J.Y., et al. 2020. Recent progress in two-dimensional nanomaterials for photocatalytic carbon dioxide transformation into solar fuels. *Materials Today Sustainability* **9**, 100037. Doi: 10.1016/j.mtsust.2020.100037.
225. Paul, S.C., Dey, S.C., et al. 2021. Nanomaterials as electrocatalyst for hydrogen and oxygen evolution reaction: Exploitation of challenges and current progressions. *Polyhedron* **193**, 114871. Doi: 10.1016/j.poly.2020.114871.
226. Wu, C., Wang, K., et al. 2020. Multifunctional nanostructured materials for next generation photovoltaics. *Nano Energy* **70**, 104480. Doi: 10.1016/j.nanoen.2020.104480.
227. Guo, R., Zhang, L., Lu, Y., Zhang, X., & Yang, D. 2020. Research progress of nanocellulose for electrochemical energy storage: A review. *Journal of Energy Chemistry* **51**, 342–361. Doi: 10.1016/j.jechem.2020.04.029.
228. Abomohra, A.E.F., Elsayed, M., Esakkimuthu, S., Sheekh, M.E., & Hanelt, D. 2020. Potential of fat, oil and grease (FOG) for biodiesel production: A critical review on the recent progress and future perspectives. *Progress in Energy and Combustion Science* **81**, 100868. Doi: 10.1016/j.pecs.2020.100868.

229. Antonacci, A., & Scognamiglio, V. 2020. Biotechnological advances in the design of algae-based biosensors. *Trends in Biotechnology* **38**(3), 334–347. Doi: 10.1016/j.tibtech.2019.10.005.

230. Wu, X., Mu, F., & Zhao, H. 2020. Recent progress in the synthesis of graphene/CNT composites and the energy-related applications. *Journal of Materials Science &Technology* **55**, 16–34. Doi: 10.1016/j.jmst.2019.05.063.

231. Dalatony, M.M.E., Zheng, Y., Ji, M.K., Li, X., & Salama, E.S. 2020. Metabolic pathways for microalgal biohydrogen production: Current progress and future prospective. *Bioresource Technology* **318**, 124253. Doi: 10.1016/j.biortech.2020.124253.

232. Zhou, R., Li, X., & Pang, H. 2021. VOx/VSx@Graphene nanocomposites for electrochemical energy storage. *Chemical Engineering Journal* **404**, 126310. Doi: 10.1016/j.cej.2020.126310.

233. Nie, B., Palacios, A., Zou, B., Liu, J., Zhang, T., & Li, Y. 2020. Review on phase change materials for cold thermal energy storage applications. *Renewable and Sustainable Energy Reviews* **134**, 110340. Doi: 10.1016/j.rser.2020.110340.

234. Zayed, M.E., Zhao, J., et al. 2020. Recent progress in phase change materials storage containers: Geometries, design considerations and heat transfer improvement methods. *Journal of Energy Storage* **30**, 101341. Doi: 10.1016/j.est.2020.101341.

235. Rubab, A., Baig, N., Sher, M., & Sohail, M. 2020. Advances in ultrathin borophene materials. *Chemical Engineering Journal* **401**, 126109. Doi: 10.1016/j.cej.2020.126109.

236. Zhang, J., Yuan, X., Si, M., Jiang, L., & Yu, H. 2020. Core-shell structured cadmium sulfide nanocomposites for solar energy utilization. *Advances in Colloid and Interface Science* **282**, 102209. Doi: 10.1016/j.cis.2020.102209.

237. Zhou, C.A., Yao, Z.J., et al. 2020. Low-strain titanium-based oxide electrodes for electrochemical energy storage devices: design, modification, and application. *Materials Today Nano* **11**, 100085. Doi: 10.1016/j.mtnano.2020.100085.

238. Rafiq, M., Shafique, M., Azam, A., & Ateeq, M. 2020. Transformer oil-based nanofluid: The application of nanomaterials on thermal, electrical and physicochemical properties of liquid insulation-A review. *Ain Shams Engineering Journal*, ISSN 2090-4479. Doi: 10.1016/j.asej.2020.08.010.

239. Yang, L., Jiang, W., et al. 2020. A review of heating/cooling processes using nanomaterials suspended in refrigerants and lubricants. *International Journal of Heat and Mass Transfer* **153**, 119611. Doi: 10.1016/j.ijheatmasstransfer.2020.119611.

240. Kotia, A. Chowdary, K., Srivastava, I., Ghosh, S.K., & Ali, M.K.A. 2020. Carbon nanomaterials as friction modifiers in automotive engines: Recent progress and perspectives. *Journal of MolecularLiquids* **310**, 113200. Doi: 10.1016/j.molliq.2020.113200.

241. Jandt, K.D., & Watts, D.C. 2020. Nanotechnology in dentistry: Present and future perspectives on dental nanomaterials. *Dental Materials* **36**(11), 1365–1378. Doi: 10.1016/j.dental.2020.08.006.

242. Makvandi, P., Gu, J.T., et al. 2020. Polymeric and inorganic nanoscopical antimicrobial fillers in dentistry. *Acta Biomaterialia* **101**, 69–101. Doi: 10.1016/j.actbio.2019.09.025.

243. Malhotra, B.D., Kumar, S., & Pandey, C.M. 2016. Nanomaterials based biosensors for cancer biomarker detection. *Physics: Conference Series* **704**, 012011–012022. Doi: 10.1088/1742–6596/704/1/012011.

244. Nasrollahzadeh, M., Sajjadi, M., Iravani, S., & Varma, V. 2021. Carbon-based sustainable nanomaterials for water treatment: State-of-art and future perspectives. *Chemosphere* **263**, 128005. Doi: 10.1016/j.chemosphere.2020.128005.

245. Malhotra, B.D., Srivastava, S., Ali, M.A., & Singh, C. 2014. Nanomaterial-based biosensors for food toxin detection. *Applied Biochemistry and Biotechnology*, **174**, 880–896. Doi: 10.1007/s12010-014-0993-0.

2 Nanomaterials' Properties, Classification, Synthesis, and Characterization[1]

Vanya Nayak
Indira Gandhi National Tribal University

Syed Muzammil Munawar
and Khaleel Basha Sabjan
C. Abdul Hakeem College

Srishti Singh
D Y Patil Deemed to be University

Kshitij RB Singh
Govt. V. Y. T. PG. Autonomous College

CONTENTS

---

[1] Vanya Nayak and Syed Muzammil Munawar have contributed equally to this work.

2.1. INTRODUCTION

Nanomaterials are the foundation of nanotechnology and nanoscience. Nanotechnology has found its profound use in various fields, emphasizing nanomaterials' shape, size, morphology, dimensions, classifications, properties, and synthesis. Nanotechnology is a wide range of interdisciplinary science explored around the globe. Nanomaterials are substances that are more than 1 nm but are less than 100 nm. A nanometer is one-millionth of a millimeter, which is one lakh times lesser than human hair width. Nanomaterials exhibit nanosized dimensions that lead to the formation of various nanostructures and have distinctive properties with huge features of nanoparticles recognized by their route of synthesis and their characterization techniques. Nanomaterials are generally classified based on their dimensionality and modulation.

Nanomaterials exhibit unique chemical, electrical, magnetic, mechanical, optical, thermal, and ionization potential, structural properties, chemical bonds, and geometric structures, which distinguish them from the bulk matter; further, these budding properties have found huge prospective in electronics, medicine, and various other fields [1]. Unique properties of nanomaterials have revolutionized nanotechnology as their products have a wide range of applications in various domains [2]. The nanomaterials are attractive and suitable for maximum impact, and hence, the investigations necessitate a broad approach. Some of the nanomaterial synthesis is performed by biological matter, namely bacteria, viruses, algae, and various plant extracts such as lotus leaves, aloevera extracts, hibiscus flowers, and lemon leaf extracts and from natural matter such as spider-mite silk, butterfly wings, volcanic ash, ocean spray, fine sand, and dust [3]. Additionally, engineered nanoparticles such as gold (Au) and silver (Ag) nanoparticles can be synthesized through top-down and bottom-up approaches. Various characterization techniques and sophisticated tools analyze physicochemical properties such as shape, size, morphology, structures, and distributions. However, these properties and structures are dependent on nanometer, atomic-scale level, and inter-particle interactions [4]. Nanotechnology builds the gap between bulk materials and molecules such as nanocrystals, quantum dots, clusters, assemblies, nanowires, nanotubes, arrays, and super-lattices in precise nanomaterials. Therefore, physicochemical properties also change from those of bulk materials or atomic molecules of the same structures. Their unique characteristics, structures, dynamics, chemistry, and energetics create a foundation of nanoscience; further, their properties and reactivity can lead to newer technologies [5]. The energy spectrum of the quantum well, wire, or dots is engineered by controlling their shape, size detention region, and the quality of the constraint potential [6]. Sophisticated direct and indirect techniques such as X-ray photoelectron spectroscopy (XPS), scanning electron microscope (SEM), transmission electron microscopy (TEM), and photoemission electron microscopy (PEEM) are essential approaches for characterization, as these characterization techniques are based on massive energy ions and electrons which mostly produce no images. The chemical information can be gathered by nanoscale secondary ion mass spectrometry (NanoSIMS) and X-rays with a spatial resolution of 50 and 100 nm or lesser, respectively, and PEEM by measuring synchrotron radiation [7]. Incidental and engineered nanoparticles have two categories of synthetic nanoparticles, in which the former is synthesized by humans by mistake or unintentionally through various heterogeneous materials and therefore shows lesser control shapes, size, and morphology. Nevertheless, humans exclusively designed the latter with exact controlled morphology and chemical compositions with various layers of heterogeneous materials made by chemists and alchemists from the past [8].

This chapter briefly describes various unique properties exhibited by the nanomaterials, which offer them a variety of applications in various fields. It will further classify various nanomaterials and their synthesis through different biological, chemical, and physical routes. In addition to the above, this chapter will also demonstrate various characterization techniques used to characterize nanomaterials, along with the conclusion and prospects of nanomaterials.

2.2 PROPERTIES

Properties are the crucial aspect of any nanomaterials because application of every nanomaterial majorly relies upon the properties. This chapter will briefly demonstrate various properties of nanomaterials, and for doing so, we will be splitting the properties into two types: the first is physiochemical properties, which will deal with nanomaterial's physical and chemical properties; and the second one is biological properties, which will deal with only the biological aspects of nanomaterials.

2.2.1 PHYSIOCHEMICAL PROPERTIES

2.2.1.1 Melting Point and Temperature

As the particle size increases in bulk atoms, the melting point decreases as the melting point depends on the surface energy. The melting temperature (Tm) in massive cluster nanoparticles decreases with a decrease in particle size, and a thermodynamic change of surface-to-volume ratio is incorrigible; one experiment proved it by using gold nanoparticles, which showed the relation between Tm and size. In the observation of nanoparticles with increased diameters of 3.8 and 2.5 nm, the Tm observed was 1,000 and 500 K, respectively, and the bulk Tm observed was 1,337 K. Further, it was observed that the melting point smoothly decreased with an increase in the size of the particle [9–13]. Therefore, it can be concluded that the physical properties have the potential to modify the chemical properties.

2.2.1.2 Wettability

A liquid's ability to retain contact with a solid surface is known as wettability and can be maintained by intermolecular interactions between cohesive and adhesive types [14]. Wettability is measured as the contact angle formed when a liquid drop is placed on a solid surface and is measured optically. In wettability, a contact point rating is used as a solid model from the fluid. Generally, if the contact angle's value is 900, it is considered as the threshold value. Similarly, if a point of contact angle is greater than 90°, it is referred to as a non-wetting frame or insufficient wettability, and if it is less than 90°, it is referred to as a wetting frame or good wettability. However, if the contact angle is measured to be zero, then complete wettability is observed. The contact angle is used to illustrate the roughness, heterogeneity, and mobility. Previously, it was believed that the contact angle of a liquid drop was formed due to the equilibrium formed by the mechanical force exerted between the drop and a solid surface with the help of three interfacial free energies. However, the liquid drop size determines the surface free energy and the contact angles of liquid metals. It was concluded that the increase in liquid metal size leads to a decrease in the contact angles [15].

2.2.1.3 Pore Size and Surface Area

Nanoparticles contain a huge explicit surface area that modifies their properties compared to the bulk materials, and therefore, particle size dramatically determines the properties of nanoparticles. The morphological influence of particle size enters the dispersion masses from the apparent surface territory. According to different portrayals, they are best in plan with the element of the same surface territory [16].

2.2.1.4 Quantum Confinement

Quantum confinement effect is a unique property of nanoparticles, as it is observed in those particles whose size is minimal and cannot be compared with the wavelength of the electrons. It is widely observed in the quantum dots, which are a part of nanomaterials. This property efficiently describes the electrons based on valence bands, conduction bands, electron energy band gaps, energy levels, and potential wells [17]. The quantum confinement increases the nanomaterial's bandgap compared to the bulk material, which results in the various fluorescent colors and makes the nanoparticles a potential semiconductor. Semiconductor nanocrystals are considered a potential semiconductor due to quantum confinement, which helps them to absorb wavelength and regulate optical emission and particle size. Quantum absorbers exhibit extraordinary luminous properties that have smaller output spectra in contrast to natural fluorophores. Due to this property, they have gained much interest in marking cells' fluorescence with cadmium selenide materials [17].

2.2.1.5 Interface Property

The charges in particles in the colloidal state are balanced emulsions, which are known as Pickering emulsions [18,19]. As the forces play a vital capacity in capturing particles at the interface, it creates an impression of identification by reducing the surface pressure at the liquid–liquid interface and the low dispersion coefficient, limiting the rate at which the molecule can propagate from the surface. Interfacial properties of naturally synthesized nanomaterials play an essential role in the fabrication of medical implants, drug delivery carriers' tissue, and engineering scaffolds [20].

2.2.1.6 Structural Properties

The decrease in molecule size with an increase in surface area and free surface energy causes variations between nuclear spaces, which can be portrayed by compressive strain driven through nanoparticle-shaped space's internal weight. However, the strength of metastable structures in bundles and nanoparticles decreases mass nuclear exposure signs. It is well known that the gold nanoparticles are perceived to take on polyhedral morphologies such as 3D-shaped octahedron, various twin icosahedrons, and various twinned decahedrons, and therefore, they are recognized as translucent twin particles in morphology. Decahedral semiprecious stones and icosahedral structures are essential for nanocluster's development and regulate the size appropriately, and they pass into additional translucent pressure plans [21]. The structural properties play an immense role in determining the nanomaterial properties. When the structure of gold nanoparticles was compared with their structural isomers: Au38T and Au38Q, it was found that they exhibited very different properties such as optical and catalytic properties. Further, it was also revealed that the nanoparticle's stability is also affected by the variation in their structural properties. This experiment has been marked as the first experiment in describing structural isomerism in the nanoparticles. Future researchers can focus on establishing the structure-stable relation in nanomaterials and finding potential applications of the nanomaterial and its structural isomers [22].

2.2.1.7 Thermal Properties

It has been observed that the melting point of the nanomaterial is lesser than that of the bulk materials because the molecules and atoms present at the surface tend to move quickly at low temperatures. In an experiment on the gold nanoparticle, it was revealed that the melting point is a unique feature that can affect the nanoparticle's size and modify the fabrication characteristic of the ceramic material [23]. When gold was brought to its melting temperature (1,336 K), a sudden reduction in size brought the bulk gold material to the nanoscale level. In nanowires, phonons execute in a technique as diverse as mass materials due to the quantum limitation in 1D. The nanowire's surface takes the modes of surface phonon, bringing about different polarizations of the phonons in addition to two longitudinal acoustics in the mass semiconductors. As the distribution of connection changes, the speed and the thickness of the band will also change. Phonon life changes are attributed to strong phonon–phonon bonds and limit the distribution within nanostructures.

2.2.1.8 Chemical Properties

Nanoparticles and nano-layers have extraordinary volume-sized surfaces and potentially displaced crystallographic structures, which cause an extreme change in synthetic responses as the use of catalyst raises the degree, selectivity, and ability of compound's responses [24]. Gold nanoparticles, which are 5 nm less, embrace icosahedron structures and cubic planes centered on the face. However, ancillary changes are increasing with the increase in the synergistic movement. Nanoscale synergist supports pore controlled sizes and can also select composite responses depending on their actual size and internal transport and response destinations. Nanoparticles show new sciences as infallible by their unmixed particle partners; new drugs are insoluble as a liquid, while in the micronized measured particle type, they will still only condense on a nanostructural structure.

2.2.1.9 Mechanical Properties

The strength of a nanomaterial depends immensely on the effectiveness of the improvement and the elegance of the material's imperfections. If the reduction in frame size and the ability to hold any deformations were becoming more confusing, the mechanical properties are adjusted. The new nanostructures are different from the massive structures based on nuclear auxiliary understanding. It clearly shows different mechanical properties as a single carbon nanotube (CNT), with many walls representing elevated mechanical qualities and versatile raised boundaries that lead to generous mechanical adaptations and minimize reversible losses. As ventures to get out of the scattering and sliding of the grain border become increasingly essential in fine material, therefore, all of these impacts were taken together to make the breaking points of the flexibility of the actual quality of custom materials and continuously increased the quality which was balanced by a flexibility-related disaster [25].

2.2.1.10 Magnetic Properties

Magnetic nanoparticles are used in various applications such as iron fluids, the formation of shady images, bioprocessing, and collection of a large and attractive memory media. Due to these large surface:volume ratios, many ion forms bond with the neighboring particles, promoting opposite charges to attract, which causes the magnetic effects. The various forms of magnetic materials are ferromagnetic, paramagnetic, diamagnetic, and antiferromagnetic. Due to their small structure, the ferromagnetic materials are found in a single magnetic domain and show the superparamagnetic property. Few particles, when they are single, exhibit ferromagnetic behavior, but collectively they exhibit paramagnetic behavior. When the external magnetic field is present, they are magnetized in the same direction as the external magnetic field, but if the external magnetic field is removed, the magnetization also fades away. The particle size determines the time of disappearance of the magnetization, as small-sized particles respond quickly to the external magnetic field compared to the larger particle size. Therefore, it can be suggested that when the material comes from bulk to its nanoform, it exhibits its magnetic behavior [26].

2.2.1.11 Optical Properties

In crystalline structures, the atomic structure and the bandgap mainly determine the optical property, whereas the grain boundaries and density determine optical transparency. The nanoscaled materials absorb light at a specific wavelength as the plasmon absorption and light transmission depending on the type of metal and particle size. It has been observed that the size of the nanomaterial plays an important role in determining the saturation of the color and tinting strength. Similarly, when present in bulk materials such as germanium and silicon, materials do not emit lights, but as the material's size is reduced to the nanoscale, it exhibits a quantum effect, which shows high light-emitting efficiency.

Semiconductor nanocrystals such as quantum dots claim the same size execution in repetition and light output strength for indirect optical properties and the updated gain for certain energy or discharge frequencies. Other properties affected by reduced dimensionality are photocatalysis, photoemission, electroluminescence, and photoconductivity. The optical characteristics of nanoparticles contrast in their structure, morphology, and size [27–30], affecting the way they communicate amid light. If the matter is exposed to light, different cycles can occur. In a similar frequency, light absorption, light scattering at the same frequency as approaching light, Mie/Rayleigh scatter or re-emit consumed light or fluorescence. The optical properties of the nanoparticle may change as a result of keratinization or delivery [28]. The chiro-optical response's tuning by the change in the nanoparticle's morphology as the circular dichroism spectra with wavelengths of the nano-helices through the right chirality methods presents a positive ellipticity, even though for the left ones they show a negative.

The nanoparticles are widely used as the raw material for the production of several electronic devices. The electrical properties and particle size play a vital role in modifying and improving product performance. For example, to develop a small and thin electronic device that exhibits a high dielectric constant, the lead(II) titanate ($PbTiO_3$) is preferred as its constant dielectric increases with the particle size decrease, whereas the minimum particle size required to maintain ferroelectric property mainly depends on the material's composition. In electronic properties, changes occur due to the decrease in the frame length scale and rise in the property of electron waves and lack of scattering foci. The frame size becomes the same as de Broglie frequency of the electrons, as the idea of discrete energy states becomes clear, while the fully discrete power range is subjected to the framework and is limited to each of the three measurements. While frame measurements are smaller than the electron-free mode for inelastic scattering, electrons can travel along with the frame without any coincidence of their wavelength period. Additional insulation is associated only with phase obstacles. Simultaneously, the frame is satisfactorily small in consequence as all the scattering foci are completely killed. Concurrently, the test boundaries are delicate; thus, the boundary reflections are speculative, and the dragging by the electron transport becomes exclusively ballistic while replacing the test as a guide for electronic wave operation [31]. The impacts of coupon barriers as a result of the conduction mass associated with single electrons are required. Accordingly, low energy is required to drive a circuit breaker, semiconductor, or memory component. Miracles have been used to make unique works for electronic, optoelectronic, and data processing applications, such as resonant tunneling transistors and single electronic transistors [32].

2.2.1.12 Vibrational Properties

Size reduction and size-induced capabilities are influenced by the vibrating properties followed by achieving nanomaterials' properties such as molecule size, morphology, and entanglement [33]. It is well known that Raman spectroscopy is a sophisticated strategy to evaluate the vibrating properties of various nanomaterials. The method depends on the scattering of inelastic light by the phonon vibrations in the crosssection and the distribution of the massive materials matching the groups of forces that decide the solitary states, $q \sim 0$ d.m.th., in which q is the phonon wave vector. In the nanocrystal, the standards of wave vector selection $q \sim 0$ are broken due to the vulnerability rule, and the expansive ways of turbulence are seen due to phones having a location in the center of the Brillouin contribution area that is $\neq 0$ and looked at Raman's profile. The upper nanocrystal profile relies on the size and bond, phonon dispersion of the mass material. The method is found to be intriguing because large number of nanoparticles are placed in the direction of the laser point, which is about 1 µm in this way, and allows the nanocrystal size to be recorded [34].

2.2.2 Biological Properties

2.2.2.1 Superoxide Dismutase (SOD) Activity

The high-impact medium digestion that burns oxygen in mammalian cells reaches some free radicals found as signaling domains are known as superoxide revolutionaries, and they perform a vital role in the scavenging of the oxidation cycle. The extent to the mass of superoxide revolutionaries is normally limited by SOD, destroying revolutionary surfaces. For example, a cerium oxide nanoparticle exists in the +3 and +4 oxidation state and greatly affects SOD's mimetic activity; they show SOD-like motions within the Ce^{3+} moiety [36]. Studies have proposed a total atomic technique for cerium oxide nanoparticles from SOD. For the mode of mechanism, please see Figure 2.1, where(4) is the original state and on (5) two Ce^{3+} ions have potential oxygen destinations to which superoxide can bind. From that point forward, the oxygen particle expands an electron by a Ce^{3+}. In (6), the two protons which are present in the solution binds with the two electronegative oxygen particles, which results in the formation of H_2O_2 molecule and is further transmitted. Further, the second superoxide molecule in (7) will bind to the residual oxygen's joining site. In (1), $2Ce^{3+}$ is oxidized to $2Ce^{4+}$ with a second H_2O_2 particle's arrival after the oxidation reaction. Regardless of how the response did not stop, as on the outside of (1), it has a site for the opening of oxygen, which contains the junction of $2Ce^{4+}$ and this binding atom H_2O_2 (2), thus offering H_2O_2 application as a declining operator. Following past reactions, protons are transferred to (3) the $2Ce^{3+}$ reduced by two electrons' exchange into two cerium particles. The oxygen supply site eventually reduced revisits to its baseline state (4) by oxygen arrival. Further, H_2O_2 exhibits a paradoxical effect in cerium oxide nanoparticles for oxidation and mass reduction. In any case, the basic properties of cerium oxide nanoparticles enable it to reproduce in its ground state and evaluate energy and find that cerium oxide nanoparticles (3–5 nm) show dizzying activity exhibiting a stable reaction rate, which is much higher than that decided on the SOD protein [35,36].

2.2.2.2 Phosphatase Mimetic Activity

Phosphate bundle enlivens the obtained RNA and DNA and coordinates development of protein, energy movement, etc. All these can be hydrolyzed to ester bonds, which can be removed by deteriorations known as phosphatases. Previously, it was observed that the cerium(IV) is responsible for the increase in synergistic reactivity

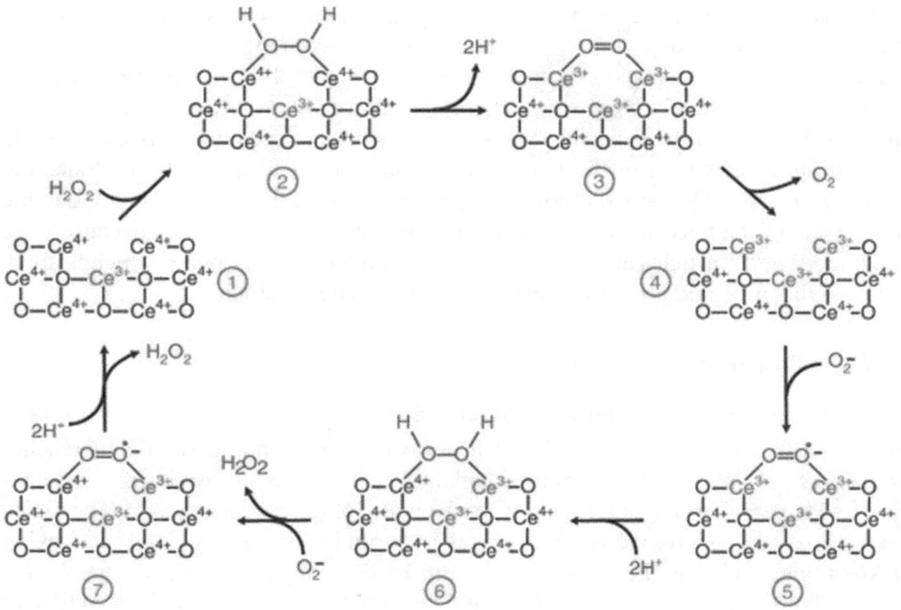

FIGURE 2.1 Diagrammatic illustration of cerium oxide nanoparticles' SOD reaction mechanism. (Reproduced from Celardo et al., *Nanoscale*, The Royal Society of Chemistry, 2011, ref. [36].)

of the synergist because of the hydrolyzation of the phosphorus–oxygen bonds, which are present in DNA and RNA. On further investigations, it was later discovered that Ce(III) could increase the synergistic activity as the phosphate's negative charge binds with the cerium oxide nanoparticle due to the presence of Lewis acidity. Therefore, it was examined and found that cerium oxide nanoparticles can disrupt the phosphate binding of O-phospho-L-tyrosine and para-nitrophenylphosphate in Ce's present (III) oxidation state. It has also been widely studied that cerium oxide nanoparticles exhibit the ability to bind with the plasmid DNA; however, no hydrolysis element has been observed. Therefore, it can be concluded that without causing DNA damage, phosphorylation of ATP and proteins can occur. Furthermore, it has been investigated that cerium oxide nanoparticles and phosphate anions affect catalase and SOD's mimetic motion by expanding and decreasing their stability individually. The catalase's mimetic motion is unique concerning the mimetic action of phosphate and mimetic action of phosphatase-specific dynamic properties; however, it follows mimetic action patterns of catalase.

2.3   NANOMATERIALS' CLASSIFICATION

According to shapes, sizes, development, and modern or biomedical applications, they are grouped according to their composite arrangement because of the extraordinary limits that nanoparticles exhibit. Natural nanoparticles have 3D structures through self-aggregation and are produced from natural particles that can be regular or processed such as total proteins, lipids, milk emulsion, and infections. These nanoparticles have an important dynamic function in the embellishment of agent enterprises and food enterprises. Some of them as food materials are chocolates, cakes, and creams that offer nanoemulsions in their portrayal. Two strategies are used to produce natural particles as top-down techniques such as mechanical processing, microfluidics, and lithography; the basic strategy creates particles from precipitation and aggregation. In this way, inorganic particles are created by the precipitation of inorganic salts and more stable than previous particles. Consequently, their safety is limited, relying on the composition/mechanics and nature of the integrated particles. Nanomaterials make bargains with zero, one, two, and three measurements as nanometers, and in Figure 2.2, these nanostructured materials are classified. Physico-synthetic characteristics change morphology, measurements, and collaborations with new materials. The allure in the semiconductor project for its endless interest in scaling is needed in the nano-domain. Due to the size reduction, quantum constraint causes properties such as electron wave capacities, which are immovably constrained, bringing about changes in the orchestrated nanomaterial's electronic and optical properties. Due to the more modest or larger size of the molecule, which brings about a more grounded or more tangible stop, improvement or reduction in the band hole changes the band of materials' structure. Electron portability, therefore, varies with mass and stable dielectric and optical properties. Their quantum

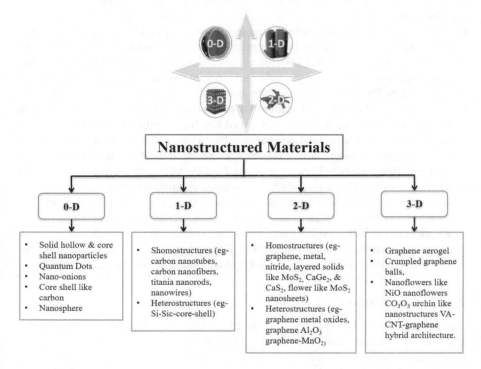

FIGURE 2.2 Classification of nanomaterials according to the dimensionality of the nanostructures.

pressure shows attractive properties such as different colloidal suspensions like molecule size, photoluminescence (PL), and bright light flux due to metal nanoparticles. Silver nanoparticles have demonstrated bright light-induced discharges and provoked nano-silver coatings as antibacterials in medical services [27]. The materials cover thin two-dimensional films with few nanometers, so their properties depend on the interface's influences and the surface and reflect the electron inhibition compared to the films. Here, the electron performs as in massive materials [37]. Nanoparticle becomes nano in size; then, de Broglie frequency and freeway of the electrons go like quantum spots. The constant thickness of states (DOS) and the band hole expansion make the materials present discrete energy levels corresponding to the molecules. As retentions shift from infrared to a visible achievement, this has a great deal of belief in nanoparticles' optical properties [38].

2.3.1   ZERO-DIMENSIONAL

The exciton states are resolved in zero-dimensional confinement by a contradiction between the influences and the electron-opening pair relationship induced by the coulomb communication. The only molecule with the degree of separation as 1/L2, where L means the quantum box's size, dictates the degree of control energy and the coulomb as l/L. In the zero dimensions, the structure expands. The mixing induced by the coulomb becomes more diligent; however, the coupon energies increase, and the electrons and openings become solidified to the particle states' lowest energy. Freezing motion in measurement occurs in quantum wells, and wires when control expands. Subsequently, the coupling-induced bond of unrestricted motion in a well or wire is improved by control [39] due to the influence of surface and quantum constraint. Regardless, the synthetic bonds, ionization potential, structure, and properties are affected by the size of the molecule at the nanoscale [40]. Nanomaterials show better properties than those of traditional coarse-grained materials, which include added quality, improved diffusion, improved flexibility, reduced thickness, reduced agility module, high electrical resistance, clear extended heat, high coefficient of warm development, lower warm conductivity, enhanced oscillator quality, blue moving assimilation, enhanced iridescence, and subtle advanced attractive properties compared to ordinary bulk materials. Nanomaterials have formed semiconductors by recording low speeds, low-current lasers, minimal circle (CD) player frames, and low noise improvement as satellite signals and hotspots in fiber optics. Favorable uses of nanomaterials are self-cleaning glass, UV-safe wood veneer, nanosize tools in clinical science for treatment and termination, disease prediction, drug delivery frameworks, attractive reversible imaging (MRI) examination, and radioactive tracers [41].

2.3.2   ONE-DIMENSIONAL

Nanostructures are the main frameworks to examine the support of different properties in size and dimensionality in 1D organized nanomaterials, which participate considerably in the interconnections, and practical segments in creating nanoscale tools. One-dimensional (1D) materials at the nanoscale level have inspired colossal detail and their importance in basic exploration and innovation applications. In addition to CNTs, nanowires or quantum wires are incomparable frameworks for predicting electric vehicle support and morphological and dimensional mechanical properties, which play an essential part in both interconnections and utility segments producing nanoscaled tools. Nanomaterials also have special attractive properties such as uncompromising durability of the repairer, higher gloss effectiveness, improved thermoelectric figure of legitimacy, and lower yield strength. Nanosols are anisotropic nanocrystals between large-angle proportions with a width of 1–200 nm and a length of several micrometers (μm); from their morphology and current properties, they differ from round nanocrystals. For examining 1D nanomaterials, their applications are essential to collect individual particles dimensionally and are applicable in the controllable strategy. Even though 1Ds are created by methods such as advanced nanolithography procedures, electronic column composition, proximal proof design, and X-ray lithography, these movements are moderate and extremely costly, improving these instruments in a way that is useful for producing extraordinary 1D materials quickly with minimal effort in an extraordinary versatility [42].

2.3.3 Two-Dimensional

2D nanomaterials are out of the nanoscale size range. Their combination is a central area in the research of materials in the coming years due to the many low-dimensional qualities not quite the same as the mass properties, while they display extraordinary structures, shape, quality dependence of size, and their use to construct nano-equipment in various application apparatuses and examinations in sensors, photocatalysts, nanocontainers, nanoreactors, and formats for 2D mathematics of different materials. They additionally occur as junctions or stable islands, glowing structures, nanoprisms, nanoplasms, nanosheets, nanowalls, and nanodisks [43].

2.3.4 3D Nanostructures

These nanomaterials have emerged from a vast surface region by implementing the properties on massive materials emerging from the influence of quantum size, which has fascinated exploration and fusion in recent years. Because of their properties, nanomaterials strongly support the morphology, the dimensionality with which they execute, and a vast range of applications. 3D nanomaterials can be morphologically orchestrated and substantially controlled and take on an important function in a wide range of applications such as traction material, battery terminal materials, and catalysis area. Due to their exceptional surface area and assimilation destinations for all atoms in the nanoscale, researchers have been attracted by research at this time and age, and nanoballs, nanocoils, nanocones, and nanopillars and nanoflowers are different 3D organized nanomaterials [44].

2.4 SYNTHESIS OF NANOMATERIALS

As particles and atoms are considered the basic foundation of an element, they have been developed with basic units necessary to understand nanomaterials' different properties and the corresponding collaborations with different atoms. Consequently, in joining nano-building blocks, a skillful control of the paths has been created, which is essential to understand different morphologies that bring forward innovation and new tools. The methodology follows two processes for the synthesis of nanoparticles. However, appropriate methodologies include the top-down methodology and the bottom-up approach. The top-down approach scales down current segments and materials, whereas the bottom-up approach is a basic approach for constructing complex subatomic structures atom by atom (Figure 2.3). These methodologies represent nanosystems' level of association as the point of intersection between subatomic objects and massive materials. In 1959, Richard Feynman, in his speech on top-down methodology, had stated: "there is much space at the bottom" which is ideal for achieving long-range structures to make relationships with the macroscopic world. Then again, the ground-up approach is the most amazingly suitable for collecting and placing short distances on the nanoscale [45].

Top-down approach

The top-down approach reduces lateral dimensions of large objects having one or two dimensions to produce fine-featured nanomaterials. This approach initiates with thin-film deposition and further proceeds for patterning features of two-dimensioned objects by serial and parallel techniques. Depending on the scaling methods, top-down approaches have several strategies such as machining, templating, or lithographic strategies.

For producing nanomaterials by a top-down approach, a large object of one or two dimensions should be fabricated, and further nanopatterning techniques should be used. This method was previously used in the microelectronic industries, but, with the advancement in thin-film deposition methods, this method has found its utility in various other fields. Thin-film deposition mainly involves the deposition of films on the substrates, divided into homogeneous and heterogeneous growth based on the relationship between the film and substrate. The deposition or growth of a

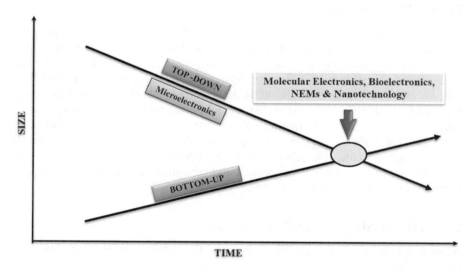

FIGURE 2.3 Convergence from top-down and bottom-up approaches in nanoelectronics.

film onto a crystalline substrate of the same materials and crystal orientations is known as homogeneous film growth, whereas the growth of thin film on a substrate consisting of various materials is known as heterogeneous growth. Several methods are found to form a thin film on substrates, such as vapor deposition, electroless plating, and electroplating. The most common method utilized for thin-film formation is the vapor deposition technique, which is further divided into physical and chemical vapor deposition methods.

To pattern the thin film deposited or grown at a small scale, nanolithography has been used, which is defined as the printing or writing on the structures which are less than 100 nm. Since the last decade, nanolithography has miraculously modernized the science and technology fields as it has been utilized for the production of various integrated circuits (IC), video screens, biochips, diffractive optical elements, micro-electro-mechanical systems (MEMS), miniaturized sensors, etc. Based on the pattern strategy, nanolithography can be categorized into serial writing and parallel replications. In the parallel replication method, predefined patterns are duplicated in a large-area patterning with high output, whereas serial writing produces arbitrary patterns with high resolution and precise registration, and no predefined patterns are required. The top-down methodology is currently driven by conventional messages such as the physical science of materials modeling, while scientific professionals in the inorganic examination improve these instruments [46].

Bottom-up approach

It produces large nanomaterials by aggregating small nanoparticles as building blocks. Here, current powers and synthetic strategies are worked out at the nanoscale level to aggregate essential things into larger structures. Materials evaluated by nanometer development of their substances suggest colloidal, supra-subatomic frameworks and undergo phase changes, such as CVD or surface smoke statement or precipitation of a strong phase from regulation. The impetus for base approaches comes from organic frameworks, from synthetic powers to include materials forever. Researchers are copying the opportunity to combine nanosized bundles of clear ions and have self-assembled them into more explained materials.

2.4.1 Physicochemical Synthesis Methods

A monodispersed size and a uniform shape are highly important to produce a highly ordered nanoparticle assembly. This section highlights several well-explored and controlled physicochemical synthesis protocols (Figure 2.4) such as sol–gel,

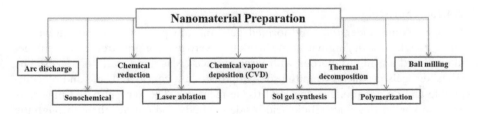

FIGURE 2.4 Different synthesis routes of nanomaterials.

sonochemical, microwave-assisted, polymerization, ball milling, cryogel, xerogel, aerogel, and chemical reduction in detail.

2.4.1.1 Chemical Reduction Method

Chemical reduction methods are completed in chemical reducing reagents and are often performed at room temperatures or moderate heating. Apart from chemical reducing reagents, the chemical reduction method also uses reduction by solvothermal methods, photocatalyst methods, and electrochemical methods. Chemical reduction is a cost-effective process and is also widely utilized for the mass production of various nanocomposites. Some of the methodology employed in reducing nanoparticles is to reduce the substances by natural/inorganic reducing agents, for example, essential hydrogen, sodium borohydride ($NaBH_4$), pyrrole, Tollens' reagent, poly(ethylene glycol)–copolymer block, ascorbate, sodium citrate, polyvinyl alcohol, polyvinylpyrrolidone, polyethylene glycol, polyacrylic methacrylate, polymethyl methacrylate, ascorbic acid, thiols, amines, and hydroquinone. This combination cycle results in the framing of colloidal aggregated nanoparticles [47].

2.4.1.2 Direct Dispersion Method

The direct dispersion method is used to achieve homogeneity in nanoparticles without the use of compression. In this technique, nanocomposites are modified synthetically to modify their affinity to polymers. It was observed that the zinc oxide nanoparticles could be produced by mixing ammonium bicarbonate and zinc sulfate [48].

2.4.1.3 Aerogel

A lightweight solid with nanometer pore size produced from a gel whose liquid component has been replaced by air is known as aerogel. The aerogels exhibit unique high porosities, open-pore structures, low densities, and large surface areas due to their three-dimensional pore structure. In aerogel production, supercritical drying of gels is considered one of the most crucial steps because it helps preserve the three-dimensional pore structures. Therefore, it is necessary to understand supercritical drying's kinetics to understand material development and improve aerogels' manufacturing. During supercritical drying conditions, no differentiation between the smoke liquid phases is observed because the density becomes equivalent, which prevents the regulation of a liquid smoke meniscus. Another promising dried gel with a similar pore volume as a wet gel is called aerogel [49,50].

2.4.1.4 Xerogel

Xerogels come under the solid-formed gels and are produced at room temperature through slow drying with fast shrinkage. Xerogels are prepared by the sol–gel method and are highly porous with large surface areas and small pore sizes. The silica xerogels produced from the sol–gel process are considered safe and biocompatible and do not produce any fatal tissue reactions. The xerogel-based materials are also known to reduce silicic acid inside the body and remove them through the kidneys [50,51].

2.4.1.5 Cryogel

Cryogels are supermacroporous gels exhibiting good osmotic stability and good mechanical strength and therefore find their potential use in the biomedical domain. They are produced at subzerotemperature by cryogelation of apposite monomers or the polymeric precursors. The various production methods of cryogels have made them suitable for several biological and biomedical domains. Cryogels can be modified in several ways, such as binding certain ligands to their surfaces or grafting polymeric chains to the surface of cryogels. The process of cryogelation is completed in the following steps: separation of phase by forming ice crystals, cross-linking, polymerization, and thawing of ice crystals, which finally forms an interconnected porous cryogel network [52].

2.4.1.6 Microemulsion

Microemulsions are clear or turbid systems consisting of hydrophilic and lipophilic phases present along with surfactants. Unlike other emulsions, microemulsions do not require high shear conditions as they are formed simply upon a mixture of compounds. Microemulsions are found in two types: direct microemulsions in which oil is dispersed in water (o/w), and reverse microemulsions in which water is dispersed on oil (w/o). Self-assembled nanostructures can be produced in a binary system, and these nanostructures formed can range from spherical and cylindrical micelles to bicontinuous microemulsions and lamellar phases. The produced spherical micelles find their potential utilization as small reactors, loaded on metal salts, and form semiconductor nanoparticles by adding organically soluble chalcogenide reagents. Cosurfactants such as the zwitterionic phospholipid L-phosphatidylcholine (lecithin) and anionic bis(2-ethylhexyl)sulfosuccinate (AOT) are used for maintaining the shape anisotropy. Triangular-shaped cadmium sulfide nanocrystals were also synthesized using a similar approach [53].

2.4.1.7 Ball Milling Process

Ball milling is used to grind powders, synthesize nanocomposites, and optimize phase composition. The ball milling process increases the reactivity of material and also uniforms the spatial distribution of elements. For nanomaterial production, the powder materials are regularly dipped for ball processing, and due to the presence of balls in the machine, the radial force is applied uniformly, which results in the smooth production of uniform-sized nanomaterials. Mechanical bonding, machining, and mechanochemical joinings are commonly used phrases that identify the creation of materials from ball machining. This strategy is ideal for creating nanomaterials because it is productive and smart [54].

2.4.1.8 Hydrothermal Process and Solvothermal Method

Solvothermal and hydrothermal processes are principal branches of inorganic synthesis, and both are used for conventional growth and synthesis of material, treatment of waste materials, etc. The chemical reaction that requires water and can be performed at both high temperature and pressure is known as the hydrothermal process, whereas the solvothermal technique is indistinguishable from the aqueous strategy, which uses various solvents except water. It is simple, easy, and direct and requires a low-temperature route to produce nanoscaled particles. The solvothermal technique exhibits good regulation of size and shape distribution and maintains crystal structure as compared to the hydrothermal strategy. It has been used to merge both TiO_2 nanoparticles and nanorods with and without the help of a surfactant guide [55].

2.4.1.9 Sol–Gel Process

The sol–gel process is a type of wet chemical technique and is also known as chemical solution deposition. The sol–gel process is when a liquid precursor is hydrolyzed and condensed into a solid in which initial precursors can be either organic species or inorganic salt. The process of sol–gel includes various steps: initially development of stable solutions of precursors or the sol, proceeding with the reaction of sol with hard, porous gel that surrounds the continuous liquid phase, and finally decomposing the gel at high temperature. Various intermediate precursors are formed during the hydrolysis and condensation reactions, and then, a suitable precursor is identified; by this, a monodispersed semiconductor nanoparticle can be synthesized. Sometimes, precursors also use stabilizing ligands such as surfactants and coordinating solvents, which provide resistance to nanoparticles from agglomeration [56].

2.4.1.10 Polymerization

Polymerization combines small molecules, called monomers, to form a large molecule, called a polymer. Monomers could be of the same compound or could be more than two different compounds. It is considered an extremely basic strategy in the synthesis of nanomaterials. The monomers are combined with stable covalent bonds, which make them distinguishable from other processes such as crystallization. Microemulsions are regulated with polymerization, which becomes the focal point for its extensive exploration globally in various applications. Some applications that use polymerizations are oil recovery, ignition, beauty care products, cutting of metals, medicines, agronomics, food technology, fats, enzymatic catalysis, natural responses, nanocapsules, etc. [57].

2.4.1.11 Microwave-Assisted Synthesis

Microwave-assisted technology arranges the dipoles of the material in the external field through microwave electromagnetic radiation. It is a more encouraging procedure than the other traditional technique, as it produces the desired shape and size of a nanoparticle. The microwave-assisted synthesis produces internal heating, which decreases the required time and energy and yields a superior item that hampers framed particles' accumulation [58].

2.4.1.12 Thermal Decomposition and Pulsed Laser Ablation

Monodispersed and metal nanoparticles are produced with the thermal decomposition technique as it is non-polluting, less time-consuming, and cost-effective compared to the conventional methods. Thermal decomposition synthesizes uniform-shaped nanoparticles, requiring low temperature and controlled concentration of the substrates with less duration of time. It has been stated that the solventless method of decomposition does not require raw material, and therefore, thermal decomposition is an easy, convenient route. However, alkyl amines and carboxylic acids can be used as capping agents, which affect the production of the monodispersed nanoparticle by the thermal decomposition method. Unlike the biological synthesis of nanoparticles, thermal decomposition produces a large number of nanoparticles [59].

Pulsed laser ablation (PLAL) is a technique that produces homogeneous nanoparticles at room temperature without the help of any reducing agents, stabilizing agents, or surfactants. The PLAL technique has been highly used to synthesize noble metal nanoparticles from their bulk metal targets, and the resulting colloids are pure and stable. High-power laser pulses of approximately $108W/cm^2$ are used to melt, evaporate, and ionize materials from the surface. This ablation process produces an ablated material collected, placed on an appropriate substrate, and grown to form a thin film. This technique has found its uses in many domains such as biomedical and electronics. Apart from this, the laser/electron beam heating is the transmission of the electrons from the electron gun to the high-temperature region using high voltage and producing a vacuum in the electronic gun. Advantageously, using a transmission electron magnification lens can heat the electron pole and illuminate materials to create diverse carbon nanomaterials, such as nanotubes, nanocapsules, and nanoparticles. There are various preferences as the heat source is outside the extinction frame, and any material, including metals, mixtures, and earthenware, cause no pollution from the heat source [60].

2.4.1.13 Template Synthesis

Mixing nanostructured materials with format strategy has gained much momentum over the last decade. Thus, the templating technique is used, which produces nanofiber from semiconductors, carbon, metals, etc., by using a nanoporous membrane as a template. Templating is an easy and simple method mainly used for the fabrication of a nanomaterial. The template-synthesized nanoparticles are highly used in electrochemical, electroanalytical, and sensing applications [61].

2.4.1.14 Sonochemical Processing

The process that uses powerful ultrasound radiations ranging from 20 kHz to 10 MHz to make molecules undergo chemical reactions is known as sonochemical processing. The high intensity of sound brings about certain chemical and physical changes in the molecule, producing and fabricating various nanoparticles. The principle that causes the modification of nanostructures in the sonochemical process is acoustic cavitation. The ultrasonic radiations cause the bubble's breakdown present in the sol, which ultimately increases pressure, temperature, heating, and cooling rate [62].

2.4.1.15 Combustion

The combustion synthesis is also known as self-propagating high-temperature synthesis (SHS), a very famous method to produce nanoparticles, and has been practiced in almost 65 countries. It is a highly effective, cost-effective, energy-saving, and high-yielding method highly used in various industrial fields. The combustion mixture produces extremely translucent particles having large surface areas. Combustion synthesis can be completed in two ways: a self-propagating mode and volume combustion synthesis. In a self-propagating mode, the preheated reactive medium is brought to an ignition temperature, which initiates this layer's point reaction. Then, the "hot" reacted layer preheats and ignites the next "cold" layer, and thus, combustion front self-propagates along the reactive mixture resulting in the formation of the desired solid product. In the volume combustion synthesis method, the entire reactive compound is heated evenly to ignition temperature with the help of an external source and the reaction at each point of the medium, which further forms the nanomaterial. The ignition temperature for the combustion synthesis ranges from 500 to 4,000 K, which proves that combustion synthesis is an energy-efficient method. It has been observed that the rate of the temperature change at the self-ignition stage is very high from 103 to 106 K/s, which shows highly non-equilibrium conditions for the material fabrication and facilitates the formation of crystalline nanomaterials with unique morphology, including one- and two-dimensional nanostructures [63].

2.4.1.16 Gas-Phase Methods

It is essential to establish a direct interaction of vapor-phase atomic metals and ligands in the gas-phase synthesis, making them suitable for creating thin films. However, the gas-phase synthesis also requires stability and high temperature, which is considered the major drawback of this method. The gas-phase synthesis is completed in a special vacuum apparatus. The zero-valent metal (the target) is vaporized with an energy source, followed by evaporated metal's movement toward the gas-phase donor ligand present in the second chamber where it condenses at low temperature. It is easy to synthesize several metal complexes with organometallic compounds, ketones, and carboxylates because the apparatus's final design can vary according to the specific metal and its nature of the ligand [64].

2.4.1.17 Arc Discharge/Plasma

The arc deposition is a method that consists of a negative cathode and positive anode of a graphitic rod placed a few millimeters apart to produce carbon products through evaporation. This procedure is widely used for producing fullerenes and other related materials. The arc discharge produces metal nanopowders of polydispersed size in both liquid and gaseous media [64].

2.4.2 BIOLOGICAL SYNTHESIS OF NANOPARTICLES

Various chemical, physical, hybrid, and biological methods have been applied for the synthesis of nanomaterials. However, physical and chemical methods are more considered to synthesize nanoparticles, but toxic chemicals greatly limit their utilization in the biomedical domain. Therefore, the development of biocompatible, non-toxic, and eco-friendly methods for synthesizing nanoparticles is now in more focus to expand their biomedical applications. Biologically synthesized nanomaterials are considered an excellent asset of cellular biocomposition. The compositions of cellulosic fibrils with 100–1,000 nm are crystalline and amorphous. The biosynthesized nanoparticles exhibit better properties such as good catalytic activity and good specific area as compared to the chemically synthesized nanoparticle. As the chemical synthesis produces hazardous and toxic wastes that could be harmful to live cells and the environment, researchers have started focusing on cleaner methods of synthesis of nanomaterials, that is, biological methods of synthesis. The biological synthesis mainly includes green synthesis from plants and microbial synthesis using microorganisms such as bacteria, yeast, fungi, and algae. Inorganic materials are made by miniature creatures inside or outside the cell. Microorganisms can meet more limited time intervals and are an obvious focus for nanoparticle recovery. This chapter has presented a brief overview of biological methods of nanoparticle synthesis, and more details related to biological synthesis will be covered in Chapter 3 of this book.

2.5 CHARACTERIZATION TECHNIQUES FOR NANOMATERIALS

Various modern methods have been continuously created to portray the size, shape, safety, and quality of nanomaterials. Portraits of nanomaterials complement two significant methodologies: spectroscopy and microscopy, which are discussed below in detail.

2.5.1 ULTRAVIOLET–VISIBLE (UV–VISIBLE) SPECTROSCOPY

UV–visible spectroscopy is a type of absorption spectroscopy, and it is one of the most used methods for the quantitative detection of various analytes such as transition metal ions, biological macromolecules, and organic compounds. When the UV light is absorbed by the molecule and excites the electrons from lower energy levels to higher energy levels, forming a distinct spectrum helps identify the compounds. UV–visible spectroscopy is used not just for sensing but also for screening due to its ambidexterity and quick results. In nanotechnology, it has found its immense role in detecting optical and luminescence properties of the prepared nanomaterial, which helps the researcher find the potential use in the perfect domain. The absorption pattern of light makes the material specific for a certain application, and UV–visible spectroscopy perfectly fits to be utilized for these analytical studies. Different materials exhibit a different interaction pattern with light; for example, when bandgap and absorption peak of TiO_2 and ZnO nanoparticles were observed under UV–visible spectroscopy, it was found to be of 0.96 eV with 354 nm and 3.26 eV with 376 nm, respectively. Therefore, it can be concluded that the absorption rate plays an important role in determining a material's properties and its potential utilization in a particular domain. With UV–visible spectroscopy, a nanomaterial's properties are detected along with the bandgap of the nanomaterial [65].

2.5.2 FOURIER TRANSFORM INFRARED SPECTROSCOPY (FT-IR)

FT-IR is a type of infrared spectroscopy that uses infrared radiations and is highly used for various analysis applications, detecting unknown materials, determining the quality of the sample, and determining components of a mixture. FT-IR works by absorbing infrared radiations by sample resulting in the formation of molecular fingerprints. FT-IR is also commonly used to build beneficial protein-binding aggregates on the surface of nanomaterials or beneficial exopolysaccharide aggregates, which act as a coating for operators and improve the strength of the operators. Distinguished biomolecules are necessary to change and improve the auxiliary compositions required to perceive the development of their properties and possible applications [66].

2.5.3 X-RAY DIFFRACTION (XRD)

XRD is an important analytical technique mainly used to detect lattice parameters, crystallinity, size, shape, and dislodgment of molecular structures. These parameters are further calculated with the help of Scherrer's equation by broadening the most intense peak of XRD. The reference patterns used for comparing the prepared nanoparticle's position and intensity are taken from the International Centre for Diffraction Data (ICDD). The ICDD usually works best for a powdered sample compared to the amorphous sample because the powder diffraction configuration provides information in the form of phases present: the phase focus, the structure, and degree of crystallinity. However, apart from these details, the crystal size and strain analysis are also analyzed in the amorphous sample. It has been observed that the large crystallinity gives rise to sharper peaks and the width extension of the roof, which indicates reduced size [66].

2.5.4 ENERGY-DISPERSIVE X-RAY SPECTROSCOPY (EDS/EDX)

EDS or EDX is a scientific instrument used to identify and quantify the elemental composition by measuring the energy and number of X-rays emitted by the sample after the excitation with an electron beam [67]. It is one of the types of X-ray fluorescence spectroscopy which uses a very small sample size, which could be as small as a few cubic micrometers. A high-energy electron beam is emitted in a properly equipped SEM that excites the atoms producing X-ray emissions unique for each element. An energy-dispersive detector is a solid-state device that analyzes these emitted X-rays according to their energies and yields the atoms' composition on the specimen surface, which will produce a signature or fingerprint spectrum used for element identification by comparing reference spectra [68]. These vivid abilities are usually exhibited because each segment has a unique nuclear structure, which allows the X-rays as the trademark for the nuclear structure of a component to be different from each other.

2.5.5 DYNAMIC LIGHT SCATTERING (DLS)

Another scientific device utilized to analyze nanoparticle size and its distribution in a suspension or the solution by using scattered light is the dynamic light scattering. The Stokes–Einstein equation calculates the nanoparticle radius. DLS is also named as photon correlation spectroscopy (PCS)/quasi-elastic light scattering. A laser passes through the colloidal solution, which scatters the light, and the intensity of scattered light is analyzed as a function of time. DLS efficiently detects the change in conformation of the DNA caused by the intercalation and is also used to study the stability [69].

2.5.6 ZETA POTENTIAL

Zeta potential is a characterization technique that is used to study the surface charge of the nanocrystals. The nanocrystal exhibits a higher positive or negative value of zeta potential value due to individual particles' electrostatic repulsion, indicating the nanoparticle's good physical stability. A zeta potential value other than −30 to +30 mV is generally considered to have sufficient repulsive force to attain better physical colloidal stability, whereas the small zeta value of a nanoparticle is caused due to the van der Waals attractive forces, leading to particle aggregation and flocculation, resulting in physical instability. Physical stability is also affected by the property of the material or the presence of any surfactants. When the surface of a charged particle is attracted to an opposite charge, it forms a thin liquid layer known as the Stern layer. An electric double layer is created when particles are diffused in a solution. Therefore, zeta potential is named for the electrical potential of the double layer and determined through velocity measurement of the charged particles moving toward the electrode across the sample solution in the presence of an external electric field [69].

The attenuation of the concentrated samples, which does not affect the incorporated nanoparticle's results, due to the movement in the zeta potential value was credited to an increase in the signal engagement by the excess molecule matter [70]. Zeta potential and hydrodynamic magnitude also vary according to the pH; for example, at a low pH, a positive surface region is observed in the nanomaterial, and on the other hand, it exhibits a negative surface charge at a high pH. Therefore, it could be concluded that the nanoparticle's condition and strength are dependent on surface load, surface coverage, and pH. From significant estimates, it is observed that the dispersion state could be changed by pH, as pH changes zeta potential, which affects the dispersion mediums ionic nature by regulating the surface electrical charge with two layers of nanomaterial [71].

2.5.7 X-Ray Photoelectron Spectroscopy (XPS)

XPS is a quantitatively sensitive surface spectroscopic method that determines the core structure, synthetic states, and electronic states of parts within the material. XPS spectra are obtained by illuminating a material with X-rays of light emission while simultaneously estimating the electron's active energy radiating from 0 to 10 nm of the primary material under consideration. XPS requires conditions with a high vacuum ($P \sim 10^{-8}$ bar) or super-high vacuum ($P < 10^{-9}$ bar). However, when used to inspect nanoparticles, the nanoparticle association's involvement should be recalled; high involvement enlivens high-level spectra. Of course, to evaluate nanoparticles, a dormant example of trade in the examination room is critical to avoid contamination and oxidation of the nanoparticle surface [72].

2.5.8 Scanning Electron Microscopy (SEM)

Scanning electron microscopy (SEM) creates an image by scanning a focused electron beam over a surface to create an image. The electrons present in the beam communicate with the sample and produce various signals that can be used to obtain information about the surface structure and composition. To produce a focused beam, firstly, the electrons are produced at the top of the column and are then passed through a combination of lenses and apertures. The sample is placed on a stage having the chamber area, which is evacuated by a combination of pumps to create the vacuum. The level of the vacuum majorly depends upon the design of the microscope. Scan coils regulate the position of the electron beam, which helps to scan the sample perfectly. The interaction between sample and electron produces various signals, which are further detected by various detectors. While having a lower purpose than TEM, SEM can easily photograph nanoparticles on mass surfaces and sometimes for a direct view of nanocrystals in larger aggregates [69,73].

The field emission SEM (FE-SEM) is a part of an electronic magnifying lens that paints the outside of the sample by examining it with high-energy light emission in a raster filter design. The field emission gun (FEG) is used as the electron source consisting of a single tungsten filament along with a pointed sharp tip as it enables the formation of a very small probe of size 0.5 nm, which helps in the formation of the high-resolution image. Electrons interact with shells on the atoms that make up the sample by making signals containing information about the surface geography of the sample, its structure, and various properties, such as electrical conductivity. Types of signals produced by an SEM include backscattered electrons (BSE), auxiliary electrons (SE), trademark X or cathodoluminescence, and current samples. Overall, the most well-known or standard location method is the SE image. The point size in a field discharge SEM is more modest than the SEM and can accordingly create high-order photographs that can reveal details in the size of 1–5 nm [69].

2.5.9 TRANSMISSION ELECTRON MICROSCOPY (TEM)

For improving the resolution limit of the optical microscope, the wave-associated microscope was developed, and within a few years, first-generation transmission electron microscopes were commercially available for the characterization of metals. A commercial TEM usually constitutes several parts such as an electron gun, electromagnetic lenses and apertures, a sample chamber, and a screen. Field emission (monochromatic electron beam) or the thermionic emission (white electrons) generates electron beams, accelerated through a bias voltage to achieve high energy. This acceleration voltage majorly defines the energy of the electron beam along with the wavelength of the electron. These electron beams are passed through the magnetic field generated by the electromagnetic lens and are bent and focused on the specimen. Once the beam transmits through the specimen, it passes through an additional set of apertures and lens to project the information onto the screen. The sample preparation is very critical as it needs to be thin enough to be transparent for electrons. However, nowadays, a sample is prepared by focused ion beam (FIB) and ion milling thinning to gain the desired thickness. TEM has found its potential use in microstructures' observation via imaging that reveals the crystallographic orientation information as a diffraction pattern and new chemical composition by the energy spectrum. TEM also records microstructures' images, but as compared to an optical microscope, it produces a much higher resolution. Generally, TEM resolution can be nanometer, which makes it an excellent tool for nanomaterials' characterization. Apart from this, the high-resolution TEM (HRTEM) can resolve a range of angstroms and even image lattice points. Currently, the resolution of TEM is also found to be astigmatism and aberration. Microscopists are conducting a huge effort to resolve these issues and further push TEM's resolution [69].

2.5.10 Auger Electron Spectroscopy (AES)

AES is considered a type of surface-specific technique that uses a high-energy and finely focused electron beam as an excitation source. AES works based on the Auger effect. In the Auger effect, a chain of radiation-less transitions occurs in an atom whose inner level is ionized, and the process ends by emitting an Auger electron. The AES analyzes the kinetic energies of the emitted Auger electrons, making the analysis of elemental composition achievable. The typical sampling depth of AES is 2–5 nm. AES is utilized in the elemental analysis of small surface features. Although AES is a destructive analysis technique, it is utilized to identify concentrated contamination areas [74].

Since the Auger cycle is certainly not a mass of three electrons, helium and hydrogen cannot be distinguished as they both have less than three electrons [75]. AES has discovered applications in evaluating oxide film structure, semiconductor synthesis, molecule investigation, oxide film structure, silicon, metallization, and surface cleaning effects. AES estimates are made in a high-vacuum climate ranging from 10^{-12} torr -10^{-10} torr to prevent the buildup of hydrocarbon decay layers on the sample surface [76].

2.5.11 PL Spectroscopy

PL spectroscopy is a valuable strategy for examining the optical characteristics of the material. The nondestructive nature and its capability to provide analytical information have made them gain much attention in the analysis techniques. The studies of nanomaterials' luminescence properties are important as they provide analysis about the quantum yields, luminescence energy transfer, decay time, etc. The PL spectroscopy uses a laser beam to catch the light produced from a substance when it comes down to the ground state from the excited state. It produces a luminescence spectrum used to study the impurities and imperfections present in the nanomaterials. The PL is not observed in metals as they show a continuous electronic state around the Fermi level [72].

2.5.12 Raman Spectroscopy

It is a spectroscopic technique that detects rotational and vibrational motions and many other properties. The analysis of the various chemical compounds present in the materials is known as Raman spectroscopy. It is a vital tool that is used in the field of vibrational spectroscopy and is similar to infrared absorption spectroscopy. Raman spectroscopy's basic principle is based on the Raman effect, which states that the different molecules reflect the different light wavelengths when exposed to incident light. These different wavelengths determine the various characteristics of the chemical compounds, which help detect and analyze various chemical reactions and properties [77]. It is an integral asset for portraying nanosized materials and structures. It also yields beneficial results in analyzing heterointerfaces and surfaces present between the constituent layers of low-dimensional structures. It is highly utilized in the medical diagnosis field due to its high chemical specificity and label-free and aseptic nature. In an experiment, it was observed that Raman spectroscopy efficiently detected the functional groups' changes, molecular composition, and variation in molecular bonding caused by skin infections. The results were analyzed in peak width in the detected spectrum and minute shifts in Raman peak location [78].

2.5.13   SCANNING PROBE MICROSCOPY (SPM)

SPM is a technique that is used for the identification of various surface properties of nanomaterials. The SPM works on a fundamental principle of quantum tunneling, comprising a probe with a very fine atomic-leveled tip that scans the desired surface at a very close distance to record the interactions between the surface and the tip. These recorded signals are rebuilt sequentially to unveil topography to analyze different properties of the nanomaterial. According to the different interaction modes between sample and probe, SPM is divided into two parts: scanning tunneling microscope (STM) and atomic force microscope (AFM). In STM, the tunneling current is first registered to reconstruct the surface information, whereas in AFM, instead of the tunneling current, the atomic force is recorded for a similar purpose [79].

2.5.13.1   Atomic Force Microscopy (AFM)

AFM is also known as scanning force microscopy (SFM)/nuclear force microscopy (NFM), which is the most modern instrument or procedure from the SPM family and takes a considerable capacity in the work of inventive exploration and parts of nanotechnology. It is a surface analysis technique widely used for nanostructured coatings and to obtain high-resolution nanoscale images. It gives quantitative and qualitative information covering many properties such as morphology, roughness, size, surface texture, and volume distributions. AFM also scans a wide range of particles ranging from 1 nm to 8 μm. It is considered a powerful technique that allows the imaging of several surfaces, such as biological samples, polymers, composites, ceramics, and glass. It is also utilized in the measurement of magnetic force, mechanical force, adhesion forces, etc. The strategy relies on measuring instantaneous forces present between a sharp metal tip and ions present on a surface, which offers high resolution and visualization in 3D images. Gerd Binning designed the AFM instrument in 1986. The instrument consists of a tip of 10–20 nm in diameter, which is mounted on the arm known as the cantilever, both the tip and the cantilever micro-fabricated from silicon (Si) or nitride (Si_3N_4). The whole arrangement is perfectly positioned in the sub-nanometer space from the sample surface. The tip–surface interactions move the tip, which is measured by focusing a laser beam with a photodiode. AFM functions in two modes: contact mode, in which the surface and the AFM tip continuously contact with each other, and tapping mode, in which the tip and the cantilever are in intermittent contact with each other. The tapping mode is the most preferred as it decreases the shear forces caused by the tip's movement. The atom present at the apex of the tip "senses" single atoms on the surface when it forms initial chemical bonds with each other as these chemical interactions slowly modify the tip's vibration frequency, making them suitable for detection and further for mapping [69].

2.5.13.2 Scanning Tunneling Microscopy (STM)

STM is considered one of the powerful types of electron microscopy used to image the sample's surface at the atomic level. It is also known as the surface-sensitive technique. The concept of quantum tunneling is the basic principle of STM. When a conducting tip is placed a few angstroms away from a metallic or semiconducting surface, a bias voltage is applied between the probe tip, and the surface enables electrons to move across the gap. The tunneling variations are recorded by the tip, which can generate topographical images of the surface. In any case, the current excavation scale is fragile and relies on the separation that separates the tip from the sample, the thickness of the electron close to the sample, just like the neighborhood boundary's stature. A couple of STM uses are nuclear imaging and electrochemical control microscopy (ECSTM), whereas scanning tunneling spectroscopy (STS) helps to image examples of inadequate conductivity [69].

2.6 CONCLUSION AND PROSPECTS

Nanotechnology is an imaginary branch among the growing sciences, which has completely transformed humanity. Nanotechnology is the study of nanoscale materials and their properties, which can be utilized in various domains. Today, many nanomaterials with enhanced properties offer them the ability to be utilized for various applications in biomedical, environmental, and agricultural domains. Further, changing the properties of the material on a subnuclear/subatomic scale gives the desire to execute updates for applications to fulfill humankind's day-to-day requirements. So, the expected nanomaterial applications in the near future are endless, with the opportunity to follow the basic advances in the fields of equipment, food, and medicine, and this is just the beginning. On the moderate front, research and development is key for improving personal satisfaction and economy, both of which can be improved only by commercializing technological advancement by nanomaterials.

Thinking along the lines of correlation, however, its impact and future proposals have been thrown over the last decade, and items using nanomaterials and nanotechnology have begun to rise dynamically. Hence, this chapter briefly describes various potential properties: physicochemical and biological, which can be utilized in various domains of science such as robotics, biomedical, electrical, chemical, material chemistry, and inorganic chemistry. This chapter further highlights various routes of nanomaterial synthesis, focusing primarily on chemical and physical methods. A brief discussion on biological synthesis is also included in this chapter as this synthesis approach is further discussed in more detail in Chapter 3 of this book. Moreover, nanomaterial classification is also briefly highlighted in this chapter, along with various characterization techniques used to determine properties of various synthesized nanomaterials. So, it can be concluded that this chapter presents the simplified data on classification, properties, synthesis, and characterization techniques of nanomaterials.

ACKNOWLEDGMENTS

Miss Vanya Nayak extends her gratitude of thanks to M.Sc. Biotechnology thesis supervisor Dr. Ravindra Pratap Singh for providing this opportunity to write this chapter. Mr. Syed Muzammil Munawar is thankful to Mrs. Syed Shabana of Thiruvalluvar University, India, and Mr. Dhandayuthabani Rajendiran of C. Abdul Hakeem College, India, for providing literature materials for writing the characterization section. Dr. Khaleel Basha Sabjan is thankful to the editor Dr. Ravindra Pratap Singh, for providing apt support for completing the chapter assignment on time. Miss Srishti Singh is thankful to the Director of DY Patil, School of Biotechnology and Bioinformatics, Mumbai, India. Kshitij RB Singh is thankful to Dr. Ajaya Singh and Principal of Govt. V. Y. T. PG. Autonomous College, Durg, India, for providing a working platform for completing this work diligently and smoothly.

REFERENCES

1. Pelletier, D.A., A.K. Suresh, G.A. Holton, C.K. McKeown, W. Wang, B. Gu, N.P. Mortensen, et al. 2010. "Effects of engineered cerium oxide nanoparticles on bacterial growth and viability." *Applied and Environmental Microbiology* 76 (24): 7981–89. Doi: 10.1128/AEM.00650-10.
2. Issa, B., I. Obaidat, B. Albiss, and Y. Haik. 2013. "Magnetic nanoparticles: Surface effects and properties related to biomedicine applications." *International Journal of Molecular Sciences* 14 (11): 21266–305. Doi: 10.3390/ijms141121266.
3. Ziemann, P., and T. Schimmel. 2012. "Physics, chemistry and biology of functional nanostructures." *Beilstein Journal of Nanotechnology* 3 (December): 843–45. Doi: 10.3762/bjnano.3.94.
4. Ramrakhiani, M. 2012. "Nanostructures and their applications." *Recent Research Science and Technology* 4(8): 14–19.
5. Dean, S.W., G.A. Mansoori, and T.A. FauziSoelaiman. 2005. "Nanotechnology — an introduction for the standards community." *Journal of ASTM International* 2 (6): 13110. Doi: 10.1520/JAI13110.
6. Kurzydlowski, K.J. 2006. "Physical, chemical, and mechanical properties of nanostructured materials." *Materials Science* 42 (1): 85–94. Doi: 10.1007/s11003-006-0060-2.
7. Winhold, M., M. Leitner, A. Lieb, P. Frederix, F. Hofbauer, T. Strunz, J. Sattelkov, H. Plank, and C.H. Schwalb. 2017. "Correlative *In-Situ* AFM & SEM & EDX analysis of nanostructured materials." *Microscopy and Microanalysis* 23 (S1): 26–27. Doi: 10.1017/S1431927617000812.
8. Lohse, S., 2013. Nanoparticles are all around us. Sustainable nano – a blog by the Center for Sustainable Nanotechnology. https://sustainable-nano.com/2013/03/25/nanoparticles-are-all-around-us/
9. Aguado, A., and M.F. Jarrold. 2011. "Melting and freezing of metal clusters." *Annual Review of Physical Chemistry* 62 (1): 151–72. Doi: 10.1146/annurev-physchem-032210-103454.
10. Couchman, P.R., and W.A. Jesser. 1977. "Thermodynamic theory of size dependence of melting temperature in metals." *Nature* 269 (5628): 481–83. Doi: 10.1038/269481a0.
11. Lai, S.L., J.Y. Guo, V. Petrova, G. Ramanath, and L.H. Allen. 1996. "Size-dependent melting properties of small tin particles: Nanocalorimetric measurements." *Physical Review Letters* 77 (1): 99–102. Doi: 10.1103/PhysRevLett.77.99.
12. Yu, X., and Z. Zhan. 2014. "The effects of the size of nanocrystalline materials on their thermodynamic and mechanical properties." *Nanoscale Research Letters* 9 (1): 516. Doi: 10.1186/1556-276X-9-516.

13. Qiu, Y., Y. Liu, L. Wang, L. Xu, R. Bai, Y. Ji, X. Wu, Y. Zhao, Y. Li, and C. Chen. 2010. "Surface chemistry and aspect ratio mediated cellular uptake of Au nanorods." *Biomaterials* 31 (30): 7606–19. Doi: 10.1016/j.biomaterials.2010.06.051.

14. Moldoveanu, S., and V. David. 2017. *RP-HPLC Analytical Columns, Selection of the HPLC Method'. Chemical Analysis.* Elsevier, Netherlands, 297–328.

15. Eustathopoulos, N. 2015. "Wetting by liquid metals—application in materials processing: The contribution of the grenoble group." *Metals* 5 (1): 350–70. Doi: 10.3390/met5010350.

16. Glover, R.D., J.M. Miller, and J.E. Hutchison. 2011. "Generation of metal nanoparticles from silver and copper objects: Nanoparticle dynamics on surfaces and potential sources of nanoparticles in the environment." *ACS Nano* 5 (11): 8950–57. Doi: 10.1021/nn2031319.

17. Neikov, O.D. 2009. "Nanopowders." In *Handbook of Non-Ferrous Metal Powders*, 80–101. Elsevier. Doi: 10.1016/B978-1-85617-422-0.00004-5.

18. Aveyard, R., B.P. Binks, and J.H Clint. 2003. "Emulsions stabilised solely by colloidal particles." *Advances in Colloid and Interface Science* 100–102 (February): 503–46. Doi: 10.1016/S0001-8686(02)00069-6.

19. Binks, B.P., J.H. Clint, P.D.I. Fletcher, T.J.G. Lees, and P. Taylor. 2006. "Growth of gold nanoparticle films driven by the coalescence of particle-stabilized emulsion drops." *Langmuir* 22 (9): 4100–4103. Doi: 10.1021/la052752i.

20. Kim, J.-K., S.C. Tjong, and Y.-W. Mai. 2018. "4.2 effect of interface strength on metal matrix composites properties." In *Comprehensive Composite Materials II*, 22–59. Elsevier. Doi: 10.1016/B978-0-12-803581-8.09982-3.

21. Sajanlal, P.R., T.S. Sreeprasad, A.K. Samal, and T. Pradeep. 2011. "Anisotropic nanomaterials: structure, growth, assembly, and functions." *Nano Reviews* 2 (1): 5883. Doi: 10.3402/nano.v2i0.5883.

22. Tian, S., Y.-Z. Li, M.-B. Li, J. Yuan, J. Yang, Z. Wu, and R. Jin. 2015. "Erratum: Structural isomerism in gold nanoparticles revealed by X-ray crystallography." *Nature Communications* 6 (1): 10012. Doi: 10.1038/ncomms10012.

23. Putnam, S.A., D.G. Cahill, P.V. Braun, Z. Ge, and R.G. Shimmin. 2006. "Thermal conductivity of nanoparticle suspensions." *Journal of Applied Physics* 99 (8): 084308. Doi: 10.1063/1.2189933.

24. "Chapter 1 nanotechnology and nanomaterials." 2006. In, 1–69. Doi:10.1016/S1383-7303(06)80002-5.

25. Guo, D., G. Xie, and J. Luo. 2014. "Mechanical properties of nanoparticles: Basics and applications." *Journal of Physics D: Applied Physics* 47 (1): 013001. Doi: 10.1088/0022-3727/47/1/013001.

26. Yokoyama, T., H. Masuda, M. Suzuki, K. Ehara, K. Nogi, M. Fuji, T. Fukui, et al. 2008. "Basic properties and measuring methods of nanoparticles." In *Nanoparticle Technology Handbook*, 3–48. Elsevier. Doi: 10.1016/B978-044453122-3.50004-0.

27. Kedia, A., H. Kumar, and P. Senthil Kumar. 2015. "Tweaking anisotropic gold nanostars: covariant control of a polymer–solvent mixture complex." *RSC Advances* 5 (7): 5205–12. Doi: 10.1039/C4RA12846J.

28. Mark, A.G., J.G. Gibbs, T.-C. Lee, and P. Fischer. 2013. "Hybrid nanocolloids with programmed three-dimensional shape and material composition." *Nature Materials* 12 (9): 802–7. Doi: 10.1038/nmat3685.

29. Mattox, T.M., X. Ye, K. Manthiram, P. James Schuck, A. Paul Alivisatos, and J.J. Urban. 2015. "Chemical control of plasmons in metal chalcogenide and metal oxide nanostructures." *Advanced Materials* 27 (38): 5830–37. Doi: 10.1002/adma.201502218.

30. Scholes, G.D. 2008. "Controlling the optical properties of inorganic nanoparticles." *Advanced Functional Materials* 18 (8): 1157–72. Doi: 10.1002/adfm.200800151.

31. Sagadevan, S. 2013. "Semiconductor nanomaterials, methods and applications: A review". *Nanoscience and Nanotechnology*, 3(3). 62–74. Doi: 10.5923/j.nn.20130303.06.
32. Edelstein, A.S, and R.C Cammaratra, eds. 1998. *Nanomaterials*. CRC Press. Doi: 10.1201/9781482268591.
33. Popov, V.N., and P. Lambin. n.d. "Vibrational and related properties of carbon nanotubes." In *Carbon Nanotubes*, 69–88. Dordrecht: Kluwer Academic Publishers. Doi: 10.1007/1-4020-4574-3_16.
34. Richter, H., Z.P. Wang, and L. Ley. 1981. "The one phonon raman spectrum in microcrystalline silicon." *Solid State Communications* 39 (5): 625–29. Doi: 10.1016/0038-1098(81)90337-9.
35. Singh, K.R.B., V. Nayak, T. Sarkar, and R.P. Singh. 2020. "Cerium oxide nanoparticles: Properties, biosynthesis and biomedical application." *RSC Advances* 10 (45): 27194–214. Doi: 10.1039/D0RA04736H.
36. Celardo, I., J.Z. Pedersen, E. Traversa, and L. Ghibelli. 2011. "Pharmacological potential of cerium oxide nanoparticles." *Nanoscale* 3 (4): 1411. Doi: 10.1039/c0nr00875c.
37. Mu, P., G. Zhou, and C.-L. Chen. 2018. "2D nanomaterials assembled from sequence-defined molecules." *Nano-Structures & Nano-Objects* 15 (July): 153–66. Doi: 10.1016/j.nanoso.2017.09.010.
38. Altavilla, C. 2016. *Inorganic Nanoparticles*. CRC Press. Doi: 10.1201/b10333.
39. Bryant, G.W. 1990. "Understanding quantum confinement in zero-dimensional nanostructures: Optical and transport properties." In, 243–54. Doi: 10.1007/978-1-4684-5733-9_24.
40. Zhang, X., X. Cheng, and Q. Zhang. 2016. "Nanostructured energy materials for electrochemical energy conversion and storage: A review." *Journal of Energy Chemistry* 25 (6): 967–84. Doi:10.1016/j.jechem.2016.11.003.
41. Mandal, G., and T. Ganguly. 2011. "Applications of nanomaterials in the different fields of photosciences." *Indian Journal of Physics* 85 (8): 1229–45. Doi: 10.1007/s12648-011-0149-9.
42. *Nanoparticle Technology Handbook*. 2018. Elsevier. Doi: 10.1016/C2017-0-01011-X.
43. Tiwari, J.N., R.N. Tiwari, and K.S. Kim. 2012. "Zero-dimensional, one-dimensional, two-dimensional and three-dimensional nanostructured materials for advanced electrochemical energy devices." *Progress in Materials Science* 57 (4): 724–803. Doi: 10.1016/j.pmatsci.2011.08.003.
44. Wang, X., M. Ahmad, and H. Sun. 2017. "Three-dimensional ZnO hierarchical nanostructures: Solution phase synthesis and applications." *Materials* 10 (11): 1304. Doi: 10.3390/ma10111304.
45. Psaro, R., Guidotti, and M., Sgobba, M., 2006. *Nanosystems: Inorganic and Bioinorganic Chemistry*. EOLSS Publishers, Oxford.
46. Guo, Z., and T. Li., 2009. *Fundamentals and Applications of Nanomaterials*. Artech House, Norwood, MA.
47. Yousefi, R., and M. Cheraghizade. 2018. "Semiconductor/graphene nanocomposites: synthesis, characterization, and applications." In *Applications of Nanomaterials*, 23–43. Elsevier. Doi: 10.1016/B978-0-08-101971-9.00002-8.
48. Naveed Ul Haq, A., A. Nadhman, I. Ullah, G. Mustafa, M. Yasinzai, and I. Khan. 2017. "Synthesis approaches of zinc oxide nanoparticles: The dilemma of ecotoxicity." *Journal of Nanomaterials* 2017: 1–14. Doi: 10.1155/2017/8510342.
49. Sakka, S. 2013. "Sol–Gel process and applications." In *Handbook of Advanced Ceramics*, 883–910. Elsevier. Doi: 10.1016/B978-0-12-385469-8.00048-4.
50. Nayak, A.K., and B. Das. 2018. "Introduction to polymeric gels." In *Polymeric Gels*, 3–27. Elsevier. Doi: 10.1016/B978-0-08-102179-8.00001-6.
51. Scherer, G.W. 2001. "Xerogels." In *Encyclopedia of Materials: Science and Technology*, 9797–99. Elsevier. Doi: 10.1016/B0-08-043152-6/01777-0.

52. Kumar, A., Mishra, R., Reinwald, Y., and Bhat, S. 2010. "Cryogels: Freezing unveiled by thawing." *Materials Today*, 13 (11), 42–44. Doi: 10.1016/S1369-7021(10)70202-9.
53. Gökçe, E.H., E.A. Yapar, S.T. Tanrıverdi, and Ö. Özer. 2016. "Nanocarriers in cosmetology." In *Nanobiomaterials in Galenic Formulations and Cosmetics*, 363–93. Elsevier. Doi: 10.1016/B978-0-323-42868-2.00014-0.
54. *Advanced Nanomaterials for Catalysis and Energy*. 2019. Elsevier. Doi: 10.1016/C2017-0-02137-7.
55. Feng, S.-H., and G.-H. Li. 2017. "Hydrothermal and solvothermal syntheses." In *Modern Inorganic Synthetic Chemistry*, 73–104. Elsevier. Doi: 10.1016/B978-0-444-63591-4.00004-5.
56. *Handbook of Nanomaterials in Analytical Chemistry*. 2020. Elsevier. Doi: 10.1016/C2018-0-00945-7.
57. Shrivastava, A. 2018. "Polymerization." In *Introduction to Plastics Engineering*, 17–48. Elsevier. Doi: 10.1016/B978-0-323-39500-7.00002-2.
58. Dahiya, M.S., V.K. Tomer, and S. Duhan. 2018. "Metal–Ferrite nanocomposites for targeted drug delivery." In *Applications of Nanocomposite Materials in Drug Delivery*, 737–760. Elsevier. Doi: 10.1016/B978-0-12-813741-3.00032-7.
59. Salavati-Niasari, M., F. Davar, and M. Mazaheri. 2008. "Synthesis of Mn_3O_4 nanoparticles by thermal decomposition of a [Bis(Salicylidiminato)Manganese(II)] complex." *Polyhedron* 27 (17): 3467–71. Doi: 10.1016/j.poly.2008.04.015.
60. Singh, R., and R.K. Soni. 2019. "Laser-induced heating synthesis of hybrid nanoparticles." In *Noble Metal-Metal Oxide Hybrid Nanoparticles*, 195–238. Elsevier. Doi: 10.1016/B978-0-12-814134-2.00011-5.
61. Nune, S.K., K.S. Rama, V.R. Dirisala, and M.Y. Chavali. 2017. "Electrospinning of collagen nanofiber scaffolds for tissue repair and regeneration." In *Nanostructures for Novel Therapy*, 281–311. Elsevier. Doi: 10.1016/B978-0-323-46142-9.00011-6.
62. Xu, H., B.W. Zeiger, and K.S. Suslick. 2013. "Sonochemical synthesis of nanomaterials." *Chemical Society Reviews* 42 (7): 2555–67. Doi: 10.1039/C2CS35282F.
63. "Methods for assessing surface cleanliness." 2019. In *Developments in Surface Contamination and Cleaning*, Volume 12, 23–105. Elsevier. Doi: 10.1016/B978-0-12-816081-7.00003-6.
64. Yang, L. 2015. "Carbon nanostructures." In *Nanotechnology-Enhanced Orthopedic Materials*, 97–120. Elsevier. Doi: 10.1016/B978-0-85709-844-3.00005-7.
65. Eyring, M.B., and P. Martin. 2013. "Spectroscopy in forensic science." In *Reference Module in Chemistry, Molecular Sciences and Chemical Engineering*. Elsevier. Doi: 10.1016/B978-0-12-409547-2.05455-X.
66. Mourdikoudis, S., R.M. Pallares, and N.T.K. Thanh. 2018. "Characterization techniques for nanoparticles: Comparison and complementarity upon studying nanoparticle properties." *Nanoscale* 10 (27): 12871–934. Doi: 10.1039/C8NR02278J.
67. Moros, M., L. Gonzalez-Moragas, A. Tino, A. Laromaine, and C. Tortiglione. 2019. "Invertebrate models for hyperthermia: What we learned from caenorhabditis elegans and hydra vulgaris." In *Nanomaterials for Magnetic and Optical Hyperthermia Applications*, 229–64. Elsevier. Doi: 10.1016/B978-0-12-813928-8.00009-0.
68. *Monitoring and Evaluation of Biomaterials and Their Performance In Vivo*. 2017. Elsevier. Doi: 10.1016/C2014-0-04050-6.
69. Shnoudeh, A.J., I. Hamad, R.W. Abdo, L. Qadumii, A.Y. Jaber, H.S. Surchi, and S.Z. Alkelany. 2019. "Synthesis, characterization, and applications of metal nanoparticles." In *Biomaterials and Bionanotechnology*, 527–612. Elsevier. Doi: 10.1016/B978-0-12-814427-5.00015-9.
70. Tantra, R., P. Schulze, and P. Quincey. 2010. "Effect of nanoparticle concentration on zeta-potential measurement results and reproducibility." *Particuology* 8 (3): 279–85. Doi: 10.1016/j.partic.2010.01.003.

71. Jiang, J., G. Oberdörster, and P. Biswas. 2009. "Characterization of size, surface charge, and agglomeration state of nanoparticle dispersions for toxicological studies." *Journal of Nanoparticle Research* 11 (1): 77–89. Doi: 10.1007/s11051-008-9446-4.
72. Telegdi, J., A. Shaban, and G. Vastag. 2018. "Biocorrosion-steel." In *Encyclopedia of Interfacial Chemistry*, 28–42. Elsevier. Doi: 10.1016/B978-0-12-409547-2.13591-7.
73. Köhler, M., and W. Fritzsche. 2004. *Nanotechnology*. Wiley. Doi: 10.1002/9783527612369.
74. "Methods for assessing surface cleanliness." 2019. In *Developments in Surface Contamination and Cleaning*, Volume 12, 23–105. Elsevier. Doi: 10.1016/B978-0-12-816081-7.00003-6.
75. Jenkins, T.E., 1998. *Semiconductor Science; Growth and Characterization Techniques*. Prentice Hall, Harlow.
76. Gao, X.-L., J.-S. Pan, and C.-Y. Hsu. 2006. "Laser-fluoride effect on root demineralization." *Journal of Dental Research* 85 (10): 919–23. Doi: 10.1177/154405910608501009.
77. Tousoulis, D. 2018. *Invasive Imaging Techniques. Coronary Artery Disease*. Elsevier, 359–376.
78. Chen, K., Y.H. Ong, C. Yuen, and Q. Liu. 2016. "Surface-enhanced raman spectroscopy for intradermal measurements." In *Imaging in Dermatology*, 141–54. Elsevier. Doi: 10.1016/B978-0-12-802838-4.00013-3.
79. Fink, H.-W. 1986. "Mono-atomic tips for scanning tunneling microscopy." In, 87–92. Doi: 10.1007/978-94-011-1812-5_10.

3 Biological Synthesis of Nanomaterials and Their Advantages

Gözde Koşarsoy Ağçeli
Hacettepe University

*Kanika Dulta, Parveen Chauhan,
and Pankaj Kumar Chauhan*
Shoolini University

CONTENTS

3.1 INTRODUCTION

Nanotechnology is a rapidly developing branch of science, especially in the recent period. Nanomaterials such as copper (Cu), zinc (Zn), titanium (Ti), gold (Au), and silver (Ag) can be synthesized with nanotechnology [1]. The synthesis process of nanoparticles (NPs) contains physical, chemical, and biological techniques. Physical and chemical processes have usually been used to synthesize NPs. However, the fact that biological synthesis produces less pollution and is environmentally friendly compared to other synthesis methods is a very important development in the field of nanotechnology. The fact that biological methods use mild reaction conditions and are cheap also supports this method to stand out [2–4].

The synthesis of NPs by biological means, that is, by living organisms, is used as a sustainable and different option for traditional production processes. Many bacteria, plants, algae, and fungi have shown the ability to synthesize NP [5,6]. These are environmentally possible, economic, and quick alternatives to the available physical and chemical synthesis process. NP production can be done by using different living creatures or their extracts by various methods that have been investigated in recent years. These NPs are not only biocompatible but also have high stability [7–9]. This chapter focuses on the biological synthesis of nanomaterials. Biological synthesis has many advantages over other synthesis methods. The mechanism of biological synthesis and how different organisms synthesize nanomaterials are discussed in detail in this section.

3.2   NANOMATERIAL SYNTHESIS METHODS

NPs can be synthesized by different methods. These are top-down and bottom-up methods using physical, chemical, or biological means [10–12]. There are different techniques for the production of NPs, and these methods are divided into three main classes as shown in Figure 3.1.

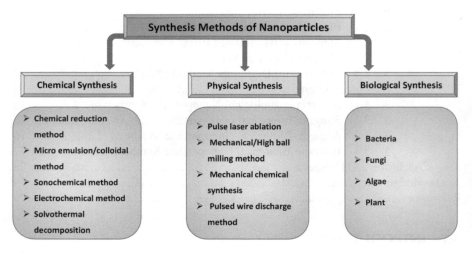

FIGURE 3.1   Synthesis methods of nanoparticles.

3.2.1 CHEMICAL SYNTHESIS

3.2.1.1 Chemical Reduction Method

The chemical reduction method was first used by Michael Faraday. Faraday reported the systematic study of the synthesis and colors of colloidal gold in 1857. Chemical reduction is the most common and simplest method used to obtain copper NPs from copper salts [13]. In addition, Haque et al. synthesized nanonickel by chemical reduction method, and Alanazi et al. synthesized gold NPs with this method [14,15].

3.2.1.2 Microemulsion/Colloidal Method

Hiral et al., with their study in 1943, obtained homogeneous and clear solutions called "microemulsions" with water, oil, and surfactant (alcohol or amine-based) substance. Microemulsions are dispersions of two or more immiscible or partly mixable fluids stabilized by added surfactants. Microemulsion method involves two immiscible liquids (such as water in oil and oil in water). Added surfactants perform a key role to reduce interfacial tension to promote thermodynamic stability of microemulsions. Microemulsions are considered to be thermodynamically stable [13,16]. The microemulsion/colloidal method is used in the preparation of NPs from a wide variety of materials, including metals (Au, Cu, Ag, Rh, Ir, Pt, Pd, Ru) [17].

3.2.1.3 Sonochemical Method

In this technique used in NP production, ultrasound is used during chemical reactions [18]. Cavitation occurs with the use of ultrasound in the sonochemical process. Cavities or microbubbles form, grow, and collapse. This happens in very short periods of time, and a large amount of energy is released [19]. The sonochemical method was originally proposed for the synthesis of Fe NPs, but is now used for the production of various metals and metal oxides [20].

3.2.1.4 Electrochemical Synthesis Method (EC)

Electrochemical reactions are processes in which chemical transformations are achieved by electron transfer. Electrochemical cell, solvent, supporting electrolyte, electrodes, and a potentiostat that generates stimulating signals according to the electrochemical technique working with this electrode are required in order to perform an electrochemical reaction. An electrochemical reaction can be done in two-electrode or three-electrode cells. This technique is inexpensive, is easy to use, and shows high flexibility. In addition, the product with higher purity compared to other chemical synthesis methods is obtained by less polluting the environment [20]. Electrochemical synthesis method can work at low temperature, it can be done with low energy, and high-purity products can be obtained with this method [21].

3.2.1.5 Solvothermal Decomposition

Solvothermal method is one of the simplest methods used in the production of nanopowder among the vacuum-free methods that do not require low temperature, high pressure, and atmospheric protection and stands out with its applicability to large scales and low cost compared to other methods [22]. Different metal oxides such as aluminum oxide, copper (II) oxide, iron oxide, nickel (II) oxide, zirconium dioxide, titanium dioxide, barium titanate, and strontium titanate can be prepared in nanostructure by both hydrothermal and solvothermal ways [23–30].

3.2.2 Physical Synthesis

3.2.2.1 Pulsed Laser Ablation (PLA)

PLA technique has been frequently used to produce noble metal NPs from their bulk metal targets [31]. It is based on the principle that the high-energy laser beam focuses on the target material surface in the vacuum chamber, providing plasma formation on the surface and concentrating the particles in the plasma on the heated substrate [32]. PLA is an easy and fast method that can effortlessly be employed to produce NPs without any primary understanding of the underlying principles governing it [33]. In the method, highly big energy is concentrated at a particular point on a solid surface to vaporize light-absorbing material. High-purity NPs can be produced by PLA because the purity of the particles is fundamentally determined by the purity of the target and ambient media (liquid or gas) without contamination from the reactor [34].

3.2.2.2 Mechanical/High Ball Milling Method

This method is widely used to grind powders into fine particles and blend materials [35]. Milling, originally used for the production of superalloys, is a solid-state processing technique for the synthesis of NPs. In this method, the raw material of micron size is fed to undergo a few changes. Various kinds of mechanical mills are available, which are frequently used for the production of NPs [20].

3.2.2.3 Mechanochemical Synthesis

In the mechanochemical synthesis method, the metal powder can be produced as a result of chemical reactions and phase changes, thanks to mechanical energy. In other words, chemical reactions occur by changing the mechanical energy. The raw material is brought to the desired content, microstructure, and grain size as a result of a mechanical effect in the presence of balls, with the help of a mill or grinder. During grinding, the desired product is obtained by chemical or reduction reactions. The product obtained is then purified by chemical resolution processes [36].

3.2.2.4 Pulsed Wire Discharge Method (PWD)

The history of this technique dates as far back as 1857, when Faraday reported that silver, copper, and aluminum powders could be prepared by heating metal wires by electric current [37]. PWD is one of the physical methods for the preparation of nanosized particles using high-density plasma/vapor [38,39]. This method has a very different mechanism from other methods. The method is not used conventionally for common industrial purposes because it is expensive and impossible to use explicitly for various metals [20]. NPs can be synthesized with these methods, but these are very expensive, environmentally incompatible, and toxic methods [40]. For this reason, in recent studies, extracts and by-products obtained from various organisms have been used in the synthesis of NPs and investigated as another bottom-up method for the synthesis of metal NPs [41].

3.2.3 BIOLOGICAL SYNTHESIS

The traditional techniques used for NP production are costly and slow and have possible biological and environmental risks [42]. Scientists have become concerned about biological synthesis for the production of NPs as alternatives to physical and chemical methods. Different biological approaches have been proposed for NP synthesis [6,43,44]. The organisms investigated and used extensively for the biosynthesis of the NPs are bacteria, fungi, algae, and plants [45–51]. The synthesis of these NPs, which can be used in biological applications, does not carry environmental risks, is inexpensive, and is easier than other methods [52]. This biological synthesis can occur with unicellular or multicellular organisms. These living creatures producing inorganic materials are either intracellular or extracellular [53,54]. In intracellular biosynthesis, the NPs will grow inside the biosource (bacteria, fungi, algae, plant, etc.) using the biomolecules present in the system (Figure 3.2).

NPs can be obtained by preparing extracts from plant parts, microorganisms, and algae in extracellular (in vivo) synthesis. In the process, the size and shape of the NPs can be changed by changing the reaction conditions as in the chemical method (Figure 3.3) [55,56]. The reduction of metal ions to NPs can be carried out extracellularly by microorganisms [57–59].

Gold (Au) and silver (Ag) are the first compounds to be synthesized biologically as NPs. In the following years, titanium dioxide (TiO_2), cerium oxide (CeO_2), copper (Cu) and copper oxide (CuO), cadmium sulfide (CdS), zinc oxide (ZnO), and iron oxide (Fe_2O_3) were added to NPs produced by biological synthesis. Palladium (Pd), metalloids such as tellurium, silicon, selenium have also been biologically obtained in nanoscale [60–63]. In addition, the reduction of graphene oxide could be accomplished by biological synthesis [64].

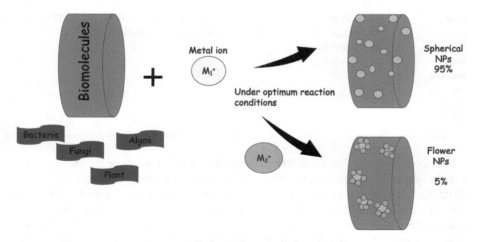

FIGURE 3.2 Intracellular synthesis of nanoparticles.

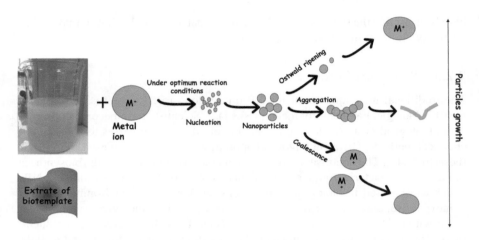

FIGURE 3.3   Extracellular synthesis of nanoparticles.

TABLE 3.1

Possible Mechanisms of Different Sources Used in Nanoparticle Biosynthesis

Biomass	Possible Mechanisms of Nanoparticle Synthesis	Reference
Bacteria	Microbial cells reduce metal ions using specific reduction enzymes such as nitrate-dependent reductase or NADH-dependent reductase.	[69]
Fungi	Intracellular and extracellular reducing enzymes and biomimetic	[70]
Algae	mineralization procedures	[71,72]
Macroalgae	Membrane-bound oxidoreductases and quinones	
Microalgae	Polysaccharides have hydroxyl groups and other functionalities that can play an important role in stabilizing and reducing nanoparticles.	
Plants	Secondary metabolites (alkaloids, flavonoids saponins, steroids,	[73]
Leaves	tannins, and other nutritional components) act as reducing and	
Stems	stabilizing agents.	
Roots		
Sprouts		
Flowers		
Shells		
Seeds		

Various shapes and sizes of NPs were produced using living organisms by varying the synthesizing parameters (oxygenation, incubation time, pH, temperature, etc.) [6,65–68]. Different organisms synthesize NPs by different mechanisms (Table 3.1) [69–73]. Environmental factors affect the shape, size, and composition of NPs. In a study in which silver NPs were synthesized from *Morganella psychrotolerans*, spherical and 2–5 nm size NPs were produced at 20°C, while the mixture of triangular and hexagonal NPs was produced at 25°C [8,74].

3.2.3.1 Bacteria in Biological Nanoparticle Synthesis

Bacteria have been used in nanotechnology since the late 1990s. The use of bacteria in nanotechnological research has gained importance day by day [75,76]. NP synthesis using bacteria was first reported by Haefeli in 1984, with the synthesis of *Pseudomonas stutzeri* AG259 silver NP. In a similar study, extracellular synthesis of gold nanoparticles (AuNPs) was performed using *Rhodomonas capsulata* [77,78]. Also, the synthesis of magnetic nanoparticles (MNPs) by magnetotactic bacteria has been informed. *Magnetospirillum magneticum* produces two types of NPs; some produce magnetic (Fe_3O_4) NPs in chain-like structures, others produce greigite (Fe_3S_4) NPs, and some produce both types of NPs [79]. There are many studies showing that bacteria are successfully used in metal NP synthesis [80]. Microorganisms are of great importance as the main employees of environmentally friendly nanofactories for the production and assembly of metal-containing nanosized particles due to their unique characteristics such as living under harsh conditions and environmental stresses, high metabolic diversity, high substrate specificity, and faster growth [81,82]. The fact that bacteria are open to gene manipulation, they do not require too much cost for reproduction, and they show a fast reproduction provides great advantages from NP synthesis [68].

Researchers continue to work for more efficient NP synthesis. Many various bacterial species, for example, *Bacillus* spp., *Acinetobacter* spp., *Escherichia coli, Pseudomonas* spp., and *Lactobacillus* spp., have been investigated for NP synthesis [68]. Das et al. isolated *Bacillus* spp. from soil contaminated with heavy metal in South India. Then, they synthesized 42–92 nm polydispersed Ag NPs with *Bacillus* spp. they isolated [83].

Bacteria can be used as a confident and not expensive organism for the production of metallic NPs such as zinc, gold, titanium, nickel, copper, silver, and palladium. The synthesis of NPs can be carried out both intracellularly and extracellularly by microorganisms [84]. The microorganismal culture filtrate is collected by centrifugation and mixed with an aqueous metallic salt solution in extracellular biosynthesis [85,86]. For the intracellular synthesis, the biomass is thoroughly washed with sterile water and incubated with metal ion solution after culturing microorganisms under optimum growth conditions. If the mixture of microorganism and metal ion solution shows NP production, it will change color. The NPs are then collected and characterized by different ways [87].

Microorganisms have a structural potential to synthesize NPs via intracellular and extracellular pathways, due to their ability to adapt to the metal-containing environment. Microorganisms convert these metal ions to their basic form through their enzymatic activities [81]. Bacteria can significantly reduce heavy metal ions to produce NPs. Throughout intracellular synthesis, solvable ions tend to insert the cytoplasm of the cell via the Mn or Mg transport chain and transform into NPs by the action of intracellular enzymes in the cytoplasm. In addition, in the synthesis of NPs, enzymes located in the cell membrane are mediated or released into the growth medium. The NPs produced in this way can hold on to the cell membrane or remain in the environment [88].

Culture conditions dictate the biogenic synthesis of NPs using bacteria. Consequently, these conditions must be standardized on a large scale for the production of NPs. With strict control over the shape, size, and combination of the particles, it is recognized that many bacteria can produce metallic NPs with characteristics similar to chemically synthesized NPs [89].

3.2.3.2 Fungi in Biological Nanoparticle Synthesis

Fungi have about 6,400 kinds of bioactive compounds, and they have a high tolerance to heavy metals [90]. Bioaccumulative ability of fungi is utilized for the reduction and stabilization of NPs in NP production. In addition, large-scale fungi cultivation is very easy, so fungi can be used for uniform NP synthesis [91–94]. Fungi, which are eukaryotic organisms, generally maintain their lives by breaking down rotting and organic compounds. Compared to plants and algae reported for biosynthesis, they can be grown easily under laboratory conditions and can reproduce rapidly, making fungi superior to other organisms. The use of fungi in the green biosynthesis of NPs has been reported in the literature. Fungi contain enzymes and proteins as ion-reducing agents. For this reason, they can always be used to produce NPs from salts of metal NPs. However, some fungi found in nature can be pathogenic. Therefore, attention must be paid when working with them during the experiment. It has been stated that fungi grow faster as biomass under the same conditions than bacteria.

Fungi have a wider surface area thanks to their micelles and produce higher amounts of protein compared to bacteria. Thus, the metal NP production speed is extremely high. Thanks to these properties, it is more advantageous compared to the synthesis of NPs from bacteria [95–101]. Gold, silver, zinc, and titanium have been informed as the metal ions used in the fungal biological synthesis of NPs [102]. Different species are involved in the synthesis of nanomaterials from fungi. For example, *Verticillium*, *Penicillium*, *Aspergillus*, *Rhizopus*, *Fusarium*, and *Trichoderma* are among the most studied fungi. The inability to synthesize large-scale NPs from fungi is due to the pathogenicity of many fungi to humans or plants [103]. For the biogenic synthesis of metallic NPs, the selection of the appropriate solvent medium is subjected to three necessary steps, which are the selection of the appropriate reducing agent (environmentally friendly) and stabilizing agents or substances for silver NP stability. This process is generally carried out by microbial enzymes found in fungi and thought to be responsible for the synthesis of NPs by organisms. Since fungi are very effective secretagogues of extracellular enzymes, it is possible for enzymes to be produced on a large scale. Economic viability and ease of use of biomass are another value for using fungi to synthesize metallic NPs [104–106].

NP synthesis by fungi is divided into intracellular and extracellular. While the NPs are synthesized in the mycelium by the intracellular approach, the fungal inert filtrate is used for NP synthesis in the extracellular pathway. Some fungi can synthesize NPs both intracellularly and extracellularly [100,107,108].

3.2.3.3 Algae in Biological Nanoparticle Synthesis

There is not much literature information on phyconanotechnology, and researchers continue to do more extensive research day by day. Algae are the largest group of photoautotrophic microorganisms. Proteins, pigments, and secondary metabolites possessed by algae are very important sources for metallic NP synthesis [109–114].

Algae are used as a model organism in biomaterial production. The most important reason for this is that they can accumulate heavy metal ions and turn them into a more harmless form. Algae extracts contain unsaturated fatty acids, minerals, carbohydrates, fats, proteins, and antioxidants. Carotenoids, phycobilins, and phycoerythrin are also included in these extracts [115,116]. These active ingredients found in extracts of algae are indicated as stabilizing or reducing agents. Nanomaterials can be synthesized from algae by uses such as algae extract, metal precursor solution, or incubation of algae extract with metal precursor solution [117]. The nanomaterial production reaction starts by mixing the metal precursor molar solution with the algae extract. The reaction mixture changes color. This color change is an indication that the reaction has begun. Next, adjacent nucleic particles come together to grow NPs, resulting in NPs of a wide variety of shapes and sizes. These NPs are thermodynamically stable [115,117,118]. NP synthesis of algae can occur in two ways: intracellular and extracellular. Although it was said that NP synthesis was intracellular at the beginning of the studies, algae were also involved in extracellular synthesis in various studies [115,119–121].

If we explain the intracellular NP synthesis mechanism from the study on silver and gold metals and *Verticillium* spp. or algal biomass, it is not fully understood, but it is based on a hypothesis. First of all, these metals attach to the surface of the cell thanks to electrostatic interactions between ions and negatively charged cell wall. This negatively charged cell wall is formed due to the carboxyl groups on the enzymes. Later, enzymes reduce the metal ions to the NP form to form a "metal core" structure [104].

3.2.3.4 Plants in Biological Nanoparticle Synthesis

Plants are ideal candidates for large-scale production of NPs onward to the early 1900s [122]. The biological synthesis of metal NPs by plants has attracted more attention as a viable alternate to chemical and physical techniques.

Plants have the potential to produce NPs at large scales due to their nature, and these NPs are highly stable. NPs synthesized from various parts of plants are obtained faster than microorganisms. In addition, compared to those synthesized by other living organisms, NPs produced from plant parts are more diverse in shape and size [123]. Extracts from plants act as both reducing and stabilizing agents in NP production. The biocompatibility of phytonanotechnology and the use of water as a solvent increase the biomedical and environmental applicability by ensuring that the synthesized NPs are nontoxic [84]. In the biosynthesis method with plants, stem, tree bark, flower, leaf, root, and fruit are used that act as a reducing and stabilizing agent [124].

The ability of plants to be part of nanosized material production is due to its components in plant extracts such as alkaloids, phenolic acids, polyphenols, proteins, and terpenoids used to reduce and stabilize metallic ions [125]. In addition to amino acids and vitamins, components such as flavonoids and polysaccharides are thought to play an important role in reducing metal salts [126]. Synthesis is a method made by mixing the metal salt solution with a plant extract [127].

One of the most important reasons for using plants for NP synthesis is their availability [128]. The biodegradation potential of plant extract is higher than microbial synthesis [129]. Plant-mediated synthesis has almost zero contamination. In addition, it is known that phytochemicals found in plant extract stabilize NPs, except that they reduce metal ions [128,130]. In particular, biological synthesis using plant extracts provides advantages in producing NPs with defined size and shape, which are significant properties for different biomedical applications [131–133]. Interestingly, when Ag ions are transformed into metal silver NPs through biogenic techniques of synthesis, their toxicity is seen to be reduced while their antibacterial activities get enhanced significantly [133–136]. This method provides all the criteria for synthesis and is suitable for synthesis as it is safe for therapeutic use, environmentally friendly, low cost, and easy to find [137–139].

Green synthesis can be done by using methods such as metal NP, metal oxide NP, MNP, and quantum dot synthesis instead of toxic reducing agents. Although NPs synthesized using biological sources such as plants and microorganisms are easier to bind to active ingredients, there are many metabolites in medicinal plants that have pharmacological activity. Many studies have indicated that these metabolites are more effective by binding to the synthesized NPs and give more properties to the NPs [140, 141].

3.3 CONCLUSION AND PROSPECTS

This chapter focuses on the biological synthesis of nanomaterials. Nanomaterials can be synthesized chemically and physically as well as biologically. Bacteria, fungi, algae, and plants can synthesize nanomaterials by the biological method. Biological synthesis has many advantages over other synthesis methods. The traditional techniques used for NP synthesis are costly and time-consuming and have eventual biological and environmental risks [42]. Toxic elements remaining on the surface of chemically and physically synthesized materials cause restrictions in clinical practice [142]. One of the goals of bionanotechnology is to design professional new materials on financial, medical, and environmental issues to solve the negative aspects of nano from the outset [143]. NP biosynthesis has taken advantage of bacteria, fungi, algae, and plants as reviewed recently [80,144–146]. Biological synthesis is sustainable, simple, clean, and economical and provides greater biocompatibility in the uses of NPs [147]. Synthesizing such materials using environmentally friendly and biocompatible reagents can reduce the toxicity of the resulting materials and the environmental impact of by-products [148]. Increasing awareness of biogenic methods has led to a desire, to develop an environmentally friendly approach to the production of nontoxic NPs [149]. As a result, in order for NPs to be used in many fields, especially in medicine, targeted NPs must be produced using advanced "green" techniques and nontoxic substances. Considering that most of the applied techniques involve low material conversion, toxic reducing agents, and waste treatments, it is important to expand the applicable green synthesis pathways [150].

In recent years, there have been great developments regarding the synthesis of nanomaterials from different organisms. However, it is not yet possible to control the particle size and morphology of the synthesized nanomaterials. However, biological synthesis can sometimes take a few hours and sometimes a few days. This is a rather slow process compared to chemical and physical synthesis. New and different studies are required to shorten the synthesis time. Biological nanomaterials whose morphology and particle size can be controlled and synthesized faster will shed light on the future. The fact that it does not leave chemical waste environmentally and is cheap increases the value of nanomaterials obtained by biological synthesis. Environmentally friendly and nontoxic NPs can be synthesized using biological methods, and it is known that the physical properties of the produced NPs can vary according to the organism type used. Researches about synthesizing biological nanomaterials from different organisms should be increased and their advantages over chemical and physical synthesis should be brought to the fore.

No matter how much we protect nature, nature will reward us. If we can produce NPs cheaper, more environmentally friendly, and faster with biological synthesis, we should use our preference as a NP production method in this direction. Different bacteria, fungi, algae, and plants can be used in future studies. The synthesis of different nanomaterials can be studied. The reaction medium (live cell/extract: NP production solution) can be optimized. For this optimization, conditions such as different pH values, temperatures, static or shaker incubation, and rpm value of shaking incubation can be investigated. Studies can be extended with different parts of the plant used. Biosynthesized nanosized materials can be used in different industrial and medical fields. The biocompatibility, antimicrobial effects, and cytotoxic effects of these nanomaterials can be investigated in detail. Producing clean and economical nanomaterials with these living organisms that nature has presented to us will become even more important in the coming years.

ACKNOWLEDGMENTS

The authors would like to extend their gratitude of thanks to Hacettepe University and Shoolini University.

REFERENCES

1. AbdelRahim, K., Mahmoud, S. Y., Ali, A. M., Almaary, K. S., Mustafa, A. E. Z. M. A., & Husseiny, S. M. 2017. "Extracellular biosynthesis of silver nanoparticles using *Rhizopus stolonifer*." *Saudi Journal of Biological Sciences*. Doi: 10.1016/j.sjbs.2016.02.025.
2. Guo, M., Li, W., Yang, F., & Liu, H. 2015. "Controllable biosynthesis of gold nanoparticles from a *Eucommia ulmoides* bark aqueous extract." *Spectrochimica Acta - Part A: Molecular and Biomolecular Spectroscopy*. Doi: 10.1016/j.saa.2015.01.109.
3. Kuppusamy, P., Ichwan, S. J. A., Parine, N. R., Yusoff, M. M., Maniam, G. P., & Govindan, N. 2015. "Intracellular biosynthesis of Au and Ag nanoparticles using ethanolic extract of *Brassica oleracea* L. and studies on their physicochemical and biological properties." *Journal of Environmental Sciences (China)*. Doi: 10.1016/j.jes.2014.06.050.
4. Ajitha, B., Ashok Kumar Reddy, Y., & Sreedhara Reddy, P. 2015. "Green synthesis and characterization of silver nanoparticles using *Lantana camara* leaf extract." *Materials Science and Engineering C*. Doi: 10.1016/j.msec.2015.01.035.
5. Willner, I., Baron, R., & Willner, B. 2006. "Growing metal nanoparticles by enzymes." *Advanced Materials*. Doi: 10.1002/adma.200501865.
6. Konishi, Y., Ohno, K., Saitoh, N., Nomura, T., Nagamine, S., Hishida, H., & Uruga, T. 2007. "Bioreductive deposition of platinum nanoparticles on the Bacterium *Shewanella algae*." *Journal of Biotechnology*. Doi: 10.1016/j.jbiotec.2006.11.014.

7. Fariq, A., Khan, T., & Yasmin, A. 2017. "Microbial synthesis of nanoparticles and their potential applications in biomedicine." *Journal of Applied Biomedicine*. Doi: 10.1016/j. jab.2017.03.004.

8. Hulkoti, N. I., & Taranath, T. C. 2014. "Biosynthesis of nanoparticles using microbes-A review." *Colloids and Surfaces B: Biointerfaces*. Doi: 10.1016/j.colsurfb.2014.05.027.

9. Bao, H., Lu, Z., Cui, X., Qiao, Y., Guo, J., Anderson, J. M., & Li, C. M. 2010. "Extracellular microbial synthesis of biocompatible CdTe quantum dots." *Acta Biomaterialia*. Doi: 10.1016/j.actbio.2010.03.030.

10. Gao, Y., & Cranston, R. 2008. "Recent advances in antimicrobial treatments of textiles. " *Textile Research Journal*. Doi: 10.1177/0040517507082332.

11. Rodríguez-Sánchez, L., Blanco, M. C., & López-Quintela, M. A. 2000. "Electrochemical synthesis of silver nanoparticles." *Journal of Physical Chemistry B*. Doi: 10.1021/jp001761r.

12. Sailaja, A. K., Amareshwar, P., & Chakravarty, P. 2011. "Different techniques used for the preparation of nanoparticles using." *International Journal of pharmacy and Pharmaceutical Science* 3: 45–50.

13. Ghorbani, H. R. 2014. "A review of methods for synthesis of Al nanoparticles." *Oriental Journal of Chemistry*. Doi: 10.13005/ojc/300456.

14. Alanazi, F. K., Radwan, A. A., & Alsarra, I. A. 2010. "Biopharmaceutical applications of nanogold." *Saudi Pharmaceutical Journal*. Doi: 10.1016/j.jsps.2010.07.002.

15. Haque, K. M. A., Hussain, M. S. Alam, S. S., & Islam, S. M. S. 2010. "Synthesis of nano-nickel by a wet chemical reduction method in the presence of surfactant (SDS) and a polymer (PVP)." *African Journal of Pure and Applied Chemistry* 31: 58–63.

16. Julian Eastoe, M. H. H., & Tabor, R. 2013. "Microemulsions." In T. Tadros (Ed.), *Encyclopedia of Colloid and Interface Science* (2013 Edition, pp. 9–16). Springer, Berlin, Heidelberg. Doi: 10.1007/978-3-642-20665-8_25

17. Capek, I. 2004. "Preparation of metal nanoparticles in water-in-oil (w/o) microemulsions." *Advances in Colloid and Interface Science*. Doi: 10.1016/j.cis.2004.02.003.

18. Fernandez Rivas, D., Stricker, L., Zijlstra, A. G., Gardeniers, H. J. G. E., Lohse, D., & Prosperetti, A. 2013. "Ultrasound artificially nucleated bubbles and their sonochemical radical production." *Ultrasonics Sonochemistry*. Doi: 10.1016/j.ultsonch.2012.07.024.

19. Padoley, K. V., Saharan, V. K., Mudliar, S. N., Pandey, R. A., & Pandit, A. B. 2012. "Cavitationally induced biodegradability enhancement of a distillery wastewater." *Journal of Hazardous Materials*. Doi: 10.1016/j.jhazmat.2012.03.054.

20. Satyanarayana, T. 2018. "A review on chemical and physical synthesis methods of nanomaterials." *International Journal for Research in Applied Science and Engineering Technology*. Doi: 10.22214/ijraset.2018.1396.

21. Parr, W. C., & Taguchi, G. 1989. "Introduction to quality engineering: Designing quality into products and processes." *Technometrics*. Doi: 10.2307/1268824.

22. Liang, X., Cai, Q., Xiang, W., Chen, Z., Zhong, J., Wang, Y., & Li, Z. 2013. "Preparation and characterization of flower-like Cu_2SnS_3 nanostructures by solvothermal route." *Journal of Materials Science and Technology*. Doi: 10.1016/j.jmst.2012.12.011.

23. Sue, K., Suzuki, M., Arai, K., Ohashi, T., Ura, H., Matsui, K., & Hiaki, T. 2006. "Size-controlled synthesis of metal oxide nanoparticles with a flow-through supercritical water method. "*Green Chemistry*. Doi: 10.1039/b518291c.

24. Hayashi, H., & Hakuta, Y. 2010. "Hydrothermal synthesis of metal oxide nanoparticles in supercritical water." *Materials*. Doi: 10.3390/ma3073794.

25. Ramesha, K., & Seshadri, R. 2004. "Solvothermal preparation of ferromagnetic submicron spinel $CuCr_2Se_4$ particles." *ChemInform*. Doi: 10.1002/chin.200444026.

26. Gupta, S. M., & Tripathi, M. 2012. "A review on the synthesis of TiO_2 nanoparticles by solution route." *Central European Journal of Chemistry*. Doi: 10.2478/s11532-011-0155-y.

27. Niederberger, M., & Garnweitner, G. 2006. "Organic reaction pathways in the nonaqueous synthesis of metal oxide nanoparticles." *Chemistry - A European Journal*. Doi: 10.1002/chem.200600313.

28. Niederberger, M., Garnweitner, G., Ba, J., Polleux, J., & Pinna, N. 2007. "Nonaqueous synthesis, assembly and formation mechanisms of metal oxide nanocrystals." *International Journal of Nanotechnology*. Doi: 10.1504/IJNT.2007.013473.

29. Niederberger, M., Garnweitner, G., Buha, J., Polleux, J., Ba, J., & Pinna, N. 2006. "Nonaqueous synthesis of metal oxide nanoparticles: Review and indium oxide as case study for the dependence of particle morphology on precursors and solvents." *Journal of Sol-Gel Science and Technology*. Doi: 10.1007/s10971-006-6668-8.

30. Chemseddine, A., & Moritz, T. 1999. "Nanostructuring titania: Control over nanocrystal structure, size, shape, and organization." *European Journal of Inorganic Chemistry*. Doi: 10.1002/(sici)1099-0682(19990202)1999:2<235::aid-ejic235>3.0.co;2-n.

31. Barcikowski, S., & Compagnini, G. 2013. "Advanced nanoparticle generation and excitation by lasers in liquids." *Physical Chemistry Chemical Physics*. Doi: 10.1039/c2cp90132c.

32. Yurtcan, M. T., Simsek, O., & Ertugrul, M. 2011. "Darbeli Lazer Yığma Sistemi ile YBCO İnce Filmlerin Büyütülmesi." *Erzincan Fen Bilimleri Enstitüsü Dergisi* 4–2: 157–167.

33. Stepanova, M., & Dew, S. 2014. *Nanofabrication: Techniques and Principles*. Doi: 10.1007/978-3-7091-0424-8.

34. Kim, M., Osone, S., Kim, T., Higashi, H., & Seto, T. 2017. "Synthesis of nanoparticles by laser ablation: A review." *KONA Powder and Particle Journal*. Doi: 10.14356/kona.2017009.

35. Moosakazemi, F., Tavakoli Mohammadi, M. R., Mohseni, M., Karamoozian, M., & Zakeri, M. 2017. "Effect of design and operational parameters on particle morphology in ball mills." *International Journal of Mineral Processing*. Doi: 10.1016/j.minpro.2017.06.001.

36. Schallehn, M., Winterer, M., Weirich, T. E., Keiderling, U., & Hahn, H. 2003. In-Situ Prepara-tion of Polymer-Coated Alumina Nanopowders by Chemical Vapor Synthesis *Chemical Vapor Deposition, WILEY-VCH Verlag GmbH & Co. KGaA, Weinheim*. 9, 40–44. Doi: 10.1002/cvde.200290006

37. Faraday, M. 1857. Experimental relations o f gold (and other metals) to light. *Philosophical Transactions of the Royal Society*. https://doi.org/10.1098/rstl.1857.0011

38. Ivanov, V., Kotov, Y. A., Samatov, O. H., Böhme, R., Karow, H. U., & Schumacher, G. 1995. "Synthesis and dynamic compaction of ceramic nano powders by techniques based on electric pulsed power." *Nanostructured Materials*. Doi: 10.1016/0965-9773(95)00054-2.

39. Jiang, W., & Yatsui, K. 1998. "Pulsed wire discharge for nanosize powder synthesis." *IEEE Transactions on Plasma Science*. Doi: 10.1109/27.736045.

40. Ahmed, S., Ahmad, M., Swami, B. L., & Ikram, S. 2016. "A review on plants extract mediated synthesis of silver nanoparticles for antimicrobial applications: A green expertise." *Journal of Advanced Research*. Doi: 10.1016/j.jare.2015.02.007.

41. Singh, J., Dutta, T., Kim, K. H., Rawat, M., Samddar, P., & Kumar, P. 2018. ""Green" synthesis of metals and their oxide nanoparticles: Applications for environmental remediation." *Journal of Nanobiotechnology*. Doi: 10.1186/s12951-018-0408-4.

42. Jacob, S. J. P., Prasad, V. L. S., Sivasankar, S., & Muralidharan, P. 2017. "Biosynthesis of silver nanoparticles using dried fruit extract of *Ficus carica* - Screening for its anticancer activity and toxicity in animal models." *Food and Chemical Toxicology*. Doi: 10.1016/j.fct.2017.03.066.

43. Khanna, P., Kaur, A., & Goyal, D. 2019. "Algae-based metallic nanoparticles: Synthesis, characterization and applications." *Journal of Microbiological Methods*. Doi: 10.1016/j.mimet.2019.105656.

44. Sharma, K., Guleria, S., & Razdan, V. K. 2020. "Green synthesis of silver nanoparticles using *Ocimum gratissimum* leaf extract: Characterization, antimicrobial activity and toxicity analysis." *Journal of Plant Biochemistry and Biotechnology.* Doi: 10.1007/s13562-019-00522-2.

45. Sekoai, P. T., Ouma, C. N. M., du Preez, S. P., Modisha, P., Engelbrecht, N., Bessarabov, D. G., & Ghimire, A. 2019. "Application of nanoparticles in biofuels: An overview." *Fuel.* Doi: 10.1016/j.fuel.2018.10.030.

46. Vahabi, K., Mansoori, G. A., & Karimi, S. 2011. "Biosynthesis of silver nanoparticles by fungus *Trichoderma reesei* (A route for large-scale production of AgNPs). "*Insciences Journal.* Doi: 10.5640/insc.010165.

47. Sahayaraj, K., Rajesh, S., & Rathi, J. M. 2012. "Silver nanoparticles biosynthesis using marine alga *Padina pavonica* (Linn.) and its microbicidal activity." *Digest Journal of Nanomaterials and Biostructures* 7 (4): 1557–1567.

48. Ankamwar, B., Damle, C., Ahmad, A., & Sastry, M. 2005. "Biosynthesis of gold and silver nanoparticles using Emblica Officinalis fruit extract, their phase transfer and transmetallation in an organic solution." *Journal of Nanoscience and Nanotechnology.* Doi: 10.1166/jnn.2005.184.

49. Chandran, S. P., Chaudhary, M., Pasricha, R., Ahmad, A., & Sastry, M. 2006. "Synthesis of gold nanotriangles and silver nanoparticles using Aloe vera plant extract." *Biotechnology Progress.* Doi: 10.1021/bp0501423.

50. Ağçeli, G. K., Hammachi, H., Kodal, S. P., Cihangir, N., & Aksu, Z. 2020. "A novel approach to synthesize TiO$_2$ nanoparticles: Biosynthesis by using streptomyces sp. HC1." *Journal of Inorganic and Organometallic Polymers and Materials.* Doi: 10.1007/s10904-020-01486-w.

51. Dulta, K., Koşarsoy Ağçeli, G., Chauhan, P., Jasrotia, R., & Chauhan, P. K. 2020. "A novel approach of synthesis zinc oxide nanoparticles by *Bergenia ciliata* rhizome extract: Antibacterial and anticancer potential." *Journal of Inorganic and Organometallic Polymers and Materials.* Doi: 10.1007/s10904-020-01684-6.

52. Morones, J. R., Elechiguerra, J. L., Camacho, A., Holt, K., Kouri, J. B., Ramírez, J. T., & Yacaman, M. J. 2005. "The bactericidal effect of silver nanoparticles."*Nanotechnology.* Doi: 10.1088/0957-4484/16/10/059.

53. Simkiss, K., & Wilbur, K. M. 1989. "*Biomineralization.*" *Cell Biology and Mineral Deposition.* Academic Press, Elsevier, Netherlands. Doi: 10.1016/C2009-0-02776-7

54. Vert, M. 1996. "Biomimetic materials chemistry. "*Biochimie.* Doi: 10.1016/0300-9084(96)89517-4.

55. Kaviya, S. 2020. "Synthesis, self-assembly, sensing methods and mechanism of bio-source facilitated nanomaterials: A review with future outlook." *Nano-Structures and Nano-Objects.* Doi: 10.1016/j.nanoso.2020.100498.

56. Ouyang, W., & Sun, J. 2016. "Biosynthesis of silver sulfide quantum dots in wheat endosperm cells." *Materials Letters.* Doi: 10.1016/j.matlet.2015.11.040.

57. Kowshik, M., Ashtaputre, S., Kharrazi, S., Vogel, W., Urban, J., Kulkarni, S. K., & Paknikar, K. M. 2003. "Extracellular synthesis of silver nanoparticles by a silver-tolerant yeast strain MKY3." *Nanotechnology.* Doi: 10.1088/0957-4484/14/1/321.

58. Bhainsa, K. C., & D'Souza, S. F. 2006. "Extracellular biosynthesis of silver nanoparticles using the fungus *Aspergillus fumigatus.*" *Colloids and Surfaces B: Biointerfaces.* Doi: 10.1016/j.colsurfb.2005.11.026.

59. Saifuddin, N., Wong, C. W., & Yasumira, A. A. N. 2009. "Rapid biosynthesis of silver nanoparticles using culture supernatant of bacteria with microwave irradiation." *E-Journal of Chemistry.* Doi: 10.1155/2009/734264.

60. Gour, A., & Jain, N. K. 2019. "Advances in green synthesis of nanoparticles." *Artificial Cells, Nanomedicine and Biotechnology.* Doi: 10.1080/21691401.2019.1577878.

61. Medina Cruz, D., Tien-Street, W., Zhang, B., Huang, X., Vernet Crua, A., Nieto-Argüello, A., & Webster, T. J. 2019. "Citric juice-mediated synthesis of tellurium nanoparticles with antimicrobial and anticancer properties." *Green Chemistry*. Doi: 10.1039/c9gc00131j.
62. Gunti, L., Dass, R. S., & Kalagatur, N. K. 2019. "Phytofabrication of selenium nanoparticles from emblica officinalis fruit extract and exploring its biopotential applications: Antioxidant, antimicrobial, and biocompatibility." *Frontiers in Microbiology*. Doi: 10.3389/fmicb.2019.00931.
63. Tiwari, A., Sherpa, Y. L., Pathak, A. P., Singh, L. S., Gupta, A., & Tripathi, A. 2019. "One-pot green synthesis of highly luminescent silicon nanoparticles using Citrus limon (L.) and their applications in luminescent cell imaging and antimicrobial efficacy." *Materials Today Communications*. Doi: 10.1016/j.mtcomm.2018.12.005.
64. Lee, G., & Kim, B. S. 2014. "Biological reduction of graphene oxide using plant leaf extracts." *Biotechnology Progress*. Doi: 10.1002/btpr.1862.
65. Shankar, S. S., Ahmad, A., Pasricha, R., & Sastry, M. 2003. "Bioreduction of chloroaurate ions by geranium leaves and its endophytic fungus yields gold nanoparticles of different shapes." *Journal of Materials Chemistry*. Doi: 10.1039/b303808b.
66. Ramkumar, V. S., Pugazhendhi, A., Gopalakrishnan, K., Sivagurunathan, P., Saratale, G. D., Dung, T. N. B., & Kannapiran, E. 2017. "Biofabrication and characterization of silver nanoparticles using aqueous extract of seaweed Enteromorpha compressa and its biomedical properties. " *Biotechnology Reports*. Doi: 10.1016/j.btre.2017.02.001.
67. Bai, H. J., & Zhang, Z. M. 2009. "Microbial synthesis of semiconductor lead sulfide nanoparticles using immobilized *Rhodobacter sphaeroides*." *Materials Letters*. Doi: 10.1016/j.matlet.2008.12.050.
68. Prasad, T. N. V. K. V., Kambala, V. S. R., & Naidu, R. 2013. "Phyconanotechnology: Synthesis of silver nanoparticles using brown marine algae *Cystophora moniliformis* and their characterisation. " *Journal of Applied Phycology*. Doi: 10.1007/s10811-012-9851-z.
69. Mishra, S., Dixit, S., & Soni, S. 2015. "Methods of nanoparticle biosynthesis for medical and commercial applications." *Bio-Nanoparticles: Biosynthesis and Sustainable Biotechnological Implications*. Doi: 10.1002/9781118677629.ch7.
70. Zielonka, A., & Klimek-Ochab, M. 2017. "Fungal synthesis of size-defined nanoparticles." *Advances in Natural Sciences: Nanoscience and Nanotechnology*. Doi: 10.1088/2043-6254/aa84d4.
71. Azizi, S., Namvar, F., Mahdavi, M., Ahmad, M. B. & Mohamad, R. 2013. "Biosynthesis of silver nanoparticles using brown marine macroalga, *Sargassum muticum* aqueous extract." *Materials*. Doi: 10.3390/ma6125942.
72. Ghodake, G., & Lee, D. S. 2011. "Biological synthesis of gold nanoparticles using the aqueous extract of the brown algae *Laminaria japonica*." *Journal of Nanoelectronics and Optoelectronics*. Doi: 10.1166/jno.2011.1166.
73. Kuppusamy, P., Yusoff, M. M., Maniam, G. P., & Govindan, N. 2016. "Biosynthesis of metallic nanoparticles using plant derivatives and their new avenues in pharmacological applications – An updated report." *Saudi Pharmaceutical Journal*. Doi: 10.1016/j.jsps.2014.11.013.
74. Ramanathan, R., O'Mullane, A. P., Parikh, R. Y., Smooker, P. M., Bhargava, S. K., & Bansal, V. 2011. "Bacterial kinetics-controlled shape-directed biosynthesis of silver nanoplates using morganella psychrotolerans." *Langmuir*. Doi: 10.1021/la1036162.
75. Klaus, T., Joerger, R., Olsson, E., & Granqvist, C. G. 1999. "Silver-based crystalline nanoparticles, microbially fabricated." *Proceedings of the National Academy of Sciences of the United States of America*. Doi: 10.1073/pnas.96.24.13611.
76. Klaus-Joerger, T., Joerger, R., Olsson, E., & Granqvist, C. G. 2001. "Bacteria as workers in the living factory: Metal-accumulating bacteria and their potential for materials science." *Trends in Biotechnology*. Doi: 10.1016/S0167-7799(00)01514-6.

77. Thostenson, E. T., Li, C., & Chou, T. W. 2005. "Nanocomposites in context. *"Composites Science and Technology*. Doi: 10.1016/j.compscitech.2004.11.003.
78. Li, S., Shen, Y., Xie, A., Yu, X., Qiu, L., Zhang, L., & Zhang, Q. 2007. "Green synthesis of silver nanoparticles using Capsicum annuum L. Extract". *Green Chemistry*. Doi: 10.1039/b615357g.
79. Nair, B., & Pradeep, T. 2002. "Coalescence of nanoclusters and formation of submicron crystallites assisted by lactobacillus strains". *Crystal Growth and Design*. Doi: 10.1021/cg0255164.
80. Saratale, R. G., Karuppusamy, I., Saratale, G. D., Pugazhendhi, A., Kumar, G., Park, Y., ... Shin, H. S. 2018. "A comprehensive review on green nanomaterials using biological systems: Recent perception and their future applications". *Colloids and Surfaces B: Biointerfaces*. Doi: 10.1016/j.colsurfb.2018.05.045.
81. Li, X., Xu, H., Chen, Z. S., & Chen, G. 2011. "Biosynthesis of nanoparticles by microorganisms and their applications." *Journal of Nanomaterials*. Doi: 10.1155/2011/270974.
82. Hagedorn, S., & Kaphammer, B. 1994. "Microbial biocatalysis in the generation of flavor and fragrance chemicals". *Annual Review of Microbiology*. Doi: 10.1146/annurev.mi.48.100194.004013.
83. Das, S. K., Shome, I., & Guha, A. K. 2012. "Surface functionalization of Aspergillus versicolor mycelia: *In situ* fabrication of cadmium sulphide nanoparticles and removal of cadmium ions from aqueous solution". *RSC Advances*. Doi: 10.1039/c2ra01273a.
84. Singh, P., Kim, Y. J., Zhang, D., & Yang, D. C. 2016. "Biological synthesis of nanoparticles from plants and microorganisms". *Trends in Biotechnology*. Doi: 10.1016/j.tibtech.2016.02.006.
85. Hossain, A., Hong, X., Ibrahim, E., Li, B., Sun, G., Meng, Y., & An, Q. 2019. "Green synthesis of silver nanoparticles with culture supernatant of a bacterium pseudomonas rhodesiae and their antibacterial activity against soft rot pathogen dickeya dadantii." *Molecules*. Doi: 10.3390/molecules24122303.
86. Ibrahim, E., Zhang, M., Zhang, Y., Hossain, A., Qiu, W., Chen, Y., & Li, B. 2020. "Green-synthesization of silver nanoparticles using endophytic bacteria isolated from garlic and its antifungal activity against wheat fusarium head blight pathogen fusarium graminearum". *Nanomaterials*. Doi: 10.3390/nano10020219.
87. Soni, M., Mehta, P., Soni, A., & Goswami, G. K. 2018. "Green Nanoparticles : Synthesis and applications." *IOSR Journal of Biotechnology and Biochemistry*, 4: 78–83.
88. Raouf Hosseini, M., & Nasiri Sarvi, M. 2015. "Recent achievements in the microbial synthesis of semiconductor metal sulfide nanoparticles. " *Materials Science in Semiconductor Processing*. Doi: 10.1016/j.mssp.2015.06.003.
89. Mubarakali, D., Gopinath, V., Rameshbabu, N., & Thajuddin, N. 2012. "Synthesis and characterization of CdS nanoparticles using C-phycoerythrin from the marine cyanobacteria". *Materials Letters*. Doi: 10.1016/j.matlet.2012.01.026.
90. Bérdy, J. 2005. "Bioactive microbial metabolites." *The Journal of Antibiotics*. Doi: 10.1038/ja.2005.1.
91. Gade, A. K., Bonde, P., Ingle, A. P., Marcato, P. D., Durán, N., & Rai, M. K. 2008. "Exploitation of *Aspergillus niger* for synthesis of silver nanoparticles." *Journal of Biobased Materials and Bioenergy*. Doi: 10.1166/jbmb.2008.401.
92. Ahluwalia, V., Kumar, J., Sisodia, R., Shakil, N. A., & Walia, S. 2014. "Green synthesis of silver nanoparticles by *Trichoderma harzianum* and their bio-efficacy evaluation against *Staphylococcus aureus* and *Klebsiella pneumonia*." *Industrial Crops and Products*. Doi: 10.1016/j.indcrop.2014.01.026.
93. Azmath, P., Baker, S., Rakshith, D., & Satish, S. 2016. "Mycosynthesis of silver nanoparticles bearing antibacterial activity." *Saudi Pharmaceutical Journal*. Doi: 10.1016/j.jsps.2015.01.008.

94. Khan, N. T., Khan, M. J., Jameel, J., Jameel, N., & Umer, S. 2017. "An overview: Biological organisms that serves as nanofactories for metallic nanoparticles synthesis and fungi being the most appropriate." *Bioceramics Development and Applications.* Doi: 10.4172/2090-5025.1000101.

95. Pantidos, N. 2014. "Biological synthesis of metallic nanoparticles by bacteria, fungi and plants." *Journal of Nanomedicine & Nanotechnology.* Doi: 10.4172/2157-7439. 1000233.

96. Taherzadeh, M. J., Fox, M., Hjorth, H., & Edebo, L. 2003. "Production of mycelium biomass and ethanol from paper pulp sulfite liquor by *Rhizopus oryzae.*" *Bioresource Technology.* Doi: 10.1016/S0960-8524(03)00010-5

97. Siddiqi, K. S., & Husen, A. 2016. "Fabrication of metal nanoparticles from fungi and metal salts: Scope and application." *Nanoscale Research Letters.* Doi: 10.1186/s11671-016-1311-2.

98. Kitching, M., Ramani, M., & Marsili, E. 2015. "Fungal biosynthesis of gold nanoparticles: Mechanism and scale up." *Microbial Biotechnology.* Doi: 10.1111/1751-7915.12151.

99. Husen, A., & Siddiqi, K. S. 2014. "Plants and microbes assisted selenium nanoparticles: Characterization and application." *Journal of Nanobiotechnology.* Doi: 10.1186/s12951-014-0028-6.

100. Durán, N., Marcato, P. D., Alves, O. L., De Souza, G. I. H., & Esposito, E. 2005. "Mechanistic aspects of biosynthesis of silver nanoparticles by several *Fusarium oxysporum* strains." *Journal of Nanobiotechnology.* Doi: 10.1186/1477-3155-3-8.

101. Sastry, M., Ahmad, A., Islam Khan, M., & Kumar, R. 2003. "Biosynthesis of metal nanoparticles using fungi and actinomycete." *Current Science* 85(2) (25 July 2003): 162–170.

102. Dorcheh, S. K., & Vahabi, K. 2017. "Biosynthesis of nanoparticles by fungi: Large-scale production." *Fungal Metabolites.* Doi: 10.1007/978-3-319-25001-4_8.

103. Vahabi, K., & Dorcheh, S. K. 2014. "Biosynthesis of silver nano-particles by trichoderma and its medical applications." *Biotechnology and Biology of Trichoderma.* Doi: 10.1016/B978-0-444-59576-8.00029-1.

104. Kalimuthu, K., Suresh Babu, R., Venkataraman, D., Bilal, M., & Gurunathan, S. 2008. "Biosynthesis of silver nanocrystals by *Bacillus licheniformis.*" *Colloids and Surfaces B: Biointerfaces.* Doi: 10.1016/j.colsurfb.2008.02.018.

105. Raveendran, P., Fu, J., & Wallen, S. L. 2003. "Completely "Green" synthesis and stabilization of metal nanoparticles." *Journal of the American Chemical Society.* Doi: 10.1021/ja029267j.

106. Kalishwaralal, K., Deepak, V., Ramkumarpandian, S., Nellaiah, H., & Sangiliyandi, G. 2008. "Extracellular biosynthesis of silver nanoparticles by the culture supernatant of *Bacillus licheniformis.*" *Materials Letters.* Doi: 10.1016/j.matlet.2008.06.051.

107. Ahmad, A., Mukherjee, P., Senapati, S., Mandal, D., Khan, M. I., Kumar, R., & Sastry, M. 2003. "Extracellular biosynthesis of silver nanoparticles using the fungus *Fusarium oxysporum.*" *Colloids and Surfaces B: Biointerfaces.* Doi: 10.1016/S0927-7765(02)00174-1.

108. Ahmad, A., Senapati, S., Khan, M. I., Kumar, R., Ramani, R., Srinivas, V., & Sastry, M. 2003. "Intracellular synthesis of gold nanoparticles by a novel alkalotolerant actinomycete, Rhodococcus species." *Nanotechnology.* Doi: 10.1088/0957-4484/14/7/323.

109. Mubarak Ali, D., Sasikala, M., Gunasekaran, M., & Thajuddin, N. 2011. "Biosynthesis and characterization of silver nanoparticles using marine cyanobacterium, oscillatoria willei ntdm01." *Digest Journal of Nanomaterials and Biostructures*, 6: 385–390.

110. Prasad, T. N. V. K. V., & Elumalai, E. K. 2013. "Marine algae mediated synthesis of silver nanopaticles using *Scaberia agardhii* greville." *Journal of Biological Sciences.* Doi: 10.3923/jbs.2013.566.569.

111. Namvar, F., Azizi, S., Ahmad, M. B., Shameli, K., Mohamad, R., Mahdavi, M., & Tahir, P. M. 2015. "Green synthesis and characterization of gold nanoparticles using the marine macroalgae Sargassum muticum." *Research on Chemical Intermediates*. Doi: 10.1007/s11164-014-1696-4.

112. Kalabegishvili, T. 2012." Synthesis of gold nanoparticles by some strains of Arthrobacter genera". *Journal of Materials* 2 (2): 164–173.

113. Patel, V., Berthold, D., Puranik, P., & Gantar, M. 2015. "Screening of cyanobacteria and microalgae for their ability to synthesize silver nanoparticles with antibacterial activity." *Biotechnology Reports*. Doi: 10.1016/j.btre.2014.12.001.

114. González-Ballesteros, N., Prado-López, S., Rodríguez-González, J. B., Lastra, M., & Rodríguez-Argüelles, M. C. 2017. "Green synthesis of gold nanoparticles using brown algae *Cystoseira baccata*: Its activity in colon cancer cells." *Colloids and Surfaces B: Biointerfaces*. Doi: 10.1016/j.colsurfb.2017.02.020.

115. Fawcett, D., Verduin, J. J., Shah, M., Sharma, S. B., & Poinern, G. E. J. 2017. "A review of current research into the biogenic synthesis of metal and metal oxide nanoparticles via marine algae and seagrasses." *Journal of Nanoscience*. Doi: 10.1155/2017/8013850.

116. Michalak, I., & Chojnacka, K. 2015. "Algae as production systems of bioactive compounds." *Engineering in Life Sciences*. Doi: 10.1002/elsc.201400191.

117. Sharma, A., Sharma, S., Sharma, K., Chetri, S. P. K., Vashishtha, A., Singh, P., & Agrawal, V. 2016. "Algae as crucial organisms in advancing nanotechnology: A systematic review." *Journal of Applied Phycology*. Doi: 10.1007/s10811-015-0715-1.

118. Prasad, R., Pandey, R., & Barman, I. 2016." Engineering tailored nanoparticles with microbes: Quo vadis?" *Wiley Interdisciplinary Reviews: Nanomedicine and Nanobiotechnology*. Doi: 10.1002/wnan.1363.

119. Lengke, M. F., Fleet, M. E., & Southam, G. 2007. "Biosynthesis of silver nanoparticles by filamentous cyanobacteria from a silver(I) nitrate complex." *Langmuir*. Doi: 10.1021/la0613124.

120. Dahoumane, S. A., Djediat, C., Yéprémian, C., Couté, A., Fiévet, F., Coradin, T., & Brayner, R. 2012. "Species selection for the design of gold nanobioreactor by photosynthetic organisms." *Journal of Nanoparticle Research*. Doi: 10.1007/s11051-012-0883-8.

121. Aboelfetoh, E. F., El-Shenody, R. A., & Ghobara, M. M. 2017. "Eco-friendly synthesis of silver nanoparticles using green algae (*Caulerpa serrulata*): Reaction optimization, catalytic and antibacterial activities." *Environmental Monitoring and Assessment*. Doi: 10.1007/s10661-017-6033-0.

122. Nasrollahzadeh, M., Sajjadi, M., Dadashi, J., & Ghafuri, H. 2020. "Pd-based nanoparticles: Plant-assisted biosynthesis, characterization, mechanism, stability, catalytic and antimicrobial activities." *Advances in Colloid and Interface Science*. Doi: 10.1016/j.cis.2020.102103.

123. Vijayakumar, S., Vaseeharan, B., Malaikozhundan, B., & Shobiya, M. 2016. "Laurus nobilis leaf extract mediated green synthesis of ZnO nanoparticles: Characterization and biomedical applications. *"Biomedicine and Pharmacotherapy*. Doi: 10.1016/j.biopha.2016.10.038.

124. Akhtar, M. S., Panwar, J., & Yun, Y. S. 2013. "Biogenic synthesis of metallic nanoparticles by plant extracts." *ACS Sustainable Chemistry and Engineering*. Doi: 10.1021/sc300118u.

125. Shah, M., Fawcett, D., Sharma, S., Tripathy, S. K., & Poinern, G. E. J. 2015. "Green synthesis of metallic nanoparticles via biological entities." *Materials*. Doi: 10.3390/ma8115377.

126. Mittal, A. K., Chisti, Y., & Banerjee, U. C. 2013. "Synthesis of metallic nanoparticles using plant extracts." *Biotechnology Advances*. Doi: 10.1016/j.biotechadv.2013.01.003.

127. Malik, P., Shankar, R., Malik, V., Sharma, N., & Mukherjee, T. K. 2014. "Green chemistry based benign routes for nanoparticle synthesis." *Journal of Nanoparticles*. Doi: 10.1155/2014/302429.

128. Iravani, S. 2011. "Green synthesis of metal nanoparticles using plants." *Green Chemistry*. Doi: 10.1039/c1gc15386b.
129. Ahmed, S., Chaudhry, S. A., & Ikram, S. 2017. "A review on biogenic synthesis of ZnO nanoparticles using plant extracts and microbes: A prospect towards green chemistry." *Journal of Photochemistry and Photobiology B: Biology*. Doi: 10.1016/j.jphotobiol.2016.12.011.
130. Dahl, J. A., Maddux, B. L. S., & Hutchison, J. E. 2007. "Toward greener nanosynthesis." *Chemical Reviews*. Doi: 10.1021/cr050943k.
131. Ethiraj, A. S., Jayanthi, S., Ramalingam, C., & Banerjee, C. 2016. "Control of size and antimicrobial activity of green synthesized silver nanoparticles." *Materials Letters*. Doi: 10.1016/j.matlet.2016.07.114.
132. Gurunathan, S., Han, J. W., Kwon, D. N., & Kim, J. H. 2014. "Enhanced antibacterial and anti-biofilm activities of silver nanoparticles against Gram-negative and Gram-positive bacteria." *Nanoscale Research Letters*. Doi: 10.1186/1556-276X-9-373.
133. Liao, C., Li, Y., & Tjong, S. C. 2019. "Bactericidal and cytotoxic properties of silver nanoparticles." *International Journal of Molecular Sciences*. Doi: 10.3390/ijms20020449.
134. Jadhav, K., Dhamecha, D., Bhattacharya, D., & Patil, M. 2016. "Green and ecofriendly synthesis of silver nanoparticles: Characterization, biocompatibility studies and gel formulation for treatment of infections in burns." *Journal of Photochemistry and Photobiology B: Biology*. Doi: 10.1016/j.jphotobiol.2016.01.002.
135. Jain, S., & Mehata, M. S. 2017. "Medicinal plant leaf extract and pure flavonoid mediated green synthesis of silver nanoparticles and their enhanced antibacterial property." *Scientific Reports*. Doi: 10.1038/s41598-017-15724-8.
136. Singh, H., Du, J., Singh, P., & Yi, T. H. 2018. "Ecofriendly synthesis of silver and gold nanoparticles by Euphrasia officinalis leaf extract and its biomedical applications." *Artificial Cells, Nanomedicine and Biotechnology*. Doi: 10.1080/21691401.2017.1362417.
137. Gan, P. P., & Li, S. F. Y. 2012. "Potential of plant as a biological factory to synthesize gold and silver nanoparticles and their applications." *Reviews in Environmental Science and Biotechnology*. Doi: 10.1007/s11157-012-9278-7.
138. Kumar, V., & Yadav, S. K. 2009. "Plant-mediated synthesis of silver and gold nanoparticles and their applications." *Journal of Chemical Technology and Biotechnology*. Doi: 10.1002/jctb.2023.
139. Mittal, A. K., Kaler, A., & Banerjee, U. C. 2012. "Free radical scavenging and antioxidant activity of silver nanoparticles synthesized from flower extract of *Rhododendron dauricum*." *Nano Biomedicine and Engineering*. Doi: 10.5101/nbe.v4i3.p118-124.
140. Sintubin, L., Verstraete, W., & Boon, N. 2012. "Biologically produced nanosilver: Current state and future perspectives." *Biotechnology and Bioengineering*. Doi: 10.1002/bit.24570.
141. Ge, L., Li, Q., Wang, M., Ouyang, J., Li, X., & Xing, M. M. Q. 2014. "Nanosilver particles in medical applications: Synthesis, performance, and toxicity." *International Journal of Nanomedicine*. Doi: 10.2147/IJN.S55015.
142. Narayanan, K. B., & Sakthivel, N. 2011. "Green synthesis of biogenic metal nanoparticles by terrestrial and aquatic phototrophic and heterotrophic eukaryotes and biocompatible agents." *Advances in Colloid and Interface Science*. Doi: 10.1016/j.cis.2011.08.004.
143. Hutchison, J. E. 2008. "Greener nanoscience: A proactive approach to advancing applications and reducing implications of nanotechnology." *ACS Nano*. Doi: 10.1021/nn800131j.
144. Siddiqi, K. S., Husen, A., & Rao, R. A. K. 2018. "A review on biosynthesis of silver nanoparticles and their biocidal properties." *Journal of Nanobiotechnology*. Doi: 10.1186/s12951-018-0334-5.

145. Vaseghi, Z., Nematollahzadeh, A., & Tavakoli, O. 2018. "Green methods for the synthesis of metal nanoparticles using biogenic reducing agents: A review." *Reviews in Chemical Engineering.* Doi: 10.1515/revce-2017-0005.
146. Rafique, M., Sadaf, I., Rafique, M. S., & Tahir, M. B. 2017. "A review on green synthesis of silver nanoparticles and their applications." *Artificial Cells, Nanomedicine and Biotechnology.* Doi: 10.1080/21691401.2016.1241792.
147. Gholami-Shabani, M., Akbarzadeh, A., Norouzian, D., Amini, A., Gholami-Shabani, Z., Imani, A., & Razzaghi-Abyaneh, M. 2014. "Antimicrobial activity and physical characterization of silver nanoparticles green synthesized using nitrate reductase from *Fusarium oxysporum.*" *Applied Biochemistry and Biotechnology.* Doi: 10.1007/s12010-014-0809-2.
148. Kharissova, O. V., Dias, H. V. R., Kharisov, B. I., Pérez, B. O., & Pérez, V. M. J. 2013. "The greener synthesis of nanoparticles." *Trends in Biotechnology.* Doi: 10.1016/j.tibtech.2013.01.003.
149. Velusamy, P., Kumar, G. V., Jeyanthi, V., Das, J., & Pachaiappan, R. 2016. "Bio-inspired green nanoparticles: Synthesis, mechanism, and antibacterial application." *Toxicological Research.* Doi: 10.5487/TR.2016.32.2.095.
150. Mckenzie, L. C., & Hutchison, J. E. 2004. "Green nanoscience." *Chimica Oggi* 22 (9): 30–33.

4 Chemistry Revolving around Nanomaterial-Based Technology

S. Saravanan and E. Kayalvizhi Nangai
K. Ramakrishnan College of Technology

C. M. Naga Sudha
Anna University MIT Campus

S. Sankar and Sejon Lee
Dongguk University

M. Velayutham Pillai
Kalasalingam Academy of Research and Education

V. Dhinakaran
Chennai Institute of Technology

CONTENTS

4.1   INTRODUCTION

Nanotechnology research is coupled to all scientific areas, namely applied chemistry and physics, engineering, and medicine with respect to their properties and a wide variety of applications. Nanochemistry is a subdiscipline of nanoscience, which elaborates chemical products and chemical reactions of the nanomaterials. It has a different approach in constructing the devices along with a molecular-scale precision. One can apply the advantages of nanodevices in medicine, scientific exploration, and electronics, which can build objects atomically [1–3]. It is essential to note that the nanostructures easily originate from the confinement effect at nanoscale dimension, which include statistical and quantum mechanical effects that can be reduced with the electronic properties of solids. However, quantum effects rule the behavior and properties of particles. The quantum size effect can start to influence the properties of matter at nanoscale and significantly change the optical, magnetic, and electrical properties [4–8].

Every material is composed of grains, which are visible or invisible to the eye, depending on their size. In simple, nanotechnology is the space at a nanoscale, which is smaller than "micro" and larger than "pico." The size of the nanoparticles ranges from 1 to 100 nm. For example, the sizes of the quantum dot, fullerene (C_{60}), and dendrimer are 8, 1, and 10 nm, respectively [9–11]. In comparison, diplomat materials such as atom, DNA (width), protein, virus, bacteria, and white blood cell, which originate naturally, have dimensions 0.1, 2, 5–50, 75–100, 1,000–10,000, and 10,000 nm, respectively [12]. Nanostructured materials such as fumed silica, sealants, adhesives, silicon rubber components, and coatings are also commercially available. It is used in car tires for providing black color and also for increasing the life of the tire [13]. In recent years, nanostructured materials have created wide applications due to their fascinating magnetic, optical, and electrical properties [15]. Magnetic nanoparticles consist of the magnetic domain and exhibit different magnetic properties like finite size effects and special crystal structure. Magnetic nanocomposites are used as electronic devices, sensors, and high-density data storage devices [14–20]. Nanostructure metal oxides have a variety of applications such as photocatalysts, wastewater purification, chemical sensors, and rechargeable batteries for cars [21]. Nanocrystalline silicon films and nanostructured titanium oxide porous films are used as a component in the preparation of dye-sensitized solar cells. Nanostructures are described in terms of dimensions under nanoscale such as zero-dimensional, one-dimensional (nanowires, rods, and tubes), two-dimensional (nanocoatings, layers, and films), and three-dimensional materials. In chemistry, cluster is defined as an ensemble of bounded atoms or molecules, which are intermediate in size between a simple molecule and a nanoparticle. A nanocrystal is defined as a nanomaterial with at least one dimension, which is less than or equal to 100 nm, known as a single crystalline.

Nanoparticles are categorized based on their size, shape, and material properties. It is also classified as organic and inorganic particles, which include liposomes, polymeric nanoparticles, and dendrimers. Polymeric nanoparticles include quantum dots, fullerenes, and gold nanoparticles [22–25]. Other classifications of nanoparticles are ceramic, semiconducting or polymeric particles, and carbon-based. Furthermore, nanoparticles are classified as hard and soft where titania, silica particles, and fullerenes belong to the hard category and liposomes, vesicles, and nanodroplets belong to the soft category [26,27]. Nanoparticles have unique material characteristics due to their size, which can be artificially produced to help in the practical applications in the field of engineering, medicine, catalysis, and environmental remediation [23–28]. In this chapter, we have explained the impact of the change in the size of the nanoparticles on the properties of the matter and the applications of this feature of nanoparticles in various fields such as medicine, drug delivery, tissue engineering, wounds, nanowire composition, and nanoenzymes.

4.2 CHEMICAL PROPERTIES

The size and shape are the general characteristics of nanomaterials. Using surface chemistry, crystalline size, shape, phase, functional groups, and elemental composition of nanomaterials are determined. Metal-containing compounds in the form of oxides and carbon nanotubes are characterized in which the surface layer should be 0–10 nm [29]. For physicochemical characterization of nanomaterials, the first step is to measure particle size and distribution, which leads nanotechnology to be utilized in core–shell structure, which is the combination of core and shell layers with novel properties and crystal structures [30,31].

It is noted that nanomaterials exhibit more effect on material shape when compared to bulk materials, and the melting, Debye temperature, and thermal expansion coefficient are size- and shape-dependent parameters [32–36]. The diameter of nanomaterials is directly proportional to the melting, Debye temperature, and thermal expansion coefficient. The size of the nanomaterial is inversely proportional to the thermal expansion coefficient value [37]. Specific heat increases with the reduction in the size of gold nanoparticles.

Spinel ferrites have unique magnetic properties and their application in memory storage devices, microwave absorbers, electromagnetic shields, magnetic strips, magnetic resonance imaging, and antenna devices with high frequency [38]. The physical properties of ferrites are controlled by the size and the shape of the nanoparticles [39]. Due to the change of pH value and density of the X-ray, there exists a small change in parameter structure such as the size of the crystal, lattice constant, and unit cell volume. When the pH value reaches 9, high clusters available in the particles and those particles in cubic spinel structure tend to be relatively in sphere shape, whereas nonuniform particle distribution is between 30 and 40 nm [40]. Whenever there occurs a change in size, shape, and distribution of the nanoparticles, the pH value increases [41]. The crystallization process has changed the distribution of ions in the lattice, which are affected by pH value. With the increase in pH value, there is an increase in uniform distribution among nanoparticles and also the size of crystals. With these changes, magnetic properties along with the optical gap of spinel ferrites are changed [42].

Spinel ferrite particle $(NiCuZn)Fe_2O_4$ is a polygonal, with a size less than 40 nm, and ferrite powder is a kind of nanoparticle and nanocrystalline where most of the particle size is normal. The dispersion rate of particles is directly proportional to the electric power generated by ultrasonic irradiation. Ultrasonic assistance of different powers influences grain size. For a reaction temperature greater than 50°C, the saturation magnetization and coercivity of ferrite nanoparticles are obtained with an increase in ultrasonic power [44–50]. This nanosized ferrite material $(NiCuZn)Fe_2O_4$ thus obtained is used for core, switching, and multilayer chip inductors [43]. Due to the large surface area, nanoparticles can be used in wastewater treatment, thereby helping to control environmental pollution. Cerium oxide nanoparticle exhibits catalytic and antioxidant properties, which act as a catalyst in fuel oxidation [51] and which are employed in solar cells, chemical–mechanical polarization, and corrosion protection [52]. Melting and boiling temperatures of nanoparticles such as 798°C and 3,424°C, respectively, for cerium oxide nanoparticles, show thermal properties. Three low-index planes, namely (100), (110), and (111), are shown in cubic fluorite, and also, plane properties are exhibited by cerium oxide nanoparticles [53–56]. Lattice expansion is greater due to the tiny size of cerium oxide nanoparticles, which results in reabsorption and decline in oxygen release. Doping of La ion with cerium oxide increases the vacancy of oxygen in the surface area [57]. Cerium oxide nanoparticles maintain catalytic behavior in harsh environments and show high catalytic mimetic activity [58].

Cerium oxide film manifests optical properties, which are applied in electro-optical and optoelectronic devices. Transition metal oxides exhibit electrochemical properties, which act as an electrode material for numerous applications, such as electrocatalysis, lithium-ion batteries, electrochemical sensors, and supercapacitor [59]. The cerium oxide nanoparticle is used in the field of biomedicines, which includes cancer treatment, antimicrobial agent, bioscaffold, and biosensor device fabrication due to their unique redox properties [60].

4.3 SYNTHESIS AND PROCESSING

Nanosynthesis and processing enhance the production of materials and new products. There are two basic techniques for synthesis, either top-down or bottom-up as shown in Figure 4.1 [61].

4.3.1 LASER ABLATION

In laser ablation, micropatterns are fabricated by removing (ablation) the small fractions of a substrate material using a focused pulsed laser beam [62]. The laser beam penetrates to the surface of a sample, which depends on the laser source wavelength and the refractive index of the aimed material. Due to the laser light beam, high

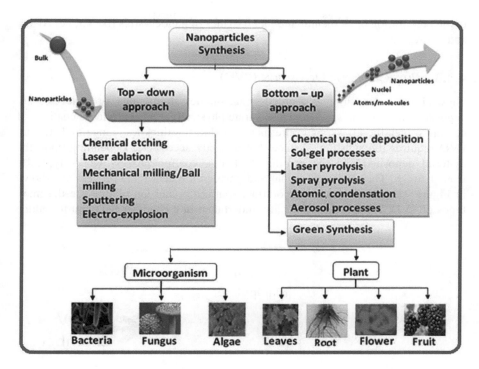

FIGURE 4.1 Nanoparticles synthesis techniques.

electric field will be generated and electrons are removed from the bulk sample. Energy transformation occurs because of collision between free electrons and atoms present on the surface of a sample. As a result, the surface gets heated, which is followed by vaporization. An intense pulse laser is most commonly used in laser ablation. Nanoparticles are ejected and completely collected in the form of colloidal solutions. Without any formation of by-products, some nanoparticles are formed in colloidal solutions. Laser ablation produces plasma, which expands considerably under strong confinement. Expansion of this kind is possible only when target is placed in solution at high temperature and high pressure. Reflections of metal surface from laser beam and laser parameters such as wavelength and pulse duration can be controlled. It is a very easy, simple, and cheap method. Laser-generated nanoparticles can be applied in many applications under the field of biophotonics and medicine.

4.3.2 BALL MILLING

Ball milling process includes reduction of particle size, blending, particle shape changes, and synthesis of nanoparticle. The tungsten ball mill with a cylindrical container turns around its axis [63]. In this technique, the container with the tungsten carbide balls along with powder of particles (50 mm) mixed with desired material is closed with tight lids. With rotation of container about the central axis, the milling balls on collision produce small grains, which are of nanosize. It helps in the preparation of elemental and metal oxides of nanocrystals like Cr, Fe, Al-Fe, and Co Ag-Fe.

4.3.3 PHYSICAL VAPOR DEPOSITION (PVD)

Physical vapor deposition is a process that implies atoms, ions, or molecules are deposited on a coating species to a substrate physically. It also produces coatings of ceramics, metallic alloys, and pure metals within the thickness range of 1–10 μm. PVD requires the substrate surface that is easily accessible. Depending upon the material deposited and the coating thickness, coating process works typically from 1 to 3 h. The schematic layout and parts of a recent PVD process are shown in Figure 4.2 [64]. Thermal evaporation, sputtering, and ion plating are the main types of PVD. All three methods are undertaken in a chamber having a forbidden

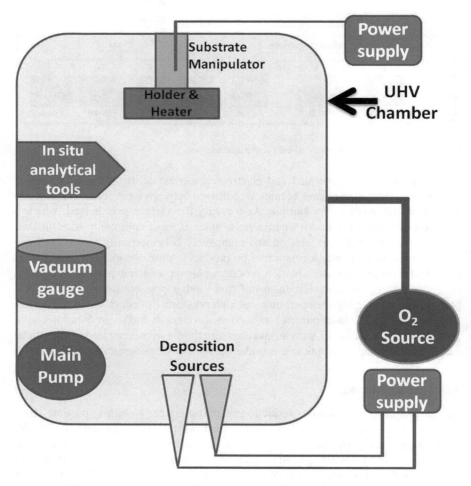

FIGURE 4.2   Schematic layout and parts of PVD process.

atmosphere at reduced pressure. These can be used for either reactive or direct deposition of a material in which atoms of the coating material and reactive gases are involved in chemical reaction at vapor/plasma phase. Substrate is being coated in the range of 200°C–400°C. In thermal evaporation method, vapor is created due to heating of the sample, and coating will be formed by condensation of substrate. Various methods that are used to achieve heating process include electrical resistance, hot filament, electron or laser beam, and electric arc. Ion plating method is the grouping of thermal evaporation and sputtering.

4.3.4 Sputtering

Sputter deposition is a kind of PVD process that works in a high-energy environment involved in the electrical generation of plasma in between the coating species and the substrate. Atoms of a particular material are displaced due to an ionized gas molecule. These atoms create a link from the atomic level to a substrate level, which makes thinner film. Sputtering processes are categorized as DC diode, radio frequency (RF), DC triode, and magnetron sputtering. In Figure 4.3, magnetron sputtering is depicted in which high voltage is applied to argon gas at low pressure, where high-energy plasma will be created [65]. Colorful light is emitted where electrons and gas ions produced by plasma are called a glow discharge. These energized plasma ions strike out the target of the desired coating material. Therefore, the atoms are ejected easily from the target material, which can bond with their respective substrate. The technique is applied for the creation of a whole new generation with smaller, lighter, and more durable products in the industrial field. In order to avoid contamination, due to residual gases, this technique starts from lower pressure and the plasma is created at relatively high pressure. Sputtering target shapes are circular, rectangular, delta, or tubular. Magnetron sputtering has its applications in microelectronics for producing metal layers.

In DC diode sputtering method, argon low-pressure plasma is ignited between target and substrate. Atoms are precipitated out of the target through positive ions and migrate to the substrate and then condense there with applied DC voltage. The only limitation is that low sputtering rates are attained due to the formation of few argon

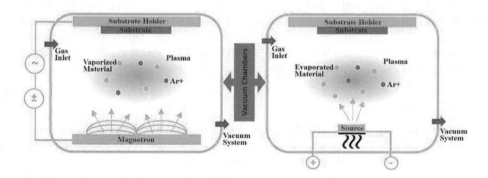

FIGURE 4.3   Magnetron sputtering process.

ions. In RF sputtering method, high-frequency alternating field is applied and connected in series along with a capacitor and plasma. Ions and electrons are accelerated in both directions with the help of alternating field. Ions are not utilized in the alternating field because of much smaller charge-to-mass ratio at a frequency of approximately 50 kHz. Positive ions move through a superimposed negative voltage along with the direction of the target and deposition, which takes place similar to DC sputtering by collision of atoms from the target material. Major advantage of this method is that insulators and semiconductors can also be sputtered. In DC triode sputtering method, plasma chamber is present outside and the target acts as a third electrode in which plasma generation and sputtering process are decoupled. Sputtering is useful in semiconductor industry and is also employed in optical applications with the help of a deposition of thin layer on glass. Another application of the sputtering process is the white material erosion. For an instance, in the process of secondary ion mass spectroscopy, at a constant speed, target is constantly sprayed.

4.3.5 Nanolithography

Lithography is a top-down fabrication technique in which huge particles are reduced to microparticles. In this technique, nanotopographical etchings are produced artificially. Various practical applications are employed in nanolithography, which include semiconductor chips in computers, photolithography, X-ray lithography, light coupling nanolithography, extreme ultraviolet lithography, scanning probe microscope lithography, electron beam, nanoimprint, and dip-pen nanolithography [66]. Dip-pen nanolithography is a technique that draws nanostructures using atomic force in the microscope scanning probe tip. A probe tip acts as a pen coated with liquid ink and in contact with a substrate. It can be used as a powerful tool to deposit soft and hard materials on a variety of surfaces.

4.3.6 Chemical Vapor Deposition (CVD)

CVD is defined as a process where coatings are formed on a heated substrate through chemical reactions from the vapor or gas phase [67]. These reactions are involved in the substrate material itself. CVD offers various advantages over other deposition processes such as thermal CVD, plasma-enhanced CVD, and laser CVD. Using CVD, extremely dense and pure materials are produced at nanoscale. Nanofilms are highly uniform and are deposited in a reasonable coverage, which has more advantages when compared to PVD technique. CVD method is able to produce a variety of coatings and chemical precursors [68].

4.3.7 Solgel

In solgel process, metal alkoxides are hydrolyzed in solutions producing dozens of colloidal particle (in mm) shaped sol, and the condensation polymerization of such particles results in a bonding, which is present in final gels [69]. Initially, sol undergoes hydrolysis and condensation polymerization, until particles consist of 1–1,000 nm. To synthesize the colloidal suspension using precursors, hydrolysis and condensation of alkoxide-based precursors such as tetraethyl orthosilicate are referred. The reactions involved in the solgel chemistry are based on the hydrolysis and condensation of metal alkoxides M (OR), which are described as follows:

$$MOR + H_2O \rightarrow MOH + ROH \text{ (hydrolysis)} \tag{4.1}$$

$$MOH + ROM \rightarrow M - O - M + ROH \text{ (condensation)} \tag{4.2}$$

Starting material forms a dispersible oxide and a sol in contact with water or dilute acid. Removing the liquid from the sol yields the gel, and the solgel transition controls the particle size and shape. Oxides are produced from the calcination of the gel. In solgel preparation, various precursors such as inorganic and nonalkoxide precursors, nitrates, carboxylates, acetylacetonates, and chloride precursors are used [70].

4.3.8 Spray Pyrolysis

In spray pyrolysis, nanoparticles are put by spraying and evaporating a droplet solution on a heated surface, in which chemical compounds are formed by constituents. These aerosol droplets are heated or diluted, which leads to solvent evaporation. Using spray pyrolysis, each droplet is converted into a single product particle in the production of microparticle as shown in Figure 4.4 [71]. In this technique, solution of metallic salts or a colloidal solution acts as precursor for the formation of aerosol. At a particular temperature, aerosol droplets are heated rapidly in a furnace. Solvent is evaporated from droplet surface and dried, which contains the precipitated solute. Annealing of the precipitate at high temperatures produces microporous particles, which are formed at defined phase composition. Porous films are made with the help of high-density packaging and high uniformity of particles. From spray pyrolysis, solar cells and solid-state lithium batteries are fabricated. This technique can be used for energy storage in electric vehicles and for load leveling in wind power production. It is carried out in the fabrication of nanofilms and Cu, TiO_2, CdS, and CdSe clusters at a large scale.

4.3.9 Coprecipitation

In chemistry, coprecipitation is defined as a process in which soluble compounds are agreed out of solution by a precipitate. Otto Hahn was provided with credits for promoting the use of coprecipitation in radiochemistry. Inclusion, occlusion, and adsorption are involved in the coprecipitation process [72]. Inclusion is the process in which crystallographic defect is caused due to impurity atoms occupied in a lattice site of the crystal structure in the carrier. Adsorbate is defined as an impurity that is weakly bound to the surface of the precipitate. Occlusion is defined as a process that occurs when an adsorbed impurity gets physically trapped on the crystal as it develops. Besides the applications in chemical analysis and in radiochemistry, coprecipitation is more important to various environmental issues, which relates to water resources, acid mine drainage, metal contaminant transport at industrial and defense

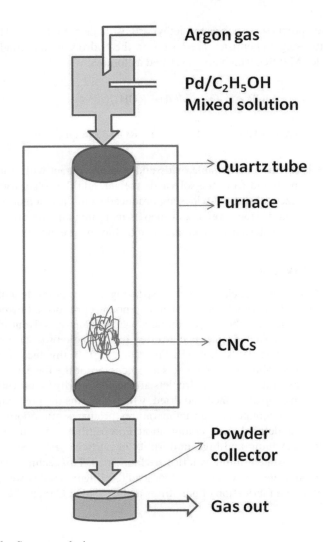

FIGURE 4.4 Spray pyrolysis processes.

sites, radionuclide migration in fouled waste repositories, metal concentrations in aquatic systems, and wastewater treatment technology.

4.3.10 ELECTROCHEMICAL DEPOSITION

Electrochemical deposition is the process by which chemical reactions are induced in an aqueous electrolyte solution when voltage is applied. Nanostructured materials are deposited, which include metal oxides and chalcogenides. It allows complex shapes, which are relatively cheap and fast. There are two different processes such as electrodeposition process and electroless (autocatalytic) deposition process [73]. Electrons are offered by an external power supply in an electrodeposition process, whereas electron source acts as a solution of reducing agents in an electroless deposition process. Both are envoy of electrochemical deposition, which is the simplest and most effective method. It can be used to fabricate one-dimensional nanomaterials. By using this method, compositions and morphologies of nanostructured materials are customized. Electroplating, electrolytic anodization, and electrophoretic deposition are the types of electrochemical deposition.

4.3.11 MOLECULAR BEAM EPITAXY

Molecular beam epitaxy is defined as a method to deposit single crystals. In the 1960s, J.R. Arthur and Alfred invented this technique at Bell Telephone Laboratories. It is broadly employed in the semiconductor device fabrication, which includes transistors and Wi-fi for mobile connections. At high or ultrahigh vacuum, molecular beam epitaxy occurs, and it is significant to note that deposition rate allows the films to grow epitaxially [74]. Compared with other deposition techniques, deposition rates require superior vacuum to achieve the same impurity levels. Because of the absence of carrier gases and the presence of ultrahigh vacuum, films at the highest purity have resulted. Ultrapure form of gallium and arsenic extraction from a solid source is heated in separate quasi-Knudsen effusion cells until it initiates to slow sublime. Gaseous elements are condensed on the wafer, which may react with each other. For an instance, gallium arsenide is formed as a single crystal. Reflection of high-energy electron diffraction is often used to monitor the crystal layers growth during operation. Shutters present in front of each furnace can be controlled by a computer. Developments of structures are allowed by the electrons confined in space and result in quantum dots. Ultrahigh vacuum environment is maintained by the combination of cryopumps and cryopanels, which are chilled using liquid or cold nitrogen gas at very low temperature within the growth chamber. Cryogenic temperatures are used as a sink for impurities in the vacuum.

4.3.12 INERT GAS CONDENSATION PROCESS

In this process, metallic source gets evaporated in reaction with inert gas and then nanoparticles are formed. Ultrafine metal particles and carbon black, which is an ink pigment, are fabricated by this method. Metallic source is evaporated by resistive heating inside a chamber and filled with inert gas at a low pressure [75]. Heating source will be taken from RF, electron beam, or laser beam, which are all equally effective methods. Due to a combination of convective flow and diffusion, metal vapor migrates from the hot source into the cool inert gas. Collision occurs between the evaporated atoms and the gas atoms, which results in the loss of kinetic energy; thereby, the particles are formed. These particles are collected by deposition on a cold surface with the help of liquid nitrogen. By this way, particles are collected in a highly concentrated and complex aggregate morphology. These structures are categorized with the help of the size of crystallites and form larger structures [76,77]. To ensure powder surface cleanliness and to minimize problems associated with trapped gas, the scrapping is done within a clean environment.

4.3.13 HYDROTHERMAL SYNTHESIS

Hydrothermal synthesis is a solution-based chemical method that helps in the synthesis of all types of nanostructure materials [78]. It has wide advantages such as crystal structure, which is easily controlled by various reaction parameters, such as reaction time, reactant source, additives, reaction temperature, and reactant solvent medium. In this technique, single crystals are synthesized at high pressure depending on the mineral solubility in hot water. An apparatus consisting of a steel vessel called autoclave is used, where nutrients are supplied along with water, and finally, the crystal growth is formed. The nutrient dissolves at the hotter end, and the cooler end increases growth by maintaining a temperature gradient.

4.3.14 GREEN SYNTHESIS

Green synthesis plays a vital role in bionanotechnology field and brings benefits in the area of economics and environment. This synthesis technique is an alternative to chemical and physical methods. Green synthesis of nanoparticles in the formation of metal and oxides is focused by many researchers [79]. This method is significantly attractive due to the reduction of the toxicity in nanoparticles. Nowadays, green synthesis of nanoparticles is extracted from plants, vitamins, and amino acids, which are of low cost, are eco-friendly, and are safe for human use. In nature, plant extracts, cyclodextrin, and chitosan are the other natural resources. In the synthesis of the nanowires and nanorods, vitamin B_2 is used as the reducing agent. Spherical-shaped nanomaterials are fabricated by microwave synthesis. In silver/gold nanoparticle synthesis, carboxyl methyl cellulose sodium acts as a reducing and capping specialist. Uniform size and shape of silver/gold nanoparticles are formed from the extraction of various antioxidant constituents such as blackberry, blueberry, turmeric, and pomegranate. These nanoparticles can be utilized for the cancer and antioxidant therapy.

4.4 APPLICATIONS

Nanochemistry provides enormous applications in the field of chemical and material sciences, which help in fabricating nanoscale materials. These materials have been investigated in wide area of electronics or nanodevice systems, composite materials, biotechnology, medicine, and even in the textile industry. Household products are made from nanoengineered materials sealing products such as self-cleaning house paints that resist dirt and marks. Nanotechnology has provided the innovative techniques such as chemical catalysis and filtration [80]. Nanotechnology offers the ability to prepare certain molecules by chemical synthesis. Clusters, nanoparticles, tailor-made molecules, polymers, etc., are provided by chemistry since it is a base for nanotechnology.

4.4.1 CATALYSIS

Extreme high surface and volume ratio in nanofabricated materials has helped to produce a better chemical catalysis process [81]. Application of nanoparticle catalysis are in fuel cell and photocatalytic devices. Catalysis provides an efficient approach for the production of various chemicals in different industries. Potential benefits of nanochemistry are wastewater management, air refinement, and energy storage device. Filtration techniques are used in mechanical or chemical methods. In the filtration process, membranes are used with proper whole sizes and the liquid is pushed via the membrane. In mechanical filtration technique, nanometer-sized porous membranes are used, which have small pores varying from 1 to 10 nm. Nanofiltration can be primarily used to remove ions, which separate different fluids. This membrane filtration technique is extensively used, which is termed ultrafiltration, and works down from 10 to 100 nm. One of the most important applications of ultrafiltration found in the medical field is real dialysis. Magnetic nanomaterials are promising materials, especially used in wastewater decontamination to remove heavy metal. Nanoscale particles are used to increase the efficiency for absorbing the contaminants and are relatively cost-effective compared to traditional precipitation and filtration methods [82].

4.4.2 MEDICINE

Medical application in nanochemistry includes sunscreen, which contains zinc oxide and titanium dioxide as nanoparticles. The skin will be protected by these nanoparticles against harmful UV light with the help of absorption or reflection of the light, which protects the skin by electrons photoexcitation in the nanoparticle and which blocks the skin cells from DNA damage [83].

4.4.3 DRUG DELIVERY

Recent drug delivery methods that involve nanotechnological process are more advantageous. It helps in increasing bodily response, efficient, nontoxic metabolism, and specific targeting. A nanomaterial carries drug contents into the body. In the research field, mesoporous silica nanoparticles have fascinated due to their large surface area and for imaging techniques, which give flexibility and high-resolution performance [84]. Activation methods are significantly different over nanoparticles drug delivery molecules. Low-intensity light and plasmonic heating can be released by nanovalve-controlled cargo. Two-photon-activated phototransducer releases the drug loads using near-infrared region, which induces disulfide bond breakage. In the case of delivery of drugs, nanodiamonds are expressed recently due to their ability to enter the blood–brain barrier and spontaneous absorption through the skin.

4.4.4 TISSUE ENGINEERING

Current trends in tissue engineering offer implant surfaces with nanoscale topographics since the cells are very sensitive. For artificial organ growth, designed 3D scaffold can be used to direct cell with appropriate conditions [85]. The 3D scaffold assimilates various nanoscale factors such as environment control for optimal and appropriate functionality.

4.4.5 WOUNDS

Nanochemistry has its applications in improving the healing process for abrasions and wounds [86]. Polymerization method like electrospinning can be utilized for drug delivery and wound dressing. It produces nanofibers under controlled environment, which encourage cell proliferation and antibacterial properties. Due to nanotopographical features, nanoscale versions show improved efficiency. Some viruses and bacteria are inhibited by silver nanoparticles. Few applications such as lithography, nanowires in sensors, and nanoenzymes are made possible through nanoparticles wound process [87].

4.4.6 NANOWIRE COMPOSITIONS

Researchers have found various nanowire compositions that have controlled length, surface structure, diameter, and doping. Nanowires that have one-dimensional structure and single crystals are applied in semiconductor nanowire devices, namely lasers, diodes, logic circuits, transistors, and sensors [88]. Electron transport efficiency makes the electrical properties more efficient in the quantum confinement effect. Nanowires are used in nanosensor elements to increase the sensitivity of electrode response. In nanolasers, semiconductor nanowires are used because of one-dimensionality and chemical flexibility. Scientists have demonstrated research in room temperature using UV nanowire nanolasers, which exhibit significant properties of nanolasers. In addition to short wavelength, nanolasers are used in a wide variety of fields such as microanalysis, optical computing, and information storage [89–94].

4.4.7 NANOENZYMES

Nanoparticle-based enzymes are attracted due to specific properties such as unique optical, magnetic, electronic, and catalytic properties, which are provided by small-sized nanoenzymes. Moreover, nanostructure of small-sized enzymes has the control of surface functionality and to create a complex structure on their surface, which in turn meets the needs of specific applications [95–100].

4.5 CONCLUSION AND PROSPECTS

Nanoparticles are one of the most focused technologies, which have various interesting applications in industry, economics, and environment. The quantum size effect can start to influence the properties of matter at nanoscale and significantly change the optical, magnetic, and electrical properties. Nanoparticles can be manufactured in different methods. One of the most important methods of such manufacturing process is the synthesis and processing. Subcategories of synthesis and processing include laser ablation, ball mining, physical vapor deposition, sputtering, nanolithography, chemical vapor deposition, solgel, spray pyrolysis, coprecipitation, electrochemical deposition, molecular beam epitaxy, inert gas condensation process, hydrothermal synthesis, and green synthesis, which are discussed in detail. Finally, applications of nanoparticles in various fields such as medicine, drug delivery, tissue engineering, wounds, nanowire composition, and nanoenzymes are explained in detail. This chapter provides knowledge about the properties of nanoparticles and their applications in various fields.

REFERENCES

1. Presting, H. and U. Konig. 2003. Future nanotechnology developments for automotive applications. *Materials Science and Engineering C* 23: 737–741.
2. Asmatulu, R. Claus, R.O. Mecham, J.B. and Corcoran, S.G. 2007. Nanotechnology: Associated coatings for aircrafts, *Materials Science* 43: 415–422.
3. Guo, K. 2012. Green nanotechnology of trends in future energy: A review. *International Journal of Energy Research* 36: 1–17.
4. Hussein, A.K. 2015. Applications of nanotechnology in renewable energies-A comprehensive overview and understanding. *Renewable and Sustainable Energy Reviews* 42: 460–476.
5. Kumar, C. 2006. *Nanomaterials: Toxicity, Health and Environmental Issues.* Wiley-VCH, 2006,351 pp; ISBN 978-3-527-31385-3.
6. Rogers, B. Pennathur, S. and Adams, J. 2008. *Nanotechnology Understanding Small Systems.* CRC Press,
7. Zhao, Q.Q. Boxman, A. and Chowdhry, U. 2003. Nanotechnology in the chemical industry-opportunities and challenges. *Journal of Nanoparticle Research* 5: 567–572.
8. Jin, W. and Maduraiveeran, G. 2018. Nanomaterial based environmental sensing platforms using state of the art electroanalytical strategies. *Journal of Analytical Science and Technology* 9(18): 1–11.
9. Khan, I. Saeed, K. and Khan, I. 2019. Nanoparticles: Properties, applications and toxicities. *Arabian Journal of Chemistry* 12: 908–931.
10. Madhumitha G. and Roopan, S.M. 2013. Devastated crops: Multifunctional efficacy for the production of nanoparticles. *Journal of Nanomaterials* 1: 1–12.
11. Zanganeh, S. Kajbafvala, A. and Zanganeh N. 2011. Hydrothermal synthesis and characterization of TiO$_2$ nanostructures using LiOH as a solvent. *Advanced Powder Technology* 22(3): 336–339.
12. Schmidt, C. and Storsberg, J. 2015. Nanomaterials-tools, technology and methodology of nanotechnology based biomedical systems for diagnostics and therapy. *Biomedicines* 3: 203–223.
13. Giftson Felix, D. and Sivakumar, G. 2014. Nano particles in automobile tires, *IOSR Journal of Mechanical and Civil Engineering* 11(4): 7–11.
14. Asmatulu, R. 2011. Toxicity of nanomaterials and recent developments in lung disease, in: P. Zobic (Ed.), *Bronchitis*, InTec, 95–108.
15. Liu, C. Burghaus, U. Besenbacher, F. and Wang, Z. 2010. Preparation and characterization of nanomaterials for sustainable energy production. *Nano Focus* 4: 5517–5526.
16. Zhou, X. 2010. New fabrication and mechanical properties of styrene-butadiene rubber/carbon nanotubes nanocomposite. *Journal of Materials Science and Technology* 26(12): 1127–1132.
17. Zhu, Y. Murali, S. Stoller, M. D. Ganesh, K. J. Cai, W. Ferreira, P.J. Pirkle, A. Wallace, R. M. Cychosz, K. A. Thommes, M. Su, D. Stach, E. A. and Ruoff, R. S. 2011. Carbon based super capacitors produced by activation of graphene. *Science* 332(6037): 1537–1541.

18. Wang, X. Ding, B. Yu, and Wang. M. 2011. Engineering biomimetic super hydrophobic surfaces of electrospun nanomaterials. *Nano Today* 6: 510.
19. Dmitruk, N.L. Borkovskaya, O.Y. Korovin, A.V. Mamontova, I.B. Romanyuk, V.R. and Sukach, A.V. 2015. Low-temperature diffused p–n junction with nano/ micro relief interface for solar cell applications, *Solar Energy Materials and Solar Cells* 137: 124–130.
20. Cohen-Karni, T. and Lieber, C.M. 2013. Nanowire nanoelectronics: Building interfaces with tissue and cells at the natural scale of biology. *Pure Applied Chemistry* 85: 883–901.
21. Zheng, G. Yang, Y. Cha, J. Hong, S. S. and Cui, Y. 2011. Hollow carbon nanofiber-encapsulated sulfur cathode for high specific capacity rechargeable lithium batteries. *Nano Letters* 11(10): 4462–4467.
22. Chen, X. and Mao, S.S. 2007. Titanium dioxide nanomaterials: Synthesis, properties, modifications, and applications, *Chemical Review* 107: 2891–2959.
23. Zeng, Z. Yin, Z. Huang, X. Li, H. He, Q. Lu, G. Boey, F. and Zhang, H. 2011. Single-layer semiconducting nanosheets: High-yield preparation and device fabrication. *Angewandte Chemie International Edition* 50: 11093–11097.
24. Wijaya, Y. N. Kim, J. Choi, W. M. Park, S. H. and Kim, M. H. 2017. A systematic study of triangular silver nanoplates: One-pot green synthesis, chemical stability, and sensing application. *Nanoscale* 9: 11705–11712.
25. Azam, A. Ahmed, A. S. Oves, M. Khan, M. S. Habib, S. S. and Memic, A. 2012. Antimicrobial activity of metal oxide nanoparticles against Gram-positive and Gram-negative bacteria: A comparative study. *International Journal of Nanomedicine* 7: 6003–6009.
26. Das, S. Dash, H. and Chakraborty, J. 2016. Genetic basis and importance of metal resistant genes in bacteria for bioremediation of contaminated environments with toxic metal pollutants. *Applied Microbiology and Biotechnology* 100: 2967–2984.
27. Lu, G. Li, H. Liusman, C. Yin, Z. Wu, S. and Zhang, H. 2011. Surface enhanced Raman scattering of Ag or au nanoparticle-decorated reduced graphene oxide for detection of aromatic molecules. *Chemical Science* 2: 1817–1821.
28. Gangopadhyay, S. Frolov, D.D. Masunov, A.E. and Seal, S. 2014. Structure and properties of cerium oxides in bulk and nanoparticulate forms. *Journal of Alloys and Compounds* 584: 199–208.
29. Torres, T. and Bottari, G. 2013. *Organic Nanomaterials: Synthesis, Characterization, and Device Applications*. Wiley ISBN: 978-1-118-01601-5, 622.
30. Rasmussen, K. Rauscher, H. Mech, A. RiegoSintes, J. Gilliland, D. González, M. Kearns, P. Moss, K. Visser, M. Groenewold, M. and Bleeker, E.A.J. 2018. Physico-chemical properties of manufactured nanomaterials - Characterisation and relevant methods. An outlook based on the OECD testing programme. *Regulatory Toxicology and Pharmacology* 92: 8–28.
31. Roming, M. Feldmann, C. Avadhut, Y.S. and Schmedt auf der Günne, J. 2008. Characterization of noncrystalline nanomaterials: NMR of zinc phosphate as a case study. *Chemistry of Materials* 20(18): 5787–5795.
32. Rawat, K. and Goyal, M. 2020 Theoretical analysis of thermophysical properties of nanomaterials. *Materials Today: Proceedings*. Doi: 10.1016/j.matpr.2020. 08.754.
33. Goyal, M. and Gupta, B.R.K. 2018 Shape and size dependent thermophysical properties of nanocrystals. *Chinese Journal of Physics* 56: 282–291.
34. Singh, M. Phantsi, T.D. 2018. Bond energy model to study cohesive energy, thermal expansion coefficient and specific heat of nanosolids. *Chinese Journal of Physics* 56: 2948–2957.
35. Qu, Y. Liu, W. Zhang, W. and Zhai, C. 2019. Theoretical study of size effect on melting entropy and enthalpy of Sn, Ag, Cu, and In nanoparticles. *Physics of Metals and Metallography* 120: 417–421.

36. Ghosh, S. and Raychaudhuri, A.K. 2013. Link between depression of melting temperature ad Debye temperature in nanowires and its implication on Lindeman relation. *Journal of Applied Physics* 114: 224313.
37. Arora, N. and Joshi, D.P. 2017. Band gap dependence of semiconducting nano-wire on cross-sectional shape and size, *Indian Journal of Physics* 91: 1493–1501.
38. Rahmayeni Putri J. Stiadi Y. and Zulhadjri Z. 2019. Green synthesis of NiFe$_2$O$_4$ spinel ferrites magnetic in the presence of Hibiscusrosa-sinensis leaves extract: Morphology, structure andactivity. *Rasayan Journal of Chemistry* 12: 1942–1949.
39. Gayathri Manju B. and Raji P. 2019. Green synthesis of nickel–copper mixed ferrite nanoparticles: Structural, optical, magnetic, electrochemical and antibacterial studies. *Journal of Electronic Materials* 48: 7710–7720.
40. Smiya, J. Geetha, V. and Francis M. 2020. Effect of pH on the optical and structural properties of SnS prepared by chemical bath deposition method. *Materials Science and Engineering* 872: 012139.
41. Ramaprasad, T. Jeevan Kumar, R. Naresh, U. Prakash, M. Kothandan, D. Chandra B. and Naidu, K. 2018. Effect of pH value on structural and magnetic properties of CuFe$_2$O$_4$ nanoparticles synthesized by low temperature hydrothermal technique. *Materials Research Express* 5: 095025.
42. Ghahfarokhi, S.M. and Shobegar, E.M. 2020. Influence of pH on the structural, magnetic and optical properties of SrFe$_2$O$_4$ nanoparticles. *Journal of Materials Research and Technology* 9(6): 12177–12186.
43. Peng, Y. Xia, C. Cui, M. Yao, Z. and Yi X. 2021. Effect of reaction condition on microstructure and properties of (NiCuZn) Fe$_2$O$_4$ nanoparticles synthesized via co-precipitation with ultrasonic irradiation. *Ultrasonics Sonochemistry* 71: 105369.
44. Nikolaev, A.L. Gopin, A.V. Severin, A.V. Rudin, V.N. Mironov, M.A. Dezhkunov, N.V. 2018. Ultrasonic synthesis of hydroxyapatite in non-cavitation and cavitation modes. *Ultrasonics Sonochemistry* 44: 390–397.
45. Kefeni, K.K. Msagati, T.A.M. and Mamba, B.B. 2017. Ferrite nanoparticles: Synthesis, characterization and applications in electronic device. *Materials Science Engineering B* 215: 37–55.
46. Somvanshi, S.B. Kharat, P.B. Khedkar, M.V. and Jadhav, K.M. 2020. Hydrophobic to hydrophilic surface transformation of nano-scale zinc ferrite via oleic acid coating: Magnetic hyperthermia study towards biomedical applications. *Ceramic International* 46: 7642–7653.
47. Kim, J.S. and Ham, C.W. 2009. The effect of calcining temperature on the magnetic properties of the ultra-fine NiCuZn-ferrites, *Materials Research Bulletin* 44: 633–637.
48. Chavan, A.R. Somvanshi, S.B. Khirade, P.P. and Jadhav, K.M. 2020. Influence of trivalent Cr ion substitution on the physicochemical, optical, electrical, and dielectric properties of sprayed NiFe$_2$O$_4$ spinel-magnetic thin films. *RSC Advances* 10: 25143.
49. Chen, Z.H. Zhou, Y. Kang, Z.T. and Chen, D. 2014. Synthesis of Mn-Zn ferrite nanoparticles by the coupling effect of ultrasonic irradiation and mechanical forces. *Journal of Alloys Compounds* 609: 21–24.
50. Arefi-Oskoui, S. Khataee, A. Safarpourd, M. Oroojia, Y. and Vatanpour, V. 2019. A review on the applications of ultrasonic technology in membrane bioreactors. *Ultrasonics Sonochemistry* 58: 104633.
51. Singh, K.R. Nayak, V. Sarkar, T. and Singh, R.P. 2020. Cerium oxide nanoparticles: Properties, biosynthesis and biomedical application. *RSC Advances* 10: 27194–27214.
52. Dhall, A. and Self, W. 2018. Cerium oxide nanoparticles: A brief review of their synthesis methods and biomedical applications. *Antioxidants* 7: 97.
53. Ciofani, G. Genchi, G.G. Liakos, I. Cappello, V. Gemmi, M. Athanassiou, A. Mazzolai, B. and Mattoli, V. 2013. Effects of cerium oxide nanoparticles on PC12 neuronal-like cells: Proliferation, differentiation, and dopamine secretion. *Pharmaceutical Research* 30: 2133–2145.

54. Manne, N.D. Arvapalli, R. Nepal, N. Thulluri, S. Selvaraj, V. Shokuhfar, T. He, K. Rice, K.M. Asano, S. and Maheshwari, M. 2015. Therapeutic potential of cerium oxide nanoparticles for the treatment of peritonitis induced by polymicrobial insult in sprague-dawley rats. *Critical Care Medicine* 43: 477–489.

55. Sack-Zschauer, M. Bader, S. and Brenneisen, P. 2017. Cerium oxide nanoparticles as novel tool in glioma treatment: An in vitro study. *Journal of Nanomedicine and Nanotechnology* 8: 474.

56. Lee, S.S. Zhu, H. Contreras, E.Q. Prakash, A. Puppala, H.L. and Colvin, V.L. 2012. High temperature decomposition of cerium precursors to form ceria nanocrystal libraries for biological applications. *Chemistry of Materials* 24: 424–432.

57. Xue, Y. Luan, Q. Yang, D. Yao, X. and Zhou, K. 2011. Direct evidence for hydroxyl radical scavenging activity of cerium oxide nanoparticles. *The Journal of Physical Chemistry C* 115: 4433–4438.

58. Hirst, S.M. Karakoti, A.S. Tyler, R.D. Sriranganathan, N. Seal, S. and Reilly, C.M. 2009 Anti-inflammatory properties of cerium oxide nanoparticles. *Small* 5: 2848–2856.

59. Xue, Y. Zhai, Y. Zhou, K. Wang, L. Tan, H. Luan, Q. and Yao, X. 2012. The vital role of buffer anions in the antioxidant activity of CeO_2 nanoparticles. *Chemistry* 18: 11115–11122.

60. Hijaz, M. Das, S. Mert, I. Gupta, A. Al-Wahab, Z. Tebbe, C. Dar, S. Chhina, J. Giri, S. and Munkarah, A. 2016. Folic acid tagged nanoceria as a novel therapeutic agent in ovarian cancer. *BMC Cancer* 16: 220.

61. Singh, J. Dutta, T. Kim, K.-H. Rawat, M. Samddar, P. and Kumar, P. 2018. Green synthesis of metals and their oxide nanoparticles: Applications for environmental remediation. *Journal of Nanobiotechnology* 16: 84.

62. Haider, A.J. Haider, M.J. and Mehde, M.S. 2018. A review on preparation of silver nano-particles. *AIP Conference Proceedings* 1968: 030086.

63. Yadav, T.P. Yadav, R.M. and Singh, D.P. 2012. Mechanical milling: A top down approach for the synthesis of nanomaterials and nanocomposites. *Nanoscience and Nanotechnology* 2(3): 22–48.

64. Wang, Y. Chen, W. Wang, B. and Zheng, Y. 2014. Ultrathin ferroelectric films: Growth, characterization, physics and applications. *Materials* 7: 6377–6485.

65. Baptista, A. Silva, F. Porteiro, J. Miguez, J. and Pinto, G. 2018. Sputtering physical vapour deposition coatings: A critical review on process improvement and market trend demands. *Coatings* 8: 402.

66. Piner, R.D. Zhu, J. Xu, F. Hong, S. Mirkin, C.A. 1999. Dip-pen nanolithography. *Science* 283: 661–663.

67. Choy, K.L. 2003. Chemical vapour deposition of coatings, *Progress in Materials Science* 48: 57–170.

68. Fay, S. Feitknecht, L. Schluchter, R. Kroll, U. Vallat-Sauvain, E. and Shah, A. Rough ZnO layers by LP-CVD process and their effect in improving performances of amorphous and microcrystalline silicon solar cells. *Solar Energy Materials and Solar Cells* 90: 2960–2967.

69. Rane, A.V. Kanny, K. Abitha, V.K. and Thomas, S. Methods for synthesis of nanoparticles and fabrication of nanocomposites. *Synthesis of Inorganic Nanomaterials* 121–139.

70. Lien, S. Wuu, D. Yeh, W. and Liu, J. 2006. Tri-layer antireflection coatings (SiO_2/SiO_2–TiO_2/TiO_2) for silicon solar cells using a sol-gel technique. *Solar Energy Materials and Solar Cells* 90: 2710–2719.

71. Huang, W.-K. Chung, K.-J. Liu, Y.-M. Ger, M.-D. Pu, N.-W. and Youh, M.-J. 2015. Carbon nanomaterials synthesized using a spray pyrolysis method, *Vacuum* 118: 94–99.

72. Chen, J. Chen, W. Tien, Y. and Shih, C. 2010. Effect of calcination temperature on the crystallite growth of cerium oxide nano-powders prepared by the co precipitation process. *Journal of Alloys and Compounds* 496: 364–369.

73. Li, Z. Meng, G. Liang, T. Zhang, Z. Zhu, X. 2013. Facile synthesis of large-scale Ag nanosheets-assembled films with Su b-10 nm Gaps as highly active and homogeneous SERS substrates. *Applied Surface Science* 264: 383–390.
74. Tardif, S. 2011. *Nanocolonnes de Ge Mn: proprieties magnetiques at structural's lummi ere du synchrotron*, Universities de Grenoble, French.
75. Suryanarayana, C. and Prabhu, B. Synthesis of nanostructure materials by inert gas condensation methods. C.C. Koch (Ed,). *Nanostructured Materials: Processing, Properties, and Applications*, 2nd Ed., William Andrew, Elsevier, Netherlands, 2007, 47–90.
76. Zhang, X. Luo, Z. Yu, P. Cai, Y. Du, Y. Wu, D. Gao, S. Tan, C. Li, Z. and Ren, M. 2018. Lithiation-induced amorphization of Pd3P2S8 for highly efficient hydrogen evolution. *Natural Catalyst* 1: 460.
77. Thovhogi, N. Diallo, A. Gurib-Fakim, A. and Maaza, M. 2015. Nanoparticles green synthesis by Hibiscus Sabdariffa flower extract: Main physical properties. *Journal of Alloys and Compounds* 647: 392–396.
78. Chaturvedi, S. Pragnesh Dave, N. and Shahb, N.K. 2012. Applications of nano-catalyst in new era. *Journal of Saudi Chemical Society* 16(3): 307–325.
79. Mathew, J. Joy, J. and George S.C. 2019. Potential applications of nanotechnology in transportation: A review. *Journal of King Saud University - Science* 31(4): 586–594.
80. Kingsley, D. Ranjan, S. Dasgupta, N. and Saha, P. 2013. Nanotechnology for tissue engineering: Need, techniques and applications. *Journal of Pharmacy Research* 7(2): 200–204.
81. Zuo, L. and Wei, C. 2007. New technology and clinical applications of nanomedicine. *Medical Clinics of North America* 91(5): 845–862.
82. Sun, R. Wang, W. Wen, Y and Zhang, X. 2015. Recent advance on mesoporous silica nanoparticles-based controlled release system: Intelligent switches open up new horizon. *Nano* 5: 2019–2053.
83. Cao, H. 2009. The application of nanofibrous scaffolds in neural tissue engineering. *Advanced Drug Delivery Review* 61(12): 1055–64.
84. Albertia, T. S. Coelhoa, D. Voytena, A. Pitz, H. de Pra, M. Mazzarino, L. Kuhnen, S. R. Ribeiro-do-Valle M. Maraschin M. and Veleirinho, B. 2017. Nanotechnology: A promising tool towards wound healing. *Current Pharmaceutical Design* 23: 1–14.
85. Witharana, C. and Wanigasekara, J. 2016. Applications of nanotechnology in drug delivery and design-an insight. *Current Trends in Biotechnology and Pharmacy* 10(1): 78–91.
86. Toimil-Molares, M.E. 2012. Characterization and properties of micro- and nanowire of controlled size, composition, and geometry fabricated by electro deposition and iontrack technology. *Journal of Nanomaterials* 3: 860–883.
87. Zhang, R. Fan, K. and Yan, X. 2020. Nanozymes: Created by learning from nature science. *China Life Sciences* 63: 1183–1200.
88. Lv, W. and Champion, J.A. 2020. Demonstration of intracellular trafficking, cytosolic bioavailability, and target manipulation of an antibody delivery platform. *Nanomedicine: Nanotechnology, Biology and Medicine* 32: 102315.
89. Soares, S. Sousa, J. Pais, A. and Vitorino, C. 2018. Nanomedicine: Principles, properties, and regulatory issues. *Frontiers in Chemistry* 6: 360.
90. Moghimi, S.M. 2014. Cancer nanomedicine and the complement system activation paradigm: Anaphylaxis and tumour growth. *Journal of Controlled Release* 190: 556–562.
91. Farokhzad, O.C. and Langer, R. 2006 Advanced drug delivery reviews nanomedicine: Developing smarter therapeutic and diagnostic *Modalities* 58(14): 1456–1459.
92. Huynh, N.T. Passirani, C. Saulnier, P. and Benoit, J.P. 2009. Lipid nanocapsules: A new platform for nanomedicine. *International Journal of Pharmaceutics* 379(2): 201–209.
93. Master, A. Livingston, M. and Gupta, A.S. 2013. Photodynamic nanomedicine in the treatment of solid tumors: Perspectives and challenges. *Journal of Controlled Release* 168(1): 88–102.

94. MoeinMoghimi, S. and ShadiFarhangrazi, Z. 2013. Nanomedicine and the complement paradigm. *Nanomedicine: Nanotechnology, Biology and Medicine* 9(4): 458–460.
95. Guo, B. Zhao, J. Zhang, Z. An, X. Huang, M. and Wang, S. 2020. Intelligent nano-enzyme for T1-weighted MRI guided theranostic applications. *Chemical Engineering Journal* 391: 123609.
96. Wu, T. Ma, Z. Li, P. Lu, Q. Liu, M. Li, H. Zhang, Y. and Yao S. 2019. Bifunctional colorimetric biosensors via regulation of the dual nanoenzyme activity of carbonized FeCo-ZIF sensors and actuators. *B: Chemical* 290: 357–363.
97. Dong, Z. Wang, Y. Yin, Y. and Liu, J. 2017. Supra molecular enzyme mimics by self-assembly. *Current Opinion in Colloid & Interface Science* 16(6): 451–458.
98. Chen, W. Fang, X. Li, H. Cao, H. and Kong, J. 2017. DNA-mediated inhibition of peroxidase-like activities on platinum nanoparticles for simple and rapid colorimetric detection of nucleic acids. *Biosensors and Bioelectronics* 94: 169–175.
99. He, X. Yang, Y. Li, L. Zhang, P. Guo, H. Liu, N. Yang, X. and Xu, F. 2020. Engineering extracellular matrix to improve drug delivery for cancer therapy. *Drug Discovery Today* 25(9): 1727–1734.
100. Wang, H. Li, J. Wang, J. Gong, Y. Xu, X. Wang, X. Li, J. Sha, Y. and Zhang, Z. 2020. Nanoparticles-mediated reoxygenation strategy relieves tumor hypoxia for enhanced cancer therapy. *Journal of Controlled Release* 319: 25–45.

5 Emergent Nanomaterials and Their Composite Fabrication for Multifunctional Applications

Karthik Kannan
Qatar University

Devi Radhika
Jain-Deemed to be University

R. Suriyaprabha
Central University of Gujarat

Sreeja K. Satheesh
Atonarp Microsystems

L. Sivarama Krishna
Universiti Malaysia Terengganu

CONTENTS

5.1 INTRODUCTION

In the past decade, metal nanoparticles (MNPs), semiconductor NPs (SCNPs), and their composites have garnered widespread attention due to their exceptional physical, optical, electrical, and biological properties. Miniaturization in terms of size, quantum confinement effects, and a large surface-to-volume ratio are some of the factors that contribute to the wide array of interesting properties that these materials tend to display. In addition to being eco-friendly, these materials also possess exceptionally high electronic and ionic conductivity, enhanced catalytic activity, and improved thermal and chemical stability, which increase their utility in many devices [1–3]. As a consequence, these materials have emerged as the backbone of technological advancement in the field of sensors, catalysis, pigments, drug delivery, and biomedicine.

Synthesis of nanomaterials with eco-friendliness is a major challenge. NPs can be synthesized via physical, chemical, and biological routes. However, these synthesis routes often yield hazardous by-products, or the synthesis procedures cannot be scaled to manufacture the NPs on the nanoscale. Thus, there is a need to enlarge ecological, facile, and cost-effective ways for the preparation of nanomaterials in laboratories at nanoscale [4]. As a result, the researchers have shifted their focus to "green syntheses" as an alternative to the conventional physical and chemical methods. The unique combination of green chemistry and nanotechnology has opened up new avenues of research, which have ultimately led to the development of nanocomposites (NCs) that can now be used in real-life applications such as energy devices, photocatalysis, antibacterial drugs, and anticancer therapy [5].

It is a well-known fact that the specific surface area and hence the surface-to-volume ratio increase rapidly on reducing the size of the material to nanoscale. The higher surface area is particularly important in the case of metal NPs, which are employed in the production of hydrogen by water splitting, odor regulation, and the extraction of dyes from contaminated water. Dyes play a major role in the pollution of water sources and are highly toxic in nature, which are used in many factories. Since dyes possess a sophisticated structure, they are difficult to degrade. Consequently, numerous photocatalytic procedures have to be used to remove them from the water supplies. The fact that water pollution would emerge as the single major threat to humanity in the coming decades serves as the major driving factor for the usage of nanomaterials for water purification, among other applications. Moreover, due to their large surface area and enhanced physicochemical properties, NPs of Ag, Au, Pd, Pt, CuO [6–10], NiO [11–16], TiO_2 [17–22], MgO [23], CdO [24], SnO_2 [25], etc., are now being used in emerging technologies such as optoelectronics, sensors, bioimaging devices, catalysis, solar cells, detectors, and biomedicine, and also, they have been identified as notable antimicrobial (foodborne pathogens) and photocatalysts (organic pollutant degradation) agents [6,16,21–25].

This chapter attempts to provide an overview of the photocatalytic, antibacterial, and anticancer capabilities of metallic, metal oxide nanoparticles (MONPs), and emerging composites for multifunctional applications prepared by green route synthesis. As potential candidates for the manufacture of extremely effective photocatalytic and pharmaceutical products, the problems associated with the use of metal oxide-based graphene composites are also addressed.

5.2 IMPORTANCE OF MIXED METAL OXIDE COMPOSITES

The doping of impurities in the semiconductor lattices is a highly efficient method to regulate the electrical conductivity and thereby improve the optical, magnetic, and biomedical properties of semiconductors. Novel materials with anticancer and anti-bacterial properties are produced by combining noble metals such as Ag, Au, Pd, and Pt and metallic oxides such as CuO [6–10], NiO [11–16], TiO_2 [17–22], MgO [23], CdO [24], and SnO_2 [25], which are the finest choices for biomedical applications. The achievement of extended absorption in the visible field is an additional emerging problem in the manufacture of semiconductor nanomaterials in photocatalytic and solar cell purposes. In comparison with pure metal NPs, NCs have gained a great deal of research interest in modern research papers on account of their superior qualities. NCs are formed by introducing nanomaterials (also called fillers) such as metal NPs and MONPs in a matrix such as ceramic, polymer, or metal. These materials exhibit interesting properties when the size of the particles ranges between 1 and 100nm. NCs can be made from different kinds of materials that can be grouped into three basic structural frames: ceramics, metals, and polymers. The NCs can have distinctly diversified electrical, thermal, electrochemical, mechanical, catalytic, and optical properties and could be further classified as zero-dimensional (0D) such as core–shell, 1D such as nanotubes and nanowires, 2D such as lamellar, and 3D such as composites of the metal matrix, keeping in view the spatial structure. NCs can also be classified as nanoparticulate, nanofilamentary, and nanolayered composites based on their structure-related properties. Researchers, scientists, and engineers are drawing even more attention to NC materials. This has led to a tremendous rise in the number of articles published in this field. Furthermore, NCs have emerged as smart materials of the 21st century, which in turn have attracted industrial and business advances in all facets of life.

5.3 PRODUCTION OF VARIOUS METAL OXIDE NANOMATERIALS USING CHEMICAL AND BIOENGINEERED ROUTES

For the preparation of nanomaterials, two methods are available: the top-down approach, which involves mechanical grinding or pulverization, and the bottom-up approach, which consists primarily of producing colloidal dispersion via wet chemical synthesis routes. When an orderly arrangement of NPs is required, the bottom-up approach is beneficial as it builds the structures atom by atom. Wet chemical pathways that are commonly used to prepare nanomaterials include coprecipitation, ultrasound synthesis, hydrothermal method, solgel method, microwave-assisted, salt-assisted aerosol decomposition method, spray pyrolysis method, and reflux synthesizing. But the green synthesis route is a sustainable procedure for the fabrication of NPs as it limits the production of hazardous wastes as by-products. Further, it has been proved that green synthesis route yields NPs with good size distribution and excellent physical and chemical properties. [10–11, 14–15, 22–25]. In the field of green route synthesis, most of the work has been focused on the fabrication of various MONPs from natural resources such as plant extracts. This is an easy and effective method to produce relatively uniform and small-sized powders, among the various other techniques that can be used effectively for the production of NPs [22]. MONPs prepared from various plant extracts are listed in Table 5.1.

TABLE 5.1
Production of Nanomaterials from Various Plant Extracts

Plant Species	Mode of Synthesis	Nanomaterials	Size (nm)	References
Guilandina moringa L.	Green route	CuO	12	[6]
Calotropis procera	Green synthesis	CuO	40-46	[7]
Oak fruit hull extract	Green route	CuO	20-35	[8]
Punica granatum	Green synthesis	CuO	10-50	[9]
Andrographis paniculata	Green route	CuO	30	[10]
Guilandina moringa L.	Green route (hot plate combustion)	NiO	8-10	[11]
Agathosma betulina	Green route	NiO	15–23	[12]
Berchemia chanetii	Bioengineered route	NiO	18	[13]
Melaleuca viminalis	Green synthesis	NiO	20-40	[14]
Citrus fruit juice	Green synthesis	NiO	20	[15]
Aloe Barbadensis	Green synthesis	TiO_2	20	[16]
Vitex negundo Linn	Green synthesis	TiO_2	10	[17]
Curcuma longa	Green synthesis	TiO_2	43	[18]
Vigna unguiculata	Green synthesis	TiO_2	11	[19]
Moringa oleifera	Green synthesis	TiO_2	12	[20]
Ledebouria revoluta	Green synthesis	TiO_2	47	[21]
Andrographis paniculata	Green synthesis (microwave-assisted)	TiO_2	19	[22]
Andrographis paniculata	Green synthesis (microwave-assisted)	MgO	35	[23]
Andrographis paniculata	Green synthesis	CdO	22	[24]
Andrographis paniculata	Green synthesis (microwave-assisted)	SnO_2	27	[25]
Ocimum tenuiflorum	Green synthesis	Ag	25-40	[26]
Syzygium cumini	Green synthesis	Ag	29-30	[27]
Psidium guajava	Green synthesis	Au	25-30	[28]
Mangifera indica	Green synthesis	Au	17-20	[29]
Pinus resinosa	Green synthesis	Pd	16-20	[30]
Glycine max	Green synthesis	Pd	15	[31]
Punica granatum L	Green synthesis	NiO/MgO NC	3	[32]
Lagerstroemia speciosa	Green synthesis	GO/Ag NC	60-100	[33]
Jatropha curcas	Hydrothermal green method	CuO/Ag NC	50-150	[34]

Species of bacteria have been broadly used for various applications in biotechnology, for example, hereditary and bioleaching purposes. Microbes can decrease metal particles and are pivotal competitors in the preparation of nanoparticles (NPs). Instead of different organisms, microorganisms can be effectively formed and controlled genetically for the biomineralization of metal particles. Microorganisms are ceaselessly presented to unforgiving and dangerous conditions coming about because of high centralizations of substantial metal particles in their environmental factors [23].

These methods can be proficiently used by the microscopic organisms for NP synthesis for various sorts of utilizations. Metal NP bacteria can combine by either extracellular or intracellular components. In ongoing reports, diverse bacterial strains, like *E. coli, B. subtilis, B. megaterium, P. aeruginosa, B. cereus, Alteromonas, and Ochrobactrum*, have been widely utilized for NPs and NCs.

Biosynthesis of fungal NPs is another basic and direct methodology that has been investigated widely for NP fabrication. MONP biosynthesis of mediated fungi is likewise a productive cycle for the monodispersed NP generation with all-around characterized morphologies. Due to the involvement of an assortment of intracellular catalysts, they also act better as organic specialists for the arrangement of MONPs and MONCs. In contrast with microorganisms, fungi have higher profitability as far as the generation of NPs and higher resilience to metals particularly in the setting of high cell divider restricting the limit of metal particles with biomass [24,25]. Metal/MONPs produced from the stem, leaf, seed, leaf, latex, etc., of plants, bacteria, and fungi have also been listed in the table. The biogenic production of NPs gives good-yield, nanosized materials with suitable properties. The biomediated process of NP synthesis can be categorized into three steps, namely activation level, growth level, and termination level. However, a very high calcination temperature is required to eliminate the impurities in the NPs.

5.4 EMERGING NANOMATERIALS AND THEIR COMPOSITES IN ENVIRONMENTAL AND BIOLOGICAL APPLICATIONS

NC materials displayed good chemical stability, improved magnetic activity, and excellent biocompatibility, and also, it has many applications in the various fields [10,15–17]. Moreover, MONPs and their composites are renowned for their contributions to enzymatic or metabolic degradation, drug delivery, pathogen identification, hyperthermia, tissue repair, and antigen diagnosis (Figure 5.1) [21–25]. Nevertheless, three major applications of NCs have been chosen for detailed analysis, namely photocatalysis, antibacterial activity, and anticancer activity. The relevance and contributions of NCs in these three spheres of technology will be elaborated in the successive sections of this chapter.

5.4.1 Photocatalytic Performance

Organic dyes from industrial effluents are major sources of water pollution with serious environmental consequences. About 0.7 million tonnes of unrefined synthetic dyes are produced annually on a pilot scale, primarily to support industries such as textiles, paints, cosmetics, synthetics, and electronics. During the dyeing process, about 20% of the overall output of dyes is lost and is released as textile effluents into the atmosphere and in the surrounding water bodies. These dyes are extremely stable and can pose as potential carcinogens when released into the atmosphere. In addition, the dyes present in the industrial effluents released into water supplies are capable of withstanding oxidation, hydrolysis, anaerobic decoloration, or other chemical reactions in wastewater and documented evidence suggests that they can have detrimental effects on human life and environmental quality [35,36].

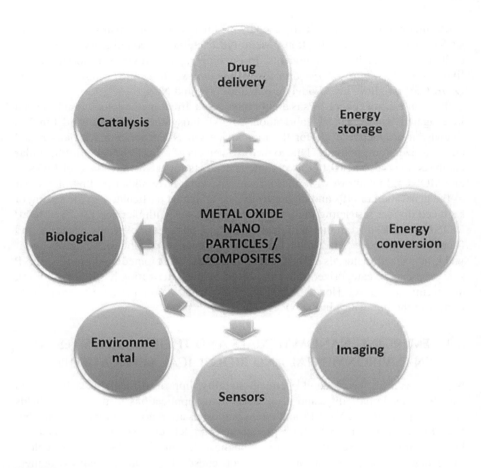

FIGURE 5.1   Applications of metal oxide nanomaterials.

Nevertheless, chemical oxidation of dyes encounters less adversities when it is performed with the aid of oxidants like H_2O_2, O_3, Cl_2, and ClO_2, although this supports only partial degradation of the dyes in water. To counter the problem, advanced oxidation processes have been developed in order to remove different toxic substances that might not be effectively removed by conventional wastewater treatment techniques. In order to address the environmental concerns associated with water purification, semiconductor photocatalysis has been introduced. In the last three decades, semiconductor photocatalysis has garnered much interest due to its reliability as an easy, straightforward, and commercial route for both energy production and water purification. Extensive work has been carried out using a variety of photocatalytic semiconductors to generate hydrogen by splitting water. Photocatalytic semiconductors are the focal point of most studies on photocatalysis available today. In photocatalytic semiconductors, the excitation of the bandgap results in the creation of electron–hole pairs. When a photocatalyst is irradiated with photos of energy identical or superior to the bandgap energy, the valence band (VB) electrons can be transmitted to the conduction band, making a positive hole in the VB as shown in Figure 5.2.

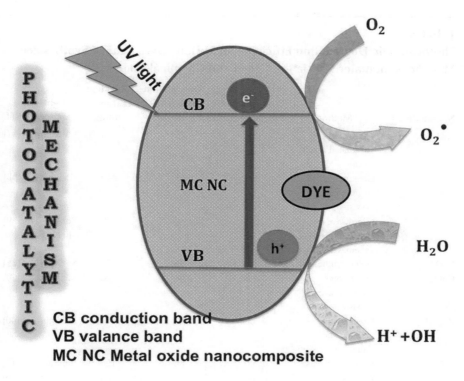

FIGURE 5.2   Photocatalytic activity of metal oxide nanocomposite.

$$\text{Photocatalyst} + (h\nu) \rightarrow e_{CB}^- + h_{VB}^+ \qquad (5.1)$$

M/MONPs catalyst behavior is another significant and overpowering region of exploration. AgNPs show superb reactant action for the decrease of different dangerous colors. For Azo colors degradation and wastewater treatment, Pd-NPs exhibit green synthesis catalytic properties [36]. NPs of metal oxides such as CuO, NiO, TiO_2 MgO, CdO, and SnO_2 are the most popular photocatalysts due to their relatively high activity, chemical stability, accessibility, low manufacture costs, and low toxicity. These materials have been researched extensively, and they have been found to completely oxidize numerous unprocessed compounds in the target contaminants, such as dyes [21–25]. Graphene-based nanoadsorbents are outstanding superior materials for the elimination of the organic pollutants from the water owing to their nanoscaled size, high surface area, and ability to cooperate via pi–pi stacking, hydrogen bonding, and electrostatic interactions. In relative adsorption studies of graphite oxide (GO) and graphite using methylene blue (MB) and malachite green (MG) as customary organic dyes, it was found that GO demonstrated much better adsorption than graphite. GO has also been employed for the removal of cationic dyes specifically MB, crystal violet (CV), and rhodamine B (RhB) from water. In the field of MONPs/ MONCs with efficient photocatalytic activity, there is plenty of literature and some of those are summarized in Table 5.2.

TABLE 5.2

Photocatalytic Degradation Efficiency of Various Dyes Against Synthesized Metallic Nanomaterials Via the Green and Chemical Routes

Nanomaterials	Method	Concentration (mg)	Degradation Efficiency (%)	Reaction Time (mins)	Light Source	Ref.
SnO_2	Microwave-irradiated green route	5.0	76	100	UV lamp	[37]
ZrO_2	Green route (microwave-assisted)	1.5	91	240	UV light	[38]
MgO	Microwave-assisted green synthesis	5.0	88	90	Solar light	[39]
SnO_2	Green synthesis	2	99	120	Sunlight	[25]
CuO	Green route (microwave-assisted)	5.0	98	120	Sunlight	[10]
$ZnFe_2O_4$	Green synthesis (sugarcane juice)	5.0	82	120	Sunlight	[40]
NiO	Green synthesis	5.0	88	120	Sunlight	[15]
NiO	Green synthesis	5.0	91	150	Sunlight	[16]
TiO_2	Green synthesis	5.0	99	80	UV light	[22]
CdO	Ultrasonic-assisted	2.5	78	150	Sunlight	[41]
CdO-MgO			82			
TMS-CM	Solvothermal route	50	95.4	180	Sunlight	[42]
ZnO-GO	Ultrasonication + hydrothermal	50	99	120	Visible light	[43]
ZnO-rGO	Hydrothermal	40	100	80	Visible light	[44]
$CeO_2/SnO_2/rGO$	Facial hydrothermal	50	90	90	Sunlight	[45]
CeO_2/SnO_2	Wet chemical	50	80	150	Visible light	[46]
$CRGO/SnO_2$	Green route	20	97.4	60	Visible light	[47]
ZnS	Coprecipitation	30	50	120	Sunlight	[48]
Cd-ZnS			96.7			
TiO_2	Solgel route	25	64.6	90	UV light 365 nm	[49]
3% bidoped TiO_2			57.14			
5% bidoped TiO_2			60.75			

5.4.2 Antibacterial Activity

M/MONPs have a unique structure, biocompatibility, intriguing redox and reactant properties, and great mechanical steadiness. Because of these properties and remarkable plasmonic properties, M/MONPs have impressive consideration in the biomedical application field. Inorganic NPs may typically interact with microorganisms that serve as an important antibacterial agent and also as an antifungal agent. However, it is important to investigate how to integrate these molecules within microorganisms in order to facilitate the transport of bioactive NPs used in targeted drug delivery systems. Gram-negative (−ve) bacteria have already been used to generate bacterial ghosts, demonstrating new distribution advances and directing to vehicles suitable for the dissemination of hydrophobic or water-soluble drugs. Various elements linked to the cell wall have been recognized as key players in fungal–host communications in pathogenic fungi such as *Candida albicans* [50].

The Gram-negative (−ve) bacterial cell wall is highly sophisticated in terms of its structure and chemistry. It consists of a thin layer of peptidoglycan (PG) and a membrane above the surface. This outer membrane often encounters hydrophobic species and covers a lipopolysaccharide as an uncommon portion, which boosts the negative (−ve) charge of cell wall membranes and is essential for the bacteria's structural integrity and feasibility [51,52]. Typically, using different forms of G− and G+ bacterial strains, the antibacterial efficiency of NPs/composites can be achieved. For metallic NPs, the concentration levels varied in intensity from 50 to 100 µg/mL.

Based on the subsequent features, the antibacterial mechanism of the fabricated MONPs/MONCs can be described as below: [53]

I. Formation of reactive oxygen species (ROS)
II. Heavy metal ions dispensation

The photogeneration of ROS in the MO/graphene NCs surface has been stated by Karthik et al. [10,16]. Normally, the antibacterial efficiency varies according to the ROS, basic surface area, particle size, etc. Through the Fenton reaction mentoring for DNA injury, lipid peroxidation, and protein oxidation, some NCs provide ROS that can kill bacteria without destroying nonbacterial cells. Another potential reason for the observed antibacterial efficiency can be described as follows: the prepared NC obstructs and attaches to the mesosome of the bacterial cell wall membrane, and the oxidative stress induced by ROS leads to cell death, as demonstrated in Figure 5.3. This in turn initiates certain intracellular functional modifications.

Nevertheless, the electrostatic attraction mechanism is explained in very few literary works. It can be understood as follows: the M^{2+} is released from the NPs substrate and interacts with the bacterial cell membrane of the microbe. Both (−ve) cell wall and (+ ve) M^{2+} are similarly intrigued and denature proteins, allowing the pathogen to die in the thrashing of the DNA's imitation competence.

$$MO\,NPs\,/\,MO-MO\,/\,Graphene\,NCs + h\upsilon \rightarrow e^- + h^+$$

$$h^+ + H_2O \rightarrow {}^\circ OH + H^+$$

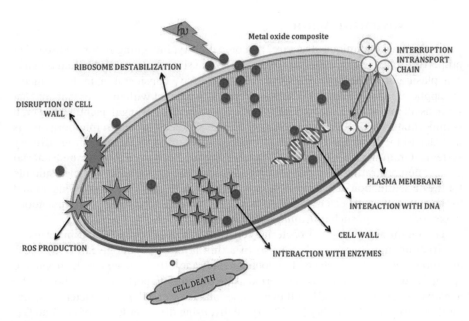

FIGURE 5.3 Antibacterial activity mechanism of metal oxide nanoparticles/ and nanocomposites

$$e^- + O_{2\rightarrow} \rightarrow {}^\circ O_{2^-}$$

$${}^\circ O_2^- + H^+ \rightarrow HO_2^\circ$$

$$HO_2^\circ + H^+ + e^- \rightarrow H_2O_2$$

The common procedure of NC antibacterial activity is explained schematically in Figure 5.3. Karthik et al. [10,16] depicted the photocatalytic and biological performances of chemically supported green route NPs, viz. CuO, NiO, ZnO, MgO, and CdO [54]. The bioengineered NC of CdO-ZnO and their photocatalytic and biological performances have been described by Rajaboopathi et al. [55]. The fascinating behavior of MONPs and their composites, in the presence of several Gram-positive and Gram-negative bacteria, has been well documented (Table 5.3). Some of these results have been tabulated below along with the antimicrobial performance observed in terms of inhibition region, in each case.

5.4.3 ANTICANCER ACTIVITY

The incorporation of green nanotechnology along with drug delivery has led to the development of a new technological era in "green nanomedicine." NPs are increasingly being used in biomedical applications in numerous processes including targeted drug delivery systems, bioimaging, and gene delivery. They can act as smart

TABLE 5.3

Antimicrobial Studies of Various Metal Oxide Nanomaterials Synthesized via Green Route Against Different Human Pathogens

Nanomaterials	Method	Tested Pathogens	Concentration	ZOI (mm)	Ref
MgO (*P. Guajava*)	Green route	*E. coli*	5 mg/mL	15	[56]
MgO (*Burm.f.*)	Green route	*E. coli*	5 mg/mL	12	[56]
MgO (*P. Guajava*)	Green route	*S. aureus*	5 mg/mL	16	[56]
MgO (*Aloe Barbadensis*)	Green route	*S. aureus*	5 mg/mL	15	[56]
CdO (*A. paniculata*)	Green route	*E. coli*	100 µg/mL	16	[24]
CdO (*A. paniculata*)	Green route	*S. aureus*	100 µg/mL	13	[24]
CuO (*A. paniculata*)	Green route (microwave-assisted)	*E. coli*	100 µg/mL	20	[10]
CuO (*A. paniculata*)	Green route (microwave-assisted)	*S. aureus*	100 µg/mL	18	[10]
MgO (*A. paniculata*)	Green route	*E. coli*	100 µg/mL	20	[23]
MgO (*A. paniculata*)	Green route	*S. aureus*	100 µg/mL	18	[23]
NiO (*Limonia acidissima Christm*)	Green route	*S. aureus*	100 µg/mL	14.6	[16]
NiO (*Limonia acidissima Christm*)	Green route	*E. coli*	100 µg/mL	15.3	[16]
TiO$_2$ (*Ledebouria revoluta*)	Green route	*E. coli*	60 µg/mL	4	[21]
TiO$_2$ (*Ledebouria revoluta*)	Green route	*C. tetani*	60 µg/mL	5.4	[21]
Ag (**Cordia dichotoma**)	Green route	*E. coli*	1 mg/mL	6	[57]
Ag (*Cordia dichotoma*)	Green route	*E. coli*	2 mg/mL	8	[57]
CdO-NiO	Microwave-assisted	*S. aureus*	10 µg/mL	20	[54]
CdO-ZnO				40	[58]
CdO-CuO			100 µg/mL	26	[59]
CdO-MgO				29	[60]
CdO-NiO-ZnO				27	[61]
N-doped CQD				16	[62]
CdO-CuO		*S. typhi*	100 µg/mL	20	[59]
CdO-ZnO-MgO				14	[63]

(*Continued*)

TABLE 5.3 (*Continued*)
Antimicrobial Studies of Various Metal Oxide Nanomaterials Synthesized via Green Route Against Different Human Pathogens

Nanomaterials	Method	Tested Pathogens	Concentration	ZOI (mm)	Ref
GO	Coreduction	*E. coli*	100 µg/mL	15	[64]
		S. aureus		18	
GO-Ag	Coreduction	*E. coli*	100 µg/mL	18	[64]
		S. aureus		28	
ZnO/Fe$_3$O$_4$/rGO	Hydrothermal	*E. coli*	1 mg/mL	18	[65]
		S. aureus		13	

weapons against many drug-resistant microorganisms and can also be used as alternatives to antibiotics. The original idea behind targeted drug delivery using NPs was the outcome of two fundamental characteristics associated with NPs. Firstly, the size of the NPs enables them to penetrate through very small capillaries and facilitates easy absorption by the cells, permitting an effective growth of drugs at the target sites. Secondly, the usage of biodegradable materials for the synthesis of NPs permits the extended release of drugs within the site targeted over a longer period of time.

Cancer, which is an uncontrolled and unbalanced growth of immature cells, is caused by genetic mutations. It contributes to unregulated cell proliferation equilibrium and advances as a small cell population attacks tissues and metastasizes to other areas of the body, producing substantial morbidity and demise. Genetics, lifestyle choices such as diet, tobacco, and lack of exercise, certain forms of diseases, and environmental exposure to distinctive types of chemicals and radiation are the most common factors that lead to genetic mutations that trigger the growth of cancerous cells.

Chemotherapy, radiation therapy, surgery, gene therapy, immunotherapy, and cryotherapy are the most widely used clinical cures for cancer. Conventional therapeutic methods have their own benefits and some drawbacks. Chemotherapy, for example, is supposed to destroy cancer cells mainly due to tumor breakdown. However, due to overcompensation from parallel signaling pathways and its side effects on the healthy cells, the clinical findings indicate tumor resistance. To resolve the demerits of traditional clinical therapies, new therapeutic strategies are therefore needed [66].

Solubility in water, nonspecificity, low permeability, and drug resistance across biological barriers are poor in chemotherapeutic agents. In the intervening years, nanomedicine has been researched thoroughly and is being promoted to address the above disadvantages of anticancer drugs. Nanomedicine essentially combines the technological advancements in nanotechnology to prevent and treat illness in the human body. It is believed that nanomedicine can transform clinical therapies into easy and cellular-level therapy, by fixing the damaged cells or entering them and replacing or supporting the defective intracellular systems as described in Figure 5.4.

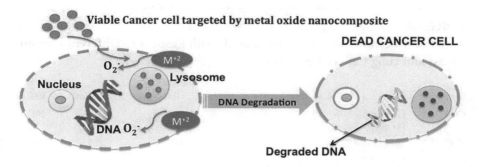

FIGURE 5.4 Anticancer activity mechanism of metal oxide nanoparticles/ and nanocomposites.

Recent advances in nanotechnology have helped to improve the contribution of nanotechnology in the field of medicine. Some of the notable contributions include:

1. Smaller nanostructures and different shapes with high surface areas tend to anchor and link with cell membranes. This in turn leads to an improvement in cell membrane mobility and communication, which enhances the cellular absorption of NPs.
2. The NPs surface can be functionalized with hydrophilic polymers that can facilitate an easy escape from macrophages.
3. NC layers can simply allow in vitro and in vivo cell-specific objectification.
4. NPs can increase intracellular drug concentrations in cancer cells, thereby avoiding excessive toxicity in the direction of healthy cells.
5. Within the NPs, chemotherapeutic drug molecules can be trapped or conjugated on the layer, which would improve the efficiency of drug loading.

Among the bulk of anticancer drugs, very few formulations based on nanomedicine have been authorized for commercial use and other formulations are still under clinical trials [67]. Table 5.4 below provides an overview of the MONPs and graphene-based composites that display anticancer properties and their behavior on different cell lines.

5.5 CONCLUSIONS, OUTLOOK, AND PERSPECTIVES

The green synthesis of MONPs is a viable and environmentally friendly substitute to physical and chemical processes. Biomediated production of NPs is a green chemistry-based approach that links nanotechnology with living organisms and offers a sustainable, resource-effective, and low-cost method for the synthesis of nanomaterials. It has great potential as the techniques used are free from hazardous and toxic by-products unlike conventional methods. The adaptability of NPs containing metal oxides presents the option of using these biocompatible, biodegradable, and safe materials in a number of uses. MONPs/MONCs are at

TABLE 5.4

Anticancer Performance of Nanomaterials with Respect to Different Human Breast and Lung Cancer Cell Lines

Sample	Method	Cancer cell lines	IC_{50} (µg/mL)	Ref.
CuO	Green route (*M. oleifera*)	MCF-7	26.7	[6]
CuO	Green route (*A. indica*)	MCF-7	25.5	[6]
MgO	Green route (*A. paniculata*)	MCF-7	15.7	[23]
NiO	Green route (*A. paniculata*)	MCF-7	44.91	[15]
ZnO	Green synthesis (*Dalbergia latifolia*)	A549	29.0	[68]
TiO$_2$	Green synthesis (*Ledebouria revoluta*)	A549	53.6	[21]
TiO$_2$	Microwave-assisted green route (*A. paniculata*)	A549	24.9	[22]
CuO	Green route (*A. paniculata*) – Microwave-assisted	A549	14.76	[10]
CuO	Green synthesis (*Ficus religiosa leaf extract*)	A549	200	[69]
Ag	Green synthesis (*Achillea biebersteinii*)	MCF-7	20	[70]
Ag	Green synthesis (*Alternanthera tenella*)	MCF-7	42.5	[71]
Ag	Green synthesis (*Cymodocea serrulata*)	A549	100	[72]
Au	Green synthesis (*Antigonon leptopus Hook. and Arn.*)	MCF-7	257.8	[73]
CdO-NiO	Microwave-assisted	MCF-7	9.98	[54]
CdO-MgO			30	[60]
CdO-MgO		A549	15.6	[60]

the focus of research and product development, as revealed by diverse types of research carried out on MONPs/MONCs. As an outcome, it can be expected that bioengineered composites of MONPs will soon be used extensively and widely in the medical, oil, electronics, gas, biotechnology, and food industries. Green-synthesized MONCs are revealed to be realistic varieties for synthetic materials in different operational applications in different industries for the imminent replacement of traditional materials and technologies. As MONPs as well as MONCs continually attract the attention of researchers around the world, the path toward advanced technology development and practical applications can rapidly expand in a way that encourages the change from laboratory to the industrial scale of novel green-synthesized MONCs.

REFERENCES

1. Artini, C, Locardi, F, Pani, M, Nelli, I, Caglieris, F, Masini, R, Plaisier, JR, and Costa, GA. 2017. "Yb-doped Gd$_2$, O$_2$, CO$_3$: Structure, microstructure, thermal and magnetic behaviour". *J. Phys. Chem. Solids* 103: 59–66. Doi: 10.1016/j.jpcs.2016.11.028.

2. Radhika, D, Karthik, K, Nesaraj, AS, and Namitha, R. 2019. "Facile low-temperature synthesis and application of LaO.85SrO.15CoO.85FeO.1503-? as superior cathodes for LT-SOFCs using C-TAB as surfactant". *Mater. Res. Innov* Doi: 10.1080/14328917.2019.1686858.
3. Liu, Y, Fan, L, and Cai, Y, 2017. "Superionic conductivity of $Sm^{3+,}$ Pr^{3+} and Nd^{3+} triple-doped ceria through bulk and surface two-step doping approach". *ACS Appl. Mater. Interfaces* 928: 23614–23623. Doi: 10.1021/acsami.7b02224.
4. Sa-nguanprang, S, Phuruangrat, A, Karthik, K, Thongtem, S, and Thongtem, T. 2020. "Tartaric acid-assisted precipitation of visible light driven Ce-doped ZnO nanoparticles used for photodegradation of methylene blue". *JAUST Ceram. Soc.* 56: 1029–1041. Doi: 10.1007/s41779-019-00447-y.
5. Rangayasami, A, Kannan, K, Joshi, S, and Subban, M. 2020. "Bioengineered silver nanoparticles using *Elytraria acaulis* (L.f.) Lindau leaf extract and its biological applications." *Biocatal. Agric. Biotech* 27: 101690. Doi: 10.1016/j.bcab.2020.101690.
6. Dilaveez Rehanaa, D, Mahendiran, R, Kumar, S, and Rahimana, AK. 2017. "Evaluation of antioxidant and anticancer activity of copper oxide nanoparticles synthesized using medicinally important plant extracts." *Biomed. Pharmacother* 89: 1067. Doi: 10.1016/j.biopha.2017.02.101.
7. Reddy, KR. 2017. "Green synthesis, morphological and optical studies of CuO nanoparticles". *J. Mol. Struct.* 1150 (15): 553–557. Doi: 10.1016/j.molstruc.2017.09.005.
8. Sorbiun, M, Mehr, ES, Ramazani, A, and Fardood, ST. 2018. "Green synthesis of zinc oxide and copper oxide nanoparticle using aqueous extract of oak fruit hull (Jaft) and comparing their photocatalytic degradation of basic violet 3". *Intl. J. Enviorn. Res.* 12: 29–37. Doi: 10.1007/s41742-018-0064-4.
9. Fuku, X, Thovhogi, N, and Maaza, M. 2018. "Photocatalytic effect of green synthesized CuO nanoparticles on selected environmental pollutants and pathogens". *AIP Conf. Proc.* 1962: 040006. Doi: 10.1063/1.5035544.
10. Kannan, K, Radhika, D, Vijayalakshmi, S, Sadasivuni, KK Ojiaku, AA, and Verma, U. 2020. "Facile fabrication of CuO nanoparticles via microwave-assisted method: Photocatalytic and antimicrobial and anticancer enhancing performance". *Int. J. Environ. Anal. Chem.* Doi: 10.1080/03067319.2020.1733543.
11. Ezhilarasi, A, Vijaya, JJ, Kaviyarasu, K, Kennedy, LJ, Jothiramalingam, R, and Al-Lohedan, HA 2018. "Green synthesis of NiO nanoparticles using *Aegle marmelos* leaf extract for the evaluation of in-vitro cytotoxicity, antibacterial and photocatalytic properties". *J Photochem. Photobiol. B* 180: 39–50. Doi: 10.1016/j.jphotobiol.2018.01.023.
12. Thema, FT, Manikandan, E, Gurib Fakim, A, and Maaza, M. 2016. "Single phase Bunsenite NiO nanoparticles green synthesis by *Aganthosma betulina* natural extract". *J. Alloy Compd.* 657: 655–661. Doi: 10.1016/j.jallcom.2015.09.227.
13. Khalil, AT, Ovais, M, Ullah, I, Ali, M, Shinwari, ZK, Hassan, D, and Maaza, M. 2018. "*Sageretia thea* (Osneck.) modulated biosynthesis of NiO nanoparticles and their in vitro pharmacognostic, antioxidant and cytotoxic potential". *Artif. Cells Nanomed Biotechnol.* 46: 838–852. Doi: 10.1080/21691401.2017.1345928.
14. Radhika, D, Kannan, K, Nesaraj, AS, Revathi, V, Sadasivuni, KK. 2020. "A simple chemical precipitation of ceria based (Sm dopedCGO) nanocomposite: Structural and electrolytic behaviour for LTSOFCs". *SN Appl. Sci.* 2: 1220. Doi: 10.1007/s42452-020-3035-2.
15. Karthik, K, Shashank, M, Revathi, V, and Tatarchuk, T. 2018. "Facile microwave assisted green synthesis of NiO nanoparticles from *Andrographis paniculata* leaf extract and evaluation of their photocatalytic and anticancer activities". *Mol. Cryst. Liq. Cryst.* 673 (1): 70–80. Doi: 10.1080/15421406.2019.1578495.
16. Kannan, K, Radhika, D, Nikolova, MP, Sadasivuni, KK, Mahdizadeh, H, and Verma, U. 2020. "Structural studies of bio-mediated NiO nanoparticles for photocatalytic and antibacterial activities". *Inorg. Chem. Commun.* 113: 10755. Doi: 10.1016/j.inoche.2019.107755.

17. Rao, KG, Ashok, CH, Rao, KV, Chakra, CS, and Tambur, P. 2015. "Green synthesis of TiO₂ nanoparticles using *Aloe vera* extract". *Asian Pac. J. Trop. Med.* 2: 28–34.
18. Jalill, RDA, Nuaman, RS, and Abd, AN. 2016. "Green synthesis of titanium dioxide nanoparticles with volatile oil of *Eugenia caryopyllata* for enhanced antimicrobial activities. *World Sci. News* 49: 204–222. Doi: 10.1049/iet-nbt.2017.0139.
19. Chatterjee, M, Ajantha, A, Talekar, N, Revathy, and Abraham, J. 2017. "Biosynthesis, antimicrobial and cytotoxic effects of titanium dioxide nanoparticles using *Vigna unguiculata* seeds. *Mater. Lett.* 9: 95–99. Doi: 10.25258/ijpapr.v9i1.8048.
20. Patidar, V, and Jain, P. 2017. "Green synthesis of TiO₂ nanoparticle using *Moringa Oleifera* leaf extract". *Int. Res. J. Eng. Technol.* 4: 470–473.
21. Aswini, R, Kannan, K, and Murugesan, S. 2020. Bio-engineered TiO₂ nanoparticles using *Ledebouria revoluta* extract: Larvicidal, histopathological, antibacterial and anti-cancer activity. *Int. J. Environ. Anal. Chem.* Doi: 10.1080/03067319.2020.1718668.
22. Karthik, K, Vijayalakshmi, S, Phuruangrat, A, Revathi, V, and Verma, U. 2019. "Multifunctional applications of microwave-assisted biogenic TiO₂ nanoparticles". *J. Clust. Sci.* 30: 965–972. Doi: 10.1007/s10876-019-01556-1.
23. Karthik, K, Dhanuskodi, S, Prabukumar, S, Gobinath, C, and Sivaramakrishnan, S. 2017. "Microwave assisted green synthesis of MgO nanorods and their antibacterial and anti-breast cancer activities". *Mat. Lett.* 206: 217–220. Doi: 10.1016/j.matlet.2017.07.004.
24. Karthik, K, Dhanuskodi, S, Gobinath, C, Prabukumar, S, and Sivaramakrishnan, S. "*Androgrphis paniculata* extract mediated green synthesis of CdO nanoparticles and its electrochemical ad antibacterial studies". *J. Mater. Sci. Mater. Electron.* 28: 7991–8001. Doi: 10.1007/s10854-017-6503-8.
25. Karthik, K, Revathi, V, and Tatarchuk, T. 2018. "Microwave assisted green synthesis of SnO₂ nanoparticles and their optical and photocatalytic properties". *Mol. Cryst. Liq. Cryst.* 67 (1): 17–23. Doi: 10.1080/15421406.2018.1542080.
26. Patil, RS, Kokate, MR, and Kolekar, SS. 2012. "Bio inspired synthesis of highly stabilized silver nanoparticles using *Ocimum tenuiflorum* leaf extract and their antibacterial activity". *Spectrochimic Acta A* 91: 234–238. Doi: 10.1016/j.saa.2012.02.009.
27. Kumar, V, Yadav, SC, and Yadav, SK. 2010. "*Syzygium cumini* leaf and seed extract mediated biosynthesis of silver nanoparticles and their characterization". *J. Chem. Technol. Biotechnol.* 85 (10): 1301–1309. Doi: 10.1002/jctb.2427.
28. Raghunandan, D, Basavaraja, S, Mahesh, B, Balaji, S, Manjunath, SY, and Venkataraman, A. 2009. "Biosynthesis of stable polyshaped gold nanoparticles from microwave0exposed aqueous extracellular anti-malignant guava (*Psidium gujava*) leaf extract". *Nanobiotechnology* 5 (1–4): 34–41. Doi: 10.1007/s12030-009-9030-8.
29. Phillip, D. 2010. "Rapid green synthesis of spherical gold nanoparticles using *Mangifera indica* leaf. *Spectrochimic Acta, Part A* 77 (4): 807–810. Doi: 10.1016/j.saa.2010.08.008.
30. Coccia, F, Tonucci, L, Bosco, D, Bressan, M, and Alessandro, N. 2012. "One pot synthesis of lignin stabilised platinum and palladium nanoparticles and their catalytic behaviour in oxidation and reduction reactions. *Green Chem.* 14 (4): 1073–1078. Doi: 10.1039/C2GC16524D.
31. Petla, RK, Vivekanandhan, S, Misra, M, Mohanty, AK, and Satyanarayana, N. 2012. "Soybean (Glycine max) J leaf extract based green synthesis of palladium nanoparticles". *Biomater. Nanobiotechnol.* 3 (1): 14–19. Doi: 10.4236/jbnb.2012.31003.
32. Fuku, X, Matinise, N, Masikini, M, Kasinathan, K, and Maaza, M. 2018. "An electrochemically active green synthesized polycrystalline NiO/MgO catalyst: Use in photocatalytic applications". *Mater. Res. Bullet.* 97: 457–465. Doi: 10.1016/j.materresbull.2017.09.022.
33. Kulshrestha, S, Qayyum, S, and Khan, AU 2017. "Antibiofilm efficacy of green synthesized graphene oxide-silver nanocomposite using *Lagerstroemia speciosa* floral extract: A comparative study on inhibition of gram-positive and gram-negative biofilms". *Microbial Pathogen* 103: 167–177. DOI: 10.1016/j.micpath.2016.12.022.

34. Nandhakumar, E, Priya, P, Selvakumar, P, Vaishnavi, E, Sasikumar, A, and Senthilkumar, N. 2019. "One step hydrothermal green approach of CuO/Ag nanocomposites: Analysis of structural, biological activities". *Mater Res Express* 6: 095036. Doi: 10.1088/2053-1591/ab2eb9.

35. World Health Statistics. 2011. *Publications of the World Health Organization*. Geneva.

36. Zollinger, H. *Color Chemistry: Syntheses, Properties, and Applications of Organic Dyes and Pigments*. 1991. 2nd rev. ed., VCH, Weinheim.

37. Sathishkumar, S, Parthibavarman, M, Sharmila V, and Karthik, M. 2017. "A facile and one step synthesis of large surface area SnO_2, nanorods and its photocatalytic activity". *J. Mater. Sci.* 28: 8192–8196. Doi: 10.1007/s10854-017-6529-y.

38. Shinde, HM, Bhosale, TT, Gavade, NL, Babar, NB, Kamble, RJ, Shirke, BS, and Garadkar, KM. 2018. "Biosynthesis of ZrO_2 nanoparticles from *Ficus benghalensis* leaf extract for photocatalytic activity". *J. Mater. Sci.* 29: 14055–14604. Doi: 10.1007/s10854-018-9537-7.

39. Karthik, K, Dhanuskodi, S, Gobinath, C, Prabukumar, S, and Sivaramakrishnan, S. 2019. "Fabrication of MgO nanostructures and its efficient photocatalytic, antibacterial and anticancer performance". *J. Photochem. Photobiol. Biol. B* 190: 8–20. Doi: 10.1016/j.jphotobiol.2018.11.001.

40. Lee, S, Mao, W, Flynn, C, and Belcher, C. 2002. "Order of quantum dots using genetically engineered viruses". *A M Sci.* 296 (5569): 892–895. Doi: 10.1126/science.1068054.

41. Karthik, K, Dhanuskodi, S, Gobinath, C, Prabukumar, S, and Sivaramakrishnan, S. 2019. "Ultrasonic-assisted CdO-MgO nanocomposite for multifunctional applications". *Mater. Technol.* 34(7): 403–414. Doi: 10.1080/10667857.2019.1574963.

42. Ali, N, Ahmad, S, Khan, A, Khan, S, Bilal, M, Din, SU Ali, N, Iqbal, HMN and Khan, H. 2020. "Selenide-chitosan as high-performance nanophotocatalyst for accelerated degradation of pollutants". *Int. J. Biol. Macromol.* Doi: 10.1016/j.ijbiomac.2020.07.132.

43. Mohamed, MM, Ghanem, MA, Khairy, M, Naguib, E, Alotaibi, NH. 2019. "Zinc oxide incorporated carbon nanotubes or graphene oxide nanohybrids for enhanced sonophotocatalytic degradation of methylene blue dye". *Appl. Surf. Sci.* 487: 539–549. Doi: 10.1016/j.apsusc.2019.05.135.

44. Liu, WM, Li, J, and Zhang, HY. 2020. "Reduced graphene oxide modified zinc oxide composites synergistic photocatalytic activity under visible light irradiation". *Optik* 207: 163778. Doi: 10.1016/j.ijleo.2019.163778.

45. Priyadharsan, A, Vasanthakumar, V, Karthikeyan, S, Raj, V, Shanavasa, S, and Anbarasan, PM. 2017. "Multi-functional properties of ternary $CeO_2/SnO_2/rGO$ nanocomposites: Visible light driven photocatalyst and heavy metal removal". *J. Photochem. Photobiol. A* 346: 32–45. Doi: 10.1016/j.jphotochem.2017.05.030.

46. Usharani, S, Rajendran, V. 2016. "Optical, magnetic properties and visible light photocatalytic activity of CeO_2/SnO_2 nanocomposites". *Eng. Sci. Technol., Int. J.* 19: 2088–2093. Doi: 10.1016/j.jestch.2016.10.008.

47. Ramanathan, S, Radhika, N, Padmanabhan, D, Durairaj, A, Selvin, SP, Lydia, S, Kavitha, S, and Vasanthkumar, S. 2020. "Eco-friendly synthesis of CRGO and CRGO/SnO_2 nanocomposite for photocatalytic degradation of methylene green dye". *ACS Omega* 5: 158–169. Doi: 10.1021/acsomega.9b02281.

48. Jabeen, U, Shah, SM, and Khan, SU. 2017. "Photo catalytic degradation of Alizarin red S using ZnS and cadmium doped ZnS nanoparticles under unfiltered sunlight". *Surf. Interf.* 6: 40–49. Doi: 10.1016/j.surfin.2016.11.002.

49. Sood, S, Mehta, SK, Umar, A, and Kansal, SK. 2014. "Photocatalytic degradation of the antibiotic levofloxacin using highly crystalline TiO_2 nanoparticles". *New J. Chem.* 38: 3220–3226. DOI: 10.1039/C3NJ01619F.

50. Chwalibog, A, Sawosz, E, Mitura, K, Grodzik, M, Orlowski P, and Sokolowska, A. 2010. "Visualization of interaction between inorganic nanoparticles and bacteria fungi". *Int. J. Nanomed.* 5: 1085–1094. Doi: 10.2147/IJN.S13532.

51. Baquero, F, Coque, MT, and Cruz, F. 2011. "Ecology and evolution as targets: The need for novel eco-evo drugs and strategies to fight antibiotic resistance". *Antimicrob. Agents Ch.* 55: 3649–3660. Doi: 10.1128/AAC.00013-11

52. Jin, A, and He, Y. 2011. "Antibacterial activities of magnesium oxide (MgO) nanoparticles against foodborne pathogens". *J. Nanopart. Res.* 13: 6877–6885.

53. Abdus Subhan, MD, Ahamed, T, Uddin, N, Azad, AK, and Begum, K. "Synthesis and characterization, PL properties, photocatalytic and antibacterial activities of nano-multi metal oxide NiO.CeO$_2$.ZnO". *Spectrochimic. Acta Part A Mol. Biomol. Spectrosc.* 136: 824–831. Doi: 10.1016/j.saa.2014.09.100.

54. Karthik, K, Dhanuskodi, S, Gobinath, C, Prabukumar, S, and Sivaramakrishnan, S. "Nanostructured CdO-NiO composite for multifunctional applications". 2018. *J. Phys. Chem. Solids* 112 (2018) 106–118. Doi: 10.1016/j.jpcs.2017.09.016.

55. Rajaboopathi, S, and Thambidurai, S. 2017. "Green synthesis of seaweed surfactant based CdO-ZnO nanoparticles for better thermal and photocatalytic activity". *Curr. Appl. Phys.* 17: 1622–1638. Doi: 10.1016/j.cap.2017.09.006.

56. Umaralikhan, L, and Jamal Mohamad Jaffar, M. 2016. "Green synthesis of MgO nanoparticles and its antibacterial activity". *Iran. J. Sci. Technol. Trans. Sci.* 1: 1–9. Doi: Doi: 10.1007/s40995-016-0041-8.

57. Devaraj, B, Seerangaraj, V, Bhuvaneshwari, V. 2018. "Green synthesis of silver nanoparticles using *Cordia dichotoma* fruit extract its enhanced antibacterial, anti-biofilm and photocatatalytic activity". *Mater. Res. Express* 5: 055404.

58. Karthik, K, Dhanuskodi, S, Gobinath, C, and Sivaramakrishnan, S. 2015. "Microwave-assisted synthesis of CdO–ZnO nanocomposite and its antibacterial activity against human pathogens." *Spectrochimic. Acta A Mol. Biomol. Spectrosc.* 139: 7–12. Doi: 10.1016/j.saa.2014.11.079.

59. Kannan, K, Radhika, D, Nikolova, MP, Andal, V, Sadasivuni, KK, and Krishna, LS. 2020. "Facile microwave-assisted synthesis of metal oxide CdO-CuO nanocomposite: Photocatalytic and antimicrobial enhancing properties". *Optik* 218: 165112. Doi: 10.1016/j.ijleo.2020.165112.

60. Kannan, K, Sivasubramanian, D, Seetharaman, P, and Sivaperumal, S. 2020. "Structural and biological properties with enhanced photocatalytic behaviour of CdO-MgO nanocomposite by microwave-assisted method". *Optik* 204: 164221. Doi: 10.1016/j.ijleo.2020.164221.

61. Karthik, K, Dhanuskodi, S, Gobinath, C, Prabukumar, S, and Sivaramakrishnan, S. 2018. "Multifunctional properties of microwave assisted CdO-NiO-ZnO nanocomposite: Enhanced photocatalytic and antibacterial activities". *J. Mater. Sci. Mater. Electron.* 29: 5459–5471. DOI: 10.1007/s10854-017-8513-y

62. Şenel, B, Demir, N, Büyükköroğlu, G, and Yıldız, M. 2019. "Graphene quantum dots: Synthesis, characterization, cell viability, genotoxicity for biomedical applications". *Saudi Pharm J* 27 (6): 846–858. Doi: 10.1016/j.jsps.2019.05.006.

63. Revathi, V, and Karthik, K. 2018. "Microwave-assisted CdO-ZnO-MgO nanocompostie and its photocatalytic and biological studies". *J. Mater. Sci. Mater. Electron.* 29: 18519–18530. Doi: 10.1007/s10854-018-9968-1.

64. Vi, TTT, Kumar, SR, Pang, J-HS, Liu, Y-K, Chen, DW, and Lue, SJ. 2020. "Synergistic antibacterial activity of silver-loaded graphene oxide towards *Staphylococcus Aureus* and *Escherichia Coli*". *Nanomaterials* 10: 366. Doi: 10.3390/nano10020366.

65. Rajan A, Khan, A, Asrar, S, Raza, H, Das, RK, and Sahu, NK. 2019. "Synthesis of ZnO/Fe$_3$O$_4$/rGO nanocomposites and evaluation of antibacterial activities towards *E. coli* and *S. aureus*". *IET Nanobiotechnol.* 13 (7): 682. Doi: 10.1049/iet-nbt.2018.5330.

66. Zimmermann, KC, and Green, DR. 2001. "How cells die: Apoptosis pathways". *J. Allergy and Clin. Immunol.* 108 (4): S99–103. Doi: 10.1067/mai.2001.117819.

67. Surendran P, Lakshmanan A, Sakthy Priya S, Balakrishnan K, Rameshkumar P, Kannan K, Geetha P, Hegde, TA, and Vinitha G. 2020. "Bioinspired fluorescence carbon quantum dots extracted from natural honey: Efficient material for photonic and antibacterial applications". *Nano-Struct. Nano-Obj.* 24: 100589. Doi: 10.1016/j.nanoso.2020.100589.

68. Elavarasana, N, Prakasha, S, Kokilaa, K, Thirunavukkarasu, C, and Sujatha, V. 2020. "The biosynthesis of a grapheme oxide based zinc oxide nanocomposites using *Dalbergia latiolia* leaf extract and its biological applications". *New J Chem.* 44: 2166. Doi: 10.1039/C9NJ04961D.

69. Sankar, R, Maheswaran, R, Karthik, S, Subramanian, and K, Shivashangari, VR. 2014. "Anticancer activity of *Ficus religious* engineered copper oxide nanoparticles". *Mater Sci Eng C* 44: 234–239. Doi: 10.1016/j.msec.2014.08.030.

70. Baharara, J, Namvar, F, Ramezani, T, Mousavi, M, and Mohamad, R. 2015. "Silver nanoparticles biosynthesized using *Achillea biebersteinii* flower extract: Apoptosis induction in MCF-7 cells via caspase activation and regulation of Bax and Bcl-2 gene expression". *Molecules* 20 (2): 2693–2706. Doi: 10.3390/molecules20022693.

71. Sathishkumar, P, Vennila, K, Jayakumar, R, Yusoff, ARM, Hadibarata, and Palvannan, T. 2016. "Phyto-synthesis of silver nanoparticle using Alternanthera tenella leaf extract: An effective inhibitor for the migration of human breast adenocarcinoma (MCF-7) cells. *Bioprocess Biosyst. Eng.* 39 (4): 651–659. Doi: 10.1007/s00449-016-1546-4.

72. Palaniappan, P, Sathishkumar, G, and Sankar, R. 2015. "Fabrication of nano-silver paticles using *Cymodocea serrulata* and its cytotoxicity effect against human lung cancer A549 cell line. *Spectrochim Acta, Part A* 138 (22): 885–890. Doi: 10.1016/j.saa.2014.10.072.

73. Balasubramani, G, Ramkumar, R, Krishnaveni, N, Pazhanimuthu, A, Natarajan, T, Sowmiya, R, and Perumal, P. 2015. "Structural characterization, antioxidant and antioxidant and anticancer properties of gold nanoparticles synthesized from leaf extract (decoction) of *Antigonon leptopus* Hook and Arn. *J. Trace. Elem. Med. Biol.* 30 (24): 83–89. Doi: 10.1016/j.jtemb.2014.11.001.

6 Current Scenario of Nanomaterials in the Environmental, Agricultural, and Biomedical Fields

Charles Oluwaseun Adetunji, Olugbemi T. Olaniyan, and Osikemekha Anthony Anani
Edo State University Uzairue

Frances. N. Olisaka
Benson Idahosa University

Abel Inobeme
Edo State University Uzairue

Ruth Ebunoluwa Bodunrinde
Federal University of Technology Akure

Juliana Bunmi Adetunji
Osun State University

Kshitij RB Singh
Govt. V. Y. T. PG. Autonomous College

Wadzani Dauda Palnam
Federal University

Ravindra Pratap Singh
Indira Gandhi National Tribal University

CONTENTS

6.1 INTRODUCTION

The application of nanotechnology in biomedical sciences and other fields has increased drastically due to elevated research output. Many fields have now positively adopted the utilization of nanotechnologies for alternative energy and environmental sources, physics, biomedicine, electronics, agriculture, and engineering. Moreover, the application of nanotechnology in biomedicine is unprecedented as cellular uptake, and encapsulation of hydrophobic photosensitizers in fluid nanocarriers may increase treatment options. Through nanotechnology, the intracellular milieu environment can be changed, and enzymatic degradation prevention and reduction in cytotoxicity of many toxicants can also be achieved [1].

It has been revealed that nanoparticles possess a unique physicochemical property that enables them to be very useful for many biomedical applications such as attachment to other biomolecules to form nanoparticle–biomolecule conjugates with tremendous capacities in drug delivery, clearance, absorption, and metabolic processes. The adoption of nanotechnology in biomedical sciences has increased therapeutic efficacy, eliminates many adverse effects, lowered wastage plus increasing bioavailability of drugs at various sites over the traditional therapeutics. Again, in biomedical imaging, nanoparticles or nanobased products are utilized for the diagnosis of a disease condition, cancer treatment, chronic diabetic wounds, intracellular mRNA delivery, drug delivery, polymicrobial infection treatment, genetic disorders plus gene therapy in clinical treatment particularly for protein replacement therapies, vaccination, plus genetic disease management [2,3]. Also, numerous studies have been performed by numerous scientists to establish the usefulness of nanomaterials in agriculture most especially as eco-friendly biopesticides, insecticides, and weedicide that could affect an increase in agricultural products [4–7].

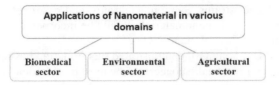

FIGURE 6.1 The various applications of nanomaterial are shown in numerous fields.

Hence, this chapter will provide detailed information on the numerous types and roles of nanomaterials in resolving several challenges encountered in the environmental, agricultural, and biomedical fields. The various applications of nanomaterial are shown in numerous fields as illustrated in Figure 6.1.

6.2 APPLICATION OF NANOMATERIAL IN BIOMEDICAL FIELDS

The nanomaterials have a vast range of applications in the field of biomedical, which is demonstrated in Figure 6.2 and is further explained in the subsections below.

6.2.1 ANTIBACTERIAL, ANTIFUNGAL, AND ANTIVIRAL ACTIVITIES

Lesley et al. [8] revealed that nanotechnology has brought about many innovative approaches to the food and agricultural sector particularly in the aspect of antimicrobial nanoparticles against pathogens in the poultry industry. The authors investigated the use of nanoparticles against *Salmonella spp.* and *Campylobacter spp.* to evaluate the in vitro activity of these nanoparticles in the poultry industry. The results revealed that the nanoparticles are very effective in a concentration-dependent manner against the pathogens. Azam et al. [9] described that nanomaterials as antibacterial agents have been utilized in the food industry against many foodborne bacterial

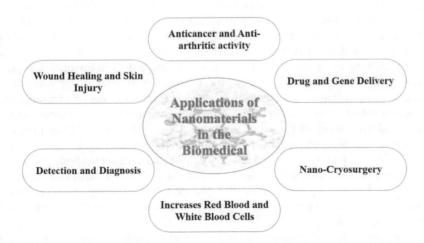

FIGURE 6.2 Schematic illustration of the nanomaterials in the biomedical domain.

such as Campylobacter strains, Salmonella strains, and gastroenteritis. The authors further suggested that these nanomaterials should be incorporated in packaging and crate design in poultry through nanoscience.

Zhen-Xing and Bin-Feng [10] demonstrated that pollution has resulted in serious environmental and human health concerns. The most troublesome of these are bacterial pollution and contamination; thus, suggestions have been given to utilize nanoinorganic metal oxide to help reduce the detrimental effects of bacterial pollutions. The authors revealed that MgO nanoparticles offer promising antibacterial mechanisms in a concentration-dependent manner that could be linked to their great resistance to harsh conditions during processing. They revealed that the proposed mechanisms by which MgO nanoparticles perform their antibacterial effects are through direct damage to the bacterial cell, development of reactive oxygen species, and interface of nanoparticles with bacteria in alkaline effect.

Tsuneo et al. [11] performed an investigation on the potential application of silver nanoparticles on virus surface receptor proteins. It is generally regarded as silver nanoparticles causing anti-inflammatory properties or as antiviral agents in allergic reactions, through blockade of viral attachment, disulfide bonds with a thiol group, binding to gp1 20, viral neuraminidase activations, degradation, and apoptosis. The silver complex formation also occurs in cancer cells destroying DNA base pairs. Benton et al. [12] reported that the fabrication of silver nanoparticles as antibacterial agents has been discovered for ages with potent efficacy against bacteria pathogen involving damage plus inhibition to the elongated peptidoglycan bacteria cell wall or damage of lipoprotein at C-, N-terminals by Ag+ ions through autolysins. Orlowski et al. [13] suggested that silver nanoparticles have been utilized as antimicrobial agents with efficient inhibitory properties against several viruses.

Lei et al. [14] showed that silver nanoparticles have many industrial applications including antiviral properties through direct interaction between extracellular virions and viral RNA with nanoparticles causing inhibitory effects. Hu et al. [15] revealed that viral amino acid enzymes like hydrolase, neuraminidase, hemagglutinin, aspartate, and glutamate are largely responsible for the catalytic action and initiation of virus degradation when activated by nanoparticles. The authors suggested that the proteasome pathway may be involved in the degradation process of the virus by nanoparticles in the host cell. Gaikwad et al. [16] demonstrated that endolysins are enzymes utilized by bacteria after replication to damage the cell membrane peptidoglycan causing the release of progeny virion. This may be a target for nanoparticle mechanisms in their antiviral activity. Mori et al. [17] revealed that silver nanoparticle interaction with the virus causes DNA damage by the induction of reactive oxygen species and silver complex formation resulting in antiviral activities. Liangpeng et al. [18] showed that silver nanoparticles have antibacterial capacity against a broad spectrum of bacterial pathogens, which is size-, shape-, and dose-dependent. The authors also suggested that nanoparticles can be combined with antibiotics to increase their antimicrobial effects by interaction with phosphorus-containing bases or regulate signal transduction mechanism of phosphotyrosine bacterial peptides causing inhibition of cell division.

Liangpeng et al. [18] revealed that the antifungal activity of nanosilver is second to none with a potent effect against *Candida* spp. The proposed mechanism involves the destruction of the cellular membrane, with the inhibition of budding activity. In another experiment done by Azam et al. [9], it was revealed that metallic nanoparticles have received massive attention from different fields due to their broad spectrum of actions and physiochemical properties. They also noted that the antibacterial and antifungal effects have been reported by different authors. Sumon and Tamalika [19] revealed that metallic nanoparticles can be produced through green technology with antibacterial properties generating reactive oxygen species and destruction of genetic materials. Wu et al. [20] demonstrated that the global threat faced with antibiotic-resistant microorganisms is becoming enormous; thus, metallic nanoparticles may be an alternative way to prevent completely the progression and development of multidrug-resistant microorganisms. Ahmed et al. [21] suggested that silver nanoparticles in combination with microbes will lead to binding of the nanoparticles with the cell membrane, membrane sulfur interaction, morphological changes, separation, damage to intracellular structures, and swelling. Humberto et al. [22] revealed that advances in bionanotechnology have brought change to the therapeutic horizons due to their nontoxic effects.

Yasutaka et al. [23] reported that silver nanoparticle composites with chitosan have potent antiviral properties against influenza A virus in a size-dependent manner as compared with the individual molecule. Verónica et al. [24] determined an in vitro antibacterial property of dental interim improved with terpenes cement combined with nanoparticles. From their analysis and results, they discovered that supplementation of the nanoparticles with terpenes increased the antibacterial activity against *S. mutans* ATCC 25175 and elevated the diametral tensile strength. Guilherme et al. [25] discovered that spread of pathogenic microbes over the years has significantly increased across the globe; thus, urgent attention is needed to develop various technological approaches as means of reducing or eliminating the cause. The authors proposed that the application of nanoparticles in the inhibition of the SARS-CoV-2 virus and other bacteria could offer a great deal of opportunity. Naeima and Hanan [26] revealed that silver nanoparticles have received great attention for their antibacterial activity. The authors produced silver nanoparticles from *Datura stramonium* plant to analyze its antibacterial characteristics against different bacterial pathogens and later confirmed its huge beneficial effects as an antibacterial agent.

Muhamed et al. [27] revealed that metal oxide nanoparticles in combination with specific bioactive molecules have been utilized in different fields with beneficial properties, particularly in biomedical sciences. Various combinations have been discovered in the past, but the authors investigated the role of functionalized curcumin aniline generated from turmeric plants combined with manganese nanoparticles produced from green synthesis against bacterial and fungal strains. From the results obtained, it was discovered that the combination displayed a significant increase in antimicrobial activity, thereby suggesting a new intervention approach to treat patients. Nurit et al. [28] briefly reported that serious attention to develop new material as an antibiotic agent has been witnessed over the past few decades due to increased resistance to many synthetic antibiotics; thus, the advancement of nanotechnology has brought significant progress to develop antimicrobial biological agents as bactericidal plus fungicidal to prevent biofilm-linked infections or drug-resistant microbes.

Mogomotsi et al. [29] briefly reported that phytomedicine and nanobiotechnology can be combined for biological purposes in a cost-effective and eco-friendly manner for pharmacological applications. Thus, the authors synthesized biogenic silver nanoparticles and investigated the antimicrobial action against different bacteria species and discovered that the silver nanoparticles displayed a higher level of antimicrobial activity. Clarence and Geoffrey [30] demonstrated that the conventional antimicrobial agents have displayed weak response to many antimicrobial-resistant strains of bacteria resulting in global concern and hence the search for alternatives such as nanoparticles with vast spectrum antimicrobial effects with tremendous physiochemical plus functionalization characteristics. Lavanya et al. [31] reported that there is a serious need to develop a potent treatment for infectious diseases due to the adverse side effects caused by prolonged use of some antibiotics and fast drug resistance development. They revealed that the relationship between microorganisms and nanoparticles has reshaped the field of biomedical sciences for therapeutic and diagnostic purposes.

Farouk et al. [32] showed that nanotechnology has created an alternative for the treatment of pathogenic fungal, viral, and bacterial as they offer prolonged treatment with minimal side effects. Zarrindokht and Pegah [33] demonstrated that ZnO nanoparticles possess antibacterial properties against many pathogenic organisms. Thus, they evaluated the bacteriological effects on *S. aureus* and *E. coli*. They discovered that the antibacterial activity of ZnO nanoparticles improved with a reduction in the size of the particles. Azam et al. [9] showed that metallic nanoparticles have wide applications in chemistry, medicine, biotechnology plus agriculture. Hence, the authors synthesized silver and copper nanoparticles with environmentally friendly materials of different plant species and evaluated the antibacterial and antifungal effects. From the results they obtained, it was observed that silver nanoparticles yielded notable positive effects. Chiriac et al. [34] demonstrated that infectious diseases caused by pathogenic organisms have resulted in a serious global burden due to the development of resistance to many synthetic drugs currently used in their management. The authors then carried out the test on different nanoparticles to establish a potent antibacterial-based nanoparticle and thus discovered that zinc-based nanoparticles have the strongest antibiotic effects and ability to cause membrane and protein damage with oxidative stress.

Carin et al. [35] reported that the antimicrobial properties of silver nanomaterials combined with other materials. The authors suggested that excellent ionic movement characteristics of the materials permit the faster release of ion plus cell damage. Bankier et al. [36] demonstrated that metallic nanoparticles have potent antimicrobial characteristics making them suitable for pharmaceutical uses. Thus, the authors evaluated the combined effects compared with single use and discovered that the combination resulted in more significant antimicrobial effects. Patcharaporn et al. [37] showed that nanotechnology possesses good antibacterial activity in vertebrates; hence, they synthesized a nanoparticle from eco-friendly *Aloe vera* leaf extract with antibacterial effect. Jo et al. [38] showed that very few studies have applied nanotechnology in controlling plant diseases; hence, the authors tested the role of silver ion nanoparticles as an antifungal against *Magnaporthe grisea* and *Bipolaris sorokiniana*. Their findings indicated that the silver nanoparticles displayed a strong antifungal effect against the two plant-pathogenic fungi by promoting the direct effect of silver with germ tubes and spores, plus inhibiting their viability.

Jana and Seth [39] revealed the application of nanoparticles in modern-day lives and thus reviewed organic- and inorganic-based nanomaterials as antimicrobial agents. They went further to and elaborated on the roles of nanomaterials-based superhydrophobic coatings in biomedical sciences particularly helping to tackle the present COVID-19 disease. Hassan [40] described how devastating the nosocomial pathogenic infection can be particularly for immunocompromised patients and the development of multidrug-resistant bacteria plus the growing concern of biofilm-linked infections has led to the search for novel treatment. Recently, the utilization of nanotechnology in clinical practice has led to significant improvement particularly in the management of infectious microbial diseases; thus, the authors reviewed the extent of multidrug resistance in nosocomial infections, mechanisms of invasion, drug resistance, and the effects of nanomaterials as antimicrobial agents. Atiqah et al. [41] described how the development of nanomaterials has been on the increase due to their wide industrial applications. The authors evaluated the antiviral and antibacterial mechanisms of silver nanoparticles and added a note on the possible toxicity of their biological applications. Tiwari et al. [42] reported that multidrug-resistant *Acinetobacter baumannii* against Carbapenems has become a significant issue that might be linked to the increased number of mortalities recorded with its application. Thus, the authors search for alternative ways to develop nanoparticles as antibacterial agents against carbapenem-resistant *A. baumannii* and the possible mechanisms of action. They proposed that the mechanisms of action could be increased membrane lipid peroxidation, elevated reactive oxygen species, DNA, and protein damage plus reduction in cell viability as zinc oxide nanoparticles showed a strong antibacterial effect against carbapenem-resistant *A. baumannii*.

Ayeshamariam et al. [43] revealed the antifungal and antibacterial of tin oxide nanoparticles with Aloe vera extract against several pathogenic bacterial and discovered that tin oxide nanoparticles with Aloe vera extract displayed a strong effect on the tested microorganisms. Hamsa et al. [44] demonstrated that silver nanoparticles utilizing *Cinnamomum zylinicum* bark extracts are a great antimicrobial agent against different pathogenic bacteria and hence recommended that they could be utilized as antibiotics against multidrug-resistant bacteria.

6.2.2 ANTI-ULCER ACTIVITY

Sreelakshmy et al. [45] briefly described that silver nanoparticles utilizing *Glycyrrhiza glabra* could be of great benefit for treating gastric ulcer disease. The authors revealed that the plant has not been documented for any in vitro anti-ulcer properties particularly for *H. pylori*; thus, they investigated silver nanoparticles obtained from *Glycyrrhiza glabra* root extract for in vitro anti-ulcer activity. They discovered that the nanoparticles showed potent cytoprotective anti-ulcer effects against *H. pylori*. Rangabhatla et al. [46] revealed the ulcer protective activity of cerium oxide nanoparticles in an animal model during their experiment on ethanol-induced gastric ulcers. They discovered that these nanoparticles were able to display strong ulcer inhibition as compared with ranitidine. The observed mechanisms were mopping of reactive oxygen species, increasing antioxidant defense system.

Fatma et al. [47] revealed that amoxicillin and doxycycline broadband antibiotics were utilized as nanocomposite in treating animal and human model experiment disease. Ethanol was utilized for ulcer induction and evaluation of wound closure rate, acute toxicity, healing percentages, protective rate, histopathological plus ulcer index were done. The findings revealed that the nanocomposites showed significant ulcer prevention and were approved for biomedical and pharmaceutical applications. Paul et al. [48] briefly reported that Uapaca staudtii Pax has strong anti-ulcer and antioxidant activities in reserpine- and indomethacin-induced ulcer models. The findings revealed that the extracts possess highly significant antioxidant and anti-ulcer properties as compared to the standard drugs and could be utilized in nanoparticle development in managing ulcer disease. Prakash et al. [49] showed that eroding the epithelial lining of the stomach, small intestine, and duodenum may lead to ulcer formation. In their study, it was revealed that many pathogenic bacteria may induce ulcer due to the release of toxins, while several drugs have been utilized in the past with little disadvantage like side effects, drug interactions, and relapses; thus, novel nanoparticles based-anti-ulcer drugs are now being suggested.

Anil et al. [50] revealed the potent advantage of cerium oxide nanoparticles as anti-ulcer, antisecretory plus antioxidant agent. In an experiment conducted by the authors, it was revealed that pyloric ligation with aspirin-induced gastric ulcer was treated with nanoceria and cerium oxide nanoparticles. It was observed that significant effects were displayed by cerium oxide nanoparticles by lowering the gastric volume, acidity, pH, plus ulcer index due to their vast surface-to-volume ratio plus physicochemical characteristics.

6.2.3 ANTIALLERGIC ACTIVITY

Sozer and Kokini [51] revealed that silver nanoparticles have attracted wide acceptance in biomedical applications particularly in treatment, diagnosis, drug delivery, and medical devices. The authors showed the physiological and physicochemical properties of silver nanoparticles in antifungal, antibacterial, antiviral, anti-inflammatory, and antiallergic reactions. The authors showed that silver nanoparticles significantly reduced pro-inflammatory cytokines, goblet cell hyperplasia, tumor necrosis factor-alpha, immunoglobulin E, IL-4, interleukin-10 plus transforming growth factor-beta expressions in dermatitis allergic reaction plus allergic rhinitis.

Saadatzadeh et al. [52] revealed that food allergenicity can be an adverse immune response to certain proteins or epitopes with severe response or effects on different organs. Many approaches have been applied to mitigate against food allergies, and the authors evaluated the adoption of novel technology such as nanotechnology as an alternative measure. Zimet et al. [53] demonstrated that soy allergens have increased drastically in Western countries as a result of increased consumption of soy proteins; hence, many strategies are put in place to reduce the allergies or immunoreactivity produced by the consumption of soy products. Nanotechnology has been identified as one of the best options for reducing the effects of food allergies. Patrignani et al. [54] revealed that higher allergenicity is produced from the milk of ruminants; hence, efforts have been put in place to reduce the allergenicity with the adoption of nano-technology. Lanciotti et al. [55] demonstrated that allergies from fruits and vegetables have been on the increase; thus, many approaches to reduce food allergy have been adopted in many Western countries. One of the fast-growing fields of science with the capacities to reduce the allergenicity of food, fruit, and vegetables is the utilization of nanotechnology such as the combination of pectinase with metallic ion nanoparticles to reduce allergenicity.

6.2.4 EFFECTS ON CENTRAL AND PERIPHERAL NERVOUS SYSTEMS

Cristina et al. [56] reported that the nervous system is a neural network connect-ing the spinal cord, brain, and other systems; thus, nanoparticles can be taken or absorbed from the diverse route in the body such as dermal, olfactory mucosa, neuro-nal, or nerve pathways. These nanoparticles may find their way into the blood–brain barrier where they may interact with many sensory nerve endings and mitochondria and then into other deeper brain structures through the blood–brain barrier. The authors have revealed that the blood–brain barrier may preferentially select the entry of cationic molecules due to its anionic nature. This route may be utilized for drug delivery into the brain and other nervous tissues; hence, the passage of nanoparticles may be influenced by the specific charge on it.

Studies have revealed that many neurodegenerative diseases are a result of the accumulation of metallic compounds and neuronal uptake of nanoparticles causing functional damage, myelin sheath breakdown due to oxidative stress, and toxicity. In the treatment protocol for managing these conditions, metal chelators and antioxi-dants may be beneficial as they can permeate through the blood–brain barrier system and initiate the therapeutic process.

6.3 APPLICATIONS OF NANOMATERIALS FOR THE TREATMENT OF THE HEAVILY POLLUTED ENVIRONMENT

The nanomaterials have a variety of applications in the environmental domain, which is highlighted in Figure 6.3, and the role of nanomaterials for the treatment of pol-lutants found in the environment such as pesticides and polyaromatic hydrocarbon is demonstrated in the below subsections.

6.3.1 PESTICIDES

In the past 50 decades, there have been indications of persistent demand for food, which has led to the sharp rise in the pressure on agricultural resources globally, as a consequence of increased human population growth. In an attempt to manage this growth and sustain the pressures placed on agricultural resources, many producers have engaged in the use of pesticides to control the invasion of pests and pathogenic agents into agricultural produce. Also, postharvest storage of crops and other agri-cultural products has utilized the adoption of pesticides and synthetic chemicals to prolong the storage rate and to prevent the invasion of pest and pathogenic organ-isms. Many studies have revealed that the adoption and utilization of these chemicals

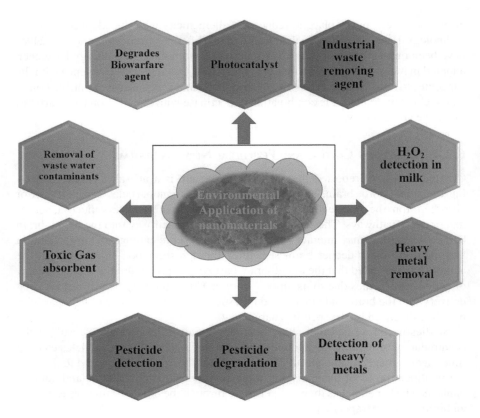

FIGURE 6.3   Applications of nanomaterials in the environmental field.

and pesticides have led to increased environmental pollution and resulted in adverse effects on human health. Thus, in an attempt to reduce the utilization of synthetic chemicals like pesticides, many novel technologies are attracting serious attention among scientists across the globe. One of these technologies suggested involves the development and adoption of nanopesticides and nanofertilizer as alternatives to synthetic chemicals with toxic chemicals, which have adversely caused serious environmental concern. Khan and Rizvi [57], in a review, looked at the utilization of nanopesticides and nanofertilizer in the improvements of crop protection, production plus ensuring environmental protection. The authors recounted that many nations have employed several measures like the green revolution in the provision of food security to meet the growing population and reduce environmental pollution. The utilization of novel technologies such as nanotechnology in the improvement and conservation of farm resources is needed at this present time to supplement the already existing technology. This technology has such great potential that can change the narrative of agriculture and associated fields in the aspect of environmental health, pest protection, and crop productivity. The authors concluded by stating that the use of nanotechnology in modern agriculture will boost food security in an emerging, challenging, and uncontrolled population setting. They also noted that nanotechnology would improve agricultural sustainability, productivity, pest protection, and delivery devices for effective improvement in crop production and the suppression of disease or pest invasions. The authors also reported that the issues of indiscriminate use and misuse of nanomaterials or utilization of pesticides and fertilizers can result to rebound toxicity in the agricultural and environmental systems. To curtail this, they recommended more research in the area of biosafety of nanoproducts for agriculture use and environmental sustainability.

Iavicolia et al. [58] did a review of occupational and toxicological risks in the utilization of pesticides in farming activities. They stated that this approach has the latency to make a sustainable influence on several ecological problems in forestry and agriculture like justifiable management of the agricultural resource, energy restraints, and urbanization. Nonetheless, they reported that human and environmental hazards may sprout up from the utilization of synthetic pesticides in an attempt to preserve and enhance agricultural productivity and thus suggested the utilization and adoption of nanopesticides and nanofertilizers as alternatives. The authors later proposed risk management and ecological assessment as well as biosafety and life cycle analysis to understudy the possible toxicity fate in the application of nanochemicals (nanopesticides and nanofertilizers) for agricultural purposes. Besides, the interaction of ecological stressors and coformulants should also be put into considerations when assessing the fate of the nanochemicals.

For the past several decades, residues from pesticides and pollutants in the environment have contributed immensely to serious health and environmental risks witnessed today. Several attempts had been made to mitigate this negative impact of pesticides on food, the environment, and human health; however, little progress has been seen particularly in developing countries. Firozjaee et al. [59] did a review on the decontamination of pesticides from the aqueous mixture using effective nanotechnology. This technology aids in the final mineralization of residues in pesticides to a less or non-noxious form. This technology offers countless hidden benefits such as uplifting or enhancing the already existing technologies (degradation, filtration, and adsorption) by employing novel nanomaterials for efficient consumption of materials plus energy. They proposed the application of nanotechnology in the decontamination of pesticides on farmland. The multitude-of-parameters method was also suggested for the enhancement of nanomaterials in aqueous solution for the deletion of residues of pesticides. This method would aid the process involved in the intensification of the surface area of nanomaterials in the solution and thus incorporates its reactivity. The authors suggested the application of nanotechnology in the elimination of pesticides in related media like wastewater and water purification to reduce the health and environmental risks linked to the media.

Urkude [60] did a review on the management of pests by the application of nanomaterials in pesticides. The author reported that the indiscriminate use of pesticides in the control of pests in agriculture has resulted in severe health and ecological problems, even though its application is beneficial in securing farm products. They stressed that nanotechnology is one of the efficient tools in resolving issues related to the pollution of soil in agriculture. The use of nanopesticides for the management and control of farm pests has yielded proper efficiency, decreased pollution, and spurred the farmers to meet up with some uncertain tasks that could not be achieved before like enhanced farmer's efficiencies. It has also improved the delivery system of farm pesticides to solve this problem of bioefficiency. They also reported that some nanopesticides formulation has shown great potential in eliminating and controlling pest rebounds as well as detectable pesticide residues in the soil. They suggested nanopesticides as an ecological footprint in the management of pests' rebound and heavily polluted environment.

Elizabeth et al. [61] in a review examined the management of flora pests using nanotechnology. The authors reported that about 20%–40% of agricultural products are destroyed by plant pathogens and pests yearly. The containment of these pests depends on existing technology like chemical usage such as synthetic pesticides, which are toxic to the plants, crops, animals, humans, and the environment. Thus, nanotechnology is believed to offer a better alternative and sustainable means of decreasing the toxicity level of these chemicals, enhance the shelf-life as well as the solubility potentials of the pesticides in the medium used, thus influencing the environment positively. The authors reported that nanoparticles can act in two ways, as a protectant and a delivery system for RNA molecules, herbicides, fungicides, and insecticides. However, there lie the benefits linked with the utilization of nanomaterials for the application and commercialization of agriculture in both field trials and farm settings in the management of pests and improvement in crop production. The authors concluded by recommending that biologists, agriculturists, environmentalists, physiologists, and other relevant stakeholders should synergize to bring expertise in resolving the inefficiency pervading the incidences of pest rebound and crop nonproductivity.

Sahithya and Nilanjana [62], in a review, examined the application of nanomaterials in the redress of pesticides in wastewater. They stressed that the rigorous application of pesticides for agricultural purposes has elicited severe ecological and health issues. Different artificial chemicals have been utilized for decades in the cleanup of pesticides from the water environment. However, it has been recognized that bionanoparticles have proven to be effective in the cleanup of pesticides, through the use of photocatalysts or adsorbent techniques. They stated that this method is cheap, less toxic, eco-friendly, and sustainable. They suggested a similar method in cleaning up polymer materials by employing biopolymers with nanomaterials on a large scale.

6.3.2 POLYAROMATIC HYDROCARBON

Hassan et al. [63] tested the use of iron oxide nanoparticles (IONPs) in the elimination of micropollutants (benzo(a)pyrene and pyrene) from wastewater through the process of adsorption. The authors stated that benzo(a)pyrene and pyrene congeners of polycyclic aromatic hydrocarbons (PAHs) even at reduced concentrations are highly carcinogenic. Several environmental factors like PAHs concentration, temperature, pH as IONPs concentration, affecting the potential remediation via adsorption of PAHs, were examined and evaluated. The outcome of the study showed that the highest capacity of IONPs against benzo(a)pyrene and pyrene was 0.029 and 2.8 mg/g, respectively. The outcome of the thermodynamic reaction revealed an adsorption exothermic process of benzo(a)pyrene and pyrene. The isotherms and kinetic investigations showed that the process of adsorption followed pseudo-2° actions and followed the Langmuir isotherm model. More so, the IONPs demonstrated to be an effective contender for the adsorption of benzo(a)pyrene and pyrene after five generation cycles at the degree of 99% and 98.5%, respectively.

Yang et al. [64] evaluated the adsorption of PAHs like pyrene, phenanthrene, and naphthalene and six chains (fullerenes) using carbon-based nanoparticles (CNs). The authors stated that CNs are newly produced particles that have great potential in the control of hydrophobic organic compounds (HOCs) through adsorption and also regulate their transformation as well as their fate in the ecosystem. It was observed that the sorption process was isothermal and not linear. It was well fitted with the Polanyi–Manes model (PMM), producing a characteristic curve that depends on the nature and properties of the CNs. It was observed that the sorption process correlated with the volume and surface area ratios of the micropore to the mesopore. However, the increased sorption capacity of the polyaromatic hydrocarbons by the CNs might lead to possible environmental hazards when released. This action can also result in a sudden change in the bioavailability and the fate of the PAHs in the ecosystem.

PAHs are environmental contaminants that have high bioaccumulation potentials, nonbiodegradable, highly noxious, and ubiquitous. For the past 50 decades, there have been several techniques used in the control of PAHs in the environment. However, a more reliable and environmentally friendly technique has been proposed by many scientists. In line with this trend, Rani and Shanker [65], in a review, evaluated the remediation potentials of nanomaterials on PAHs. The authors hinged on rapid industrialization and population burst as the chief precursors of the release of PAHs into the environment. The use of nanoparticles to control PAH's influence on the environment is one of the novel technologies invoked. The application of nanotechnology uses the principles of redox breakdown, photocatalytic, and adsorption in the removal of PAHs in the environment. Among all these techniques, the adsorption method was singled out as one of the best because it is efficient and equally cheap to use. The authors also suggested the utilization of native methods like photolysis and microbial degradation as well as some conventional technique like activated carbon and biological wastes from agriculture. They stated some of the benefits of using nanomaterials as an indelible tool for PAHs removals like their low cost, green technology, high adsorbent quality under sunlight and UV irradiations, fast degradation, and efficient biogenic properties. The authors also recommended the use of hexacyanoferrate metal, ZnO, and TiO_2 for the removal of PAHs from the environment.

Mahgoub [66] did a review of the utilization of PAH extraction from the environment using nanoparticles. The authors listed some commonly used nanoparticles (magnetized and magnetic NPs, metal oxides, metals, and mesoporous silica) in the sorption of PAHs in environmental media. These nanomaterials are applied in many sectors like in a space ship as fuel additives, optics, mechanics, feed and food, biomedical, pharmaceutical as a drug delivery tool, the environment, and health industries. This is because NPs have high sorption and surface area capacities. The successful application of NPs in the decontamination of PAHs on different matrices was highlighted and discussed. Some problems of the adsorbent benefits of NPs of PAHs like their fate and disposal were pointed out. However, some biological, green, and nonhazardous techniques were recommended by the authors that will improve the affinity and removal efficiency of PAHs. This will avert protracted exposure of nanomaterial on the environment and humans. The incidences of health risks on the brain, lungs, and skin will be eliminated by the introduction of these eco-friendly NPs.

6.3.3 HEAVILY POLLUTED SOIL WITH HEAVY METALS

Mohamed et al. [67] tested and evaluated the application of NPs in the elimination of heavy metals in the soils. The authors stated that nanotechnology is a novel technology that offers a variety of benefits like highly effective immobilization of pollutants converting them to less toxic forms as it is a cheap technology. Nanocarbon (NanoC), nanoalginite, nanoZVI bentonite, nanoZVI, and nanovalentFe are potential nanomaterials that can be applied for the potential immobilization and sorption of heavy metals like Pb and Cd in contaminated soils. These nanomaterials are produced ex situ using bottom-up and top-down techniques. Thereafter, the transmission electron microscope (TEM) was used to characterize them. The results from the nanocharacterization showed that the cation capacity exchange, surface area, and particle sizes were observed to be 42.5–47.7 cmolc/kg, 194.2–259.7 m^2/g, and < 70 nm correspondingly. The capacity of adsorption for Cd (17,850–25,970 mg/kg) and Pb (37,450–93,450 mg/kg) was observed to be high as well as the adsorbate retention potential. Also, a small percentage of Cd (10.8%–33.4%) and Pb (13.7%–35.6%) were desorbed, respectively, exempting that of the nanoC. The examined nanoparticles showed high efficacy in the immobilization of Pb and Cd in contaminated soils. The plant treated with nanomaterials via diethylenetriaminepentaacetic acid (DTPA) extraction of Cd and Pb from it showed a reduction in Pb and Cd concentration at 2.88 and 0.06 mg/kg, respectively. The findings from the study showed that nanoC is an effective pollutant cleanser, and it is recommended for soil reclamation of heavily polluted metals.

The influence of toxic metals in the aquatic environment has been long-fingered for their negative impacts on the resources therein. However, there is a clarion call to address this environmental menace. On this note, Yang et al. [68] did a review of the elimination of toxic metals from sewer water utilizing nanoparticles. The authors stated that the removal of metal contaminants in wastewater is geared toward sustainable management of the ecosystem. In recent times, various types of technologies have been employed in the elimination of heavy metal from effluents. However, they have been spotted to be ineffective and also capable of generating additional wastes during the process. A novel technology that involves the use of nanomaterials that have been proven to be green, sustainable, nontoxic, eco-friendly, and cheap can be used to substitute these traditional methods. This new technology can be used to remove heavy metal from polluted water because of the nanometer effects it possesses. Examples of these nanomaterials are nanocomposites, metal oxides, zerovalent metal, and nanocarbon. They stated that given the benefits of nanomaterials in the utilization for environmental cleaning of heavily polluted environments, especially heavy metal from wastewater, a comprehensive nanotechnique should be considered in the reduction of cost of production, their synthesis, separation, reusability, and capability of the process. This will reduce the impact and risk of the fate of nanomaterials in the ecosystem. In conclusion, the application in the environmental and health sectors, advantages, and limitations, as well as their future potentials, were discussed.

Baragaño et al. [69] tested and evaluated the environmental elimination of heavy metals like P, As, Cd, Pb, and Cu using graphene oxide nanomaterials (nGOx) as related to its counterpart zerovalent iron nanoparticles (nZVI). The nanoparticles were identified by microscopy (atomic force), dynamic light scattering, CHNS-O examination, and X-ray methods. Samples of soils were treated with nGOx and nZVI to determine the ability of the materials to immobilize and break down pollutants of concentration range 0.2%, 1.0%, and 5%. It was noticed that nGOx immobilized efficiently Cd, Pb, and Cu but mobilized P and As at both high and low concentrations, respectively. On the other hand, nZVI indorsed prominent immobilization outcome for Pb and As, a lesser outcome for Cd, and elicited the accessibility for Cu. The soil electrical conductivity and pH have been affected slightly by nGOx. In general, nGOx appears as an effective immobilization and/or mobilization option for the nanobreakdown of soil pollutants if combined with other methods of bioremediation.

Medina-Pérez et al. [70] did a review of the use of nanoparticles in the cleanup of contaminated soil with metals. The authors stated that nanomaterials are generally developed for use by humans in the industrial sector to combat by-products and waste chemicals generated in the environment. Of recent, the management of adulterated soils using nanomaterials has become an evolving remediation technique in modern times to enhance the performance of environmental sustainability. However, there has been serious concern about the use of this technology on human health and the environment as the fate of the wastes released cannot be trusted. Be that as it may, nanotechnology has immense benefits in the removal of toxic wastes from soils like metals when compared to the traditional methods. They recommended that this technology should be standardized and evaluated by decision-makers and scientists for effective deployment in large-scale use while considering the social well-being, health, and environmental fitness.

Zand et al. [71] tested and evaluated the coapplication of titanium dioxide nanomaterials (TiO_2 NMs) and biochar to enhance remediation of antimony (Sb) from the soil via uptake and flora response of Sorghum bicolor (millet). The millet seedlings were exposed to the varying concentrations: 0, 100, 250, and 500 mg/kg of titanium dioxide nanomaterials and 0%, 2.5%, and 5% of biochar concentration to ascertain the influence of flora growth on the physiological retort, accumulation, and absorption to Sb in polluted soil. The combined application of the biochar and the TiO_2 NMs demonstrated a positive influence on the tested plant and established growth in the polluted soil. It was noticed that an ample amount of Sb was found in the shoots of the plant when related to the roots of the same plant in all the treated groups. The use of the biochar elicited Sb immobilization in the soil. However, the utilization of the titanium dioxide nanomaterials significantly improved the ability of the tested plant for Sb with the highest accumulation size (1624.1 µg) in each pot attained in 250 mg/kg titanium dioxide nanomaterials per 2.5% biochar treatment. The association of the biochar and the TiO_2 NMs significantly enhanced the chlorophyll contents of the tested plant (S. bicolor) when related to the titanium dioxide nanomaterial treatment and amendments. The findings of the study presented a novel method of combined utilization of the biochar and the titanium dioxide nanomaterials in the flora remediation of antimony in polluted soils, which is an intelligent collaborated technique for future application of heavy metal remediation in different environmental settings.

Haijiao et al. [72] in a review examined the treatment of wastewater polluted with excessive Zn, Fe, and Ag by using nanomaterials. They stated that nanomaterials such as zerovalent metal nanomaterials have been used to improve the adsorption and catalysis properties of metals in a polluted environment in recent decades. They opined that studies have shown the effectiveness of the elimination of contaminants from sewer water by the treatment techniques of nanomaterials. They stressed the benefits gained in the use of iron oxides, nanocomposites, carbon nanotubes, ZnO, and TiO_2 in the cleanup of the polluted environment with excessive ZN, Fe, and Ag contaminants. However, there are worries about the potential toxicity nanomaterials portend in the extensive use. Nonetheless, there is also a paucity of insufficient standards for the evaluation of the toxicity potentials of nanomaterials. So, an all-inclusive comparison and evaluation of nanoparticle toxicity are urgently needed to safeguard their actual applications. Thus, the evaluation of the performance mode of action of nanoparticles in wastewater and water cleanup should be looked upon in subsequent researches and investigations in different sectors of usage.

6.4 APPLICATION OF NANOMATERIALS IN AGRICULTURE

The agricultural field is very vast, and the nanomaterials have a great potential application (Figure 6.4) in this domain for enhancing the crop yield, stress tolerance, antipesticidal activity, etc.

6.4.1 NANOMATERIALS AS SEED ENHANCERS/GROWTH STIMULATOR

Nanotechnology has proven to be a very effective means of promoting food safety by the enhancement of crop growth and production, thereby enabling sustainable agriculture. Nanomaterials that are used as fertilizers in agriculture possess some physicochemical characteristics, which are known to significantly increase the metabolism of plants. Boutchuen et al. [73] investigated the use of low and high concentrations of hematite nanoparticle fertilizer on the growth and final yield of four various leguminous plants: mung bean (*Vigna radiata*), chickpea (*Cicer arietinum*), red beans, and black (*Phaseolus vulgaris*), using the modified seed presoaked method. They recorded a 230%–830% significant improvement in the growth of all legumes with

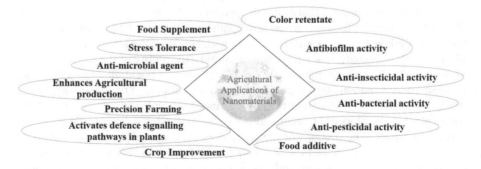

FIGURE 6.4 Potential applications of nanomaterials in the agricultural field.

the mung bean having the maximum growth impact. It was concluded that hematite nanofertilizers are a promising nanotechnology approach for sustainable agricultural development with low levels of environmental toxicity on the soil physiology. Janmohammadi and Sabaghnia [74] soaked seeds of sunflower (Helianthus annuus) before sowing in various concentrations of 0, 0.2, 0.4, 0.6, 0.8, 1.0, and 1.2 mM of nanosilicon solutions for 8 h. The result obtained indicated that at a minimal concentration of 0.2–0.4 mM, there was a significant upsurge in the germination rate of the seeds by 50% and enhanced root length and also a better seed potency. Ghazal et al. [75] tested the capability of biomass and secondary metabolite production of *Stevia rebaudiana* by immersing metallic alloys of gold(Au) and copper(Cu) into root cultures of the plant. When 30 μg/L of nanoparticles was added to the root cultures, there was an increase in biomass production on the 27th day of the experiment. The authors also recorded a significant increase in phenolic and flavonoid production, both of which had a direct correlation to dry biomass.

Carbon nanotubes have been shown to aid seed sprouting and development of *Solanum lycopersicum* (tomato) by improving the water absorbance by the plant [76]. Pandey et al. [77] investigated the biomass production and germination rate of switchgrass and sorghum using multiwalled carbon nanotubes and graphene. They recorded an increase in the rate of germination of switchgrass by 28% when compared with the control and also prompt germination of sorghum. Carbon-based nanoparticles were also reported to increase salt stress tolerance when NaCl was added to the medium of growth. It was concluded that carbon-based nanoparticles are key to sustainable agriculture. Also, Nair et al. [78] inspected the influence of the dimension of C60, carbon nanotubes, and graphene on the sprouting of rice seeds. It was revealed that carbon nanotubes significantly increased the sprouting rate of rice seeds. Thus, it was recorded that there was an increase in moisture content at the sprouting time of the seeds treated with carbon nanomaterials. The authors reported healthier plants with well-established roots, which showed the effect of a more advanced seedling when previously sprouted seeds were grown in a basal medium containing carbon nanomaterial. They concluded that carbon nanomaterials are good development boosters of rice seedling. Juhel et al. [79] assessed the biological impact of alumina nanoparticles (Al_2O_3) on the photosynthesis, development, and morphology of *Lemna minor.* They discovered a considerable improvement in the biomass with an increase in alumina concentration. It was reported that the mechanism involved in the use of alumina nanoparticles in *Lemna minor* is novel and is known for its enhancement of photosynthesis in the weed.

Li et al. [80] compared the influence of different concentrations of 20 mL nanoscale zerovalent iron (NZVI) (10–320 μmol) and a single concentration of 20 mL of ethylene diamine tetraacetate-iron (EDTA-Fe) solution (40 μmol/L), for its stimulation ability on the germination and subsequent growth of peanut (*Arachis hypogaea*) seedlings. They discovered that at a low concentration of 40 μmol/L NZVI, there was an early seed sprout and a longer stem length when compared to 40 μmol/L EDTA-Fe solution. TEM showing the cotyledon of *Arachis hypogaea* treated by NZVI revealed that NZVI can loosen and open the seed coat, and this, in turn, aids the intake of water helping early germination. They also recorded that at higher concentrations of 300–320 μmol/L, NZVI was lethal to *Arachis hypogaea* seedlings. The use and importance of silicon nanoparticles as pesticides have been studied and emphasized [81–83]. Rouhani et al. [84] discovered that SiO_2 nanoparticles were toxic to the adult and larvae stage of cowpea seed beetle at 100% and 83%, respectively, and therefore could be used in the control of *Callosobruchus maculatus* agricultural pest.

6.4.2 NANOMATERIALS AS BIOPESTICIDE

Among the current development in agricultural innovative science, nanoparticles play a crucial part in protecting crops due to their exceptional physical and chemical features. Nanoparticles attach themselves to the membrane of pathogenic microbes, thereby causing damage/death due to its high rate of transfer of energy [85]. Viruses, insects, bacteria, pests, and fungi were identified to infect, lessen yield, and quality of farm produce. Researchers stated that the use of silver NPs has significantly caused the suppression of pathogenic diseases in the agricultural sector. Nanomanipulation of nanosized particles with fluconazole was effective in protecting *Trichoderma* sp, *P. glomerata*, and *C. albicans* [86]. Zinc oxide was discovered to lower the number of cells in *S. typhimurium* and *Staphylococcus aureus* to the barest minimum within 4–8 h of administration [87]. They reported that Ag NPs, Ti O_2, and Au NPs helped in protecting protein synthesis by bacteria, protecting tobacco plants from *Vibrio harveyi*, pathogen, and the NPs act on *E. coli* by reducing ATPase action, thus disrupting the adenosine triphosphate (ATP) activity at the membrane level [88,89]. Utilizing conventional pesticides as well as herbicides has been limited, due to the problem of resistance and biological diversity in the soil resulting in the biological accumulation of pesticides and herbicides in the soil with the resultant soil pathophysiology [90]. The work of Nair et al. [91] revealed the inhibitory potential of Ag nanoparticles against some fungi species that resulted in the death of oak trees. Indicating that nanoencapsulated pesticides are easily taken up through the surface of plants, this assists in the extended-release period of active agents compared to commercially produced pesticides that could be easily washed off by rain [92]. Thus, using conventional chemicals could cause a grievous adverse effect on both the human and immediate environment [93].

Nanoparticles help to improve the dissolution of partially soluble crucial ingredients and help in discharging the useful ingredients gradually to the plants. Chitosanderived NPs embedded with some amino groups and complexes of commercial pesticides showed that incorporating commercial pesticides (rotenone) into NPs was amplified to 13,000 times powerful when compared to the single use of rotenone [94,95]. Nanomaterials impregnated with essential oil from garlic were found to be very effective against *T. castaneum* (Herbst) [96]. Aluminosilicate in nanotube can glue to the surface of plants applied, whereby the active components in the nanotube possess the capacity to glue to the hairs of the insect pest surface and diffuse into their body to cause physiological disruption and damage [97–99]. Nanoparticles give a wider range of effectiveness in the agricultural application when compared to other chemicals, owing to the enormous superficial area, quick mass transmission, and easy attachment to the target area of the plant [87]. They can let out important complexes to the plants due to the presence of a modified control system when compared to the conventional type of pesticides. The presence of the control transport system has made it easy for the release of active component to the plant and also to meet the targeted site for the expected period of time, thereby proving effective control against insects and pests [100], increasing solubility, the dispersal of dissolved fat in water [101], enhancing biological effectiveness, lowering the toxicity that would have been caused by the application of normal pesticides on plants, removal of organic diluents, and amplified effectiveness [102].

The use of nanotechnology has reduced the dosage of conventional pesticides to just the needed components to reach the target sites. Through nanoencapsulation, insecticides are gradually effectively released in a regulated pattern to control the damage usually caused by pests and insects. Nanoformulation has been found to possess various advantages over conventional pesticides owing to its great target distribution and modified mechanism of release. Also, the dosage needed to exhibit a pesticide action is very minimal, thereby limiting the rate of resistance by plants to the nanomaterials and the environmentally friendly nature [68]. Nanocrystalline oxides of metals are greatly used for broad-spectrum pesticides. Metals like ferric oxides, aluminum oxides, cerium oxide, titanium oxides, and Mg oxides are very efficient [59]. They are applied owing to their greater rate of adsorption, quick migration due to higher definite surface area, and superior amount of reaction on the surface compared to those conventional pesticides [103–107]. These NPs also help to convert hazardous chemicals into forms that are eco-friendly and help to reduce the issue of toxicity [108].

Elbeshely et al. [109] carried out a study by spraying Faba beans plants already infected with the mosaic virus with various concentrations of AgNPs and discovered that significant change was recorded within 24 h and thus suggested that silver NPs possess insecticidal activity. The administration of poly detached AuNPs was found to suppress mosaic virus on the barley plant, thereby bringing about resistance development in the barley plant [89]. Studies have shown that chitosan could induce some level of protection on different plants against viral infections. It controls the rot in the roots of the tomato plant, in grapes, in a bunch of Botrytis and *P. grisea* in rice plant [110,111]. Solid lipid NPs make available the medium to trap lipid-linked active materials without utilizing carbon-based solvents [112,113].

Insecticides that possess a reduced ability to dissolve in the presence of aqueous need carbon-based solvents to assist in the proper dissolution of the pesticides/insecticides leading to reduced toxicity of the insecticides [114]. Thus, nanoparticles could be utilized in increasing the rate of solvent and also to reduce the degree to which the insecticides could become toxic [115]. Rani et al. [116] preserved castor leaves with encapsulated silica NPs scattered in acetone. The authors stated that *S. litura* and leaves of castor plant possess reduced feeding potential, resulting in death owing to malnourishment; thus, it was concluded from these studies that NPs-encapsulated active components improved the issues of instability [117]. Nanotechnology has greatly helped in the agricultural and food industry, like helping in plant growth, production, and yield [118]. Nanotechnology has also assisted the environment to renew itself and to manage some wastes that would have constituted environmental hazards [119]. The implementation and adoption of nanotechnology have also helped in the reduction of toxic chemicals on farm products [120,121]. Agriculturally produced bionanocomplexes have been applied to upgrade production and eradicate microorganisms causing diseases in plants [122]. During their research, El-Benday and El-Helaly [81] found out that silica nanoparticles could be effectively used as plant pesticides. The administration of nanosilica particles in the larva of *Spodoptera littoralis* showed great lethal potency at all the administered concentrations [123].

The antimicrobial potency of several nanoparticles, mostly copper and silver NP, has been assayed against different types of microorganisms causing plant diseases. Cioffi et al. [124] revealed the antifungal effectiveness of polymer-NPs in agriculture particularly on farm produce, thus bringing about greater water and nutrient administration and retention. Nanotechnology has also been helpful in the areas of plant production and the transfer of genetic materials [120]. Park et al. [125] assayed the potency of silica-NPs in controlling plant pathogens such as *Botrytis cinerea, Colletotrichum gloeosporioides, Pythium ultimum,* and *Rhizoctonia solani.* This was done in others to conquer the parasitic disease in pumpkin by spraying the NPs on the contaminated leaves for about 3 days; upon this, there were positive results in the eradication of the pathogen. Kim et al. [126] conducted a study on the fungus causing disease state in oak trees (*Raffaelea* sp.); this fungus sometimes causes mortality and stunts the development of the oak trees. These NPs were very effective in stopping the activity of the fungus. There was also another discovery on the elimination of the activity of some fungi such as *Phoma and Fusarium* [127]. The use of NPs as pesticides, herbicides, and insecticides has widely been part of people involved in agricultural business and in others to help their yield and possible maximum production [128]. There has been an improvement in the production of agro-friendly material with the utilization of NPs, thus assisting in breaking down some solvents much more than those that existed before NPs. The majority of farmers have utilized several suspensions of nanosized materials, either the oil- or water-based ones comprising of accurate pesticides or herbicides materials in them, which have been used in the treatment and/or preservation during postharvest [87].

Nanoparticles have been used to transport DNA and some important chemical agent into tissues of the plant in others to secure the plants from plant pathogens [2012]. Permeable silica NPs could be loaded with validamycin and used in the production of liquid pesticides. This formulation had quick actions on the elimination of plant pathogen [129]. Oil–water nanosuspensions were a great formulation viable against insects and pests while farming [130]. Nanosilica was found to have an outstanding nanoparticle, gotten from silica. It was used for various applications in liquid treatment and prophylactic application. Lately, researchers discovered nanoparticles to be beneficial as a catalyst and as nanopesticides [123]. Researchers investigated the application of nanosilica as pesticides and found out that the effectiveness of nanosilica depends on the form/manner in which the pests utilize the collection of lipids found in their cuticle for confirming the lethal state of pest via dryness [89]. Studies revealed the insecticidal efficacy of polyethylene glycol coated with nanomaterials seeded with garlic used against the growth of *Tribolium castaneum*; this nanoformulation was found to have 80% effectiveness against the insect [131].

During their work, they performed bioexamination where they set up the liquid formulation of an aqueous suspension of NPs, applied the technique on rice, and kept it under observation for 7 days. A great result was recorded of the mortality of rice weevil upon application of NPs on the grasserie infection from silkworm; there was a great reduction in the viral load of the leaves treated with the ethanol suspension of the NPs aluminum silicate. There was an investigation on the insecticidal effectiveness of nanoaluminum against *Sitophilus oryzae* and *R. dominica,* which were known to cause the largest percentage of spoilage harvested crops worldwide. There was a substantial lethal rate after 3 days observation of nanocapsulated alumina when infected wheat plants were treated with it, the result obtained when compared with economically available insecticides, showed that nanocapsulated alumina brought about some promising advances to the world as a pest control agent. Zhang et al. [138] reported that wings of cicada found from insects were gotten from a field and were discovered to structure the pillars of nanomaterials. The cicada used the light-shiny features of the nanoparticles to elude the effect of predators [132]. They then concluded that nanotechnology has a great deal to contribute to the agricultural world [131].

Different studies have investigated *Talaromyces flavus* isolated from strawberry in others to determine their inhibitory potency. *Glomerella cingulata* is known to cause anthracnose and *C. acutatum* in plant nursery [133]. There was a great reduction in the disease-infected plant after the administration of the modified strains. Nanotechnology has been able to add more value to the development of biological pesticides that are not toxic to plants, humans, and the environment [134]. Studies have revealed that utilizing nanomaterial in preventing *Azadirachta indica* oil from quick dilapidation, provides plant's lasting protection against pest [134]. This technology has been so helpful in improving the sustainability and efficacy of natural produce [135]. It also provides measured molecules at the site of the target, reduces toxicity effects, and avoids digestion of the active compounds by microbes [135,121].

Studies have examined and highlighted different experiments done on the use of NPs as nanofungicides; this was done by integrating the fungicides in solid woods [136]. Several fungal isolates were utilized in determining the effectiveness of the nanofungicides. The results obtained from the investigation showed a reduced level of evaporation of the NPs at ambient temperature, enhanced stability, and increased rate of stability [137]. The experimental work of Damalas and Koutroubas [138] explained that the use of synthetic insecticides could lead to resistance of pests to the applied pesticides while trying to manage *Oleander aphid* [139]. Their study revealed the biological assay of that insecticidal potency of Ag-Zn nanomaterials on *A.* Merrie and thus concluded that it could be very efficient as an agent that controls pest [140].

6.5 CONCLUSION AND FUTURE RECOMMENDATIONS

This chapter has provided detailed information on the various types of nanomaterials in resolving several challenges encountered in the environmental, agricultural, and biomedical fields.

Therefore, there is a need to search for novel nanomaterial that could be applied in future functional research and development of all the highlighted sectors above. Also, there is a need to intensify more effort in the utilization of natural or biological nanomaterials that could be commercialized as well as that could be patented. Moreover, there is a need for the government, stakeholders, and scientists to integrate and encourage the study of nanotechnology into various education systems most especially in developing countries.

ACKNOWLEDGMENT

Charles Oluwaseun Adetunji, Olugbemi T. Olaniyan, Osikemekha Anthony Anani, and Abel Inobeme are thankful to the administration of Edo University Iyamho, Nigeria. Frances. N. Olisaka is thankful to Benson Idahosa University, Nigeria, for extending facilities to work on this manuscript. Ruth Ebunoluwa Bodunrinde thanks the Federal University of Technology Akure, Nigeria. Juliana Bunmi Adetunji is thankful to Osun State University, Nigeria. Wadzani Dauda Palnam is thankful to the Federal University, Nigeria. Ravindra Pratap Singh is thankful to Hon'ble Vice-chancellor, Indira Gandhi National Tribal University, Amarkantak, M.P., India, for providing financial assistance to work smoothly and diligently. Kshitij RB Singh is thankful to Principal, Govt. V. Y. T. PG. Autonomous College, Durg, Chhattisgarh, India, for providing apt support throughout the writing process of this manuscript draft.

REFERENCES

1. Ulatowska-Jarża A, Pucińska J, Wysocka-Król K, Hołowacz I, and Podbielska H. (2011) Nanotechnology for biomedical applications – enhancement of photodynamic activity by nanomaterials. *Bulletin of the Polish Academy of Sciences Technical Sciences*, 59(3). Doi: 10.2478/v10175-011-0031-0.

2. Editorial. (2019) The role of nanoparticles for biomedical application. *Asian Biomedicine (Res Rev News)*, 13(4): 121–122. Doi: 10.1515/abm-2019-0050.

3. Jain N, Jain R, Thakur N, Gupta BP, Jain DK, Banveeri J, and Jain S. (2010) Nanotechnology: A safe and effective drug delivery system. *Asian Journal of Pharmaceutical and Clinical Research*, 3: 159–164.

4. Rai M, and Ingle A. (2012) Role of nanotechnology in agriculture with specia reference to management of insect pests. *Applied Microbiology and Biotechnology*, 94(2): 287–293. Doi: 10.1007/s00253-012-3969-4.

5. Adetunji CO. (2019) Environmental impact and Eco-toxicological influence of biofrabricated and Inorganic nanoparticle on soil activity. In *Nanotechnology for Agriculture*. Doi: 10.1007/978-981-32-9370-0_12.

6. Adetunji CO, Panpatte D, Bello OM, and Adekoya MA (2019) Application of nanoengineered metabolites from beneficial and eco-friendly microorganisms as a biological control agents for plant pests and pathogens. In *Nanotechnology for Agriculture: Crop Production & Protection*. Doi: 10.1007/978-981-32-9374-8_13.

7. Adetunji CO, and Ugbenyen AM (2019) Mechanism of action of Nanopesticide derived from microorganism for the alleviation of abiotic and biotic stress affecting crop productivity. In *Book: Nanotechnology for Agriculture: Crop Production & Protection*. Doi: 10.1007/978-981-32-9374-8_7.

8. Duffy LL, Osmond-McLeod MJ, Judy J, and King T. (2018) Investigation into the antibacterial activity of silver, zinc oxide and copper oxide nanoparticles against poultry-relevant isolates of salmonella and campylobacter. *Food Control*, 92: 293–300.

9. Jafari A, Pourakbar L, Farhadi K, Mohamadgolizad L, and Goosta Y. (2015) Biological synthesis of silver nanoparticles and evaluation of antibacterial and antifungal properties of silver and copper nanoparticles. *Turkish Journal of Biology*, 39: 556–561 © TÜBİTAK Doi: 10.3906/biy-1406-81.

10. Tang Z-X and Lv B-F. (2014) MgO nanoparticles as antibacterial agent: Preparation and activity. *Brazilian Journal of Chemical Engineering*, 31(03): 591–601. Doi: 10.1590/0104-6632.20140313s00002813.

11. Ishida T. (2018) Anti-viral activity of Ag+ ions for viral prevention, replication, cell surface receptors, virus cleavage, and DNA damage by Ag+-DNA interactions. *Archives of Immunology and Allergy*, 1(1): 29–40.

12. Benton JJ, Wharton SA, Martin SR, and McCauley JW. (2017) Role of neuraminidase in influenza A(H7N9) virus receptor binding. *Journal of Virology*, 91(11): 1–10.

13. Piotr O, Tomaszewska E, Gniadek M, Baska P, Nowakowska J, Sokolowska J, Nowak Z, Donten M, Celichowski G, Grobelny J, and Krzyzowska M. (2014) Tannic acid modified silver nanoparticles show antiviral activity in HSV type 2 infection. *PLOS One*, 9: 1–15.

14. Lu L, Sun RW-Y, Chen R, Hui C-K, Ho C-M, Luk JM, Lau GKK, and Che C-M (2000) Silver nanoparticles inhibit HBV replication. *Antiviral Therapy*, 13(28): 253–262.

15. Hu RL, Li SR, Kong FJ, Hou RJ, Guan XL, and Guo F. (2014) Inhibition effect of silver nanoparticles on HSV 2. *Genetics and Molecular Research*, 13(3): 7022–7028.

16. Swapnil G, Ingle A, Gade A, Rai M, Falanga A, Incoronato N, Russo L, Galdiero S, and Galdiero M. (2013) Antiviral activity of mycosynthesized silver nanoparticles against HSV and human parainfluenza virus type 3. *International Journal of Nanomedicine*, 8: 4303–4314.

17. Yasutaka M, Ono T, Miyahira Y, Nguyen VQ, Matsui T, and Ishihara M. (2013) Antiviral activity of silver nanoparticle/chitosan composites against H1N1 influenza A virus. *Nanoscale Research Letters*, 8: 1–6.

18. Ge L, Li Q, Wang M, Ouyang J, Li X, and Xing MMQ. (2014) Nanosilver particles in medical applications: Synthesis, performance, and toxicity. *International Journal of Nanomedicine*, 9: 2399–2407.

19. Das S, and Chakraborty T. (2018) A review on green synthesis of silver nanoparticle and zinc oxide nanoparticle from different plants extract and their antibacterial activity against multi-drug resistant bacteria. *Journal of Innovations in Pharmaceutical and Biological Sciences*, (JIPBS) ISSN: 2349–2759 5(4): 63–73.

20. Wu D, Fan W, Kishen A, Gutmann JL, and Fan B. (2014) Evaluation of the antibacterial efficacy of silver nanoparticles against Enterococcus faecalis biofilm. *Journal of Endodontics*, 40(2): 285–290.

21. Ahmed S, Ahmad M, Swami BL, and Ikram S. (2016) A review on plants extract mediated synthesis of silver nanoparticles for antibacterial applications: A green expertise. *Journal of Advanced Research*, 7(1): 17–28.

22. Lara HH, Garza-Treviño EN, Ixtepan-Turrent L, and Singh DK. (2011) Silver nanoparticles are broad-spectrum bactericidal and virucidal compounds. *Journal of Nanobiotechnology*, 9: 30 http://www.jnanobiotechnology.com/content/9/1/30. 1–8.

23. Mori Y, Ono T, Miyahira Y, Nguyen VQ, Matsui T, and Ishihara M. (2013) Antiviral activity of silver nanoparticle/chitosan composites against H1N1 influenza A virus. *Nanoscale Research Letters*, 8: 93 http://www.nanoscalereslett.com/content/8/1/93. 1–6.

24. Andrade V, Martínez A, Rojas N, Bello-Toledo H, Flores P, Sánchez-Sanhueza G, and Catalán A. (2018) Antibacterial activity against Streptococcus mutans and diametrical tensile strength of an interim cement modified with zinc oxide nanoparticles and terpenes: An in vitro study. *The Journal of Prosthetic Dentistry*, 119(5): 1–7.

25. Tremiliosi GC, Simoes LGP, Minozzi DT, Santos RI, Vilela DB, Durigon EL, Machado RRG, Medina DS, Ribeiro LK, Viana Rosa IL, Assis M, Andrés J, Longo E, and Freitas-Junior LH. (2020) Ag nanoparticles-based antimicrobial polycotton fabrics to prevent the transmission and spread of SARS-CoV-2. bioRxiv preprint bioRxiv 2020.06.26.152520. Doi: 10.1101/2020.06.26.15252.

26. Yousef NMH, and Temerk HA. (2017) Enhancement the antibacterial potential of the biosynthesized silver nanoparticles using hydrophilic polymers. *Microbiology Research Journal International*, 19(3): 1–9; Article no. MRJI.31738. Doi: 10.9734/MRJI/2017/31738.

27. Muhamed Haneefa M, Jayandran M, and Balasubramanian V (2017) Evaluation of antimicrobial activity of green-synthesized manganese oxide nanoparticles and comparative studies with curcuminaniline functionalized nanoform. *Asian Journal of Pharmaceutical and Clinical Research*, 10(3): 347–352.

28. Beyth N, Houri-Haddad Y, Domb A, Khan W, and Hazan R. (2015) *Alternative Antimicrobial Approach: Nano-Antimicrobial Materials.* Hindawi Publishing Corporation Evidence-Based Complementary and Alternative Medicine Volume 2015, Article ID 246012, 16 p. Doi: 10.1155/2015/246012.
29. Mogomotsi Kgatshe DC, Aremu OS, Katata-Seru L, and Gopane R. (2019) Characterization and antibacterial activity of biosynthesized silver nanoparticles using the ethanolic extract of pelargonium sidoides. *Hindawi Journal of Nanomaterials,* 2019, Article ID 3501234, 10 p. Doi: 10.1155/2019/3501234.
30. Yah CS, and Simate GS. (2015) Nanoparticles as potential new generation broad spectrum antimicrobial agents. *DARU Journal of Pharmaceutical Sciences,* 23: 43. Doi: 10.1186/s40199-015-0125-6.
31. Singh L, Kruger HG, Maguire GEM, Govender T, and Parboosing R. (2017) The role of nanotechnology in the treatment of viral infections. *Therapeutic Advances in Infectious Disease,* 4(4): 105–131. Doi: 10.1177/ 2049936117713593.
32. Farouk SN, Muhammad A, and Aminu MA. (2018) Application of nanomaterials as antimicrobial agents: A review. *Archives of Nanomedicine: Open Access Journal,* 1(3). ANOAJ.MS.ID.000114. Doi: 10.32474/ANOAJ.2018.01.000114.
33. Emami-Karvani Z, and Chehrazi P. (2011) Antibacterial activity of ZnO nanoparticle on grampositive and gram-negative bacteria. *African Journal of Microbiology Research,* 5(12): 1368–1373, Available online http://www.academicjournals.org/ajmr Doi: 10.5897/AJMR10.159.
34. Chiriac V, Stratulat DN, Calin G, Nichitus S, Burlui V, Stadoleanu C, Popa M, and Popa IM. (2016) Antimicrobial property of zinc based nanoparticles. *International Conference on Innovative Research 2016- ICIR Euroinvent 2016 IOP Publishing IOP Conf. Series: Materials Science and Engineering,* 13: 1–7. 012055 Doi:10.1088/1757-899X/133/1/012055.
35. Batista CCS, Albuquerque LJC, de Araujo I, Albuquerque BL, da Silvaa FD, and Giacomelli FC. (2018) Antimicrobial activity of nano-sized silver colloids stabilized by nitrogen-containing polymers: The key influence of the polymer capping. *RSC Advances,* 8: 10873. Doi: 10.1039/c7ra13597a.
36. Bankier C, Matharu RK, Cheong YK, Ren GG, Cloutman-Green E, and Ciric L. (2019) Synergistic antibacterial effects of metallic nanoparticle combinations. *Scientific Reports,* 9: 16074. Doi: 10.1038/s41598-019-52473-2.
37. Tippayawat P, Phromviyo N, Boueroy P, and Chompoosor A. (2016) Green synthesis of silver nanoparticles in aloe vera plant extract prepared by a hydrothermal method and their synergistic antibacterial activity. *Peer Journal,* 4: e2589; Doi: 10.7717/peerj.2589. 1–15.
38. Jo Y-K, Kim BH, and Jung G. (2009) Antifungal activity of silver ions and nanoparticles on phytopathogenic fungi. *Plant Disease,* 93: 1037–1043.
39. Jana S, and Seth M. (2020) Nanomaterials based superhydrophobic and antimicrobial coatings. *NanoWorld Journal,* 6(2): 26–28.
40. Hemeg HA. (2017) Nanomaterials for alternative antibacterial therapy. *International Journal of Nanomedicine,* 12: 8211–8225.
41. Salleh A, Naomi R, Utami ND, Mohammad AW, Mahmoudi E, Mustafa N, and Fauzi MB. (2020) The potential of silver nanoparticles for antiviral and antibacterial applications: A mechanism of action. *Nanomaterials,* 10: 1566. Doi: 10.3390/nano10081566. 1–20.
42. Tiwari V, Mishra N, Gadani K, Solanki PS, Shah NA, and Tiwari M (2018) Mechanism of anti-bacterial activity of zinc oxide nanoparticle against carbapenem-resistant acinetobacter baumannii. *Frontiers in Microbiology,* 9: 1218. Doi: 10.3389/fmicb.2018.01218.
43. Ayeshamariam A, Tajun Meera Begam M, Jayachandran M, Praveen Kumar G and Bououdina M. (2013) Green synthesis of nanostructured materials for antibacterial and antifungal activities. *International Journal of Bioassays,* 02(01): 304–311.

44. Almalah HI, Alzahrani HA, and Abdelkader HS. (2019) Green synthesis of silver nanoparticles using cinnamomum zylinicum and their synergistic effect against multi-drug resistance bacteria. *Journal of Nanoparticle Research*, 1(3): 095–107. Doi: 10.26502/jnr.2688-8521008.

45. Sreelakshmy V, Deepa MK, and Mridula P (2016) Green synthesis of silver nanoparticles from *Glycyrrhiza glabra* root extract for the treatment of gastric ulcer. *Journal of Developing Drugs*, 5: 152. Doi: 10.4172/2329–6631.1000152.

46. Prasad RGSV, Davan R, Jothi S, Phani AR, and Raju DB. (2013) Cerium oxide nanoparticles protects gastrointestinal mucosa from ethanol induced gastric ulcers in in-vivo animal model. *Nano Biomedicine and Engineering*, 5(1): 46–49. Doi: 10.5101/nbe.v5i1.

47. Abo El-Ela FI, Farghali AA, Mahmoud RK, Mohamed NA, and Abdel Moaty SA. (2019) New approach in ulcer prevention and wound healing treatment using doxycycline and amoxicillin/ LDH nanocomposites. *Scientific Reports*, 9: 6418 1–15. Doi: 10.1038/s41598-019-42842-2.

48. Thomas P, Umoh U, Udobang J, Bassey A, Udoh A, and Asuquo H. (2019) Determination of antiulcer and antioxidant activities of the ethanol leaf extract of Uapaca staudtii Pax (Phyllanthaceae). *Journal of Pharmacognosy and Phytochemistry*, 8(4): 927–932.

49. Rao SP, Amrit I, Singh V, and Jain P. (2015) Antiulcer activity of natural compounds: A review. *Research Journal of Pharmacognosy and Phytochemistry*, 7(2): 124–130.

50. Kumar SA, Ramesh A, and Phani AR. (2015) Evaluation of antiulcer activity of nanoceria by pyloric ligation in rats. *Proceedings of 28th IRF International Conference*, 7th June 2015, Pune, India, ISBN: 978-93-85465-29-1, 68–70.

51. Sozer N and Kokini JL (2009) Nanotechnology and its applications in the food sector. *Trends in Biotechnology*, 27 (2): 82–89.

52. Saadatzadeh A, Atyabi F, Fazeli MR, Dinarvand R, Jamalifar H, Abdolghaffari AH, Mahdaviani P, Mahbod M, Baeeri M, Baghaei A, Mohammadirad A and Abdollahi M. (2011) Biochemical and pathological evidences on the benefit of new biodegradable nanoparticles of probiotic extract in murine colitis. *Fundamentals of Clinical Pharmacology*, Doi: 10.1111/j.1472–8206.2011.00966.x.

53. Zimet P, Rosenberg D and Livney YD. (2011) Re - assembled casein micelles and casein nanoparticles as nano - vehicles for u -3 polyunsaturated fatty acids. *Food Hydrocolloid*, 25 (5): 1270–1276.

54. Patrignani F, Burns P, Serrazanetti D, Vinderola G, Reinheimer J, Lanciotti R and Guerzoni ME. (2009) Suitability of high pressure - homogenized milk for the production of probiotic fermented milk containing *Lactobacillus paracasei* and *Lactobacillus acidophilus*. *Journal of Dairy Research*, 76(1): 74–82.

55. Lanciotti R, Vannini L, Pittia P and Guerzoni ME. (2004) Suitability of high dynamic - pressure - treated milk for the production of yoghurt. *Food Microbiology*, 21(6): 753–760.

56. Buzeaa C, Pachecob II, and Robbie K. (2007) Nanomaterials and nanoparticles: Sources and toxicity. *American Vacuum Society*. Biointerphases 2: 4.

57. Khan MR and Rizvi TF (2017) Application of nanofertilizer and nanopesticides for improvements in crop production and protection. *Nanoscience and Plant–Soil Systems, Soil Biology*, 48: 405–427. Doi 10.1007/978-3-319-46835-8_15.

58. Iavicolia I, Lesoa V, Beezholdb DH and Shvedova AA. (2017, August 15) Nanotechnology in agriculture: Opportunities, toxicological implications, and occupational risks. *Toxicology and Applied Pharmacology*, 329: 96–111. Doi: 10.1016/j.taap.2017.05.025.

59. Firozjaee TT, Mehrdadi N, Baghdadi M, and Nabi Bidhendi GR. (2018) Application of nanotechnology in pesticides removal from aqueous solutions - A review. *International Journal of Nanoscience and Nanotechnology*, 14(1): 43–56.

60. Rashmi U. (2019) Application of nanotechnology in insect pest management, *International Research Journal of Science & Engineering*, 7(6): 151–156.

61. Worrall EA, Hamid A, Mody KT, Mitter N, and Pappu HR. (2018) Nanotechnology for plant disease management. *Agronomy*, 8: 285. Doi: 10.3390/agronomy8120285.
62. Sahithya K, and Das N. (2015) Remediation of pesticides using nanomaterials: An overview. *International Journal of Chem Tech Research*, 8(8): 86–91.
63. Hassan SSM, Abdel-Shafy HI, and Mansour MSM (2018) Removal of pyrene and benzo(a) pyrene micropollutant from water via adsorption by green synthesized iron oxide nanoparticles. *Advances in Natural Sciences: Nanoscience and Nanotechnology* 9: 015006.
64. Yang K, Zhu,L, and Xing B. (2006) Adsorption of polycyclic aromatic hydrocarbons by carbon nanomaterials. *Environmental Science & Technology*, 40: 1855–1861. Kun. Doi: 10.1021/es052208w.
65. Rani M, and Shanker U. (2018) Remediation of polycyclic aromatic hydrocarbons using nanomaterials. In: Crini G., Lichtfouse E. (eds) *Green Adsorbents for Pollutant Removal. Environmental Chemistry for a Sustainable World*, Vol 18. Springer, Cham. Doi: 10.1007/978-3-319-92111-2_10.
66. Mahgoub HA. (2019) Nanoparticles used for extraction of polycyclic aromatic hydrocarbons. *Journal of Chemistry*. 2019, Article ID 4816849, 20 p. Doi: 10.1155/2019/4816849.
67. Helal MID, Khater HA, and Marzoog A. (2016) Application of nanotechnology in remediation of heavy metals polluted soils. *Journal of Arid Land Studies*, 26(3): 129–137. Doi: 10.14976/jals.26.3129.
68. Yang J, Hou B, Wang J, Tian B, Bi J, Wang N, Li X, and Huang X. (2019) Nanomaterials for the removal of heavy metals from wastewater *Nanomaterials*, 9: 424. Doi: 10.3390/nano9030424.
69. Baragaño D, Forján R, Welte L, and Gallego JLR. (2020) Nanoremediation of As and metals polluted soils by means of graphene oxide nanoparticles. *Scientific Reports*, 10: 1896. Doi: 10.1038/s41598-020-58852-4.
70. Medina-Pérez G, Fernández-Luqueño F, Vazquez-Nuñez E, López-Valdez F, Prieto-Mendez J, Madariaga-Navarrete A, and Miranda-Arámbula M. (2019) Remediating polluted soils using nanotechnologies: Environmental benefits and risks. *Polish Journal of Environmental Studies*, 28(3): 1013–1030.
71. Zand AD, Tabrizi AM, and Heir AV. (2020) Co-application of biochar and titanium dioxide nanoparticles to promote remediation of antimony from soil by Sorghum bicolor: Metal uptake and plant response. *Heliyon*, 6: e04669.
72. Lu H, Wang J, Stoller M, Wang T, Bao Y, and Hao H. (2016) An overview of nanomaterials for water and wastewater treatment. *Advances in Materials Science and Engineering*, 2016, Article ID 4964828, 10 p. Doi: 10.1155/2016/4964828.
73. Boutchuen A, Zimmerman D, Aich N, Masud AM, Arabshahi A, and Palchoudhury S. (2019) Increased plant growth with hematite nanoparticle fertilizer drop and determining nanoparticle uptake in plants using multimodal approach *Journal of Nanomaterials*, 2019: 7–9 Article ID 6890572. Doi: 10.1155/2019/6890572 2019.
74. Janmohammadi M, and Sabaghnia N. (2015) Effect of pre-sowing seed treatments with silicon nanoparticles on germinability of sunflower (*Helianthus annuus*) *Botanica Lith*, 21(1): 13–21.
75. Ghazal B, Saif S, Farid K, Khan A, Rehman S, Reshma A, Fazal H, Ali M, Ahmad A, Rahman L, and Ahmad N. (2018) Stimulation of secondary metabolites by copper and gold nanoparticles in submerge adventitious root cultures of *Stevia rebaudiana* (Bert.) *IET Nanobiotechnology*, 12(5): 569–573. Doi: 10.1049/iet-nbt.2017.0093.
76. Khodakovskaya M, Dervishi E, Mahmood M, Xu Y, Li Z, Watanabe F, and Biris AS. (2009) Carbon nanotubes are able to penetrate plant seed coat and dramatically affect seed germination and plant growth. *ACS Nano*, 3: 3221–3227.
77. Pandey K, Lahiani MH, Hicks VK, Hudson MK, Green MJ, and Khodakovskaya M. (2018) Effects of carbon-based nanomaterials on seed germination, biomass accumulation and salt stress response of bioenergy crops. *PLoS One*, 13(8): 1–17. e0202274. Doi: 10.1371/journal.pone.0202274.

78. Nair R, Mohamed MS, Gao W, Maekawa T, Yoshida Y, Ajayan PM, and Kumar DS. (2012) Effect of carbon nanomaterials on the germination and growth of rice plants effect of carbon nanomaterials on the germination and growth of rice plants. Doi: 10.1166/jnn.2012.5775.
79. Juhel G, Batissea E, Huguesa Q, Dalya D, van Pelt FNAM, O'Hallorana J, and Jansen MAK. (2011) Alumina nanoparticles enhance growth of Lemna minor. *Aquatic Toxicology*, 105: 328–336. Doi: 10.1016/j.aquatox.2011.06.019.
80. Li X, Yang Y, Gao B, and Zhang M. (2015) Stimulation of peanut seedling development and growth by zero-valent iron nanoparticles at low concentrations. *PLoS One*, 10(4): e0122884. Doi: 10.1371/journal.pone.0122884.
81. El-Bendary HM, and El-Helaly AA (2013) First record nanotechnology in agricultural: Silica nano-particles a potential new insecticide for pest control. *Applied Scientific Reports*, 4(3): 241–246.
82. Shukla P, Chaurasia P, Younis K, Qadri OS, Faridi SA, and Srivastava G. (2019) Nanotechnology in sustainable agriculture : Studies from seed priming to post - harvest management. *Nanotechnology for Environmental Engineering*, 4: 11 Doi: 10.1007/s41204-019-0058-2 2, 1–15.
83. Ziaee M, and Ganji Z (2016) Insecticidal efficacy of silica nanoparticles against *Rhyzopertha dominica* F. and *Tribolium confusum* Jacquelin du Val. *Journal of Plant Protection Research*, 56: 250–256.
84. Rouhani M, Samih MA, and Kalamtari S. (2012) Insecticidal effect of silica and silver nanoparticles on the cowpea seed beetle, *Callosobruchus maculatus* F. (Col.: Bruchidae). *Journal of the Entomological Research Society*, 4: 297–305.
85. Ashish K and Ritika J. (2018) Synthesis of nanoparticles and their application in agriculture. *Acta Scientific Agriculture*, 2(3): 10–13.
86. Kasprowicz MJ, Kozioł M, and Gorczyca A. (2010) The effect of silver nanoparticles on phytopathogenic spores of *Fusarium culmorum*, *Canadian Journal of Microbiology*, 56: 247–253.
87. Ahmed AT, Wael FT, Shaaban MA, and Mohammed FS. (2011) Antibacterial action of zinc oxide nanoparticle agents' foodborne pathogens. *Journal of Food Safety*, 31(2): 211–218.
88. Kamran S, Forogh M, Mahtab E, and Mohammad A. (2011) In vitro antibacterial activity of nanomaterialsfor using in tobacco plants tissue culture. *World Academy of Science, Engineering and Technology*, 79: 372–373.
89. Gopinath K, Karthika V, Sundaravadivelan C, Gowri S, and Arumugam A. (2015) Mycogenesis of cerium oxide nanoparticles using *Aspergillus niger* culture filtrate and their applications for antibacterial and larvicidal activities. *Journal of Nanostructure in Chemistry*, 5: 295–303.
90. Gruere G, Clare N, and Linda A. (2011) *Agricultural, Food and Water Nanotechnologies for the Poor Opportunities, Constraints and Role of the Consultative Group on International Agricultural Research*. The International Food Policy Research Institute, Washington, 1–35.
91. Nair R, Varghese SH, Nair BG, Maekawa T, Yoshida Y, and Kumar DS. (2010) Nanoparticulate material delivery to plants, *Plant Science*, 179: 154–163.
92. Scrinis G and Lyons K. (2007) The emerging nano-corporate paradigm: Nanotechnology and the transformation of nature, food and agri-food systems. *International Journal of Sociology of Food and Agriculture*, 15(2): 22–44.
93. Singh A, Singh NB, Hussain I, Singh H, and Singh SC. (2015) Plant-nanoparticle interaction: An approach to improve agricultural practices and plant productivity surfaces A. *Physicochemical and Engineering Aspects*, 372: 66–72.
94. Chowdappa P, and Gowda S. (2013) Nanotechnology in crop protection: Status and Scope. *Pest Management in Horticultural Ecosystems*, 19(2): 131–151.
95. Shui-Bing L, Zhang Z-X, Xu H-H, and Jiang G-B. (2010) Novel amphiphilic chitosan derivatives: Synthesis, characterization and micellar solubilization of rotenone. *Carbohydrate Polymers*, 82(4): 1136–1142.

96. Mohamed R, and Al-kazafy HS. (2014) Nanotechnology for insect pest control. *International Journal of Science, Environment and Technology*, 3(2): 528–545.

97. Allan GG, Chopra CS, Neogi AN, and Wilkins RM. (1971) Design and synthesis of controlled release pesticide-polymer combinations. *Nature*, 10; 234(5328): 349–351. Doi: 10.1038/234349a0.

98. Patil SA. (2009) *Economics of Agri Poverty: Nano-Bio Solutions*. Indian Agricultural Research Institute, New Delhi, Indian, 56. Physiology 92: 83–91.

99. Gopal M, Chaudhary SR, Ghose M, Dasgupta R, Devakumar C, Subrahmanyam B, Srivastava C, Gogoi R, Kumar R, and Goswami A. (2011) Samfungin: A novel fungicide and the process for making the same. Indian Patent Application No. 1599/DEL/2011.

100. John H, Lucas J, Clare W, and Dusan L. (2017) Nanopesticides: A review of current research and perspectives. *New Pesticides and Soil Sensors*, 193–225.

101. Anjali CH, Sharma Y, Mukherjee A, and Chandrasekaran N. (2012) Neem oil (Azadirachta indica) nanoemulsion - a potent larvicidal agent against Culex quinque-fasciatus. *Pest Management Science*, 68: 158–163.

102. Yu M, Yao J, Liang J, Zeng Z, Cui B, Zhao X, Sun C,Wang Y, Liu G, and Cui H. (2017) Development of functionalized abamectin poly(lactic acid) nanoparticles with regulatable adhesion to enhance foliar retention. *RSC Advances*, 7: 11271–11280.

103. Armaghan M, and Amini M. (2012) Adsorption of diazinon and fenitrothion on nanocrystalline alumina from non-polar solvent. *Colloid Journal*, 74: 427–433.

104. Moradi Dehaghi S, Rahmanifar B, Moradi AM, and Azar PA. (2014) Removal of permethrin pesticide from water by chitosan-zinc oxide nanoparticles composite as an adsorbent. *Journal of Saudi Chemical Society*, 18: 348–355.

105. Tavakkoli H, and Yazdanbakhsh M. (2013) Fabrication of two perovskite-type oxide nanoparticles as the new adsorbents in efficient removal of a pesticide from aqueous solutions: Kinetic, thermodynamic, and adsorption studies. *Microporous and Mesoporous Materials*, 176: 86–94.

106. Cheng Y. (2013) Effective organochlorine pesticides removal from aqueous systems by magnetic nanospheres coated with polystyrene. *Journal of Wuhan University of Technology (Materials Science) Editorial Department*, 29: 168–173.

107. Bardajee G, and Hooshyar Z. (2013) Degradation of 2-Chlorophenol from wastewater using γ-Fe_2O_3 Nanoparticles. *International Journal of Nanoscience and Nanotechnology*, 9: 3–6.

108. Fryxell GE, and Cao G. (2012) Environmental applications of nanomaterials: Synthesis, sorbents and sensors. *World Scientific*, 89: 236–244 Doi: 10.1142/P814.

109. Elbeshehy EKF, Elazzazy AM, and Aggelis G. (2015) Silver nanoparticles synthesis mediated by new isolates of Bacillus spp., nanoparticle characterization and their activity against Bean Yellow Mosaic Virus and human pathogens. *Frontiers in Microbiology*, 6: 453.

110. Kochkina Z, Pospeshny G, and Chirkov S. (1994) Inhibition by chitosan of productive infection of T-series bacteriophages in the *Escherichia coli* culture. *Mikrobiologiia*, 64: 211–215.

111. Kashyap PL, Xiang X, and Heiden P. (2015) Chitosan nanoparticle based delivery systems for sustainable agriculture. *International Journal of Biological Macromolecules*, 77: 36–51.

112. Ekambaram P, Sathali AAH, and Priyanka K. (2012) Solid lipid nanoparticles: A review. *Scientific Reviews & Chemical Communications*, 2: 80–102.

113. Borel T, and Sabliov C. (2014) Nanodelivery of bioactive components for food applications: Types of delivery systems, properties, and their effect on ADME profiles and toxicity of nanoparticles. *Annual Review of Food Science and Technology*, 5: 197–213.

114. Lu W, Lu ML, Zhang QP, Tian YQ, Zhang ZX, and Xu HH. (2013) Octahydrogenated retinoic acid-conjugated glycol chitosan nanoparticles as a novel carrier of azadirachtin: Synthesis, characterization, and in vitro evaluation. *Journal of Polymer Science Part A: Polymer Chemistry*, 51: 3932–3940.

115. Campos EV, Proença PL, Oliveira JL, Melville CC, Vechia JF, Andrade DJ, and Fraceto LF. (2018) Chitosan nanoparticles functionalized with-cyclodextrin: A promising carrier for botanical pesticides. *Scientific Reports*, 8: 2067.

116. Rani PU, Madhusudhanamurthy J, and Sreedhar B. (2014) Dynamic adsorption of α-pinene and linalool on silica nanoparticles for enhanced antifeedant activity against agricultural pests. *Journal of Pest Science*, 87: 191–200.

117. Wang Y, Cui H, Sun C, Zhao X, and Cui B. (2014) Construction and evaluation of controlled-release delivery system of Abamectin using porous silica nanoparticles as carriers. *Nanoscale Research Letters*, 9: 2490.

118. Rickman D, Luvall J, Shaw J, Mask P, Kissel D, and Sullivan D. (2003) Precision agriculture: Changing the face of farming. *Geotimes* 48(11): 28–33.

119. Tungittiplakorn W, Cohen C, and Lion LW. (2004) Engineered polymeric nanoparticles for bioremediation of hydrophobi contaminants. *Environmental Science & Technology* 39(5): 1354–1358. Doi: 10.1021/es049031a.

120. Tomey F, Trenyn BG, Lin VSY, and Long K. (2007) Mesoporous silica nanoparticles deliver DNA and chemicals into plants. *Nature Nanotechnology*, 2: 295–300.

121. Gogos A, Knauer K, and Bucheli TD. (2012) Nanomaterials in plant protection and fertilization: Current state, foreseen applications, and research priorities. *Journal of Agricultural and Food Chemistry*, 60 (39): 9781–9792.

122. Alemdar A, and Sain A. (2008) Mesoporous silica nanoparticles deliver DNA and chemicals into plants. *Nature Nanotechnology*, 99: 1664–1671.

123. Dimetry, N and Hussein, H. (2016) Role of nanotechnology in agriculture with special reference to pest control. *International Journal of PharmTech Research*, 9(10): 121–144.

124. Cioffi N, Toisi L, Ditaranto N, Sabbatini L, Zambonin PG, Tantillo G, Ghibelli L, D'Alessio M, Bleve-Zacheo T, and Traversa E. (2004) Antifungal activity of polymer-based copper nanocomposite coatings. *Applied Physics Letters*, 85: 2417. Doi: 10.1063/1.1794381.

125. Park H-J, Kim S-H, Kim H-J, and Choi S-H (2006) A new composition of nanosized silica silver for control of various plant diseases. *The Plant Pathology Journal*, 22(3): 295–302.

126. Kim JK (2009) An in vitro study of the antifungal effect of silver nanoparticles on oak wilt pathogen Raffaelea sp. *Journal of Microbiology and Biotechnology*, 19: 760–764.

127. Esteban-Tejeda L, Malpartida F, Esteban-Cubillo A, Pecharroman C, and Moya J. (2009) Antibacterial and antifunfal activity of a Soda-lime glass containing copper nanoparticales. *Nanotechnology*, 20(50): 505701.

128. Owolade OF, and Ogunleti DO. (2008) Effects of titanium dioxide on the diseases, development and yield of edible cowpea. *Journal of Plant Protection Research*, 48(3): 329–336.

129. Liu F, Wen L-X, Li Z-Z, Yu W, Sun H-Y, and Chen J-F. (2006) Porous hollow silica nanoparticles as controlled delivery system for water-soluble pesticide. *Materials Research Bulletin*, 41(12): 2268–2275.

130. Wang L, Li X, Zhang G, Dong J, and Eastoe J (2007) Oil-in-water nanoemulsions for pesticide formulations. *Journal of Colloid and Interface Science*, 314(1): 230–235.

131. Yang F-L, Li X-G, Zhu F, and Lei C-L. (2009) Structural characterization of nanoparticles loaded with garlic essential oil and their insecticidal activity against *Tribolium castaneum* (Herbst) (Coleoptera: Tenebrionidae). *Journal of Agricultural and Food Chemistry*, 57(21): 10156–10162.

132. Zhang J and Liu Z. (2006) Insects make nanotech impression. *Royal Society of Chemistry*. On line, Doi: 10.1002/smll.200600255.

133. Ishikawa, S. (2013) Integrated disease management of strawberry anthracnose and development of a new biopesticide. *Journal of General Plant Pathology*, 79: 441–443.

134. Mishra S, Keswani C, Abhilash PC, Fraceto LF, and Singh HB. (2017) Integrated approach of agri-nanotechnology: Challenges and future trends. *Frontiers in Plant Science*, 8: 471.

135. Perlatti B, de Souza Bergo PL, Fernandes da Silva, MF, Fernandes JB, and Forim MR. (2013) Polymeric nanoparticle-based insecticides: A controlled release purpose for agrochemicals. In *Insecticides-Development of Safer and More Effective Technologies*; Trdan, S., Ed., InTech: Rijeka, 523–550.
136. Liu Y, Laks P, and Heiden P. (2002) Controlled release of biocides in solid wood. III. Preparation and characterization of surfactant-free nanoparticles. *Journal of Applied Polymer Science*, 86: 615–621.
137. Worrall EA, Hamid A, Mody KT, Mitter N, and Pappu NP. (2018) Nanotechnology for plant disease management agronomy. 8: 285. Doi: 10.3390/agronomy8120285.
138. Damalas CA, and Koutroubas SD. (2018) Current status and recent development in Biopesticides use. *Agriculture*, 8: 13. Doi: 10.3390/agriculture8010013.
139. Eleka N, Hoffmanb, R, Ravivb, U, Reshb, R, Ishaayac, I, and Magdassi, S. (2010) Novaluron nanoparticles: Formation and potential use in controlling agricultural insect pests. *Colloids and Surfaces A: Physicochemical and Engineering Aspects*, 372: 66–72.
140. Rouhani M, Samih MA, Aslani A, and Beiki, KH. (2011) Side effect of nano-Zno-Tio$_2$-Ag mix-oxide nanoparticles on Frankliniella occidentalis Pergande (Thys.: Thripidae). In *Proceedings Symposium: Third International Symposium on Insect Physiology, Biochemistry and Molecular Biology*, 2–5 July 2011. East China Normal University, Shanghai, China, 51.

7 Nanomaterials for Environmental Hazard
Analysis, Monitoring, and Removal

S. Sreevidya
Kalyan PG College

Kirtana Sankara Subramanian
University of Melbourne

Yokraj Katre
Kalyan PG College

Ajaya Kumar Singh
Govt. V.Y.T. PG. Autonomous College

CONTENTS

7.1 INTRODUCTION

7.1.1 ENVIRONMENTAL SITUATION

Lack of immunity, a major factor to combat the present notorious pandemic situation of COVID-19, can be attributed to various environmental effects. On the onset of the pandemic break, the research community globally is completely focused on typical investigating platforms, for the chemical analysis of water and effluents to find the archetypal agents to get supplementary inputs and fundamental cause for this dreadful COVID-19 [1,2]. Our environment, the main source of potent minerals in different forms, undergoing continuous drastic alteration by countless disasters, is either man-made due to variations in technological innovations and advancement, selfishness for sustainability, and existence of the human race or by natural disasters [3]. Continuous advancement of easy lifestyle in this fast-moving world, rapidity in commercial and economic sectors, industrialization and pharmaceutical development, and imperfection in ecological management and policies lead to an enormous large-scale catastrophe, jeopardizing our mother Earth [4].

Potable water, a potent source of livelihood, with its availability privileged for the human race and other living species present only in our planet, is at the stake of depletion due to mismanagement [5]. Land (soil), another major component for survival for all terrestrial living species under continual stress, by ongoing deliverance of urbanization (construction/transportation) and industrial development by new technological expansions is at the pitch of exhaustion with hazardous potentials and hence needs immediate attention [6]. Air, last but not the least the most important vital component required for survival, found to be impure with toxic pollutant gasses engulfed in it, by emission of smokes and fumes from transportation and industrial expansion, is at a tremendous pressure for purification and fortification [7].

Depletion of nature's natural components is to be handled in a better way for sustainability and protection of all species, especially the endangered ones [8]. The toxic components (hydrocarbons and fertilizers, pesticides and herbicides, phenolics and oil trips, heavy metals and their compounds, oxides of S, C, and N, halogens, volatile organics, suspended particulate air particles, microbial harmful pathogens, industrial, pharmaceutical, and sewage effluents, etc.) in these vital bodies are to be exterminated for the environmental security and safety [9]. Inhalation, assimilation, ingestion, inclusion, and absorption are some pathway entry points of the harmful toxins that hamper the physiological stream of every individual species [10]. Accumulation of contaminants on food and agricultural produce and aqua bodies is the major source of risks to be managed for protecting all living bodies [11]. Figure 7.1 underlines the importance of the removal of various contaminants present in the environment.

7.1.2 NANOTECHNOLOGICAL THREE-WAY CYCLIC APPROACH

Henceforth, in light of the growing exigent dictates, to protect our ecosystem against the threat of the toxic environmental impurities, new sustainable, competent, and resourceful, less toxic or non-toxic, cost-effective innovations are mandatory. With vibrant systems for pollution removal available [12,13], nanotechnology, though

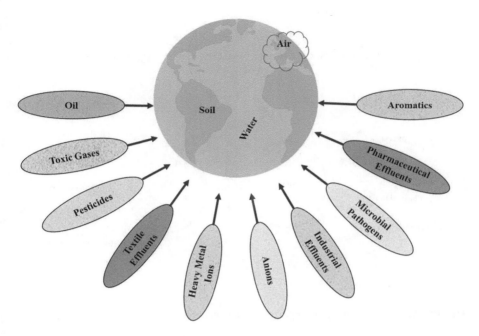

FIGURE 7.1 Contaminants in the environment.

possessing particles with a very tiny array, offers a better solution in a bigger module for the betterment and protection of the ecosystem in a broader angle [14,15]. Momentous noteworthy "nanomaterials (NMs)" that have caught the sight of the scientific society for the past few decades are approached by different techniques such as coprecipitation [16], hydrothermal [17], sol–gel [18], green [19], microwave [20], sonification [21], and RF sputtering (RF – radio-frequency magnetron) [22]. NMs, although in a very small range of morphological dimension with remarkable potentials, find their place in a noteworthy way in diversified sectors. Significance of a better surface:volume ratio of the NMs when equated with their counterparts (bulky materials) show a versatile prolific property [23]. Thus, the surface interaction, when contrasted with other standard approaches, targets the particular modules of interest (noxious waste) for well-organized remediation. The premeditated fine-tuning of the characteristic features such as particle size, geometrical shape, morphological dimensions, material porosity, and chemical constitutions confers an additive beneficial advantage that can primarily influence the operation of the NM for a consistent noxious waste remediation [24], water decontamination [25], soil remediation, and air purification.

About 2.2 billion individuals are unable to have a clean and natural potable drinking water globally (UN/WHO) [26]. An indication of pollution by lifesaving pharmaceuticals and other self-care items is prominently visible in a number of water bodies [27]. Water, an essential component for a sustainable livelihood in a better way, can be effectually provided to every individual habitat, in a clean manner by the utility of NMs in a simple three-way cyclic approach of analysis, monitoring, and removal,

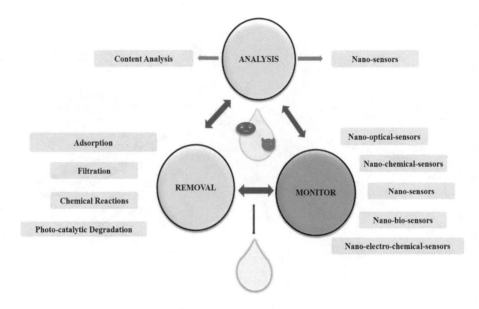

FIGURE 7.2   Three-way cyclic approach of nanotechnologies used for water management.

where the water-management infrastructures are to be well exploited. New innovative approaches are emphasized and modulated as an alternative approach to utilize and reutilize water from various resources such as salty seawater, dirty domestic wastewater, and rainy stormwater, in areas with less availability of potable water.

The performances of NMs that are fabricated by using various building blocks, limited not only to synthesis of nano-absorbents [28], nanofilters [29], and nano-membranes [30], as nanoparticles (NPs), nanocomposites (NCs) [31], or nano-functionalization [32], in nanosensors (NSs) [33,34], shows a promising note for a laboratory and full-scale technological environmental analysis, monitoring, and removal of toxic components in water technological management. A sample representation of nanotechnological developments for water management using nanoengineered NMs employed in a three-way cyclic approach is depicted in Figure 7.2. Analysis: Qualification and quantification of environmental hazardous (contents/checking) materials present in water. Monitoring: Continuous checkup if the water sample is within the range of limits permissible. Removal: Exclusion of pollutants by innovative means.

7.2   OPERATIONAL MANAGEMENT

A speedy increase in the rising population and expanding standards of the lifestyle demand the prerequisite for a clean potable drinking water. Recent studies indicate that 34% of publication deals with the utilization of NMs for water management [35]. Pollutants, a non-intermittent universal impurity, can easily drift from the spot location of contaminant to pollute the non-affected zone to contaminate it too [36,37]. Presently and more frequently, novel NMs are primarily focused for their applications and, hence, studied in depth for environmental water pre-/post-treatment as a solid-phase absorbent. The breakthrough and innovative utilization of novel NMs have permitted the advances in improving the treatment required for the water sample management, with potential profits with rapidity, consistency, and sensitivity, especially in water analysis, monitoring, and removal of contaminants. However, an important key factor is the difficulty that crops up when complex matrices in the environmental water bodies formed, that leads to interferences in analysis, monitoring, and removal. Hence, it is mandatory to eliminate these interfering points that lead to complexity for simplicity and rapidity of procedures by pre-treatment protocols [38].

Providing potable clean and hygienic H_2O is of great challenge to be met in an interdisciplinary forum; hence, answers to the existing problem sought must cross the different margins of disciplines [39]. The three-way cyclic approach would cut across the borders of bounties with new revolutions in the nano-world. Carbonaceous materials and their derivatives, with modification as nanostructural derivatives as carbon nanotubes (CNTs), G-materials, carbon fibers (CFs) have been successfully manipulated for their activities in sensing, adsorption, membrane, monitoring, removal, and decontamination technologies [40]. Nano-based polymers and metal and metal oxide (MO) nanostructures are prudential for exploitation in fields of sensing for content analysis, monitoring, and removal as adsorbents, photocatalysts, and disinfectants [41]. Detecting and sensing the presence and quantifying the contents of environmentally hazardous materials present in water system can be framed as environmental analysis of water bodies.

7.3 NMs FOR ENVIRONMENTAL WATER ANALYSIS

Many a time, the components that are getting accumulated in the environmental water segments in spite of their toxicity at a very low level of concentration are limited to their accessibility for detection levels with the present technological analytical systems. Hence, it is an essential and vital need to expand the operational analytical practices for detection and quantification that leads to accuracy and reproducibility, so as to get rid of the micro-pollutants at a macroscopic level when present even at a negligible low concentration in environmental water system [42]. Consequently, the new developments are in a great demand to identify and structurally qualify and quantify the environmental contaminants with sensitivity and rapidity in analytical approaches for focusing on the removal of contaminants from water samples [43].

The prominence of NMs as nanosorbents and NSs that play a crucial role in environmental water analysis (E-WA) can be well noticed as supporters in pseudo-stationary and stationary levels in analytical methods [44]. Isolation, elimination, and transformation into a right and proper analyte enable easy analysis in the nano-structural field with high specific area and modification in its chemical features. The trace residues as impurities that form complexities even at the microlevel can effectively be retarded by suitable sample preparation of the analytes [45,46]. This recuperates the procedural activities in the right direction, for improvement that favors sensitivity, precision, and reproducibility for qualification and quantification protocols. Precipitation, digestion, extraction, concentration, coprecipitation, solid-phase extraction (SPE), etc., are distinctive pre-treatment methodologies for E-WA, where SPE occupies a better position in E-WA. Time-saving procedures to pick up the objective analytes for analysis are an essential prerequisite for E-WA measures [47].

7.3.1 Content Analysis: Recognition of Hazardous Pollutants

Effluents as contaminants are dispersed quickly into the environment as wastewater segments, wherein pharmaceutical wastes and textile wastes can be detected without difficulty, with well-ascertained, selective, and sensitive methodologies. The molecular components and chemical structures help in using the developed philosophies in an amenable way for the management of pollutant drifts [48]. Chromatographic methodologies offer a better solution to tackle the existing challenges that are coprecipitated in some analytical technique, while handling trace impurities LC-MS/MS [49,50], GC-MS [51], HPLC [52] (LC – liquid chromatography; MS – mass spectrometry; GC – gas chromatography; HPLC – high-performance liquid chromatography) are some of the versatile approaches that responds with accuracy in qualification and quantification of the unwanted deliverables in waste-water, without any interference and intrusions of the constituents associated in it [53]. Moreover, as these methodologies are exceedingly sensitive to very low levels of components at parts per million/parts per billion/parts per trillion (ppm/ppb/ppt), environmental contaminants can be effectively detected and quantified to precision and accuracy [54]. The detection of environmental wastes (EWs) in the ecosystem is complexed with the presence of naturally appearing colloidal particle, economics, and finance required as supports [55,56]. At very high-to-very low concentrations, relatively more hindrances are visible, hence leading to complications in results for biological, chemical, and nuclear debris. Advanced high-tech, analytical overflows with microextraction procedures are essential to examine the targeted analytes for a single-scale detection of a wide segment of pollutants [57]. Chromatographic evaluations generally depend upon high-resolution (HR) and precise systems. HR-MS, an indispensable part, provides salient features such as isotopic arrangement, fragmental daughter ions, and data (mass) of molecular species with the time-of-flight (ToF) not narrowing its output [58].

In one of their recent experiments, the experimentalists however worked on the identification of the detrimental organic toxins of polycyclic aromatic hydrocarbons (PAHCs) in real water samples, where detection was by the precise GC using He as a carrier gas, and elimination was at higher levels by the micro-SPE method using Sod. Alg-MWCNT (multiwalled carbon nanotube) beads [59]. Later, elsewhere polyurethane as sieve, with functionalized modifications using polypyrrole–ZnO as a nanocomponent, was used for refining the properties of water. The work for detection delivered supports using GC, spectrophotometric, and biological oxygen demand (BOD) methods. Potential and powerful chromatographic detector HPLC was coupled for the identification of phenolics, wherein the workers extracted the adaptability with supports of photo-Fenton for their product as hybridized NC of $MgAC-Fe_3O_4/TiO_2$ [60]. Herein, the coworkers adapted reversed-phase high-performance liquid chromatography–ultraviolet (RP-HPLC/UV) analysis for the quantification of the concentration in the pharmaceutical target pollutant diclofenac. Later, it was proven for its reduction using GO-FeO [graphene oxide (GO)–iron oxide)] composites by ultrasonification methodologies [61]. Quantification measurements of organophosphorus used as pesticide were done using GC for the supernatant liquid, after a gestation period of 15 min, for the pesticide that was effectively adsorbed over the nanogold bioconjugate [62]. Photolytically cleavable NMs of poly(ethylene glycol)-b-poly(lactic acid), found effective in elimination of PAHCs, were detected by HPLC detectors and identified as toxic pollutants in the water samples [63].

7.3.2 NSs for Uncovering Water Contaminants

Recent advances in nanotechnology have enabled us to foresee and overcome the fundamental challenges that crop up while formulating and testing. Hence, advances in nanoproduct as NSs for analyzing the presence and quantifying not only the contaminated metals but also other noxious components in unalike matrices, unambiguously in H_2O, play a vital part in any research work [64]. NMs as chemical [65], electrochemical [66] and optical [67], biosensors and nanobiosensors [68] have been expansively exploited for the uncovering of heavy inorganic metals, organo-pesticides, other organic/inorganic contaminants, dyes, drugs, etc., in water bodies. With the improvements and feasibility in on-site identification, pronounced sensitivity, distinct selectivity, transportability, and enhanced functioning, NSs stands in a better position for evaluations of toxins [69]. The assimilation of toxicants, at an extremely low ppb, delivers innumerable adverse health problems [70]. The restrictions in sampling techniques, usage of high-priced gadgets, preconcentration practices, and the need for professional technicians for handling the situation in the regular conventional circumstances can be effectively overdone by NS protocols [71]. NMs of a wide category varies from organic (carbon derivatives, polymeric) to inorganic (metallic/nonmetallic) as NPs, NCs, and functionalized materials with versatile characteristics such as huge surface:volume ratio, good responsiveness, and enhanced sensitivity and selectivity and hence, are perfect candidates [72].

Electro-optical NSs with an eclectic range of potentials and applicability on an on-site platform employ electromagnetic particle emission that generates the required analytical magnetic pulse indicative of the contaminant present. A surfeit of NSs testified for validity in the past by exploiting the fluorescence swapping of quantum dots (QDs), carbon quantum dots (CQDs), and QDs in the existence of an analyte delivers optical signal to testify the water pollutant [73]. Fluorescence of the material can be attributed to quantum confinement [74], molecular fluorescence [75], quenching [76], and external oxidation on the outer layer.

In one of their recent studies, the authors used plant-based nanobionic fluorescent NSs in the near-infrared (NIR) region for the detection of As (water pollutant), which was later proved for its selectivity and sensitivity [77]. The authors and their coworkers integrated an Au probe with Ag enhancement for the identification and quantification of Cd in water, with a detection limit of 5 mg/L [78]. However, identification of pesticides and metallic toxicities present in freshwater by NSs was reviewed by the authors in one of their works [79]. In a similar trial, Au NMs in a non-modified phase were found to be very sensitive to Pb^{2+} ions with the color tint variation from blue to pinkish red (detection limit 0.2×10^{-9} mol/L) in natural water samples [80].

Hg^{2+} ions from 50 different sampling on-site locations were reported by coworkers using self-built smartphones, with integrated Au NM chips, found to be sensitive to ~ 3.5 ppb quantification limits [81]. Overuse of pesticides by agriculturist, a great concern globally, could be effectively detected and degraded by fluorescence NM ZnO QDs. The authors revealed that aldrin and tetradifon had a good binding capacity with ZnO QDs when compared with glyphosate and atrazine [82]. A novel platform of AuNPs supported on reduced graphene oxide (rGO) layers with an e-tongue system was authenticated for the detection of organophosphates and was proven for high sensitivity and selectivity [83]. Later, in one of the experiments, the research team investigated the presence of carbaryl pesticides in lab water samples with high potentials (recognition limits 10–12 g/L) using single-walled carbon nanotubes (SWCNTs)/acetylcholinesterase (AChE) [84]. The presence of toxicants has been unwrapped by new inventive, simpler, inexpensive NMs with fluorescence, which were the later frills segmented as polymeric, functionalized, metal/MO, or carbon/silicon-based materials [85].

7.4   NMs FOR ENVIRONMENTAL WATER MONITORING

The second stage of the cyclic progression is in conjunction with the monitoring of water, either natural or effluent. Environmental water monitoring is an essential component to meet the needs for delivering a safe and good potable quality water free from undesirable materials that favor unhealthy issues to the livelihood. More cohesive approaches are required to readjust the monitoring parameters that access and assess the status of the chemical components in the ecosystem in terms of ground and surface water bodies. A widespread extensive surveillance to meet the needs is in progress technically and socially.

Traditional approaches (reverse osmosis, HPLC, LC/GC-MS), despite delivering templates of accuracy and reproductivity at on-site levels, need voluminous levels of water for purification. Hence, they demand the need for professional technicians or expensive practices [86]. The harmful detriments of effluent water flow to the environments can be minimized to a maximum extent, provided monitoring of health risks is supplemented with regular pollution checkups and controls of the toxicants, with strategic protocols as per standards of the dependability [87]. Constant complexity in the matrices, arises due to a seemingly low level of concentrations of the contaminants, hence, regular monitoring of water is indispensable to ensure good quality, though it is a challenging task at variable scales [88]. More organized approaches and innovative efforts are well addressed to qualify and quantify the unwanted pollutants [89]. Identification of biological and chemical components and newly improved sampling schemes, with safety and purposeful achievements, in amalgamation with regular monitoring of on-site and off-site platforms, are essential to meet the frames delivered by WHO and UN [90]. Traditional columns and their counterparts that focus primarily on individual components are challenged for multiple components. NMs and their composites offer better achievements in both platforms.

Although the mighty challenge of potential environmental risk by NMs is a question mark, its efficiency and significant technical approaches as NSs, nano-chemical sensors, nanobiosensors, nano-optical sensors, and nano-electromagnetic sensors offer a better status for water monitoring [91]. Quantum dots [MQDs – metal quantum dots CQDs], carbon derivatives [mesoporous C, CNTs, fullerenes, graphene (G), and its derivatives], graphene family members (GFMs), metal [MOs, metal–organic frameworks (MOFs), and polymeric NMs], with notable properties, are employed in a holistic pattern, for fabricating water monitoring accessories [92]. Significant features and methodological principles of fluorescence, quantum confinement, nano-small size, a large surface area/energy that help in binding with the pollutants, such as surface-enhanced Raman spectroscopy (SERS) [93], UV–Vis spectroscopy [94], colorimetry [95], electrochemistry [96], photoluminescence [97], chemiluminescence [98], and electrochemiluminescence [99] are by and large exercised in exploitation of monitoring the quality of water samples (real/lab). NMs exercised in different nanometric size, shape, and structure show a commendable performance for identifying and monitoring various classes of undesirable and undeniable pollutants (organic/inorganic/biological pathogens) arising from industrial, mining, pharmaceutical, and domestic fronts [100].

7.4.1 MONITORING OF CONSTITUENTS AND CONTENTS PRESENT

NMs have shown a signature note, with reliability, rapidity, reproducibility, and sensitivity for content monitoring, so as to enable a quick implementation of policies to retard and discard the unworthy toxins [101]. Some noteworthy pollutants contaminating water systems to be mentioned here are heavy metal ions of transition, metalloids, rare earths, alkaline, and some p-blocks, residual oxide radicals of N, P, S, and Cl$^-$, organic C derivatives of simple and polymeric pharmaceutical effluents, industrial dyes, agricultural pesticides, pathogenic microorganisms, biotoxins, and other surfacing chemical components. Similarly, NMs have been remarkably utilized for chemical oxygen demand (COD) and BOD for water-quality monitoring [102].

A comparative analysis in monitoring COD and BOD from real water samples (coal-mine wastewater and Mzingazi river water) by Fe@Cu-NPs, Fe-NPs, and Cu-NPs indicated a wide range of contaminants present from ~3.30 to ~842 mg/L in all, which was proved for its removal efficacy from 48% to 94% by the same NMs [103]. In another trial, the authors monitored crude wastewater samples from San-Elhagr, Al-Sharqia, Egypt, by using nano-Fe in the batch methods for the evaluation of the pollutants in pre-/post-condition in mg/L (COD: 600/76, BOD: 365/31, Pb: 0.216/0.003, CN$^-$: 0.265/0.006, NO^{3-}: 0.62/0.06) [104]. The existence of the toxins as a residual organophosphorus agricultural pesticide (fenamiphos/malathion/isocarbophos/chlorpyrifos/profenofos) in an array of 89.0%–99.8% in Qinghe river water and 89.0%–97.8% in agricultural ground water at Beijing was effectively monitored using graphene@SiO$_2$@Fe$_3$O$_4$ NCs by the research team in a specific test [105]. In one of their experiments, the scientific workers used G-Fe$_3$O$_4$ to monitor the presence of five carbamate agricultural pesticides (isoprocarb/metolcarb/pirimicarb/carbofuran/diethofencarb) in water samples collected from reservoir (Wangkuai), river (Yimu), and pool from Baoding (China), where the carbamates were found to be in the range of 87.0%–97.3% [106].

NO^{3-}/NO^{2-} levels of contamination (10–500 mg/L) in groundwater and wastewater in areas of Taiwan, effectively monitored by zero-valent Fe NPs (ZVIN), were proven for operative elimination with efficiencies as 65%–83% and 51%–68%, respectively [107]. Nevertheless, recently some authors have reviewed the functionalities of the bimetallic NMs and their applicability in monitoring wastewater and other water systems. Arsenic, a noxious element found at high levels in ground and natural waters of eastern India and Bangladesh, was significantly monitored for its concentration, from specific sources at Kolkata, and was eliminated by different trials using nanostructured bimetallic Fe(III)-Ti(IV) [108]. Swine wastewater gathered from the site (hoggery) at Fuzhou, China, was monitored for the removal of COD, total P, and total N as 84.5%, 30.4%, and 71.7%, respectively, by Fe-NPs [109]. In a similar trial, wastewater from the same site was tested to monitor the concentration of NO^{3-} by comparative analysis between Fe(0) and Fe_3O_4 NPs prepared using different leaf extracts (green tea, eucalyptus) for the efficient takeaway of the toxin being 159.7/1.7 and 41.4%, respectively [110]. Textile industrial original wastewater samples were successfully monitored for COD (87%) and decrease in color (40%) using hydrothermally synthesized Fe_2O_3/MgO NMs [111]. Wastewaters from olive mill at Palestine with a high COD varies between (153 to 10) 10^3 mg/L together with other metallic contaminants ($Na^+/K^+/Ca^{2+}/Cu^{2+}/Fe^{3+}/Cr^{2+}$) were effectively proved for the removal (79%–44%) by γ-Fe_2O_3 NPs [112].

7.4.2 NSs as Examiners of Water Toxicants

NSs, as nano-sized particles, provide an innovative and new inventive airway for water-quality monitoring, rapidity in identification, and precision with accuracy in measurements [113]. Topical momentum in chemical, electrochemical, optical, electro-optical, and field-effect transistors has given rise to sensors in nano-range minuscule models, which have gained their importance in monitoring the analyte in an ultralow scale in environmental requests [114]. Traditional methods, unable to cater for the quick demands for on-site work, with high cost, have been efficaciously replaced by the miniature easy-handling, handy models of NSs. Innovative advancement in nanotechnology has paved a platform for a potential output for NSs with high sensitivity, good reproducibility, and explicit selectivity [115]. Optical micro-NSs' utility intercepts vividly employing basic tools such as optical fiber, surface plasmon resonance (SPR), and fluorescence capitalization, with absorbance and luminescence, supplemented with the characteristic changes according to the reaction needs of the system [116]. Optical sensors focus on fluorescence with a very quick response, SERS, and SPR protocols, while chemical sensors are based on molecular analyte recognition and transducers resulting in physical signals by the chemical change [117]. Interestingly, Paolo and his coworkers have reviewed AgNPs as probes for the detection of water contaminants [118].

7.4.3 Vehicles for Monitoring the Unwanted Analytes

NSs with functionalities of fluorescence have physicochemical features that enable the analyte to have a charge transference with the NMs to induce and enhance the binding portals. QDs, carbon dots (CDs), and CQDs have challenging fluorescence intensity, with a wide excitation outline, and fluorescence emission signals that depend upon the particle size, constricted emission bands, greater quantum efficacies, and a better photoinduced stability for a multiplexed recognition and competency of the contaminants [119]. Hence, a favorable outlook is noticed for pollutant sensing, by NMs in conjunction with adaptive large surface area and florescence. Doped and undoped NMs are classic metaphors for contaminant identification for small- and large-scale trials. The next suitable vehicle for contaminant export that monitors to broom and clean the restricted site of water bodies is carbon and their derives such as CNTs (hollow/multilayered dimensional structures), fullerenes (C-60: icosahedral), graphene, and GO. They are some active photocatalytic monitors of metallics, pesticides, and eminent degraders of dyes [120]. An alternative significant carrier of pollutant that has seemingly increased the attention of the research candidates for the shipment of cargo pollutants is the metal and MOs as NMs due to their inherent cohesiveness, modification of size requirement, good deliveries of adsorption, and suitable enhanced recoveries [121,122]. Table 7.1 delivers some platforms used for monitoring unwanted segments that pollute water samples by the potential NMs, detected at a very low level.

7.5 UNDESIRED TOXICANTS IN WATER SEGMENTS: REMOVAL

The third and the most vital segment in the three-way cyclic process, i.e., in conjunction with the monitoring of water pollutants, is toxin removal and is immensely sought after for a healthy living [137]. Toxicant removal, though a challenging part globally, can be encountered in a securely way, better than the existing conventional procedures (non-cost-effective/laborious) by adopting nano-range supplies using nanotechnology. However, utmost care has to be hired for the proper handling of NMs in environmental remediation approaches. The groundwater, surface water, river water, and wastewater are invariably wedged with undesirable contaminants of inorganic/organic (heavy metals/pesticides/textile materials/lifesaving medicinal components/healthcare and sanitization products) [138]. Nanotechnology offers a rapid sustainability, cost-effectivity, reproducibility, and recyclability, supported by the assets of copious protocols, for environmental water remediation. NMs such as NPs, NCs, CNTs, QDs, and M/MOs occupy a highest position in the podium of toxin exclusion. Filtration by nanofilters/nanomembranes, utilizing different techniques such as chemical reactions by NPs, adsorption by NPs, NCs, and C-derivatives, sorbents by MOs and its associates, and photocatalysis by NCs, NPs, C-derivatives, or polymers, render a very effective arena for remediation of contaminants in both platforms on- and off-site locations [139]. NMs, basically belonging either to inorganic or organic, can be classified into NSs/NPs/NCs, are synthesized individually or in fusion so as to be modulated for a better wanted result in terms of morphology (size, shape), porosity, surface area, and volume capacity [140]. NMs such as metallics – Ag, Au, Cu, and Fe [141] – and oxides of metallics – Zn, Cu, Fe, Ti, Al, and Si as NPs [142], NCs as oxyacids – with Bi, Ti, W, Mo, V, Fe, etc. [143], sulfides/selenides as QDs of Cd, Zn-functionalized oxides of Ag-Ti [144], nanostructures as carbon derivatives – C60, C, G, GO, organic polymers either form naturals such as sodium

TABLE 7.1

Platforms for Monitoring Unwanted Segments

NMs/ Nanosensors	On-/Off-Sites: Platforms (Water)	Pollutant Selectivity	Limits of Detection	Ref
MoS_2 QDs	River Qinhuai: Nanjing, (Jiangsu)/lake: Nanjing University. Sci. and Tech./tap: lab	FOX-7: explosive (1,1-diamino-2,2-dinitroethylene)	0.19 µM	[123]
ZnO QDs	River: Hooghly, West Bengal (India)	$Cr^{6+}/Fe^{3+}/Cu^{2+}$	16 pM (Cu^{2+}) and 0.18 nM (Cr^{6+})	[74]
CdS QDs/NMPPY	Channel: Yazd (Iran/Persian Gulf)	Picric acid	4.6×10^{-7} M	[124]
SnO_2 QDs	Deionized water/reclaimed water/seawater: Xinghai Bay: Yellow Sea, Dalian (China)	$Cd^{2+}/Fe^{3+}/Ni^{2+}/Pb^{2+}$	0.01 ppm (Ni^{2+})	[125]
GSH-Mn-ZnS QDs	Lake: Beihang University	$Pb^{2+}/Cr^{3+}/Hg^{2+}$	9.3×10^{-7} M	[126]
N-CQDs	Lake: South Lake: Changchun, Jilin Province (China) and tap (lab – Changchun)	Hg^{2+}	0.23 µM	[127]
P-CQDs	Raw/filtered/expelled/IISER, Tirupati (India)	Fe^{3+}	9.5 nM	[128]
S/N-CQDs	DI/mineral/tap/river/Jinan (China)	Cr^{3+}	6 µM	[129]
CDs (potato)	River: Brahmaputra – tannery water	Cr^{6+}/Fe^{3+}	0.012 µM (Cr^{6+})/0.000549 µM (Fe^{3+})	[130]
CDs–RhB	Tap: Central South University, *Hunan (China)*	Free Cl_2	4 µM	[131]
AgNPs–cysteine	Lake: Weiming Peking University/bottled mineral water: Wahaha	Cu^{2+}/Hg^{2+}	10 pM (Cu^{2+})/10 pM (Hg^{2+})	[132]
AuNPs/rGO	River: Yingtao/tap: lab/drinking pure-water Wahaha/Shanghai (China)	Pb^{2+}	1 nM	[133]
R-GO/TiO_2 NT	Lake: Qarun, Egypt	Hg(II)/Cu(II)/Mn(II)	4×10^{-11} M	[134]
Fe_3O_4@m-SiO_2/ PSA@Zr-MOF	Pond: Fudan University/lake: Jiading, Shanghai (China)	Bifenthrin: insecticide	0.5 µg/L	[135]
Cantilever nanobiosensor (Au, Si, enzyme)	River: northern Rio Grande do Sul/RS (Brazil)	Pb/Ni/Cd/Zn/Co/Al	0.32 (Pb)/0.87 (Ni)/0.33 (Cd)/0.48 (Zn)/0.42 (Co)/0.39 (Al): in ppb	[136]

NMPPY, N-methylpolypyrrole; GSH, glutathione; N-CQDs, nitrogen-doped CQDs; P-CQDs, phosphorus-doped CQDs; PSA, N-(n-propyl)ethylenediamine.

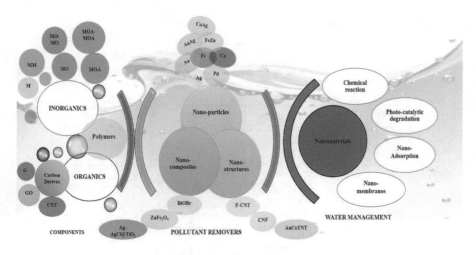

FIGURE 7.3 A perpetual visualization of NMs for water management.

alginate, chitin, starch, cellulose, chitosan, or polyvinyl glycol, poly(methyl methacrylate), polyaniline, etc. [145], to mention a few, are extensively deployed in the exclusions of microbial [146], radioactive nuclides, undesirables as organics/inorganics, dyes and drugs, pesticides and heavy metals, when percolated into the ecosystem. Figure 7.3 is a simple diagrammatic representation of some NMs that are actively utilized for water management.

Structural modifications of NMs install them to be a perfect candidate for environmental water reparation by specific schematic philosophies. Modulating the surface area with a nano-range size, altering the porosity for absorbency, and transformation in the band gap for electronic excitation to bring quick radical effect for an oxidative reduction render NMs for selectivity, sensitivity, reproducibility, recyclability, and removability of toxicants. The physisorption/chemisorption mechanisms that support the removability of the dyes, drugs, and heavy metals in many instances have been reported in many ongoing and off-going platforms for an intrinsic trial. A simple schematic philosophy practiced in environmental water redressal is pictured in Figure 7.4.

7.5.1 NPs for Water Remediation

NPs, a foremost subdivision in NMs, have M/MOs/M-oxoacids (MOAs) with the components in the atomic or molecular level, and the atoms that are not completely saturated resides at the outer surface level thus, chemically bind with other reacting species [147]. This situation significantly leads to adsorption, degradation, and bioremediation of contaminants in water ecosystem, for environmental clean-ups [148].

7.5.2 M/MOs/MOAs as Photocatalytic Degraders/Nano-Adsorbents

MOs/MOAs such as TiO_2 [149], Bi_2WO_6 [150], and Fe-MOF [151] have numerous roles in photocatalytic degradation (PCD) to disintegrate or transfigure dangerous contaminants

WATER / WASTE-WATER MANAGEMENT

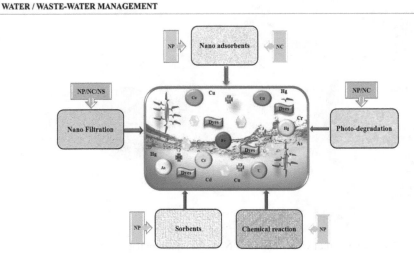

FIGURE 7.4   Schematic philosophies in environmental water redressal.

into unhazardous components [152]. The photoexcitation of electrons (e⁻) from valence band to conduction band, which results in e⁻/h⁺ (holes), undergoes suitable transformation to oxidize and degrade the pollutants by the formation of OH/O_2^- radicals, as the case may be. In a similar trend, the ratio of surface to volume and the porosity of NPs and/or NCs, magnetites [153], and other NMs [154] have been exploited for the capabilities as photocatalytic degraders and nano-absorbents with worthy cost-effectiveness and increased potentials for immobilizing the analytes for carrying and recuperating the removal proficiencies of the pollutants from water samples. Nano-MOAs with good removal efficiencies have different scales of degradation kinetic isotherms in photocatalytic (sunlight/UV–Vis energy) and adsorption mechanisms in different modes [155].

7.5.3   NCs as Photocatalytic Degraders/Nano-Adsorbents/Nanofilters

NCs with an inner organic core supported by outer inorganic shell or in reverse modes or conjunction of two or more (organic/inorganic) of nm range favor for an appropriate utilization in on-site/off-site sampling water-management technologies. In the past few years, appropriate amalgamation of NPs with polymers/carbon derivatives to form NCs has delivered constructive functionalities with remarkable properties, for water technologies. Bimetallics as photocatalysts offer an improved charge split-up and provide more active spots for redox reaction [156]. However, the cost-effective approaches are rendered by coupling noble metals with other inexpensive alternatives [157]. The advantage of low density with increased surface area primarily enhances the availability for absorbency, hence, for a good removal proficiency. Doped/undoped combination of Mn-Co doped [158], and CdS-Nylon6 [159], are available in the literature, for sustaining stability, thus, avoiding aggregation of the nanoclusters for easy accessibility. Vacancies that promote photocatalytic activity by electron transference are aided by surface alteration.

7.5.4 CARBON DERIVATIVES AS PHOTOCATALYTIC DEGRADERS/ NANO-ADSORBENTS/NANOFILTERS

Exceptional assemblies and powerful electronic features render NSs of carbon derivatives such as CNTs [160], g-C_3N_4 (graphitic nitride), and G [161]/rGO [162] for a specialized status in water nanoremediation factories, with advantages of speedy reaction kinetics for a quick removal, wide pertinence and non-restricted usability, enhanced surface area, rapid photocatalytic activity, and profound absorptivity. CNTs and cellulose nanofibers (CNFs) with unique characteristics are further modulated with metallic/other supports, to enrich the absorptivity for a wider segment of contaminants [163] with boosted mechanical and increased optical and electrochemical outputs. The surface functionalization ensures a rapid increase in N/O/S/H for its quick disposal for adsorption/degradation, with improved surface availability [164]. f-MWCNTs had agreeable nano-properties for a swift adsorption (118.41 mg/g) of harmful Cu^{2+} and removability of 93% [165]. In another study, Au sequined on Co/TNT (titanium nanotubes) was proven for an effective adsorption of 53%, with the remaining 47% removed by PCD of methylene blue (MB) dye in water with a profuse efficacy [166]. Nevertheless, ultrafine CNFs had an effective adsorption of 167 mg/g, for an operative elimination of the noxious radioactive component UO_2^{2+} that drifts into water causing contamination [167].

GO, an oxidated derivative of G, with functional water attracting parts imparts an exceptional dispersibility, with robust property for nanomembranes in nanofiltration and desalination techniques. Responsive sites are achieved by surface alteration, with the functional oxygen moiety responsible for functionalization, and hence, PCD/ filtration takes place for an easy removal/retention of pollutant in the membrane. The experimental studies using GO-NiO-β-CD functionalities were evidenced to fit PCD, and this had a good removal ability for MB (90%) and MV (~69.5) dye [168]. In one of the studies, it was found that GO/ethylene glycol (EG) membranes had an effective H_2O permeance, with a good rejection capacity of ~99.0% for the dye rhodamine B (RhB) [169]. In another trial, it was found that G-membranes were highly stable with a non-degrading capacity for more than ~3 months, with the good retention capacity of ~99.0% for dyes [170]. Similarly, ultrathin rGO nanomembranes that were efficient for nanofiltration and decontamination had a retention capacity of ~99% of the organic pollutant – MB dye that was tested [171]. Nanofilters with GO-Py-PDEAEMA (poly(N,N-diethylaminoethyl methacrylate)-pyrene) supported by CO_2/Ar had a good elimination capacity of 98.9% for RhB and 96.5% for methyl orange (MO), and hence, were evidenced as better alternatives for water treatment [172]. The nanomembrane platelet GO synthesized by vacuum filtration had a very good competency for isotopic D_2O/H_2O separation, using uncomplicated modules [173]. In an analogous style, GO-PVA-chitosan NC (PVA – polyvinyl alcohol) was verified for its efficiency for the takeaway of MB (97.5%) by the researchers in their work [174]. Maintenance of the quality of potable water is an essential task for a healthy living. Hence, in the trio cycle, the next crucial target is the removal of toxicants. Numerous efforts with unique novelties have been established for the removal of pollutant. Table 7.2 recapitulates some NMs that are used effectively and efficiently in water remediation technologies as nano-adsorbents (NADs) and for PCD of unwanted noxious components.

TABLE 7.2

NMs for Effective and Efficient Water Remediation Technologies

Removed Fixtures	Limits of Removal Efficiency	Removing Fixtures	Remarks	Ref
Methylene blue (MB)	99.53%	Fe_3O_4/AC/CD/Alg	NADs: Langmuir's isotherm	[175]
Anthraquinonic acid blue 25	89%	$ZnMoO_4$/BiFeWO$_6$/ rGO	PCD: Z-scheme	[176]
Acid orange II (AO II)	AO II (90%), COD (72%)	H_2O_2-meso-$ZnFe_2O_4$	Photocatalytic D/pseudo-first order	[177]
Micrococcus lylae, MB, Cr(VI)	Cr(VI) (90%), MB (55%), Micrococcus lylae (50%)	BiOBr	Adsorption, desorption, type IV isotherms – Langmuir	[178]
Methyl orange (MO)	(90, 94, 98)% –variable	TiO_2-γ-alumina	Photolysis – nanofiber membrane	[179]
Auramine O (AO)	AO (97–98.5%)	GO	PCD	[180]
Cd^{2+}/Zn^{2+}/Pb^{2+}/Cu^{2+}	Metal ions (95%)	Fe_3O_4@APS@ AA-co-CA: NCs	Adsorption: Freundlich/ Langmuir	[181]
As(III), As(V)	Water (91%–96%) and soil (78%)	Fe–Mn binary oxide/CMC	Adsorption: Langmuir	[182]
Acid Black-194	AB-194 100% and TOC 87%	Fe–P NPs	Pseudo-first order	[183]
Norfloxacin (NOX)	86%	Mn:ZnS QDs	PCD	[184]
Tetracycline (TC), total organic carbon	TC 83%, TOC 37%	$MWCNTs/TiO_2$: NCs	PCD, pseudo-first order	[185]
Levofloxacin (LEVO)	(99.2 – 99.6)% variable	ZnO NP/GO nanosheets	PCD/adsorption	[186]
Nitenpyram (NTP) and rhodamine B (RhB)	95% NTP and 100% (RhB)	g-C_3N_4/Ag_3PO_4/ AgI: NCs	PCD: Z-scheme	[187]
Methyl violet (MV) Brilliant Green (BG) NOX Ciprofloxacin (CPX) Cu^{2+}	MV 98.43% BG 93.53% NOX 96.57% CPX 94% Cu^{2+} 96.08%	Fe_3O_4/CD/AC/SA: NCs	Adsorption: Freundlich, Langmuir, Temkin	[188]
Cd^{2+}	94.5%	MgS@CNF	Adsorption: Langmuir, pseudo-first order	[189]
MB F$^-$	98.4% (MB) >80% (F$^-$)	FAT: NCs	PCD: MB, pseudo-first and -second order Adsorption: Langmuir (F$^-$)	[190]
CPX	87.70%	Bio-Pd: NPs	Adsorption: pseudo-second order	[191]
Celestine blue	81.3%	ZnO@ACP	Adsorption: Langmuir, pseudo-second order	[192]

(Continued)

TABLE 7.2 (*Continued*)
NMs for Effective and Efficient Water Remediation Technologies

Removed Fixtures	Limits of Removal Efficiency	Removing Fixtures	Remarks	Ref
Hg^{2+}	96%	Ag/Q: NPs (Ag/quartz)	Adsorption: Freundlich, Langmuir, Dubinin–Radushkevich, Redlich–Peterson, pseudo-second order	[193]
Congo Red	96.49%	Pb@ZnFe$_2$O$_4$: NCs	Adsorption: Langmuir, Freundlich, Temkin, Dubinin–Radushkevich, Harkins–Jura	[194]
Ibuprofen (Ibu) Naproxen (Nap) Diclofenac (Dic)	74.4% (Ibu) 86.9% (Nap) 91.4% (Dic)	Cu: NPs	Adsorption: Langmuir, Freundlich, Temkin, Dubinin–Radushkevich, Lagergren's first order, McKay–Ho second order, Elovich, Weber and Morris	[195]
Zn^{2+}	90%	CMMC	Adsorption: Langmuir	[196]
Pb^{2+} Cd^{2+} Ni^{2+}	99.5% (Pb^{2+}) 55.9% (Cd^{2+}) 24.3% (Ni^{2+})	CM-β-CD- Fe$_3$O$_4$	Adsorption: Langmuir, Freundlich, pseudo-second order	[197]

AC, activated carbon, CD, cyclodextrin; APS, 3-aminopropyltriethoxysilane; AA, acrylic acid; CA, crotonic acid; CMC, carboxymethyl cellulose; Fe–P NPs, Fe–polyphenol NPs; SA, sodium alginate; FAT, Fe/Al/Ti oxide; ACP, Ananas comosus waste; CMMC, cross-linked magnetic modified chitosan.

7.6 CONCLUSIONS AND PIONEERING OUTLOOK

Legitimate apprehension for a safe and pure water accessibility with boundaries of approachability in the diverged lifestyle can be well achieved eventually by nanotechnological revolutions. The occurrence of noxious waste as an obstacle in the path of water management, segregated into three frames as (1) analysis, (2) monitoring, and (3) removal, is flawlessly converged into a single wall by the efficacies of nano-metric scales. Simple and complex nanostructures in nano-metric range with interlinked bridges as NSs, nano-adsorbents, nanofilters, and nanomembranes or photodegraders clubbed with versatile chromatographic methodologies catalyze the path for an effective integrated three-way cyclic approach. The essential unavoidable wants (drugs, dyes, metallics, pesticides) but unwanted (as it pollutes), can be well handled and controlled through an eye of nanotechnological lines for remediating the water system, so as to protect the ecosystem. The discarded effluents from the industrial sectors such as – the heavy metal ions the drugs, the colorful dyes, the harsh pesticides, the avoidable microbiological colonies that are detrimental, can be best handled with proper utilization on NMs as NPs, NCs, NSs, and carbon derivatives to protect the depleting water system. Rising population and the future generation, henceforth, can be provided with a better-quality water for a good healthy longevity and sustainability in these present COVID times and others.

REFERENCES

1. S Wang, HC Green, ML Wilder, Q Du, BL Kmush, MB Collins, DA Larsend and T Zeng. 2020. "High-throughput wastewater analysis for substance use assessment in central New York during the COVID-19 pandemic." *Environment Science: Processes and Impacts.* 22, 2147. Doi: 10.1039/d0em00377h.
2. TM Rawson, LSP Moore, N Zhu, N Ranganathan, K Skolimowska, M Gilchrist, G Satta, G Cooke and A Holmes. 2020. "Bacterial and fungal co-infection in individuals with coronavirus: A rapid review to support COVID-19 antimicrobial prescribing." *Clinical Infectious Disease* 71 (9) 2459–2468. Doi: 10.1093/cid/ciaa530.
3. G Halkos and A Zisiadou. 2019. "An overview of the technological environmental hazards over the last century." *Economics of Disasters and Climate Change* 4 411–428. Doi: 10.1007/s41885-019-00053-z.
4. N Savage and MS. Diallo. 2005. "Nanomaterials and water purification: opportunities and challenges." *Journal of Nanoparticle Research* 7 331–342. Doi: 10.1007/s11051-005-7523-5.
5. R Chakraborty, A Asthana, AK Singh, S Yadav, MABH Susan and SAC Carabineiro. 2020. "Intensified elimination of aqueous heavy metal ions using chicken feathers chemically modified by a batch method." *Journal of Molecular Liquids* 312 113475. Doi: 10.1016/j.molliq.2020.113475.
6. Y Qian, C Qin, M Chen and S Lin. 2020. "Nanotechnology in soil remediation – applications vs. implications." *Ecotoxicology and Environmental Safety* 201 110815. Doi: 10.1016/j.ecoenv.2020.110815.
7. Q Zhao. 2020. "Nanomaterials developed for removing air pollutants." In *Advanced Nanomaterials for Pollutant Sensing and Environmental Catalysis*, 203–247. Elsevier. Doi: 10.1016/B978-0-12-814796-2.00006-X.
8. B Jain, AK Singh, S Banchhor, SB Jonnalagadda and MABH Susan. 2020. "Treatment of pharmaceutical wastewater by heterogeneous Fenton process: An innovative approach." *Nanotechnology for Environmental Engineering* 5 13. Doi: 10.1007/s41204-020-00075-z.
9. MM Khin, A Sreekumaran Nair, VJ Babu, R Murugan and S Ramakrishna. 2012. "A review on nanomaterials for environmental remediation." *Energy and Environmental Science* 5 8075–8109. Doi: 10.1039/C2EE21818F.
10. T Weldeslassie, H Naz, B Singh and M Oves. 2018. "Chemical contaminants for soil, air and aquatic ecosystem." In *Modern Age Environmental Problems and their Remediation*, 1–22. Springer International Publishing. Doi: 10.1007/978-3-319-64501-8_1.
11. R Chakraborty, A Asthana, AK Singh, B Jain and ABH Susan, 2020. "Adsorption of heavy metal ions by various low-cost adsorbents: A review." *International Journal of Environmental Analytical Chemistry.* Doi: 10.1080/03067319.2020.1722811.
12. R Chakraborty, R Verma, Anupama Asthana, SS Vidya and AK Singh. 2019. "Adsorption of hazardous chromium (VI) ions from aqueous solutions using modified sawdust: Kinetics, isotherm and thermodynamic modelling." *International Journal of Environmental Analytical Chemistry.* Doi: 10.1080/03067319.2019.1673743.
13. R Chakraborty, A Asthana, AK Singh, R Verma, S Sankarasubramanian, S Yadav, SAC Carabineiro and MABH Susan. 2020. "Chicken feathers derived materials for the removal of chromium from aqueous solutions: Kinetics, isotherms, thermodynamics and regeneration studies." *Journal of Dispersion Science and Technology.* Doi: 10.1080/01932691.2020.1842760.

14. RK Ibrahim, M Hayyan, MA AlSaadi, A Hayyan and S Ibrahim. 2016. "Environmental application of nanotechnology: Air, soil, and water." *Environment Science Pollution Research International* 23 (14) 13754–13788. Doi: 10.1007/s11356-016-6457-z.

15. A Hashmi, AK Singh, B Jain and SAC Carabineiro. 2020. "Chloramine-T/N-Bromosuccinimide/FeCl$_3$/KIO$_3$ decorated graphene oxide nanosheets and their antibacterial activity." *Nanomaterials* 10 105. Doi: 10.3390/nano10010105.

16. MF Zawrah, ESE El Shereefy and AY. Khudir. 2019. "Reverse precipitation synthesis of ≤ 10 nm magnetite nanoparticles and their application for removal of heavy metals from water." *Silicon* 11 85–104. Doi: 10.1007/s12633-018-9841-0.

17. R Dewangan, A Asthana, AK Singh and SAC Carabineiro. 2020. "Synthesis, characterization and antibacterial activity of a graphene oxide based NiO and starch composite material." *Journal of Dispersion Science and Technology.* Doi: 10.1080/01932691.2020.1844014.

18. X Jaramillo-Fierro, S González, HA Jaramillo and F Medina. 2020. "Synthesis of the ZnTiO$_3$/TiO$_2$ nanocomposite supported in ecuadorian clays for the adsorption and photocatalytic removal of methylene blue dye." *Nanomaterials* 10 1891. Doi: 10.3390/nano10091891.

19. A Hashmi, AK Singh, B Jain and A Singh. 2020. "Muffle atmosphere promoted fabrication of graphene oxide nanoparticle by agricultural waste". *Fullerenes, Nanotubes and Carbon Nanostructures.* Doi: 10.1080/1536383X.2020.1728744.

20. S Shad, M-FA Belinga-Desaunay-Nault, N Bashir and I Lynch. 2020. "Removal of contaminants from canal water using microwave synthesized zero valent iron nanoparticles." *Environmental Science: Water Research and Technology* 6 3057–3065. Doi: 10.1039/D0EW00157K.

21. T Pasinszki and M Krebsz. 2020. "Synthesis and application of zero-valent iron nanoparticles in water treatment, environmental remediation, catalysis, and their biological effects." *Nanomaterials* 10 917. Doi: 10.3390/nano10050917.

22. J Rashid, MA Barakat, N Salah and SS Habib. 2014. "ZnO-nanoparticles thin films synthesized by RF sputtering for photocatalytic degradation of 2-chlorophenol in synthetic wastewater." *Journal of Industrial and Engineering Chemistry* 23 134–139. Doi: 10.1016/j.jiec.2014.08.006.

23. JN Sharma, DK Pattadar, BP Mainali, and FP Zamborini. 2018. "Size determination of metal nanoparticles based on electrochemically measured surface-area-to-volume ratios." *Analytical Chemistry* 90(15) 9308–9314. Doi: 10.1021/acs.analchem.8b01905.

24. MT Amin, AA Alazba and U Manzoor. 2014. "A review of removal of pollutants from water/wastewater using different types of nanomaterials." *Advances in Materials Science and Engineering* 24. Doi: 10.1155/2014/825910.

25. R Soni, AK Pal, P Tripathi, JA Lal, K Kesari and V Tripathi. 2020. "An overview of nanoscale materials on the removal of wastewater contaminants." *Applied Water Science* 10 189. Doi: 10.1007/s13201-020-01275-3.

26. "UN World Water Development Report 2020 'Water and Climate Change'." 21 March 2020. https://www.unwater.org/world-water-development-report-2020-water-and-climate-change/.

27. AS Adeleye, JR Conway, K Garner, Y Huang, Y Su and AA Keller. 2016. "Engineered nanomaterials for water treatment and remediation: Costs, benefits, and applicability." *Chemical Engineering Journal* 286 640–662. Doi: 10.1016/j.cej.2015.10.105.

28. M Wen, G Li, H Liu, J Chen, T An and H Yamashita. 2019. "Metal–organic framework-based nanomaterials for adsorption and photocatalytic degradation of gaseous pollutants: recent progress and challenges." *Environmental Science: Nano* 6 1006–1025. Doi: 10.1039/c8en01167b.

29. TA Siddique, NK Dutta and NR Choudhury. 2020. "Nanofiltration for arsenic removal: Challenges, recent developments, and perspectives." *Nanomaterials* 10 1323. Doi: 10.3390/nano10071323.

30. M Anjum, R Miandad, M Waqas, F Gehany and MA Barakat. 2019. "Remediation of wastewater using various nano-materials." *Arabian Journal of Chemistry* 12 (8) 4897–4919. Doi: 10.1016/j.arabjc.2016.10.004.

31. HD Beyene and TG Ambaye. 2019. "Application of sustainable nanocomposites for water purification process." In *Sustainable Polymer Composites and Nanocomposites*, 387–412. Springer Nature. Doi: 10.1007/978-3-030–05399-4_14.

32. VB Cashin, DS Eldridge, A Yu and D Zhao. 2018. "Surface functionalization and manipulation of mesoporous silica adsorbents for improved removal of pollutants: A review." *Environmental Science: Water Research and Technology* 4 110–128. Doi: 10.1039/C7EW00322F.

33. S Liu, L Yuan, X Yue, Z Zheng and Z Tang. 2008. "Recent advances in nanosensors for organophosphate pesticide detection." *Advanced Powder Technology* 19 419–441. Doi: 10.1163/156855208X336684.

34. J Patel, B Jain, AK Singh, MABH Susan and L Jean-Paul. 2020. "Mn-doped ZnS quantum dots-an effective nanoscale sensor." *Microchemical Journal* 155 104755. Doi: 10.1016/j.microc.2020.104755.

35. SC Bernardo, ACA Sousa, MC Neves and MG Freire. 2019. "Use of nanomaterials in the pretreatment of water samples for environmental analysis." In *Nanomaterials for Healthcare, Energy and Environment. Advanced Structured Materials* 118 103–142. Springer Nature. Doi: 10.1007/978-981-13-9833-9_6.

36. C Lee Goi. 2020. "The river water quality before and during the Movement Control Order (MCO) in Malaysia." *Case Studies in Chemical and Environmental Engineering* 2 100027. Doi: 10.1016/j.cscee.2020.100027.

37. ML Brusseaul, L Pepper and CP Gerba. 2019. "The extent of global pollution." In *Environmental and Pollution Science* (3rd Edition), 3–8. Elsevier. Doi: 10.1016/B978-0-12-814719-1.00001-X.

38. WJ Cosgrove and DP Loucks. 2015. "Water management: Current and future challenges and research directions." *Water Resources Research* 51 (6) 4823–4839. Doi: 10.1002/2014WR016869.

39. A Nagar and T Pradeep. 2020. "Clean water through nanotechnology: Needs, gaps, and fulfilment." *ACS Nano* 14 (6) 6420–6435. Doi: 10.1021/acsnano.9b01730.

40. A Baruah, V Chaudhary, R Malik and VK Tomer. 2019. "Nanotechnology based solutions for wastewater treatment." In *Nanotechnology in Water and Wastewater Treatment. Theory and Applications Micro and Nano Technologies*, 337–368. Elsevier. Doi: 10.1016/B978-0-12-813902-8.00017-4.

41. GN Hlongwane, PT Sekoai, M Meyyappan and K Moothi. 2019. "Simultaneous removal of pollutants from water using nanoparticles: A shift from single pollutant control to multiple pollutant control." *Science of the Total Environment* 656 808–833. Doi: 10.1016/j.scitotenv.2018.11.257.

42. IS Yunus, Harwin, A Kurniawan, D Adityawarman and A Indarto. 2012. "Nanotechnologies in water and air pollution treatment." *Environmental Technology Reviews* 1 (1) 136–148. Doi: 10.1080/21622515.2012.733966.

43. M Liang and L-H Guo. 2009. "Application of nanomaterials in environmental analysis and monitoring." *Journal of Nanoscience and Nanotechnology* 9 (4) 2283–2289. Doi:10.1166/jnn.2009.SE22.

44. SR Beeram, E Rodriguez, S Doddavenkatanna, Z Li, A Pekarek, D Peev, K Goerl, G Trovato, T Hofmann and DS Hage. 2017. "Nanomaterials as stationary phases and supports in liquid chromatography: A review." *Electrophoresis* 38 (19) 1–48. Doi: 10.1002/elps.201700168.

45. "Water sampling and analysis." 51–72. www.who.int › dwq. Water Sampling and Analysis- World Health Organisation.

46. I Liška. 2006. "Pesticides in water: Sampling, sample preparation, preservation." In *Encyclopaedia of Analytical Chemistry: Applications, Theory and Instrumentation*, 1–16. John Wiley & Sons. Doi: 10.1002/9780470027318.a1723.

47. S Büyüktiryaki, R Keçili and CM Hussain. 2020. "Functionalized nanomaterials in dispersive solid phase extraction: Advances & prospects." *Trends in Analytical Chemistry* 127 115893. Doi: 10.1016/j.trac.2020.115893.

48. SN Zulkifli, HA Rahim and W-J Lau. 2018. "Detection of contaminants in water supply: A review on state-of-the-art monitoring technologies and their applications." *Sensors and Actuators B* 255 2657–2689. Doi: 10.1016/j.snb.2017.09.078.

49. A Medina, FA Casado-Carmona, AI Lopez-Lorente and S Cardenas. 2020. "Magnetic graphene oxide composite for the microextraction and determination of benzophenones in water samples." *Nanomaterials* 10 168. Doi: 10.3390/nano10010168.

50. J Hollender, J Rothardt, D Radny, M Loos, J Epting, P Huggenberger, P Borer and H Singer. 2018. "Comprehensive micropollutant screening using LC-HRMS/MS at three riverbank filtration sites to assess natural attenuation and potential implications for human health." *Water Research* 1 (1) 100007. Doi: 10.1016/j.wroa.2018.100007.

51. S Ozcan, A Tor and ME Aydin. 2012. "Application of magnetic nanoparticles to residue analysis of organochlorine pesticides in water samples by GC/MS." *Journal of AOAC International* 95 (5) 1343–1349. Doi: 10.5740/jaoacint.SGE_Ozcan.

52. ZA ALOthman and SM Wabaidur. 2019. "Application of carbon nanotubes in extraction and chromatographic analysis: A review." *Arabian Journal of Chemistry* 12 (5) 633–651. Doi: 10.1016/j.arabjc.2018.05.012.

53. KM Giannoulis, DL Giokas, GZ Tsogas and AG Vlessidis. 2014. "Ligand-free gold nanoparticles as colorimetric probes for the non-destructive determination of total dithiocarbamate pesticides after solid phase extraction." *Talanta* 119 276–283. Doi: 10.1016/j.talanta.2013.10.063.

54. K Godlewska, P Stepnowski and M Paszkiewicz. 2020. "Application of the polar organic chemical integrative sampler for isolation of environmental micropollutants – a review." *Critical Reviews in Analytical Chemistry* 50 (1) 1–28. Doi: 10.1080/10408347.2019.1565983.

55. H Lu, J Wang, M Stoller, T Wang, Y Bao and H Hao. 2016. "An overview of nanomaterials for water and wastewater treatment." *Advances in Materials Science and Engineering* 10. Doi: 10.1155/2016/4964828.

56. J de Jong, T van Buuren and JPA Luiten. 1996. "Systematic approaches in water management: Aquatic outlook and decision support systems combining monitoring, research, policy analysis and information technology." *Water Science and Technology* 34 (12) 9–16. Doi: 10.1016/S0273-1223(96)00848-7.

57. A Spietelun, Ł Marcinkowski, M de la Guardia and J Namieśnik. 2014. "Green aspects, developments and perspectives of liquid phase microextraction techniques." *Talanta* 119 34–45. Doi: 10.1016/j.talanta.2013.10.050.

58. J Aceña, S Stampachiacchiere, S Pérez and D Barceló. 2015. "Advances in liquid chromatography-high-resolution mass spectrometry for quantitative and qualitative environmental analysis." *Analytical and Bioanalytical Chemistry* 407 (21) 6289–99. Doi: 10.1007/s00216-015-8852-6.

59. AS Abboud, MM Sanagi, WAW Ibrahim, AS Abdul Keyon and HY Aboul-Enein. 2018. "Calcium alginate-caged multiwalled carbon nanotubes dispersive microsolid phase extraction combined with gas chromatography-flame ionization detection for the determination of polycyclic aromatic hydrocarbons in water samples." *Journal of Chromatographic Science* 56 (2) 177–186. Doi: 10.1093/chromsci/bmx095.

60. VK Hoang Bui, D Park, TN Pham, Y An, JS Choi, HU Lee, O-H Kwon, J-Y Moon, K-T Kim and Y-C Lee. 2019. "Synthesis of MgAC-Fe$_3$O$_4$/TiO$_2$ hybrid nanocomposites via sol-gel chemistry for water treatment by photo-Fenton and photocatalytic reactions." *Science Reports* 9 11855. Doi: 10.1038/s41598-019-48398-5.

61. S Mura, Y Jiang, I Vassalini, A Gianoncelli, I Alessandri, G Granozzi, L Calvillo, N Senes, S Enzo, P Innocenzi and L Malfatti. 2018. "Graphene oxide/iron oxide nano-composites for water remediation." *ACS Applied Nano Materials* 1 (12) 6724–6732. Doi: 10.1021/acsanm.8b01540.

62. SK Das, AR Das and AK Guha. 2009. "Gold nanoparticles: Microbial synthesis and application in water hygiene management." *Langmuir* 25 (14) 8192–8199. Doi: 10.1021/la900585p.

63. F Brandl, N Bertrand, EM Lima and R Langer. 2015. "Nanoparticles with photoin-duced precipitation for the extraction of pollutants from water and soil." *Natures Communication* 6 7765. Doi: 10.1038/ncomms8765.

64. C Wang and C Yu. 2013. "Detection of chemical pollutants in water using gold nanopar-ticles as sensors: A review." *Review in Analytical Chemistry* 32 (1) 1–14. Doi: 10.1515/revac-2012-0023.

65. S Yaqub, U Latif, and FL Dickert. 2011. "Plastic antibodies as chemical sensor material for atrazine detection." *Sensors and Actuators B* 160 227–233. Doi: 10.1016/j.snb.2011.07.039.

66. N-U-A Babar, KS Joya, MA Tayyab, MN Ashiq and M Sohail. 2019. "Highly sensi-tive and selective detection of arsenic using electrogenerated nanotextured gold assem-blage." *ACS Omega* 4 (9) 13645–13657. Doi: 10.1021/acsomega.9b00807.

67. SBD Borah, T Bora, S Baruah and J Dutta. 2015. "Heavy metal ion sensing in water using surface plasmon resonance of metallic nanostructures." *Groundwater for Sustainable Development* 1 1–11. Doi: 10.1016/j.gsd.2015.12.004.

68. C Steffens, J Steffens, AM Graboski, A Manzoli and FL Leite. 2017. "Nanosensors for detection of pesticides in water." In *New Pesticides and Soil Sensors*, 595–634. Elsevier. Doi: 10.1016/B978-0-12-804299-1.00017-5.

69. N Ullah, M Mansha, I Khan and A Qurashi. 2018. "Nanomaterial-based optical chemi-cal sensors for the detection of heavy metals in water: Recent advances and challenges." *Trends in Analytical Chemistry* 100 155–166. Doi: 10.1016/j.trac.2018.01.002.

70. AA Yaqoob, T Parveen, K Umar and MNM Ibrahim. 2020. "Role of nanomaterials in the treatment of wastewater: A review." *Water* 12 495. Doi: 10.3390/w12020495.

71. B Kaur, R Srivastava and B Satpati. 2015. "Ultratrace detection of toxic heavy metal ions found in water bodies using hydroxyapatite supported nanocrystalline ZSM-5 modified electrodes." *New Journal of Chemistry* 39 5137–5149. Doi: 10.1039/c4nj02369b.

72. MR Willner and PJ Vikesland. 2018. "Nanomaterial enabled sensors for environmental contaminants." *Journal of Nanobiotechnology* 16, 95. Doi: 10.1186/s12951-018-0419-1.

73. SK Tuteja, DAN Dilbaghi and E Lichtfouse. 2020. *"Nanosensors for Environmental Applications"*. Springer Nature.

74. MMR Khan, T Mitra and D Sahoo. 2020. "Metal oxide QD based ultrasensitive micro-sphere fluorescent sensor for copper, chromium and iron ions in water." *RSC Advance* 10 9512. Doi: 10.1039/c9ra09985a.

75. A Boruah, M Saikia, T Das, RL Goswamee and BK Saikia. 2020. "Blue-emitting fluo-rescent carbon quantum dots from waste biomass sources and their application in fluo-ride ion detection in water." *Journal of Photochemistry & Photobiology, B: Biology* 209 111940. Doi: 10.1016/j.jphotobiol.2020.111940.

76. P Devi, P Rajput, A Thakur, K-H Kim and P Kumar. 2019. "Recent advances in carbon quantum dot-based sensing of heavy metals in water." *Trends in Analytical Chemistry* 114 171e195. Doi: 10.1016/j.trac.2019.03.003.

77. TTS Lew, M Park, J Cui and MS Strano. 2020. "Plant nanobionic sensors for arsenic detection." *Advanced Materials* 33(1) 2005683. Doi: 10.1002/adma.202005683.

78. C Xing, H Kuang, C Hao, L Liu, L Wang and C Xu. 2014. "A silver enhanced and sen-sitive strip sensor for cadmium detection." *Food and Agricultural Immunology* 25 (2) 287–300. Doi: 10.1080/09540105.2013.781140.

79. H Xiang, Q Cai, Y Li, Z Zhang, L Cao, K Li and H Yang. 2020. "Sensors applied for the detection of pesticides and heavy metals in freshwaters." *Journal of Sensors* 22. Doi: 10.1155/2020/8503491.

80. AG Memon, X Zhou, Y Xing, R Wang, L Liu, M Khan and M He. 2019. "Label-free colorimetric nanosensor with improved sensitivity for Pb^{2+} in water by using a truncated 8–17 DNAzyme." *Frontiers of Environmental Science and Engineering* 13 (1) 12. Doi: 10.1007/s11783-019-1094-7.

81. Q Wei, R Nagi, K Sadeghi, S Feng, E Yan, S Jung Ki, R Caire, D Tseng and A Ozcan. 2014. "Detection and spatial mapping of mercury contamination in water samples using a smart-phone." *ACS Nano* 8 (2) 1121–1129. Doi: 10.1021/nn406571t.

82. D Sahoo, A Mandal, T Mitra, K Chakraborty, M Bardhan, and AK Dasgupta. 2018. "Nanosensing of pesticides by zinc oxide quantum dot: An optical and electrochemical approach for the detection of pesticides in water." *Journal of Agricultural and Food Chemistry* 66 414–423. Doi: 10.1021/acs.jafc.7b04188.

83. MHM Facure, LA Mercante, LHC Mattoso and DS Correa. 2017. "Detection of trace levels of organophosphate pesticides using an electronic tongue based on graphene hybrid nanocomposites." *Talanta* 167 59–66. Doi: 10.1016/j.talanta.2017.02.005.

84. S Firdoz, F Ma, X Yue, Z Dai, A Kumar and B Jiang. 2010. "A novel amperometric biosensor based on single walled carbon nanotubes with acetylcholine esterase for the detection of carbaryl pesticide in water." *Talanta* 83 269–273. Doi: 10.1016/j.talanta.2010.09.028.

85. JA Buledi, S Amin, SI Haider, MI Bhanger and AR Solangi. 2020. "A review on detection of heavy metals from aqueous media using nanomaterial-based sensors." *Environmental Science and Pollution Research.* Doi: 10.1007/s11356-020-07865-7.

86. R Das, CD Vecitis, A Schulze, B Cao, AF Ismail, X Lu, J Chene and S Ramakrishna. 2017. "Recent advances in nanomaterials for water protection and monitoring." *Chemical Society Review* 46 6946. Doi: 10.1039/c6cs00921b.

87. L Madhura, S Singh, S Kanchi, M Sabela, and K Bisetty. 2019. "Nanotechnology-based water quality management for wastewater treatment." *Environmental Chemistry Letters* 17 65–121. Doi: 10.1007/s10311-018-0778-8.

88. Vikesland PJ 2018. "Nanosensors for water quality monitoring." *Nature Nanotechnology* 13 651–660. Doi: 10.1038/s41565-018-0209-9.

89. J Plazas-Tuttle, FM Giraldo and A Avila. 2020. "Nano-enabled technologies for wastewater remediation." In *Nanomaterials for the Detection and Removal of Wastewater Pollutants. Micro and Nano Technologies*, 1–17. Elsevier. Doi: 10.1016/B978-0-12-818489-9.00001-3.

90. "Developing drinking-water quality regulations and standards, 14 October 2018 | Report." https://www.who.int/publications/i/item/9789241513944, https://www.un.org/sustainabledevelopment/water-and-sanitation/

91. M Li, H Gou, I Al-Ogaidi and N Wu. 2013. "Nanostructured sensors for detection of heavy metals: A review." *ACS Sustainable Chemistry and Engineering* 1 (7) 713–723. Doi: 10.1021/sc400019a.

92. M Ghadimi, S Zangenehtabar and S Homaeigohar. 2020. "An overview of the water remediation potential of nanomaterials and their ecotoxicological impacts." *Water* 12 1150. Doi: 10.3390/w12041150.

93. D Song, R Yang, H Wang, W Li, H Wang, H Long and F Long. 2017. "A label-free SERRS-based nanosensor for ultrasensitive detection of mercury ions in drinking water and wastewater effluent." *Analytical Methods* 9 154–162. Doi: 10.1039/C6AY02361D.

94. Y Guo, C Liu, R Ye and Q Duan. 2020. "Advances on water quality detection by UV-Vis spectroscopy." *Applied Sciences* 6874. Doi: 10.3390/app10196874.

95. M Annadhasan, T Muthukumarasamyvel, VR Sankar Babu and N Rajendiran. 2014. "Green synthesized silver and gold nanoparticles for colorimetric detection of Hg^{2+}, Pb^{2+}, and Mn^{2+} in aqueous medium." *ACS Sustainable Chemistry and Engineering* 2 (4) 887–896. Doi: 10.1021/sc400500z.

96. SA Jame and Z Zhou. 2016. "Electrochemical carbon nanotube filters for water and wastewater treatment". *Nanotechnology Reviews* 5 (1). Doi: 10.1515/ntrev-2015-0056.

97. Q Zheng, DT Tan and D Shuai. 2016. "Research highlights: Visible light driven photocatalysis and photoluminescence and their applications in water treatment." *Environmental Science: Water Research and Technology* 2 13–16. Doi: 10.1039/c5ew90026c.

98. KV Ragavan and S Neethirajan. 2019. "Nanoparticles as biosensors for food quality and safety assessment." In *Nanomaterials for Food Applications Micro and Nano Technologies*, 147–202. Elsevier. Doi: 10.1016/B978-0-12-814130-4.00007-5.

99. Z Han, Z Yang, H Sun, Y Xu, X Ma, D Shan, J Chen, S Huo, Z Zhang, P Du and X Lu. 2019. "Electrochemiluminescence platforms based on small water-insoluble organic molecules for ultrasensitive aqueous-phase detection." *Angewendte Chemie International Edition*. Wiley VCH. 58 (18) 5915–5919. Doi: 10.1002/anie.201814507.

100. X-y Xue, R Cheng, L Shi, Z Ma and X Zheng. 2020. "Nanomaterials for monitoring and remediation of water pollution." In *Nanoscience in Food and Agriculture 2. Sustainable Agriculture Reviews* 21 207–233. Springer. Doi: 10.1007/978-3-319-39306-3_6.

101. L Falciola, V Pifferi and A Testolin. 2020. "Detection methods of wastewater contaminants: State of the art and role of nanotechnology." In *Nanomaterials for the Detection and Removal of Wastewater Pollutants. Micro and Nano Technologies*, 47–68. Elsevier. Doi: 10.1016/B978-0-12-818489-9.00003-7.

102. Y Lu, L Doan, A Bafana, G Yu, C Jeffryes, T Benson, S Wei and EK Wujcik. 2019. "Multifunctional nanocomposite sensors for environmental monitoring." In *Polymer-Based Multifunctional Nanocomposites and Their Applications*, 157–174. Elsevier. Doi: 10.1016/B978-0-12-815067-2.00006-8.

103. NG Dlamini, AK Basson and VSR Pullabhotla. 2020. "A comparative study between bimetallic Iron@copper nanoparticles with iron and copper nanoparticles synthesized using a bioflocculant: Their applications and biosafety." *Processes* 8 1125 Doi: 10.3390/pr8091125.

104. AS Mahmoud, RS Farag and MM Elshfai. 2020. "Reduction of organic matter from municipal wastewater at low cost using green synthesis nano iron extracted from black tea: Artificial intelligence with regression analysis." *Egyptian Journal of Petroleum* 19 9–20. Doi: 10.1016/j.ejpe.2019.09.001.

105. P Wang, M Luo, D Liu, J Zhan, X Liu, F Wang, Z Zhou and P Wang. 2018. "Application of a magnetic graphene nanocomposite for organophosphorus pesticide extraction in environmental water samples." *Journal of Chromatography A* 1535 9–16. Doi: 10.1016/j.chroma.2018.01.003.

106. Q Wu, G Zhao, C Feng, C Wang and Z Wang. 2011. "Preparation of a graphene-based magnetic nanocomposite for the extraction of carbamate pesticides from environmental water samples." *Journal of Chromatography A*, 1218 7936–7942. Doi: 10.1016/j.chroma.2011.09.027.

107. K-S Lin, N-B Chang and T-D Chuang. 2008. "Fine structure characterization of zero-valent iron nanoparticles for decontamination of nitrites and nitrates in wastewater and groundwater." *Science and Technology of Advanced Materials* 9 (2) 025015-9 p. Doi: 10.1088/1468-6996/9/2/025015.

108. K Gupta and UC Ghosh. 2009. "Arsenic removal using hydrous nanostructure iron(III)-titanium(IV) binary mixed oxide from aqueous solution." *Journal of Hazardous Materials* 161 (2–3) 884–892. Doi: 10.1016/j.jhazmat.2008.04.034.

109. T Wang, X Jin, Z Chen, M Megharaj and R Naidu. 2014. "Green synthesis of Fe nanoparticles using eucalyptus leaf extracts for treatment of eutrophic wastewater." *Science of the Total Environment* 466–467 210–213. Doi: 10.1016/j.scitotenv.2013.07.022.

110. T Wang, J Lin, Z Chen, M Megharaj and R Naidu. 2014. "Green synthesized iron nanoparticles by green tea and eucalyptus leaves extracts used for removal of nitrate in aqueous solution." *Journal of Cleaner Production* 83 413–419. Doi: 10.1016/j. jclepro.2014.07.006.

111. HR Mahmoud, SA El-Molla and M Saif. 2013. "Improvement of physicochemical properties of Fe_2O_3/MgO nanomaterials by hydrothermal treatment for dye removal from industrial wastewater." *Powder Technology* 249 225–233. Doi: 10.1016/j. powtec.2013.08.021.

112. NN Nassar, LA Arar, NN Marei, MM Abu Ghanim, MS Dwekat and SH Sawalha. 2014. "Treatment of olive mill based wastewater by means of magnetic nanoparticles: Decolourization, dephenolization and COD removal." *Environmental Nanotechnology, Monitoring and Management* 1–2 14–23. Doi: 10.1016/j.enmm. 2014.09.001.

113. Y Dahman. 2017. "Nanosensors." In *Nanotechnology and Functional Materials for Engineers. Micro and Nano Technologies*, 67–91. Elsevier. Doi: 10.1016/ B978-0-323-51256-5.00004-6.

114. CC Bueno, PS Garcia, C Steffens, DK Deda and F de Lima Leite. 2017. "Nanosensors." In *Nanoscience and its Applications Micro and Nano Technologies*, 121–153. Elsevier. Doi: 10.1016/B978-0-323-49780-0.00005-3.

115. B Kuswandi, D Futra and LY Heng. 2017. "Nanosensors for the detection of food contaminants." *Nanotechnology Applications in Food Flavor, Stability, Nutrition and Safety* 307–333. Doi: 10.1016/B978-0-12-811942-6.00015-7.

116. RM Balakrishnan, P Uddandarao, K Raval and R Raval. 2019. "A perspective of advanced biosensors for environmental monitoring." *Tools, Techniques and Protocols for Monitoring Environmental Contaminants* 19–51. Doi: 10.1016/ B978-0-12-814679-8.00002-9.

117. I Yaroshenko, D Kirsanov, M Marjanovic, PA Lieberzeit, O Korostynska, A Mason, I Frau and A Legin. 2020. "Real-time water quality monitoring with chemical sensors." *Sensors* 20 3432. Doi: 10.3390/s20123432.

118. P Prosposito, L Burratti and I Venditti. 2020. "Silver nanoparticles as colorimetric sensors for water pollutants." *Chemosensors* 8 26. Doi: 10.3390/chemosensors8020026.

119. M Bhatt, S Bhatt, G Vyas, IH Raval, S Haldar and P Paul. 2020. "Water-dispersible fluorescent carbon dots as bioimaging agents and probes for Hg^{2+} and Cu^{2+} ions." *ACS Applied Nano Material* 3 (7) 7096–7104. Doi: 10.1021/acsanm.0c01426.

120. G Asgari, A Seidmohammadi, A Esrafili, J Faradmal, MN Sepehrgh and M Jafarinia. 2020. "The catalytic ozonation of diazinon using nano-MgO@Cnt@Gr as a new heterogenous catalyst: The optimization of effective factors by response surface methodology." *RSC Advance* 10 7718. Doi: 10.1039/c9ra10095d.

121. C Ray and T Pal. 2017. "Recent advances of metal-metal oxide nanocomposites and their tailored nanostructures in numerous catalytic applications." *Journal of Material Chemistry A* 5 9465–9487. Doi: 10.1039/C7TA02116J.

122. S Ashraf, A Siddiqa, S Shahida and S Qaisar. 2019. "Titanium-based nanocomposite materials for arsenic removal from water: A review." *Heliyon* 5 (5) e01577. Doi: 10.1016/j.heliyon. 2019.e01577.

123. S Feng, J Lv, F Pei, X Lv, Y Wu, Q Hao, Y Zhang, Z Tong and W Lei. 2020. "Fluorescent MoS_2 QDs based on IFE for turn-off determination of FOX-7 in real water samples." *Spectrochimica Acta Part A: Molecular and Biomolecular Spectroscopy* 231 118131. Doi: 10.1016/j.saa.2020.118131.

124. F Abbasi, A Akbarinejad and N Alizadeh. 2019. "CdS QDs/N-methylpolypyrrole hybrids as fluorescent probe for ultrasensitive and selective detection of picric acid." *Spectrochimica Acta Part A: Molecular and Biomolecular Spectroscopy* 216 230–235. Doi: 10.1016/j.saa.2019.03.032.

125. J Liu, Q Zhang, W Xue, H Zhang, Y Bai, L Wu, Z Zhai and G Jin. 2019. "Fluorescence characteristics of aqueous synthesized tin oxide quantum dots for the detection of heavy metal ions in contaminated water." *Nanomaterials (Basel)* 9 (9) 1294. Doi: 10.3390/nano9091294.
126. J Liu, G Lv, W Gu, Z Li, A Tang and L Mei. 2017. "A novel luminescence probe based on layered double hydroxides loaded with quantum dots for simultaneous detection of heavy metal ions in water." *Journal of Materials Chemistry C* 5 5024–5030. Doi: 10.1039/C7TC00935F.
127. R Zhang and W Chen. 2014. "Nitrogen-doped carbon quantum dots: Facile synthesis and application as a "Turn-Off" fluorescent probe for detection of Hg^{2+} ions." *Biosensors and Bioelectronics* 55 83–90. Doi: 10.1016/j.bios.2013.11.074.
128. G Kalaiyarasan, J Joseph and P Kumar. 2020. "Phosphorus-doped carbon quantum dots as fluorometric probes for iron detection." *ACS Omega* 5 (35) 22278–22288. Doi: 10.1021/acsomega.0c02627.
129. C Wang, J Xu, H Li and W Zhao. 2020. "Tunable multicolour S/N co-doped carbon quantum dots synthesized from waste foam and application to detection of Cr^{3+} ions." *Luminescence* 35 (8) 1373–1383. Doi: 10.1002/bio.3901.
130. R Sinha, AP Bidkar, R Rajasekhar, SS Ghosh and TK Manda. 2020. "A facile synthesis of nontoxic luminescent carbon dots for detection of chromium and iron in real water sample and bio-imaging." *The Canadian Journal of Chemical Engineering* 98 (1) 194–204. Doi: 10.1002/cjce.23630.
131. Y Ding, J Ling, J Cai, S Wang, X Li, M Yang, L Zha and J Yan. 2016. "A carbon dot-based hybrid fluorescent sensor for detecting free chlorine in water medium." *Analytical Methods* 8 1157–1161. Doi: 10.1039/C5AY03143E.
132. F Li, J Wang, Y Lai, C Wu, S Sun, Y He and H Ma. 2013. "Ultrasensitive and selective detection of copper (II) and mercury (II) ions by dye-coded silver nanoparticle-based SERS probes." *Biosensors and Bioelectronics* 39 82–87. Doi: 10.1016/j.bios.2012.06.050.
133. L Zhao, W Gu, C Zhang, X Shi and Y Xian. 2016. "*In situ* regulation nanoarchitecture of Au nanoparticles/reduced graphene oxide colloid for sensitive and selective SERS detection of lead ions." *Journal of Colloid and Interface Science* 465 279–285. Doi: 10.1016/j.jcis.2015.11.073.
134. IH Abdullaha, N Ahmed, MA Mohamed, FMA Ragab, MTA Abdel-Wareth and NK Allam. 2018. "Engineered nanocomposite for sensitive and selective detection of mercury in environmental water samples." *Analytical Methods* 10 2526–2535. Doi: 10.1039/C8AY00618K.
135. M Xu, K Chen, C Luo, G Song, Y Hu and H Cheng. 2017. "Synthesis of Fe_3O_4@m-SiO_2/PSA@Zr-MOF nanocomposites for bifenthrin determination in water samples." *Chromatographia* 80 463–471. Doi: 10.1007/s10337-017-3253-y.
136. AA Rigo, AM De Cezaro, DK Muenchen, J Martinazzo, A Manzoli, J Steffens and C Steffens. 2019. "Heavy metals detection in river water with cantilever nanobiosensor." *Journal of Environmental Science and Health, Part B* 55 (3) 239–249. Doi: 10.1080/03601234.2019.1685318.
137. B Nowack. "Pollution prevention and treatment using nanotechnology." In *Nanotechnology. Volume 2: Environmental Aspects.* 2008 Wiley-VCH Verlag GmbH & Co. Doi: 10.1002/9783527628155.nanotech010.
138. F Lu and D Astruc. 2018. "Nanomaterials for removal of toxic elements from water." *Coordination Chemistry Reviews* 356 147–164. Doi: 10.1016/j.ccr.2017.11.003.
139. FD Guerra, MF Attia, DC Whitehead and F Alexis. 2018. "Nanotechnology for environmental remediation: Materials and applications." *Molecules* 23 1760. Doi: 10.3390/molecules23071760.

140. S Taghipour, SM Hosseini and B Ataie-Ashtiani. 2019. "Engineering nanomaterials for water and wastewater treatment: Review of classifications, properties and applications." *New Journal of Chemistry* 43 7902–7927. Doi: 10.1039/C9NJ00157C.

141. S Das, J Chakraborty, S Chatterjee and H Kumar. 2018. "Prospects of biosynthesized nanomaterials for the remediation of organic and inorganic environmental contaminants." *Environmental Science: Nano* 5 2784. Doi: 10.1039/c8en00799c.

142. G Ghasemzadeh, M Momenpour, F Omidi, MR Hosseini, M Ahani and A Barzegari. 2014. "Applications of nanomaterials in water treatment and environmental remediation." *Frontiers of Environmental Science and Engineering* 8 471–482. Doi: 10.1007/s11783-014-0654-0.

143. H Park, A Bak, YY Ahn, J Choi and MR Hoffmannn. 2012. "Photoelectrochemical performance of multi-layered BiOx–TiO₂/Ti electrodes for degradation of phenol and production of molecular hydrogen in water." *Journal of Hazardous Materials* 211–212 47–54. Doi: 10.1016/j.jhazmat.2011.05.009.

144. DT Sass, ESM Mouele and N Ross. 2019. "Nano silver-iron-reduced graphene oxide modified titanium dioxide photocatalytic remediation system for organic dye." *Environments* 6 106. Doi: 10.3390/environments6090106.

145. N Pandey, SK Shukla and NB Singh. 2017. "Water purification by polymer nanocomposites: An overview." *Nanocomposites* 47–66. Doi: 10.1080/20550324.2017.1329983.

146. SY Rikta. 2019. "Application of nanoparticles for disinfection and microbial control of water and wastewater." In *Nanotechnology in Water and Wastewater Treatment. Theory and Applications Micro and Nano Technologies*, 159–176. Elsevier Inc. Doi: 10.1016/B978-0-12-813902-8.00009-5.

147. N Pourreza, S Rastegarzadeh and A Larki. 2014. "Nano-TiO₂ modified with 2-mercaptobenzimidazole as an efficient adsorbent for removal of Ag(I) from aqueous solutions." *Journal of Industrial and Engineering Chemistry* 20 127–132. Doi: 10.1016/j.jiec.2013.04.016.

148. PS Goh, CS Ong, BC Ng and AF Ismail. 2019. "Applications of emerging nanomaterials for oily wastewater treatment." In *Nanotechnology in Water and Wastewater Treatment*, 101–113. Elsevier Inc. Doi: 10.1016/B978-0-12-813902-8.00005-8.

149. S Kanan, MA Moyet, RB Arthur and HH Patterson. 2019. "Recent advances on TiO₂-based photocatalysts toward the degradation of pesticides and major organic pollutants from water bodies." *Catalysis Reviews* 62 (1). Doi: 10.1080/01614940.2019.1613323.

150. VP Singh, M Sharma and R Vaish. 2020. "Enhanced dye adsorption and rapid photo catalysis in candle soot coated Bi₂WO₆ ceramics." *Engineering Research Express* 1 025056. Doi: 10.1088/2631-8695/ab5e93.

151. K Li, YdR de Mimérand, X Jin, J Yi and J Guo. 2020. "Metal oxide (ZnO and TiO₂) and Fe-based metal-organic-framework nanoparticles on 3D-printed fractal polymer surfaces for photocatalytic degradation of organic pollutants." *ACS Applied Nano Materials* 3 (3) 2830–2845. Doi: 10.1021/acsanm.0c00096.

152. U Shanker, M Rani and V Jassal. 2017. "Degradation of hazardous organic dyes in water by nanomaterials". *Environmental Chemistry Letters* 15 623–642. Doi: 10.1007/s10311-017-0650-2.

153. R Verma, A Asthana, AK Singh, S Prasad and MABH Susan. 2016. "Novel glycine-functionalized magnetic nanoparticles entrapped calcium alginate beads for effective removal of lead." *Microchemical Journal* 130 168–178. Doi: 10.1016/j.microc.2016.08.006.

154. KS Varma, RJ Tayade, KJ Shah, PA Joshi, AD Shukla and VG Gandhi. 2020. "Photocatalytic degradation of pharmaceutical and pesticide compounds (PPCS) using doped TiO₂ nanomaterials: A review." *Water-Energy Nexus* 3 46–61. Doi: 10.1016/j.wen.2020.03.008.

155. A Tadesse, D RamaDevi, M Hagos, G Battu and K Basavaiah. 2018. "Synthesis of nitrogen doped carbon quantum dots/magnetite nanocomposites for efficient removal of methyl blue dye pollutant from contaminated water." *RSC Advance* 8 8528–8536. Doi: 10.1039/C8RA00158H.

156. R Rajendran, K Varadharajan, V Jayaraman, B Singaram and J Jeyaram. 2018. "Photocatalytic degradation of metronidazole and methylene blue by PVA-assisted Bi_2WO_6-CdS nanocomposite film under visible light irradiation." *Applied Nanoscience* 8 61–78. Doi: 10.1007/s13204-018-0652-9.

157. Y Feng, J Yin, S Liu, Y Wang, B Li and T Jiao. 2020. 'Facile synthesis of Ag/Pd nanoparticle-loaded poly(ethylene imine) composite hydrogels with highly efficient catalytic reduction of 4-nitrophenol." *ACS Omega* 5 (7) 3725–3733. Doi: 10.1021/acsomega.9b04408.

158. M Abdullah Iqbal, S Irfan Ali, F Amin, A Tariq, MZ Iqbal and S Rizwan. 2019. "La- and Mn-codoped bismuth ferrite/Ti_3C_2 MXene composites for efficient photocatalytic degradation of congo red dye." *ACS Omega* 4 8661–8668. Doi: 10.1021/acsomega.9b00493.

159. H Sereshti, H Gaikani and HR Nodeh. 2017. "The effective removal of mercury ions (Hg^{2+}) from water using cadmium sulfide nanoparticles doped in polycaprolactam nanofibers: Kinetic and equilibrium studies." *Journal of the Iranian Chemical Society* 15 (3). Doi: 10.1007/s13738-017-1274-y.

160. H Filik and AA Avan. 2020. "Review on applications of carbon nanomaterials for simultaneous electrochemical sensing of environmental contaminant dihydroxybenzene isomers." *Arabian Journal of Chemistry* 13 (7) 6092–6105. Doi: 10.1016/j.arabjc.2020.05.009.

161. Y Jiang, P Biswas and JD Fortner. 2016. "A review of recent developments in graphene enabled membranes for water treatment." *Environmental Science Water Research and Technology* 2 915. Doi: 10.1039/c6ew00187d.

162. A Hashmi, AK Singh, AAP Khan and AM Asiri. 2020. "Novel and green reduction of graphene oxide by capsicum annuum: Its photo catalytic activity." *Journal of Natural Fibers* Doi. 10.1080/15440478.2020.1819930.

163. B Arora and P Attri. 2020. "Carbon nanotubes (CNTs): A potential nanomaterial for water purification." *Journal of Composite Science* 4 135. Doi: 10.3390/jcs4030135.

164. A Rehman, M Park and S-J Park. 2019. "Current progress on the surface chemical modification of carbonaceous materials." *Coatings* 9 103. Doi: 10.3390/coatings9020103.

165. VK Gupta, S Agarwal, AK Bharti and H Sadegh. "Adsorption mechanism of functionalized multi-walled carbon nanotubes for advanced Cu (II) removal." *Journal of Molecular Liquids* 230 667–673. Doi: 10.1016/j.molliq.2017.01.083.

166. WMA El Rouby. 2018. "Selective adsorption and degradation of organic pollutants over Au decorated Co doped titanate nanotubes under simulated solar light irradiation." *Journal of the Taiwan Institute of Chemical Engineers* 88 201–214. Doi: 10.1016/j.jtice.2018.04.003.

167. H Ma, BS Hsiao and B Chu. 2012. "Ultrafine cellulose nanofibers as efficient adsorbents for removal of UO_2^{2+} in water." *ACS Macro Letters* 1 213–216. Doi: 10.1021/mz200047q.

168. R Dewangan, A Hashmi, A Asthana, AK Singh and MABH Susan. 2020. "Degradation of methylene blue and methyl violet using graphene oxide/NiO/β-cyclodextrin nanocomposites as photocatalyst." *International Journal of Environmental Analytical Chemistry.* Doi: 10.1080/03067319.2020.1802443.

169. Y Zhang, L-J Huang, Y-X Wang, J-G Tang, Y Wang, M-M Cheng, Y-C Du, K Yang, MJ Kipper and M Hedayati. 2019. "The preparation and study of ethylene glycol-modified graphene oxide membranes for water purification." *Polymers* 11 188. Doi: 10.3390/polym11020188.

170. P Su, F Wang, Z Li, CY Tang and W Li. 2020. "Graphene oxide membranes: The controlling of transport pathways." *Journal of Materials Chemistry A* 8 15319–15340. Doi: 10.1039/D0TA0224.

171. Y Han, Z Xu and C Gao. 2013. "Ultrathin graphene nanofiltration membrane for water purification." *Advanced Functional Materials* 23 3693–3370. Doi: 10.1002/adfm.2012026010.

172. L Dong, W Fan, X Tong, H Zhang, M Chen and Y Zhao. 2018. "CO_2-responsive graphene oxide/polymer composite nanofiltration membrane for water purification." *Journal of Materials Chemistry A* 6 6785–6791. Doi: 10.1039/C8TA00623G.

173. A Mohammadi, MR Daymond and A Docoslis. 2020. "Graphene oxide membranes for isotopic water mixture filtration: Preparation, physicochemical characterization, and performance assessment." *ACS Applied Materials and Interfaces* 12 (31) 34736–34745. Doi: 10.1021/acsami.0c04122.

174. VF. Medina, CS Griggs, B Petery, J Mattei-Sosa, L Gurtowski, SA Waisner, J Blodget, and R Moser. 2017. "Fabrication, characterization, and testing of graphene oxide and hydrophilic polymer graphene oxide composite membranes in a dead-end flow system." *Journal of Environmental Engineering* 143 (11) 04017072–04017078. Doi: 10.1061/(ASCE)EE.1943–7870.0001268. American Society of Civil Engineers.

175. S Yadav, A Asthana, R Chakraborty, B Jain, AK Singh, SAC Carabineiro and MABH Susan. 2020. "Cationic dye removal using novel magnetic/activated charcoal/β-cyclodextrin/alginate polymer nanocomposite." *Nanomaterials* 10 170. Doi: 10.3390/nano10010170.

176. PJ. Mafa, B Ntsendwana, BB Mamba and AT Kuvarega. 2019. "Visible light driven $ZnMoO_4$/BiFeWO$_6$/rGO Z-scheme photocatalyst for the degradation of anthraquinonic dye." *Journal of Physical Chemistry C* 123 20605–20616. Doi: 10.1021/acs.jpcc.9b05008.

177. M Su, C He, VK Sharma, MA Asi, D Xia, X-z Li, H Deng and Y Xiong. 2012. "Mesoporous zinc ferrite: Synthesis, characterization, and photocatalytic activity with H_2O_2/visible light." *Journal of Hazardous Materials* 211–212 95–103. Doi: 10.1016/j.jhazmat.2011.10.006.

178. D Zhang, M Wen, B Jiang, G Li and JC Yu. 2012. "Ionothermal synthesis of hierarchical BiOBr microspheres for water treatment." *Journal of Hazardous Materials* 211–212 104–111. Doi: 10.1016/j.jhazmat.2011.10.064.

179. G Em. Romanos, CP Athanasekou, FK Katsaros, NK Kanellopoulos, DD Dionysiou, V Likodimos and P Falaras. 2012. "Double-side Active TiO_2-modified nanofiltration membranes in continuous flow photocatalytic reactors for effective water purification." *Journal of Hazardous Materials* 211–212 304–316. Doi: 10.1016/j.jhazmat.2011.09.081.

180. B Jain, A Hashmi, S Sanwaria, AK Singh, MABH Susan and SAC Carabineiro. 2020. "Catalytic properties of graphene oxide synthesized by a "Green" process for efficient abatement of auramine-O cationic dye." *Analytical Chemistry Letters* 10 (1) 21–32. Doi: 10.1080/22297928.2020.1747536.

181. F Ge, M-M Li, H Ye and B-X Zhao. 2012. "Effective removal of heavy metal ions Cd^{2+}, Zn^{2+}, Pb^{2+}, Cu^{2+} from aqueous solution by polymer-modified magnetic nanoparticles." *Journal of Hazardous Materials* 211–212 366–372. Doi: 10.1016/j.jhazmat.2011.12.013.

182. B An and D Zhao. 2012. "Immobilization of As(III) in soil and groundwater using a new class of polysaccharide stabilized Fe-Mn oxide nanoparticles." *Journal of Hazardous Materials* 211–212 332–341. Doi: 10.1016/j.jhazmat.2011.10.062.

183. Z Wang, C Fang and M Megharaj. 2014. "Characterization of iron-polyphenol nanoparticles synthesized by three plant extracts and their Fenton oxidation of azo dye." *ACS Sustainable Chemistry and Engineering* 2 (4) 1022–1025. Doi: 10.1021/sc500021n.

184. J Patel, AK Singh and SAC Carabineiro. 2020. "Assessing the photocatalytic degradation of fluoroquinolone norfloxacin by Mn:ZnS quantum dots: Kinetic study, degradation pathway and influencing factors." *Nanomaterials* 10 964. Doi: 10.3390/nano10050964.

185. M Ahmadi, HR Motlagh, N Jaafarzadeh, A Mostoufi, R Saeedi, G Barzegar and S Jorfi. 2017. "Enhanced photocatalytic degradation of tetracycline and real pharmaceutical wastewater using MWCNT/TiO$_2$ nano-composite." *Journal of Environmental Management* 186 55e63. Doi: 10.1016/j.jenvman.2016.09.088.

186. CM El-Maraghy, OM El-Borady and OA El-Naem. 2020. "Effective removal of levofloxacin from pharmaceutical wastewater using synthesized zinc oxide, graphen oxide nanoparticles compared with their combination." *Scientific Reports* 10 5914. Doi: 10.1038/s41598-020-61742-4.

187. M Tang, Y Ao, C Wang and P Wang. 2020. "Facile synthesis of dual Z-scheme g-C$_3$N$_4$/ Ag$_3$PO$_4$/AgI composite photocatalysts with enhanced performance for the degradation of a typical neonicotinoid pesticide." *Applied Catalysis B: Environmental* 268 118395. Doi: 10.1016/j.apcatb.2019.118395.

188. S Yadav, A Asthana, AK Singh, R Chakraborty, S Sree Vidya, MABH Susan and SAC Carabineiro. 2020. "Adsorption of cationic dyes, drugs and metal from aqueous solutions using a polymer composite of magnetic/β-cyclodextrin/activated charcoal/Na alginate: isotherm, kinetics and regeneration studies." *Journal of Hazardous Materials.* Doi: 10.1016/j.jhazmat.2020.124840.

189. N Sankararamakrishnan, R Singh and I Srivastava. 2019. "Performance of novel MgS doped cellulose nanofibres for Cd(II) removal from industrial effluent-mechanism and optimization." *Scientific Reports* 9 1263. Doi: 10.1038/s41598-019-49076-2.

190. A Mukherjee, MK Adak, S Upadhyay, J Khatun, P Dhak, S Khawas, UK Ghorai and D Dhak. 2019. "Efficient fluoride removal and dye degradation of contaminated water using Fe/Al/Ti oxide nanocomposite." *ACS Omega* 4 9686–9696. Doi: 10.1021/acsomega.9b00252.

191. P He, T Mao, A Wang, Y Yin, J Shen, H Chen and P Zhang. 2020. "Enhanced reductive removal of ciprofloxacin in pharmaceutical wastewater using biogenic palladium nanoparticles by bubbling H$_2$." *RSC Advance* 10 26067. Doi: 10.1039/d0ra03783d.

192. KG Akpomie and J Conradie. 2020. "Synthesis, characterization, and regeneration of an inorganic–organic nanocomposite (ZnO@biomass) and its application in the capture of cationic dye." *Scientific Reports* 10 14441. Doi: 10.1038/s41598-020-71261-x.

193. RS El-Tawil, ST El-Wakeel, AE Abdel-Ghany, HA Abuzeid, KA Selim and AM Hashem. 2019. "Silver/quartz nanocomposite as an adsorbent for removal of mercury (II) ions from aqueous solutions." *Heliyon* 5 e02415. Doi: 10.1016/j.heliyon.2019. e02415.

194. G Jethave, U Fegade, S Attarde, S Ingle, M Ghaedi and MM Sabzehmeidani. 2019. "Exploration of the adsorption capability by doping Pb@ZnFe$_2$O$_4$ nanocomposites (NCs) for decontamination of dye from textile wastewater." *Heliyon* 5 e02412. Doi: 10.1016/j.heliyon.2019.e02412.

195. DZ Husein, R Hassanien and MF Al-Hakkani. 2019. "Green-synthesized copper nanoadsorbent for the removal of pharmaceutical pollutants from real wastewater samples." *Heliyon* 5 (8) e02339. Doi: 10.1016/j.heliyon.2019.e02339.

196. L Fan, C Luo, Z Lv, F Lu and H Qiu. 2011. "Preparation of magnetic modified chitosan and adsorption of Zn^{2+} from aqueous solutions." *Colloids and Surfaces B: Biointerfaces* 88 574–581. Doi: 10.1016/j.colsurfb.2011.07.038.

197. AZM Badruddoza, ZBZ Shawon, TWJ Daniel, K Hidajat and MS Uddin. 2013. "Fe$_3$O$_4$/Cyclodextrin polymer nanocomposites for selective heavy metals removal from industrial wastewater." *Carbohydrate Polymers* 91 322–332. Doi: 10.1016/j. carbpol.2012.08.030.

8 Recent Development in Agriculture Based on Nanomaterials

Elaine Gabutin Mission
University of Valladolid

CONTENTS

8.1 INTRODUCTION

The agriculture industry is the world's largest industry and a globally significant sector to produce raw materials for food purposes and chemical industries. It is estimated to have a global gross value of about $5 trillion by 2028 and engage more than a billion people for employment [1]. However, agriculture is highly vulnerable to the sweeping consequences of climate change. Being one of the most pressing issues of our times, it has brought about tremendous changes in weather patterns, massive rains and droughts, and pests that threatened the productivity of the agricultural sector and employment of rural households [2–4]. Coupled with the population boom that is projected to reach around 9.6 billion in 2050 and rapid urbanization, this may adversely affect fundamentals of food and raw material supply that may lead to shortages, instabilities, and escalating prices [5,6]. Thus, various mitigation and adaptation techniques, such as crop and soil science, mechanization, and land and water management, are being explored to improve the productivity of the agriculture sector [7–9].

With an apparent challenge to produce 40% more food by 2030 despite restricted land and water resource and lesser intake of energy, fertilizer, and pesticide to curtail global greenhouse gas emissions, the agriculture sector can widely benefit from the advancement of emerging and innovative technologies such as nanotechnology. Nanotechnology allows the synthesis of nanomaterials, commonly termed nanoparticles, at nanoscale level, which means having a dimension within 100 nm range [10].

At present, nanoparticles abound in various fields and industries such as in energy and electronics, food and drugs, pharmaceuticals and cosmetics, and biology and engineering [11]. The market for nanotechnology has an estimated value of around $50 billion in 2018 and has an annual projected growth rate of 18% [12]. The current market size and growth rate are dominated by the major players, namely the medical and healthcare, energy, electronics, and information technology infrastructures and environmental sectors. The technology can trickle down to the agricultural sector, which can further intensify the growth projections and thus the market size. In agriculture, nanoparticles have been investigated for soil fertility management to improve soil quality, nanofertilizers, and nutrient delivery system for the delivery of agrochemicals and nutrients and stimulate plant growth, precision farming, water purification, and nanosensors and diagnostic devices to detect chemicals or food-borne pathogens [13–18]. This has been evident based on the compilation illustrated in Figure 8.1 where articles published on "nanoparticles and agriculture" have tripled in the last 5 years alone.

Nanoparticles are classified into natural or synthetic, based on their origin. Naturally occurring nanoparticles found in the environment can be synthesized by biological species and take the form of bacterial products, clays, and minerals. On the other hand, synthetic or engineered nanoparticles are synthesized through various biophysicochemical techniques or a combination thereof, under controlled or artificial environments [19]. Nanoparticles may also be classified depending on their morphologies (nanosheets, nanofibers, nanowires, nanorods, nanospheres) [20]. Nanoparticles can also be classified into four classes, based on material components:

    i. organic-based (dendrimers, liposomes, polymers)
   ii. inorganic-based (metal and metal oxides, quantum dots)
  iii. carbon-based (carbon nanotubes, fullerenes, graphenes)
  iv. composite-based (multiphase)

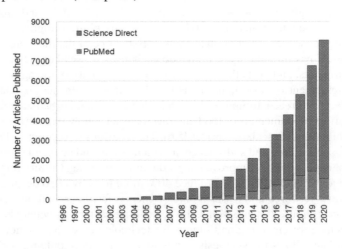

FIGURE 8.1 Evolution of publication numbers on "nanoparticle and agriculture" in ScienceDirect and PubMed in the last 25 years.

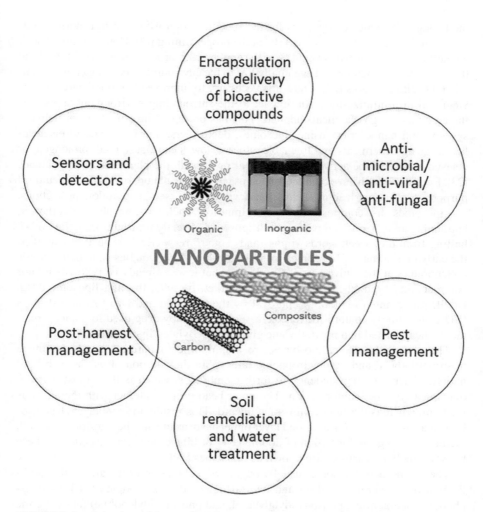

FIGURE 8.2 Overview of the nanoparticles discussed in this chapter and their applications.

In this chapter, the recent developments in nanomaterials according to material component classification for agricultural applications will be discussed, as illustrated in Figure 8.2. The design, synthesis, properties, characterization, and application of the nanomaterials will be discussed, including the trends, opportunities, and challenges on nanomaterial design and synthesis. In addition, perspectives regarding the social and ethical repercussions of nanomaterial uses in agriculture are provided.

8.2 ORGANIC NANOPARTICLES

The stimulation of plant growth and the management of pests can be best achieved through the application of various agrochemicals such as insecticides, herbicides, fungicides, and nematicides at a specific target part of the plant. Thus, it is important that the biologically active compounds for the specific action would be transported, protected, and released at the intended site at the proper timing [21]. However, this is such an inefficient process, with over 90% of the applied crop protection agents not reaching the target due to leaching and eventually lost to the environment, causing environmental and health concerns [22]. Thus, it is necessary to improve plant uptake to enhance yield and to minimize the toxic impact to the surrounding environment as well as the amounts of agrochemicals applied. This has prompted investigation for delivery systems and nanocarriers using liposomes, dendrimers, and polymers, to encapsulate these agrochemicals and to enable either a slow, steady, sustained, prolonged, or delayed release of the active chemicals near or through the plant tissues and cuticles [23]. Recent research developments on the use of organic nanoparticles for agricultural purposes are summarized in Table 8.1 and will be discussed in the following sections.

Liposomes are composed of one or multiple enclosed phospholipid bilayers, entrapping an aqueous compartment. "Liposome" directly translates into "fat body," hailing from the Greek words "lipos" and "soma," respectively. The 1960s marked the early beginnings of liposomes from the work of Alec Douglas Bangham and his collaborators at the University of Cambridge, but it was already in 1986 when Dior commercially brought it to the cosmetic market. In 1995, the first liposome drug, Doxil, was granted regulatory approval by the US Food and Drug Administration (FDA). It actually contains doxorubicin, a chemotherapy drug used for cancer treatment, encapsulated in a poly(ethylene glycol) (PEG)ylated liposome.

The phospholipids in liposomes are amphiphilic in nature, usually having a hydrophilic head and two hydrophobic tails. With the addition of water, such lipid moieties have a strong tendency to form membranes, which is their most thermodynamically stable conformation. The polar head groups interact with the aqueous medium, whereas the lipid chains turn inward. This results in monolayer structures known as micelles. Then, a second layer of heads turns toward the interior of the cell, thereby favoring the formation of double layer or bilayer lamellar structures. These bilayer lamellar structures are known as liposomes [29].

The synthesis of liposomes usually begins with cholesterol (that allows the modification of membrane rigidity) and natural nontoxic phospholipids (such as phosphatidylethanolamine, phosphatidylglycerol, and phosphatidylcholine) following the thin film hydration or Bangham's method [30,31]. Due to the limitation of Bangham's method such as low encapsulation ability and difficulty for batch scale-up, emerging methods have been developed [32]. Some of these are reversed-phase evaporation, detergent dialysis, and solvent injection techniques. More importantly, several technologies such as ultrasonication, high-pressure extrusion, and microfluidization have been developed to help decrease the size of vesicles and facilitate increase of concentration at the target area [33–38]. Some of the most important parameters in the design and manipulation of liposomes include lipid physicochemical properties, compatibility of compounds being confined, polydispersity (or the extent by which the sizes of liposomes vary), zeta potential (where high values are preferred signifying stability toward aggregation of particles in a solution or dispersion), stability and shelf life (to prevent leakage of encapsulated compound), particle size (smaller liposomes have higher tendency toward efficient delivery of the encapsulated compound at the target location), and reproducibility (to maintain quality and effectiveness).

TABLE 8.1

Organic Nanoparticles as Applied to Agriculture

	Size, nm	Purpose	Findings	References
Liposome	100	Delivery of agricultural ingredients to seedling and fully grown tomato plants through leaves	30% uptake, much higher than the previous report of 0.1%	[22]
Fluorescent cationic dendrimers	1.1–3.2	Thiamethoxam pesticide delivery	Consists of fluorescent perylene-3,4,9,10-tetracarboxydiimide chromophore (PDI) in the center, which allows the detection by fluorescence microscopy Cytotoxic assay revealed up to 75% live cell mortality at 3 μL dosage	[24]
Starch-based polymers	5,000–300,000	Encapsulation of natural essential oils and delivery against Varroa mite in honeybee colonies	Encapsulation efficiency 65% and 100% for camphor and clove, respectively.	[25]
Polylactic acid (PLA)	600–800	Encapsulation of metazachlor herbicide	Encapsulation efficiency as high as 73% 56%–92% metazachlor release for 600 and 800 nm nanoparticles, respectively	[26]
Poly(amidoamine) (PAMAM)	2–20	Removal of metal ions from contaminated water	Designed with EDA core and terminal NH_2 groups 100% of Cu(II) ions can be bound onto the dendrimers at pH = 7	[27]
methoxy poly (ethylene glycol)-poly (lactide-co-glycolide) (mPEG-PLGA)	90–126	Metolachlor pesticide delivery system	Enhanced solubility of hydrophobic metolachlor and positive effect to *Oryza sativa*	[28]

Since the liposome's bilayer structure can be made to mimic the lipid fraction in cell membranes, they were turned into model membranes of plant organelles and have enabled researches on the physiological processes involving the effects of pesticides and toxins as well as plant aging, drying, and freezing tolerance [39]. Due to the amphiphilic nature of phospholipids, various lipophilic and hydrophilic drugs can be encapsulated within these vesicles. This has been particularly welcome in aquaculture for the administration of immunostimulants and vaccines to a huge number of fishes. In the work of Ruyra et al. [40], immunostimulant cocktail has been encapsulated in unilamellar liposomes with a diameter of about 125 nm and was found to elicit explicit proinflammatory and antiviral reaction from zebrafish hepatocytes and trout macrophages. Liposomes with an average size of 100 nm were used for supplement delivery to planted tomato seedlings and full-grown plants, as described by Karny et al. [22]. First, they found out that the liposomes can creep into the plant leaves and can be carried throughout the whole plant structure according to the natural pathways by which water and nutrients are carried across and without any toxic effect. Moreover, the applied liposomes resulted in 33 times higher penetration in the leaves than when the nutrient in free form (without encapsulation) was applied in a similar approach. Having naturally occurring components, liposomes are so far considered the most successful drug carrier system. This reduces or eliminates hurdles from regulatory agencies that may discourage their use in food and drug systems. On the other hand, a major difficulty in developing liposomes is the capability to construct homogeneous molecular structures with satisfactory properties and acceptable encapsulation performance [41].

Dendrimers, also called "arboroles" or "cascade polymers," are a class of artificial, three-dimensional hyperbranched, monodisperse macromolecules resembling tree branches with defined molecular weights [42]. The term directly translates into "tree part," originating from the Greek words "dendron" and "meros," respectively. It was first introduced in 1978 by Fritz Vogtle and coworkers, whereas Donald A. Tomalia and his coworkers synthesized the first family of dendrimers at Dow chemicals in 1985 [43,44]. "Starburst" or polyamidoamines (PAMAM) were the first dendrimers synthesized by the Dow chemicals company, using ammonia as the starting molecule. Polypropylene imine (PPI) is also a widely used dendrimer.

Structurally, dendrimers bear a central nucleus, branches, and terminal functional groups scattered around the macromolecule's exterior sections and are laid out in a distinct molecular architecture. There is a distinguishing aspect between linear polymers and dendrimers; that is, in a linear polymer, the molecule chain resembles coils, whereas dendrimers consist of chain of molecules that branch out from a communal core without entanglement [45]. The vast number of functional groups on the surface contributes to the ability to covalently or noncovalently bind molecules of different properties (i.e., solubility, viscosity, and thermal behaviors).

In general, dendrimers are synthesized via the divergent or convergent method. The divergent method was introduced by Tomalia et al., wherein dendrimers are constructed in a stepwise manner from a small core molecule. As an illustrative example, nitrogen as core would react with monomer molecules such as carbon and other elements that consist of one reactive and two nonreactive groups, leading to the initial-generation dendrimer. The edge of this initial generation will react accordingly in the presence of more monomers. Through the successive reiteration of a sequence of reaction, dendrimers are built layer after layer, or generation after generation, and then, the structure can be expanded to a certain desired size. The size of the resulting spherical macromolecular structure can be similar to that of blood albumin and hemoglobin [46]. This method, however, suffers from some drawbacks such as (1) occurrence of trailing generations and surface defects due to the huge number of branches as generation number increases; (2) difficulty in product purification; and (3) usage of excess reagents and eventually elimination of the excess. Conversely, in the convergent method, proposed by Hawker and Frechet in 1990, dendrimer growth starts from end groups and progresses inward owing to the reactivity of their core [47]. While this approach has made the removal of impurities easier, monodisperse dendrimers are also obtained. Moreover, the steric constraints limit its size.

Dendrimers have become one of the most interesting multivalent scaffolds due to the variety of functionalities that can be incorporated into them. Drugs or active substances can be incorporated into dendrimers through complexation and encapsulation, which successfully led to the solubilization of poorly soluble drugs. The hydrophobic cores can hold hydrophobic drugs, and hydrophilic drugs may be connected along their periphery involving covalent bonding or electrostatic interaction [48]. Having mostly strong positive charges, this enables functionalization of a wide array of target molecules, including, but not limited to, antibodies, peptides, and folate [49]. It is possible to control the release of only the active substances attached to the periphery by controlling pH or utilizing enzymes. Alternatively, the full structure can be broken down through the dissociation of covalent bonds to release the same [50]. Despite their biocompatibility and nonimmunogenicity, the application of single dendrimers is quite limited due to size constraints and surface properties. Thus, more recently, supramolecular structures with tunable size, shape, structure, and functions have been widely investigated. Designing dendrimers where the structure, size, aggregation number, and colloidal stability can be controlled while maintaining nanosize and branched assembly remains to be a major challenge in this research field [51].

Polymeric nanoparticles are often found within the colloidal regions, having sizes between 10 nm and 1 μm. They can be naturally found in nature, the most common of which are chitosan, albumin, and heparin that are generally hydrophilic. Meanwhile, a number of synthetic polymers that include N-(2-hydroxypropyl)-methacrylamide copolymer, poly(glycolic acid), poly(lactide-co-glycolide), poly (ε-caprolactone), poly(cyanoacrylate), poly(lactic acid–glycolic acid), and poly(lactic acid) (PLA) are hydrophobic in nature. Polymeric nanoparticles can be prepared using solvent displacement or emulsion polymerization method. Depending on the preparation method, they can form two types of structures: nanosphere (polymeric matrix-type structure) and nanocapsule (shell type with an oil or aqueous core) [52]. Polymeric nanoparticles can be synthesized in aqueous media via emulsion polymerization from monomers of polyacrylamide, poly(alkyl acrylates), poly(methyl methacrylate), and poly(ethyl cyanoacrylate) [53]. Polymeric nanoparticles are structurally stable with varying particle distribution, electrostatic potential, and drug release profiles. Modification of its properties through synthesis could be achieved by changing the length of resulting polymer, choice of surfactants/emulsifier/wetting agent or detergent, and organic solvents used [54]. Furthermore, polymeric nanoparticles have surface edges available for chemical modification through functionalization as applied to targeted antimicrobial delivery.

There is a surge of interest in the use of natural polymers because they are perceived as biodegradable, thereby alleviating environmental pollution [17]. As a consequence, they have found extensive applications as controlled-release fertilizers, as pesticide carrier and delivery, and as water-retaining agents. A long list of biopolymers has been employed in controlled-release fertilizers including starch [55], chitosan [56] gelatin [57], lignin [58] pectin, cellulose, and kappa-carrageenan [59]. Water-insoluble chitosan and cellulose as well as chemically modified starch are ideal for agricultural uses [60]. BASF has launched Seltima in 2016 for pyraclostrobin pesticide encapsulation. It aims to safeguard rice plants by using a humidity-responsive controlled-release technology to selectively discharge the pesticides onto rice leaves instead of into the water in paddy fields, which will guarantee appropriate pesticide application and reduce water contamination in the paddy fields [61]. Microencapsulation technologies have also been employed in several commercially available products, such as Centium 360 CS from FMC UK, which contains the herbicide clomazone, and Syngenta Australia's Karate Zeon, which contains the insecticide lambda-cyhalothrin.

Polymeric materials have also been explored for water retention applications in drought-prone and semiarid regions. Guo et al. [62] reported on the synthesis of an extremely absorbent polymer from poly-γ-glutamic acid (γ-PGA SAP) and its ability to preserve water in the soil for a prolonged duration. Also, its addition to the soil leads to the reduction of the permeation volume of water, avoiding the deep penetration of soil water and substantially favoring the extent of evaporation. Cheng et al. have prepared hydrogels, or superabsorbent polymers, by cross-linking acrylic acid and urea with potassium sulfate using N,N'-methylenebis acrylamide as the initiator and cross-linker. Using the as-prepared hydrogels for 192 h, relative water contents of 42%, 56%, and 45% were measured for the treated sandy loam, loam, and paddy soil, respectively [63].

Chitosan is second to cellulose in abundance, which can be sourced from shellfishes, insects, and in some fungi and bacteria [64]. This polysaccharide, consisting of D-glucosamine and N-acetyl-D-glucosamine monomers adjoined in β-1,4-configuration, has been widely studied for the development of nanocapsules because of its biocompatibility. Chitosan is widely acknowledged in agricultural systems due to its antimicrobial properties, capability to impede the growth of fungal and bacterial pathogens, and ability to stimulate resistance in plants [64,65]. It has also been reported to improve the photosynthetic rate and overall plant growth. In addition, chitosan can help alleviate the toxic effects of some heavy metals such as cadmium, lead, and zinc by binding and forming molecular complexes, thus avoiding bioaccumulation in plants [66].

A chitosan/alginate system was developed for encapsulating the herbicide paraquat. With a mean diameter of 635 nm, the encapsulation efficiency was found to be around 74% [67]. On the other hand, a chitosan/tripolyphosphate system with an average size of 300 nm was used to carry paraquat at an encapsulation efficiency of about 62%. Moreover, it was found that encapsulated herbicide has the same effectiveness as the nonencapsulated form but with an advantage due to a slower discharge profile [68]. Meanwhile, Maruyama et al. (2016) developed alginate/chitosan system and chitosan/tripolyphosphate system to encapsulate the herbicides imazapic and imazapyr. The nanoparticles, with an average size of 400 nm, both demonstrated encapsulation efficiencies of between 50% and 70% and demonstrated 30-day persistent stability at ambient temperature [69]. Starch nanoparticles produced via water-in-oil microemulsion at ambient temperature, then coated with poly-L-lysine, were successfully used as plant transferring vehicle [25].

Nanoparticles can also play a crucial role in wastewater treatment and soil remediation in order to make contaminated water and soil resources useful. Amphiphilic polyurethane (APU) nanoparticles were reported in remediating polynuclear aromatic hydrocarbons (PAHs) – contaminated soils, a class of hydrophobic organic groundwater contaminants. The synthesized APU particles have 17–97 nm size. Moreover, the APU particles were found tunable and can be tailored to have hydrophobic internal sections that could enable particle mobility in soil [70].

8.3 INORGANIC NANOPARTICLES

Metal nanoparticles refer to particles of metal atoms having diameters from 1 nm to about a few hundreds of nanometers. The metal nanoparticles have a substantial difference in terms of properties to the associated bulk materials due to their special characteristics, such as having substantial fraction of their atoms on their surface, high surface energy (which is synonymous with having strong molecular attraction), spatial confinement (which results in a change in optical and electronic properties), and controlled morphology (that corresponds to an increase in active area and uniformity). These unique properties of nanoparticles make them applicable in the fields of catalysis, agriculture, electronics, biomedical research, and even groundwater purification [71]. Metal ions form metal oxides when reacted with oxygen. The most commonly reported metal oxide nanoparticles include oxides of aluminum (Al), bismuth (Bi), cerium (Ce), copper (Cu), cobalt (Co), iron (Fe), magnesium (Mg), nickel (Ni), tin (Sn), titanium (Ti), and zinc (Zn) [72]. The metal oxides can be synthesized in a variety of techniques including hydrothermal, microwave solvothermal, chemical precipitation, thermal vapor transport [73,74].

Among different nanoparticles, silver nanoparticles (AgNPs) draw more attention, thanks to their antimicrobial and antiviral properties. For instance, AgNPs have shown antifungal inhibition effect on Ambrosia fungus *Raffaelea spp.*, which was accountable for the death of a significant number of oak trees in Korea [75]. Similarly, spherical AgNPs have demonstrated potent action against *Trichophyton mentagrophytes* strains and *Candida* species, which are responsible for dermatophytosis on hair, nails, and the outer skin layer of human skin, pets, and farm animals [76]. AgNPs also exhibited antiviral efficacy against the deadly and highly infectious peste des petits ruminants virus (PPRV), a prototype of the Morbillivirus that infects goats and sheep in Asia and Africa. The *in vitro* study of Khandelwal et al., 2014, demonstrated that the PPRV replication was impaired as AgNPs inhibited the entry of virus into the cells [77]. AgNPs at 24 µg/mL have also been used *in vitro* for the prevention of bovine herpes virus type 1 (BoHV-1) that brought about diseases in buffaloes and cattle worldwide [78]. AgNPs have also found a way to be directly applied onto plants. Jasim et al., 2016, reported that 200–800 nm AgNP size can enhance plant growth while 35–40 nm could positively affect the root and shoot growth of different plants [79,80].

In a similar manner, 200 µg/mL of gold nanoparticles (AuNPs) was used to inhibit bacterial growth of *C. pseudotuberculosis,* which causes Caseous lymphadenitis (CLA) disease wherein infected goats and sheep develop abscess in their lymph nodes and other tissues [81]. Fungicidal activity of AuNPs has been reported for fungal pathogens such as *Candida spp.* Wani and Ahmad, 2013, prepared gold nanodisks and obtained an average diameter and surface area of 25 nm and 179.5 m^2/g, respectively, using tin chloride ($SnCl_2$) as a reducing agent. On the other hand, polyhedral assemblies with a mean particle size of 30 nm and 150.5 m^2/g surface area were formed using sodium borohydride ($NaBH_4$). The antifungal properties were observed to be size-dependent, with the 25-nm strongly inhibiting more fungal growth than the 30-nm particles [82].

AuNPs are antifungal, but their antibacterial function has contradictory effects [83]. However, Lopez-Miranda et al. [84] have successfully demonstrated that antibacterial properties of the AuNPs are synthesized using aqueous extracts of *Aloysia triphylla* and obtained mostly spherical nanoparticles between 40 and 60 nm. The plant extracts acted as reducing agents and stabilizers for the synthesis. As synthesized AuNPs showed antibacterial activity against Gram-positive and Gram-negative bacteria, the corresponding inhibition sections for *S. Aureus* and *E. coli* were 11.3 and 10.6 mm, respectively. AuNPs have been used as biosensors for the rapid and sensitive detection of multiple veterinary pathogens such as swine reproductive and respiratory syndrome virus, bluetongue virus, and foot and mouth disease in ruminants. AuNPs have also been used for the detection and monitoring of toxin levels, heavy metals, and pollutants in the environment and food sources [85]. AuNPs have also been explored as a potential nanofertilizer at 6.25 µg/mL and were found to accelerate the plant growth-promoting rhizobacteria (soil microbes), more specifically, of *P. fluorescens, P. elgii, and B. subtilis* [86].

Zerovalent iron (ZVI) has been widely reported for the treatment of various contaminants, including heavy metals, dioxins, and organic solvents in soil, groundwater, and wastewater. In a study by Singhal et al., ZVI was used to remediate soil that has been contaminated by the insecticide malathion due to leaching. The oxidation of malathion took place in less than 10 min [87]. Pawlett and coworkers investigated the effect of ZVI on soil microbial community, namely mycorrhizal fungi and Gram-negative bacteria in sandy, loam, and clay soil. They deduced that the effect of ZVI is dependent on other factors, such as organic matter content of soil and soil mineral type, rather than applied nanoparticle concentration and duration of application [88].

Titanium dioxide nanoparticles (TiO_2) at 100 mg/L demonstrated positive effect on the growth of Moldavian balm (*Dracocephalum moldavica L.*) plants. Moreover, it has mitigated the negative effects due to salinity levels, which indicates that plants can thrive at higher salinity levels. Furthermore, the essential oil content in the plants was enhanced [89]. TiO_2 was also found to promote seed germination rate, shoot length, and root length in wheat [90]. A 60% increase in fresh and dry weight for spinach was under the influence of nanoanatase TiO_2 [91]. Zinc oxide nanoparticles (ZnO) at 1 mg/L were found to have bactericidal activity against *Aeromonas veronii,* a predominant fish pathogen causing hemorrhage and red sore disease in various fishes [92]. ZnO also inhibited the growth of other pathogens such as *Aeromonas hydrophila, Aeromonas salmonicida, Aphanomyces invadans, Edwardsiella tarda, Flavobacterium branchiophilum, Vibrio spp.*, and *Yersinia ruckeri* [93].

Metal nanoparticles have found application involving direct contact with plants, perhaps during germination of seeds and direct application onto leaves and roots. Thus, recent trends in metal nanoparticles include deeper analysis of the plant uptake, translocation, accumulation rate, possible transformation, biocompatibility, and phytotoxicity properties [94–96]. There are a significant number of studies devoted to metal nanoparticle synthesis using plant extracts and microorganisms as an effective, powerful, low-cost, and environment-friendly technique for synthesizing nanoparticles with defined properties. The emphasis is on the role played by natural plant biomolecules in the reduction of metal salts during synthesis [96]. Rapid synthesis has also been conducted using microwaves [97]. Metal nanoparticles have been frequently conjugated or cosynthesized with other nanoparticle types, such as carbons, to give rise to hybrid nanoparticles with new properties [98].

Quantum dots (QDs) are colloidal, semiconductor particles that are of nanometer size, typically within 1–20 nm, containing merely 10–50 atoms. It was first produced by Alexi Ekimov in 1977. Structurally, they bear a core–shell structure. QDs are man-made nanocrystals with semiconductor behavior; hence, the term exclusively refers to semiconductor nanocrystals, while the rest of the inorganic substances in the nanosystem are denoted as a "nanocrystal" [99]. QD could confine charged carriers (maybe electrons and/or holes) firmly in all three spatial directions so that these charged carriers would occupy only the set of discrete energies [100]. Typical examples of QD are CdS, CdSe, CdTe, CdS@ZnS, CdSe@ZnS, and CdSeTe@ZnS. The elements from the periodic table (from groups III–V, II–VI, or IV–VI) can be used to prepare QD [101]. QDs have unique physical, electronic, and optical properties, which can be tuned to affect energy flow. They produce energy in the form of light through photoluminescence when electrically excited or illuminated. Therefore, they found wide applications as energy conversion architectures for fluorescent probes and light-capture systems. Since then, QDs have been used in a wide range of optoelectronic devices, and QD suspension has found applications in *in vivo* and *in vitro* imaging, labeling, and sensing techniques [102]. Some key parameters affecting the properties of QD include precise sizing and shaping, core composition (alloys, semiconductor/metal boundaries), surface configuration (ligand properties), and film thickness and composition (such as electronic coupling, bulk heterojunction, biological boundaries) [103].

QDs demonstrate electronic properties that are intermediary to those of bulk semiconductors and individual molecules, owing to their specific size. Their optoelectronic characteristics can therefore be dictated upon by any modification of their shapes and sizes. They possess a color tuning ability that can be demonstrated when QDs are excited by a photon of energy; those that are about 5–6 nm will emit energy in the orange or red spectrum, whereas the smaller QDs emit shorter wavelengths producing blue or green color. This would result in avoiding the need to produce a different set of materials in order to achieve a certain spectrum; this can be simply resolved by amending the size and/or the shape of the QD.

Since QD can emit specific wavelengths of light when supplied with energy, this property has been harnessed to shift incoming light into a specific part of the spectrum that plants can most easily use for photosynthesis [104]. Nanoco Technologies has developed sheets of QDs that can be installed on ceilings of greenhouse farms that convert UV and visible light to the deep red/magenta spectrum that promotes plant growth twice as fast that of standard LED lights. The Merck Innovation Center also developed a fluorescent carrier nanostructure to enhance the specific radiation spectrum that favors and accelerates crop growth [105].

Liu et al. have established a QD-based immunochromatographic strip (ICS) assay to identify acetamiprid, a commercial insecticide with a strong propensity to collect in soil and potentially contaminate air and groundwater. The QD-ICS demonstrated satisfactory reliability and high agreement with the performance of the liquid chromatography–tandem mass spectrometry. Moreover, the full detection of a sample could finish within an hour. The results specifically demonstrated that QD-ICS was well capable of a fast and sensitive screening for acetamiprid deposits in an agricultural produce and paved the way for on-site detection of other analytes [106].

Due to its nephrotoxicity, their absorption, and accumulation in the kidneys, *in vivo* analysis of QDs is still being pursued. Some researchers consider QDs as inert and biocompatible, which merited its use for biomedical applications. However, there is still a cloud of doubt as to their safety since the most common QD cores, selenium and cadmium, may cause chronic and acute cytotoxicity in vertebrates and possess significant environmental and health concerns [107,108].

Table 8.2 provides a summary of the researches carried out on inorganic particles and their applications to the field of agriculture.

TABLE 8.2
Inorganic Nanoparticles as Applied to Agriculture

	Size, nm	Purpose	Findings	References
Ag (silver)	7–21	Antifungal against plant pathogen *Fusarium oxysporum*	Antifungal activity was demonstrated at 8 µg/mL	[109]
Ag (silver)	70–40	Antiparasitic against bovine ticks *Rhipicephalus (Boophilus) microplus*	High mortality rates were observed at an acaricidal activity LC_{50} value of 3.44 mg/L	[110]
Au (gold)	Colloidal	Biosensor for porcine reproductive and respiratory syndrome virus (PRRSV)	Detection limit at three particles/µL	[111]

(Continued)

TABLE 8.2 (*Continued*)
Inorganic Nanoparticles as Applied to Agriculture

	Size, nm	Purpose	Findings	References
Au (gold) coated with citrate	3–50	Foliar uptake in wheat	Accumulation in leaf impaired photosynthesis; 10%–30% accumulated in shoots; 10%–25% root accumulation; 5%–15% exuded into rhizosphere soil.	[112]
Zinc oxide	~50	Germination of lettuce seeds	No significant effect on germination rate for 25 and 50ppm but mean seedling length improved with 25 ppm	[113]
TiO_2/FeO-NP	22–33	Antibacterial against *Streptococcus iniae and Edwardsiella tarda*	Requires 120 min of visible fluorescent light exposure for effect	[114]
Multicomponent	< 20	Micronutrient deposition in plant tissues	Synthesis of $Ni_{0.4}Cu_{0.2}Zn_{0.4}Nd_{0.05}$ $Y_{0.05}Fe_{1.9}O_4$ novel nanoparticle via solgel method. Treatment with 500 mg/L increased macroelements K, Ca, Mg, and P in roots while Ca decreased in leaves.	[115]

8.4 CARBON-BASED NANOPARTICLES

In general, carbon-based nanomaterials contain carbon and its allotropes. The most popular carbon allotropes in recent years comprised of low-dimensional carbon forms that include fullerenes (C60), carbon nanotubes (CNT), carbon black, carbon onions, and graphene. Carbon-based materials exist in morphologies such as sheets, hollow tubes, ellipsoids, or spheres. Carbon-based materials are fabricated using electrolytic approach, laser ablation, hydrothermal, arc discharge, chemical vapor deposition (CVD), and template [116]. These materials have been widely investigated due to their myriad properties and novel features such as superconductivity and super strength. So, it is not impossible that their applications have also been extended toward increasing and optimizing agricultural production. The results of these applications are summarized in Table 8.3.

In the carbon nanoallotropes family, sp^2 nanocarbons can be assembled as a two-dimensional (2D) structure known as graphene. Graphene is a honeycomb crystal lattice of sp^2-C-atoms and with sp^3-C-atoms at the defect sites. When graphene is wrapped into a sphere, the zero-dimensional (0D) fullerene can be formed. When graphene is rolled along the horizontal, one-dimensional (1D) carbon nanotubes can be formed. Fullerenes, also referred to as buckyballs, consist of hexagonal carbon

TABLE 8.3

Carbon-Based Nanoparticles as Applied to Agriculture

	Size, nm	Purpose	Findings	References
MWCNT	18.8±4.1	Plant growth	Soybean plants flowered earlier	[117]
Graphene nanoplatelets	350±320	Plant growth	and produced 60% more flowers	
			Reduced whole plant N₂ fixation	
Carbon black	36±8	Plant growth	potential as much as 91%	
Graphite	400	Plant growth of *Arabidopsis thaliana*	Accumulation in leaves > stem > stem tissues; induced early flowering	[118]
Multiwalled carbon nanotubes	500–1,000	Seed germination and plant growth Callus tobacco cells	Average diameter 20 nm Enhanced tobacco cell growth by 55% to 65% with 5–500 µg/mL	[119]
Multiwalled carbon nanotubes	13,100–24,000	Pesticide uptake by lettuce (*Lactuca sativa L.*)	Pristine – decrease root and shoot pesticide content by 88% and 78%, respectively. Amine functionalized – decrease by 57% and 23% in roots and shoots, respectively.	[120]
Fullerene	Not indicated	Effect on p,p'-DDE estrogenic pollutant	Coexposure with fullerene has significantly increased plant uptake of DDE: 29% and 48% in zucchini and soybean shoots, respectively. Contaminant uptake increased from 30 to 65%.	[121]
Carbon nanotubes	15–40 nm outer diameter varying lengths	CO₂ assimilation via photon influx (photosynthetic photon flux density)	Enhancement of photosynthesis after 20-week exposure at 300 µmol/m²/s; no toxic issue observed	[122]

atoms connected with single and double bonds assembled into a hollow sphere with <1 nm diameter [123]. Meanwhile, carbon nanotubes (CNTs) can consist of single-walled (SWCNTs) or several concentric tubes (MWCNTs), with different diameters of up to 100 nm. Accordingly, if SWCNT would be cut along its length and unrolled, graphene would be obtained [124].

Fullerene and carbon nanotubes. Being carbon-based, fullerenes and CNTs found direct application onto plants. Husen and Siddiqi, 2014, have reported that C_{60} can improve water retention volume, biomass yield, and fruit yield in plants by up to ~118%. For example, bitter melon seeds treated with C60 increased the levels of their active components such as cucurbitacin B (74%), lycopene (82%), charantin (20%), and insulin (91%) [125]. Lahiani and coworkers (2011) reported that the MWCNT activated seed sprouting, development, and growth of barley, soybean, and corn when deposited on seed surface at low doses of 5–200 µg/mL [122]. Meanwhile, Jiang and coworkers (2013) have demonstrated that CNT at appropriate concentrations of < 100 µg/mL encouraged seed sprouting and root propagation of rice plant [126]. As low as 50 µg/mL of CNT has doubled the yield of tomatoes. Therefore, CNT can be used to improve agricultural yields in the future.

While there are several positive impacts brought by fullerene and CNT on plants, several studies showed that both do harm on plants as well, such as reduced biomass, reduced root length, delayed flowering, and decreased yield. For instance, Liu et al., 2010, have observed that fullerenes can inhibit plant growth, with specific effects such as shortened length of seedling roots and change in the direction of its growth in response to gravity (phenomenon also known as gravitropism) [127]. Aside from the particle size and concentration, the negative effects may also be attributed to the synthesis method for the carbon nanoparticles. It has also been suggested that molecular configuration has an effect on the interactions [121]. This coincides with the report that pristine multiwalled carbon nanotubes are lethal to both plants and animals, which can be resolved by surface modification or functionalization [125]. Shrestha et al. also evaluated the effect of MWCNT on soil microbial community on sandy loam soil for 90 days. MWCNT was found not to affect soil metabolic properties (which include soil respiration, enzymatic activities, and composition at quantities closely resembling actual usage levels until 1,000 mg/kg soil). However, when the dosage was increased to 10,000 mg/kg, the composition of the soil microbial community has changed and favored only those that can tolerate the huge quantity such as *Rhodococcus, Cellulomonas, Nocardioides, and Pseudomonas*. While it is not lethal altogether, huge doses drive changes in the soil microecosystem [128]. CNT structure has also provided a potential for antimicrobial delivery due to its possible carrying capacity and capability to freely infiltrate membranes [129].

Graphene. Graphene was isolated as a single atomic layer of carbon in 2004 by Sir Andre Geim and Sir Kostya Novoselov for which they have won the 2010 Nobel Prize in Physics. Since then, the research on graphene and its popular derivative graphene oxide (GO) has exploded, such that it found application on almost anything. The oxidized form of graphene, known as GO, contains few layers of graphene stacked together with abundant oxygenated moieties on its surface and periphery. With a very high surface area of 2,600 m^2/g and special 2D structure, graphene and its derivatives offer an excellent medium for carrying nutrients and slow-release fertilizer [130]. For instance, KNO_3 pellets have been encapsulated in GO films using a simplified heat treatment in air at 90°C for 6 h. Potassium ions were found to act as a binder to the adjacent graphene sheets, which enabled GO films to act as a covering around the KNO_3 pellets and prevented its immediate release. The as-prepared GO-encapsulated KNO_3 pellets displayed slow-release activity within the first 7 h, where only 34.5% of potassium ions have been liberated into the water. Within the next hour, the burst release of potassium ions took place such that roughly 93.8% of potassium ions have been liberated from the fertilizers [131]. GO has also served as carrier of Zn and Cu for controllable slow release of the micronutrients. About 10% of the micronutrients can be loaded onto the GO sheets partly attributed to its high surface area and abundant oxygen-binding sites on its surface and along its edges. The GO-based vehicle demonstrated both the capacity to supply micronutrients in quick-release scenario (around 40% in 5 h) and the ability to support gradual sustained release [132].

GO may also be used as a regulatory mechanism to improve plant growth and stability. Meanwhile, graphene has fast-tracked the seed germination process (accelerated sprouting) and greatly shortened the germination time in tomatoes. The mean germination rates for control seeds were 33.3% in 4 days and 76.7% in 6 days, whereas the mean germination rates for those treated with graphene were reported as 76.7% in 4 days and 90% in 6 days [133]. It should be noted, however, that plant responses have wide variations and may not be easily extrapolated to different plant types. This is most likely due to the high heterogeneity of nanomaterials in terms of chemical composition, size and form, surface structure, solubility, aggregation, and application modes. GO has also been applied to the culture media of *Arabidopsis thaliana L.* while being introduced into the stems of the watermelon plant. It was found that at the recommended quantity, GO has had a desirable influence on plant development in terms of growing the root length, the area and number of leaves, and the flower bud formation. In addition, GO influenced the watermelon ripeness and increased the perimeter and sugar content of the fruit [134]. Graphene and CNT are also favorable components for biosensor transducers due to their high sensing properties and sensitivity to environmental changes [135,136].

Graphene nanosheets have displayed toxicity on bacteria (both Gram-positive and Gram-negative) models. GO nanosheets, resembling nanowalls, have been synthesized by electrophoretic deposition of Mg^{2+} on GO nanosheets produced via chemical exfoliation. The edges of the nanowalls were found to be incredibly sharp and could cause damage to the cell membrane bacteria when in close contact. This was an important mechanism in the bacterial inactivation that could affect the Gram-positive *Staphylococcus aureus* (that lacks an outer membrane) unlike the Gram-negative *Escherichia coli* bacteria (with an external membrane). In comparison, hydrazine-reduced GO has become more harmful to bacteria than pristine ones. The increased antibacterial behavior of the reduced nanowalls was due to a smoother flow of charge between the bacteria and the finer edges of the reduced nanowalls due to fewer oxygenated functionalities, during the contact [137].

Graphene, GO, and modified/functionalized GO also found application in postharvest management, particularly as catalyst for the conversion of agricultural residues into useful platform compounds. Graphene-based carbons have been used as catalysts in the conversion of cellulose, hemicellulose, and lignin components found in waste agricultural residues. The complex molecules are broken down into sugar monomers or converted into acids and furans that can be used as starting molecules to synthesize a wide variety of chemical commodities [138,139].

One of the main weaknesses of the graphene material is its receptivity toward oxidative attack, making it transform into various structures at relatively mild conditions. Such structural changes affect its functionality in the system and may eventually result in unlikely toxic qualities. In fact, scientists from Brown University have found out that the zigzag corners and armchair edges of graphene are rather sharp, which can easily pierce through cell membranes. It can then be ferried into the cell and impair normal cellular functioning [140].

Other carbon-based nanostructures, such as single-walled carbon nanohorns (SWCNHs) that bear conical shape, have also been utilized to enhance the growth, as well as accelerate seed germination for several crops, namely corn, tomato, rice, and soybean [141].

8.5 COMPOSITE NANOPARTICLES

Composite nanoparticles, oftentimes referred to as hybrid nanosystems, are nano-materials with composite structures and can consist of two or more nanoscale components with unique physicochemical properties [142]. The combination of different organic–organic and organic–inorganic materials or polymer–hybrid nanoparticles based on the combination of individual constituents can yield superior features through the combination of the desirable properties [143]. The synthesis of these materials was driven by the need to maximize the strengths of individual component materials, to bring about new properties, and, to some extent, minimize their defects. The advent of more sensitive imaging, sensing, and analytical capabilities has also revolutionized composite nanoparticle research and development. Composites are typically found in three geometries: (1) core–shell materials with concentric stacking over each other; (2) composite materials encompassing separate, diversified domains of various substances; and (3) hybrid materials covering diversified domains that are beyond distinction above the molecular/macromolecular level [144]. As shown in Figure 8.3, composite nanoparticles can be synthesized from component materials in either parallel or sequential manner. The preparation of the individual components may include self-assembly, grafting, immobilization, functionalization, or heat treatment.

As discussed in an earlier section, metal nanomaterials have been primarily exploited to enhance plant growth and protect crops from diseases and insects. Unfortunately, these metal-based nanomaterials have a huge tendency to

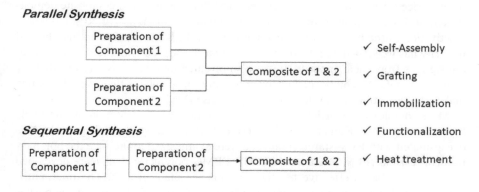

FIGURE 8.3 Synthesis routes for composite nanoparticles.

bioaccumulate on the plant, originating from the roots and eventually translocated at lesser concentration, within the shoots and leaves. In this context, controlled-release fertilizers have been enabled through encapsulation. Martins et al. reported that ZnO can be immobilized onto natural polymers, such as cellulose, alginate, and chitosan to address Zn deficiencies in maize. Zn deposition efficiencies can range between 81% and 94%. ZnO encapsulated in alginate beads dosed at 100 mg of Zn per 1 kg of soil has slightly improved plant growth without any signs of Zn deficiency or intoxication [145].

Andelkovic et al. synthesized functionalized GO/iron (GO-Fe) as phosphate ions (P) carrier for improved plant nutrient delivery. They reported that ferric ions have supported the incorporation of phosphate ions onto the GO-Fe composite, yielding 48 mg P/g. When the as-prepared GO-Fe composite with phosphate (GO-Fe-P) was tested, it was found that P has been released slowly when compared with the commercial monoammonium phosphate (MAP) fertilizer. These findings are crucial to address the leaching issue of soluble P onto surface water as well as groundwater. Through visualization and chemical analysis of the diffusion paths, it was found that >99% of the P discharged from GO-Fe-P persisted in the composite or within localized area of the application site, indicating the sluggish release characteristics [146].

Atrazine is a common herbicide for weed control, which may potentially contaminate water supplies by leaching, runoff, and spray drift. Nsibande et al. reported on the fabrication of a sensor consisting of L-cysteine-capped CdSeTe/ZnS QDs functionalized with MIPs for selective atrazine detection. CdSeTe/ZnS@MIP sensor displayed a quick reaction time (5 min) when dealing with atrazine, and the fluorescence intensity was linearly extinguished within the 2–20 mol/L atrazine range. The detection limit of 0.80×10^{-7} mol/L[1] is comparable to the published environmental standards. Finally, the sensor was applied in actual water samples and has demonstrated adequate recoveries (92%–118%) in spiked samples, making it a promising candidate for use in water monitoring [147].

GO-Ag composite was applied to combat *Xanthomonas oryzae pv. oryzae*, which induces bacterial leaf blight, a highly damaging disease in the growth and development of rice (*Oryza sativa L.*). It was observed that the prepared composite demonstrated an improvement of up to four times relative to pristine Ag NPs. Complete inactivation of the bacteria was gained at 2.5 μg/mL. It was elucidated that the cell integrity has weakened and the intercellular material seeped with the slow release of the Ag ions from the composite. It was also revealed in a phytotoxicity test that the as-prepared GO-Ag composite had lower toxicity for the plants than Ag NPs, which makes it more attractive for deterring crop disease [148].

More complex nanoparticle systems may also involve three or more nanoparticle types. In a study by Kumar et al., bimetallic (Cu/Zn) nanoparticle-dispersed CNFs encapsulated with PVA–starch composite were synthesized to produce the polymer-bi-metal-carbon (PBMC) composites. The PBMC polymeric composite developed was shown to be an effective fertilizer to enhance plant growth. The micronutrients (Cu/Zn) have been precisely released from the CNFs and polymeric composites [149]. The application of composite nanoparticles in agriculture is summarized in Table 8.4.

TABLE 8.4

Composite Nanoparticles as Applied to Agriculture

	Size, nm	Purpose	Findings	References
ZnO immobilized on microcrystalline cellulose	120–400	Controlled release of zinc and timed delivery for maize	Deposition efficiency = 94%	[130]
ZnO immobilized on chitosan	113 ± 33		Deposition efficiency = 81%	
ZnO immobilized on alginate	87 ± 24		Deposition efficiency = 88%	
Poly(amidoamine) (PAMAM) grafted onto poly(styrene-divinylbenzene-glycidyl methacrylate)	600	Removal of glyphosate from aqueous solution	Maximum glyphosate adsorption was achieved at pH = 3 with minimum adsorption equilibrium time = 5 min. Adsorption efficiency in contaminated drinking water = 95% Can be reused for about five cycles	[150]
Urea-hydroxyapatite	15–20 diameter, 100–200 length	Nutrient delivery (nitrogen)	63 min to release 86% of N, vs 99% in pure urea, remaining 14% about 1 week	[151]
Silver on graphene oxide	$Ag = 18 \pm 3$ and 5 ± 1.8; aggregate not stated	Plant Disease Management: against *Xanthomonas perforans* in tomatoes	Optimal antibacterial activity at 20 ppm Ag (18 nm) and 16 ppm (5 nm) after 1 h incubation. Applied 100 ppm solution (26.9 ppm Ag) to tomato transplants	[152]
Superhydrophobic biopolymer-coated slow-release fertilizer	20–500	Slow-release fertilizer	Longer N release longevity ~100 days	[153]

8.6 ISSUES, CHALLENGES, AND OPPORTUNITIES

Nanotechnology indeed has made significant changes and continues to hold key to various advancements and developments in agriculture. However, there are numerous problems and concerns that need to be tackled in order to shift technology toward adoption and acceptance.

- **Design**: Most nanoparticles are insoluble in water, which poses limitations with their interaction in biological environment. Therefore, one of the major challenges is to enhance nanoparticle solubility in water. This is particularly promising for developing novel nontoxic drug carriers and minimizing the use of organic solvents during synthesis. One way to do so is to modulate the structure and nanoparticle design through well-identified architectures, porous internal cavities, and variety of surface functionalities [154].

- **Green and sustainable synthesis of nanoparticles**: Nanoparticle synthesis is quite tedious. Oftentimes, it requires a myriad of chemicals that produce waste streams in solid, liquid, and gaseous forms. These waste streams require additional handling and may impact the environment, if not properly treated. In consequence, many researchers have attempted to synthesize nanoparticles in one pot [155,156]. Moreover, many studies have synthesized nanoparticles using natural products or using components derived from waste biomass or industrial by-products. For instance, Ag-nanoparticles have been synthesized using the extracts of *P. nigrum* derived from its leaves and stem. The synthesis of Ag nanoparticles using *P. nigrum* leaf and stem extracts started at 10 min and ended at 2 h for leaf and 4 h for stem extracts. The transmission electron microscopy (TEM) photographs of the as-prepared Ag NP revealed twin-shaped nanoparticles. Usage of agricultural waste (coconut shell) for the synthesis of Ag NP and assessment of their antibacterial action against selected human pathogens has also been reported [157].

  Nanotechnology truly holds a promising potential in agricultural systems to support the plant growth and to improve the input of pesticides and fertilizers into plant systems in order to generate more food. However, the production of engineered nanomaterials also poses some risks that need to be investigated in the future.

- **Nanotoxicity**: Nanoparticles are quite small and can easily elude the naked eye. It can be unintentionally inhaled or ingested. Hence, potential exposure cannot be readily assessed and potential toxic effects may accumulate due to repeated exposure over time. Therefore, it is crucial to elucidate and unveil mechanisms and pathways on the interaction between nanoparticles and living organisms for safety reasons. As discussed earlier, not all synthesized and tested nanoparticles were successful. There were several conflicting studies on the impact of nanoparticles on plant growth [121,122,125–127]. These are highly influenced by their size, structure, properties, and synthesis method. However, the rapid development of composite nanoparticles poses higher risks since new properties are elicited when they are formed, which merits further studies particularly on their pathways to humans and the environment. Furthermore, it would eventually be necessary that the chemicals, solvent medium, and the reducing and stabilizing agents used during nanoparticle preparation be evaluated for toxicity and environmental effects.

- **Bioaccumulation**: Engineered nanoparticles have been reported in the environment and its detrimental impact is largely unknown. Some studies have indicated that the amounts of metal nanoparticles in the environment can range from 3.1 to 31 µg/kg in the soil to 76–760 µg/L in water, as in the case of ZnO [158,159]; thus, an in-depth and comprehensive ecological risk assessment must be necessary. Global life cycle assessments pointed out that 63%–91% of the 260,000–309,000 metric tons engineered nanomaterials produced end up in landfills, with 8%–28% accrued in soil, 0.4%–7% in water bodies, and 0.1%–1.5% in the atmosphere [160]. With global annual production of metal nanoparticles reaching 585,000 metric tons in 2019, the trace quantities reaching environmental compartments could be critical.

- **Absence of policies or regulations**: There is currently no international oversight of nanoproducts or the underlying nanotechnology, and there are no globally accepted definitions or terms for nanotechnology. As a result, there are no internationally agreed guidelines for the toxicity monitoring of nanoparticles and no formal protocols for the estimation of the environmental impacts of nanoparticles. Furthermore, nanomaterials do not come under the framework of any current international conventions on toxic chemicals [161].

It should be noted, however, that there are several agencies and organizations that have instituted regulatory guidelines, which are currently limited in scope. For example, the Food and Agriculture Organization of the United Nations has several roundtable discussions with the World Health Organization on nanotechnology for food safety and food security issues. The US Food and Drug Administration regulates new products including pharmaceuticals and food additives, whereas the US Environmental Protection Agency covers regulatory review for nanoscale materials as a chemical substance that might pose risk to human health and the environment. The European Commission and its Joint Research Center have identified nanotechnology as a key enabling technology and have thus issued legislative frameworks and regulatory reviews on how to classify or treat nanomaterials. They have been evaluating relevant and applicable legislations through Registration, Evaluation, Authorization, and Restriction of Chemicals (REACH) and Classification, Labeling, and Packaging (CLP) as well as product-specific categories (cosmetics, toys, pesticides, etc.). In Canada, nanomaterials are covered under existing legislative and regulatory frameworks for chemicals. Unless a nanomaterial is not reflected in the Domestic Substances List (DSL), it will be considered as a new substance and will be covered by the New Substances Notification Regulations (Chemicals and Polymers). In China, the National Nanotechnology Standardization Technical Committee is the main coordinator for the methods, safety, protocols, and technical standards for nanotechnology. With respect to standardization, the International Standard Organization (ISO), American Standard for Testing Materials (ASTM), the British Standards Institution, and the European Committee for Standardization are some of the major players who have technical committees that review and address issues on nanomaterials and its evolving needs [162–164].

- **Incomplete risk assessment**: Without any regulation, the environmental fate of any newly synthesized nanoparticles is not necessary even until commercial production levels. Therefore, the end of life cycle is also uncertain. Thus, there is an increasing concern about the unregulated exposure of the nanoparticles. For example, nanoparticle release or addition onto soil can cause unfavorable impact toward the microbes (fungi, bacteria, protozoa, etc.) inhabiting the soil, and the varied responses of soil fungi and soil bacteria have been previously reported [155]. In several instances, the presence of nanoparticles induces damage on the physicochemical characteristics of soil and toxicity. These merit further research and in-depth studies.

- **Social and ethical aspects**: Research and development on nanoparticles are quite concentrated on a select number of countries. Thus, understanding and acceptance of the public also vary. This may also impact vulnerable communities because they might not be aware of their exposure levels. In contrast, since only major economies have initiated nanomaterial research centers and heavily invested on them, the positive benefits will also be enjoyed by these countries. Therefore, there is a widening gap in terms of knowledge economy and commercial exploitation of nanotechnology.

8.7 CONCLUSION AND PROSPECTS

The use of nanoparticles is an emerging technology to alleviate agricultural issues and address growing demands. Nanoparticles of various forms and compositions have found extensive application as water-retaining agents, plant growth enhancer, pest management, enhanced nutrient and fertilizer delivery, postharvest management, sensors, and ambient condition modulator, with almost equal advantages and disadvantages. Much of the research studies mentioned are in their early and exploratory phases; thus, it is expected that the next 5 years would usher development in this field. Composite nanoparticles encompass all possible applications with improved properties from its constituent materials, so there will be countless structures and architectures that will be developed and reported in the immediate future. Nonetheless, the current report has outlined the importance of interdisciplinary and multidisciplinary collaboration in various fields, namely the physical sciences for the design and synthesis, life sciences for the application and effectiveness, health sciences for the safety and risk assessments, and social sciences for acceptance and regulatory framework. Thus, it is expected that more studies in the future will be more comprehensive and encompassing, and there will be multiple touchpoints with respect to nanotechnology. At this point, major nanotechnology research and development are carried out and funded by big economies such as China, United States, and Europe. Therefore, there will be plenty of opportunities for other economies to take part. Amidst the COVID-19 global pandemic, swine flu and avian flu have also wreaked havoc in the food industry. Thus, the developments in nanotechnology would be crucial to avert further impacts in the current future outbreaks that would otherwise affect the food supply chain.

REFERENCES

1. Food and Agriculture Organization of the United Nations. 2019. "OECD-FAO Agricultural Outlook 2019–2028". http://www.agri-outlook.org/Outlook-Summary-ENG.pdf.
2. Rial-Lovera, K., W.P. Davies, and N.D. Cannon. 2016. "Implications of climate change predictions for UK cropping and prospects for possible mitigation: A review of challenges and potential responses." *Journal of the Science of Food and Agriculture* 97 (1): 17–32. Doi: 10.1002/jsfa.7767.
3. Elijah, V.T., and J.O. Odiyo. 2019. "Perception of environmental spillovers across scale in climate change adaptation planning: The case of small-scale farmers' irrigation strategies, Kenya." *Climate* 8 (1): 3. Doi: 10.3390/cli8010003.

4. Gornall, J., R. Betts, E. Burke, R. Clark, J. Camp, K. Willett, and A. Wiltshire. 2010. "Implications of climate change for agricultural productivity in the early twenty-first century." *Philosophical Transactions of the Royal Society B: Biological Sciences* 365 (1554): 2973–2989. Doi: 10.1098/rstb.2010.0158.
5. Fischer, G., K. Frohberg, M.L. Parry, and C. Rosenzweig. 1996. "Impacts of potential climate change on global and regional food production and vulnerability." *Climate Change and World Food Security* 115–159. Doi: 10.1007/978-3-642-61086-8_5.
6. Schnitter, R., and P. Berry. 2019. "The climate change, food security and human health nexus in Canada: A framework to protect population health." *International Journal of Environmental Research and Public Health* 16 (14): 2531. Doi: 10.3390/ijerph16142531.
7. Beddington, J. 2010. "Food security: Contributions from science to a new and greener revolution." *Philosophical Transactions of the Royal Society B: Biological Sciences* 365 (1537): 61–71. Doi: 10.1098/rstb.2009.0201.
8. Lu, F.-M. 2009. "The role of agricultural mechanization in the modernization of Asian agriculture: Taiwan's experience." *Engineering in Agriculture, Environment and Food* 2 (4): 124–131. Doi: 10.1016/s1881-8366(09)80003-7.
9. Ali, M.H., and M.S.U. Talukder. 2008. "Increasing water productivity in crop production—A synthesis." *Agricultural Water Management* 95 (11): 1201–1213. Doi: 10.1016/j.agwat.2008.06.008.
10. Auffan, M., J. Rose, J.Y. Bottero, G.V. Lowry, J.P. Jolivet, and M.R. Wiesner. 2009. "Towards a definition of inorganic nanoparticles from an environmental, health and safety perspective. "*Nature Nanotechnology* 4: 634–641. Doi: 10.1038/nnano.2009.242.
11. Enamala, M.K., B. Kolapalli, P.D. Sruthi, S. Sarkar, C. Kuppam, and M. Chavali. 2019. "Applications of nanomaterials and future prospects for nanobionics." *Plant Nanobionics* 177–197. Doi: 10.1007/978-3-030-16379-2_6.
12. Industry Arc. 2019. "Nanotechnology market overview". https://www.industryarc.com/Report/15022/ nanotechnology-market.html.
13. Ditta, A., S. Mehmood, M. Imtiaz, M.S. Rizwan, and I. Islam. 2020. "Soil fertility and nutrient management with the help of nanotechnology." *Nanomaterials for Agriculture and Forestry Applications* 273–287. Doi: 10.1016/b978-0-12-817852-2.00011-1.
14. Kopittke, P.M., E. Lombi, P. Wang, J.K. Schjoerring, and S. Husted. 2019. "Nanomaterials as fertilizers for improving plant mineral nutrition and environmental outcomes." *Environmental Science: Nano* 6 (12): 3513–3524. Doi: 10.1039/c9en00971j.
15. Duhan, J.S., R. Kumar, N. Kumar, P. Kaur, K. Nehra, and S. Duhan. 2017. "Nanotechnology: The new perspective in precision agriculture." *Biotechnology Reports* 15: 11–23. Doi: 10.1016/j.btre.2017.03.002.
16. Hesni, M.A., A. Hedayati, A. Qadermarzi, M. Pouladi, S. Zangiabadi, and N. Naqshbandi. 2020. "Using *Chlorella vulgaris* and iron oxide nanoparticles in a designed bioreactor for aquaculture effluents purification." *Aquacultural Engineering* 90: 102069. Doi:10.1016/j.aquaeng.2020.102069.
17. Sampathkumar, K., K.X. Tan, and S.C. Loo. 2020. "Developing nano-delivery systems for agriculture and food applications with nature-derived polymers." *iScience* 23 (5): 101055. Doi: 10.1016/j.isci.2020.101055.
18. Vamvakaki, V., and N.A. Chaniotakis. 2007. "Pesticide detection with a liposome-based nano-biosensor." *Biosensors and Bioelectronics* 22 (12): 2848–2853. Doi: 10.1016/j.bios.2006.11.024.
19. Khan, I., K. Saeed, and I. Khan. 2019. "Nanoparticles: Properties, applications and toxicities." *Arabian Journal of Chemistry* 12 (7): 908–931. Doi: 10.1016/j.arabjc.2017.05.011.
20. Jeevanandam, J., A. Barhoum, Y.S. Chan, A. Dufresne, and M.K. Danquah. 2018. "Review on nanoparticles and nanostructured materials: History, sources, toxicity and regulations." *Beilstein Journal of Nanotechnology* 9: 1050–1074. Doi: 10.3762/bjnano.9.98.

21. Vega-Vásquez, P., N.S. Mosier, and J. Irudayaraj. 2020. "Nanoscale drug delivery systems: From medicine to agriculture." *Frontiers in Bioengineering and Biotechnology* 8. Doi: 10.3389/fbioe.2020.00079.

22. Karny, A., A. Zinger, A. Kajal, J. Shainsky-Roitman, and A. Schroeder. 2018. "Therapeutic nanoparticles penetrate leaves and deliver nutrients to agricultural crops." *Scientific Reports* 8 (1). Doi: 10.1038/s41598-018-25197-y.

23. Pérez-de-Luque, A., and D. Rubiales. 2009. "Nanotechnology for parasitic plant control." *Pest Management Science* 65 (5): 540–545. Doi: 10.1002/ps.1732.

24. Liu, X., B. He, Z. Xu, M. Yin, W. Yang, H. Zhang, J. Cao, and J. Shen. 2015. "A functionalized fluorescent dendrimer as a pesticide nanocarrier: Application in pest control." *Nanoscale* 7 (2): 445–449. Doi: 10.1039/c4nr05733c.

25. Liu, J., F.-h. Wang, L.-l. Wang, S.-y. Xiao, C.-y. Tong, D.-y. Tang, and X.-m. Liu. 2008. "Preparation of fluorescence starch-nanoparticle and its application as plant transgenic vehicle." *Journal of Central South University of Technology* 15 (6): 768–773. Doi: 10.1007/s11771-008-0142-4.

26. Stloukal, P., P. Kucharczyk, V. Sedlarik, P. Bazant, and M. Koutny. 2012. "Low molecular weight poly(lactic acid) microparticles for controlled release of the herbicide metazachlor: Preparation, morphology, and release kinetics." *Journal of Agricultural and Food Chemistry* 60 (16): 4111–4119. Doi: 10.1021/jf300521j.

27. Diallo, M.S., S. Christie, P. Swaminathan, J.H. Johnson, and W.A. Goddard. 2005. "Dendrimer enhanced ultrafiltration. 1. Recovery of Cu(II) from aqueous solutions using PAMAM dendrimers with ethylene diamine core and terminal NH_2 groups." *Environmental Science & Technology* 39 (5): 1366–1377. Doi: 10.1021/es048961r.

28. Tong, Y., Y. Wu, C. Zhao, Y. Xu, J. Lu, S. Xiang, F. Zong, and X. Wu. 2017. "Polymeric nanoparticles as a metolachlor carrier: Water-based formulation for hydrophobic pesticides and absorption by plants." *Journal of Agricultural and Food Chemistry* 65 (34): 7371–7378. Doi: 10.1021/acs.jafc.7b02197.

29. Daraee, H., A. Eatemadi, E. Abbasi, S.F. Aval, M. Kouhi, and A. Akbarzadeh. 2014. "Application of gold nanoparticles in biomedical and drug delivery." *Artificial Cells, Nanomedicine, and Biotechnology* 44 (1): 410–422. Doi: 10.3109/21691401. 2014.955107.

30. Bangham, A.D. 1978. "Properties and uses of lipid vesicles: An overview." *Annals of the New York Academy of Sciences* 308 (1): 2–7. Doi: 10.1111/j.1749-6632.1978.tb22010.x.

31. Lee, S.-C., K.-E. Lee, J.-J. Kim, and S.-H. Lim. 2005. "The effect of cholesterol in the liposome bilayer on the stabilization of incorporated retinol." *Journal of Liposome Research* 15 (3–4): 157–166. Doi: 10.1080/08982100500364131.

32. Ko, Y.T., and U. Bickel. 2012. "Liposome-encapsulated polyethylenimine/oligonucleotide polyplexes prepared by reverse-phase evaporation technique." *AAPS PharmSciTech* 13 (2): 373–378. Doi: 10.1208/s12249-012-9757-8.

33. Deamer, D.W. 2019. "Preparation of solvent vaporization liposomes." *Liposome Technology* 29–35. Doi: 10.1201/9781351074100-3.

34. Zumbuehl, O., and H.G. Weder. 1981. "Liposomes of controllable size in the range of 40 to 180 nm by defined dialysis of lipid/detergent mixed micelles." *Biochimica et Biophysica Acta (BBA) - Biomembranes* 640 (1): 252–262. Doi: 10.1016/0005-2736(81)90550-2.

35. Hussain, F.H.S. and H. Abdulla. 2018. "Synthesis and characterization of some new Bis-(dihydropyrimidinones-5-carboxamide) by using ultrasonic irradiation." *Eurasian Journal of Science & Engineering* 3 (3). Doi: 10.23918/eajse.v3i3p132.

36. Mui, B. and M.J. Hope. 2018. "Formation of large unilamellar vesicles by extrusion." In *Liposome Technology: Liposome Preparation and Related Techniques,* 77–88. Doi: 10.1201/9781420005875-8.

37. Schubert, R. 2003. "Liposome preparation by detergent removal." *Methods in Enzymology,* 46–70. Doi: 10.1016/s0076-6879(03)67005-9.

38. Cheung, C.C., G. Ma, A. Ruiz, and W.T. Al-Jamal. 2020. "Microfluidic production of lysolipid-containing temperature-sensitive liposomes." *Journal of Visualized Experiments* 157. Doi: 10.3791/60907.
39. Taylor, T.M., J. Weiss, P.M. Davidson, and B.D. Bruce. 2005. "Liposomal nanocapsules in food science and agriculture." *Critical Reviews in Food Science and Nutrition* 45 (7–8): 587–605. Doi: 10.1080/10408390591001135.
40. Ruyra, A., M. Cano-Sarabia, S.A. MacKenzie, D. Maspoch, and N. Roher. 2013. "A novel liposome-based nanocarrier loaded with an LPS-dsRNA cocktail for fish innate immune system stimulation." *PLoS One* 8 (10): e76338. Doi: 10.1371/journal.pone.0076338.
41. Maja, L., K. Željko, and P. Mateja. 2020. "Sustainable technologies for liposome preparation." *The Journal of Supercritical Fluids* 165: 104984. Doi: 10.1016/j.supflu.2020.104984.
42. Boas, U., and P.M. Heegaard. 2004. "Dendrimers in drug research." *Chemical Society Reviews* 33 (1): 43. Doi: 10.1039/b309043b.
43. Buhleier E., W. Wehner, and F. Vogtle. 1978. "Cascade and Nonskid-chain-like synthesis of molecular cavity topologies." *Synthesis* 2: 155–158.
44. Tomalia D.A., H. Baker, J. Dewald, M. Hall, G. Kallos, S. Martin, J. Roeck, J. Ryder, and P. Smith. 1985. "A new class of polymers: Starburst-dendritic macromolecules." *Polymer Journal* 17: 117–132.
45. Astruc, D., E. Boisselier, and C. Ornelas. 2010. "Dendrimers designed for functions: From physical, photophysical, and supramolecular properties to applications in sensing, catalysis, molecular electronics, photonics, and nanomedicine." *Chemical Reviews* 110 (4): 1857–1959. Doi: 10.1021/cr900327d.
46. Caminade, A.-M., R. Laurent, M. Zablocka, and J.-P. Majoral. 2012. "Organophosphorus chemistry for the synthesis of dendrimers." *Molecules* 17 (11): 13605–13621. Doi: 10.3390/molecules171113605.
47. Hawker, C.J., and J.M. Frechet. 1990. "Preparation of polymers with controlled molecular architecture. A new convergent approach to dendritic macromolecules." *Journal of the American Chemical Society* 112 (21): 7638–7647. Doi: 10.1021/ja00177a027.
48. Choudhary, S., L. Gupta, S. Rani, K. Dave, and U. Gupta. 2017. "Impact of dendrimers on solubility of hydrophobic drug molecules." *Frontiers in Pharmacology,* 8. Doi: 10.3389/fphar.2017.00261.
49. Palmerston Mendes, L., J. Pan, and V. Torchilin. 2017. "Dendrimers as nanocarriers for nucleic acid and drug delivery in cancer therapy." *Molecules* 22 (9): 1401. Doi: 10.3390/molecules22091401.
50. Turrin, C.-O., and A.-M. Caminade. 2011. "Dendrimer conjugates for drug delivery." *Dendrimers,* 437–461. Doi: 10.1002/9781119976530.ch18.
51. Wang, J., I.K. Voets, R. Fokkink, J. Van der Gucht, and A.H. Velders. 2014. "Controlling the number of dendrimers in dendrimicelle nanoconjugates from 1 to more than 100." *Soft Matter* 10 (37): 7337–7345. Doi: 10.1039/c4sm01143k.
52. Sharma, M. 2019. "Transdermal and intravenous nano drug delivery systems." *Applications of Targeted Nano Drugs and Delivery Systems,* 499–550. Doi: 10.1016/b978-0-12-814029-1.00018-1.
53. Siracusa, V. 2019. "Microbial degradation of synthetic biopolymers waste." *Polymers* 11 (6): 1066. Doi: 10.3390/polym11061066.
54. Rai, M., S. Bansod, M. Bawaskar, A. Gade, C.A. Dos Santos, A.B. Seabra, and N. Duran. 2015. "Nanoparticles-based delivery systems in plant genetic transformation." *Nanotechnologies in Food and Agriculture,* 209–239. Doi: 10.1007/978-3-319-14024-7_10.
55. Naz, M.Y., and S.A. Sulaiman. 2014. "Testing of starch-based carbohydrate polymer coatings for enhanced urea performance." *Journal of Coatings Technology and Research* 11 (5): 747–756. Doi: 10.1007/s11998-014-9590-y.

56. Perez, J.J., and N.J. Francois. 2016. "Chitosan-starch beads prepared by ionotropic gelation as potential matrices for controlled release of fertilizers." *Carbohydrate Polymers* 148: 134–142. Doi: 10.1016/j.carbpol.2016.04.054.

57. Tang, J., J. Hong, Y. Liu, B. Wang, Q. Hua, L. Liu, and D. Ying. 2017. "Urea controlled-release fertilizer based on gelatin microspheres." *Journal of Polymers and the Environment* 26 (5): 1930–1939. Doi: 10.1007/s10924-017-1074-6.

58. Jiao, G.-J., Q. Xu, S.-L. Cao, P. Peng, and D. She. 2018. "Controlled-release fertilizer with lignin used to trap urea/hydroxymethylurea/ urea-formaldehyde polymers." *BioResources* 13 (1). Doi: 10.15376/biores.13.1.1711-1728.

59. Liu, L.S., J. Kost, M.L. Fishman, and K.B. Hicks. 2008. "A review: Controlled release systems for agricultural and food applications." *ACS Symposium Series* 265–281. Doi: 10.1021/bk-2008-0992.ch014.

60. Akalin, G.O., and M. Pulat. 2019. "Controlled release behavior of zinc-loaded carboxymethyl cellulose and carrageenan hydrogels and their effects on wheatgrass growth." *Journal of Polymer Research* 27 (1). Doi: 10.1007/s10965-019-1950-y.

61. Huang, B., F. Chen, Y. Shen, K. Qian, Y. Wang, C. Sun, Z. Xiang, et al. 2018. "Advances in targeted pesticides with environmentally responsive controlled release by nanotechnology." *Nanomaterials* 8 (2): 102. Doi: 10.3390/nano8020102.

62. Guo, J., W. Shi, L. Wen, X. Shi, and J. Li. 2019. "Effects of a super-absorbent polymer derived from poly-γ-glutamic acid on water infiltration, field water capacity, soil evaporation, and soil water-stable aggregates." *Archives of Agronomy and Soil Science*, 1–12. Doi: 10.1080/03650340.2019.1686137.

63. Cheng, D., Y. Liu, G. Yang, and A. Zhang. 2018. "Water- and fertilizer-integrated hydrogel derived from the polymerization of acrylic acid and urea as a slow-release N fertilizer and water retention in agriculture." *Journal of Agricultural and Food Chemistry* 66 (23): 5762–5769. Doi: 10.1021/acs.jafc.8b00872.

64. El Knidri, H., R. Belaabed, A. Addaou, A. Laajeb, and A. Lahsini. 2018. "Extraction, chemical modification and characterization of chitin and chitosan." *International Journal of Biological Macromolecules* 120: 1181–1189. Doi: 10.1016/j.ijbiomac.2018.08.139.

65. Arunkumar, R., K.V. Harish Prashanth, and V. Baskaran. 2013. "Promising interaction between nanoencapsulated lutein with low molecular weight chitosan: Characterization and bioavailability of lutein *in vitro* and *in vivo*." *Food Chemistry* 141 (1): 327–337. Doi: 10.1016/j.foodchem.2013.02.108.

66. Hidangmayum, A., P. Dwivedi, D. Katiyar, and A. Hemantaranjan. 2019. "Application of chitosan on plant responses with special reference to abiotic stress." *Physiology and Molecular Biology of Plants* 25 (2): 313–326. Doi: 10.1007/s12298-018-0633-1.

67. Silva, M.D., D.S. Cocenza, R. Grillo, N.F. Melo, P.S. Tonello, L.C. Oliveira, D.L. Cassimiro, A.H. Rosa, and L.F. Fraceto. 2011. "Paraquat-loaded alginate/chitosan nanoparticles: Preparation, characterization and soil sorption studies." *Journal of Hazardous Materials* 190 (1–3): 366–374. Doi: 10.1016/j.jhazmat.2011.03.057.

68. Grillo, R., Z. Clemente, V. Chalupe, C. Jonsson, R. Lima, G. Sanches, C. Nishisaka, K. Oehlke, R. Greiner, and L. Fraceto. 2014. "Effect of the presence of aquatic humic substances on the toxicity of chitosan/tripolyphosphate nanoparticles containing paraquat." *Toxicology Letters* 229: 191. Doi: 10.1016/j.toxlet.2014.06.648.

69. Maruyama, C.R., M. Guilger, M. Pascoli, N. Bileshy-José, P.C. Abhilash, L.F. Fraceto, and R. De Lima. 2016. "Nanoparticles based on chitosan as carriers for the combined herbicides imazapic and imazapyr." *Scientific Reports* 6 (1). Doi: 10.1038/srep19768.

70. Tungittiplakorn, W., L.W. Lion, C. Cohen, and J.-Y. Kim. 2004. "Engineered polymeric nanoparticles for soil remediation." *Environmental Science & Technology* 38 (5): 1605–1610. Doi: 10.1021/es0348997.

71. Watanabe, K. 2018. "Photochemistry on nanoparticles." *Encyclopedia of Interfacial Chemistry*, 563–572. Doi: 10.1016/b978-0-12-409547-2.13211-1.

72. Wang, S., and L. Gao. 2019. "Laser-driven nanomaterials and laser-enabled nanofabrication for industrial applications." *Industrial Applications of Nanomaterials*, 181–203. Doi: 10.1016/b978-0-12–815749-7.00007-4.

73. Kharissova O.V., B.I. Kharisov, C.M. Oliva González, Y.P. Méndez, and I. López. 2019. "Greener synthesis of chemical compounds and materials." *Royal Society Open Science* 6(11): 191378. Doi: 10.1098/rsos.191378.

74. Wojnarowicz, J., T. Chudoba, S. Gierlotka, K. Sobczak, and W. Lojkowski. 2018. "Size control of cobalt-doped ZnO nanoparticles obtained in microwave solvothermal synthesis." *Crystals* 8 (4): 179. Doi: 10.3390/cryst8040179.

75. Kim, S.-W., J.-H. Jung, K. Lamsal, Y.-S. Kim, S.-J. Sim, H.-S. Kim, S.-J. Chang, J.-K. Kim, K.-S. Kim, and Y.-S. Lee. 2011. "Control efficacy of nano-silver liquid on oak wilt caused by Raffaelea sp. in the field." *Research in Plant Disease* 17 (2): 136–141. Doi: 10.5423/rpd.2011.17.2.136.

76. Kim, K.-J., W.S. Sung, B.K. Suh, S.-K. Moon, J.-S. Choi, J.G. Kim, and D.G. Lee. 2008. "Antifungal activity and mode of action of silver nano-particles on Candida albicans." *BioMetals* 22 (2): 235–242. Doi: 10.1007/s10534-008-9159-2.

77. Khandelwal, N., G. Kaur, K.K. Chaubey, P. Singh, S. Sharma, A. Tiwari, S.V. Singh, and N. Kumar. 2014. "Silver nanoparticles impair Peste des petits ruminants virus replication." *Virus Research* 190: 1–7. Doi: 10.1016/j.virusres.2014.06.011.

78. El-Mohamady, R.S., T.A. Ghattas, M.F. Zawrah, and Y.G.M. Abd El-Hafeiz. 2018. "Inhibitory effect of silver nanoparticles on bovine herpesvirus-1." *International Journal of Veterinary Science and Medicine* 6 (2): 296–300. Doi: 10.1016/j.ijvsm. 2018.09.002.

79. Jasim, B., R. Thomas, J. Mathew, and E.K. Radhakrishnan. 2017. "Plant growth and diosgenin enhancement effect of silver nanoparticles in Fenugreek (Trigonella foenum-graecum L.)." *Saudi Pharmaceutical Journal* 25 (3): 443–447. Doi: 10.1016/j. jsps.2016.09.012.

80. Pallavi, C. M. Mehta, R. Srivastava, S. Arora, and A.K. Sharma. 2016. "Impact assessment of silver nanoparticles on plant growth and soil bacterial diversity." *Biotech* 6 (2). Doi: 10.1007/s13205-016-0567-7.

81. Mohamed, M.M., S.A. Fouad, H.A. Elshoky, G.M. Mohammed, and T.A. Salaheldin. 2017. "Antibacterial effect of gold nanoparticles against *Corynebacterium pseudotuberculosis*." *International Journal of Veterinary Science and Medicine* 5 (1): 23–29. Doi: 10.1016/j.ijvsm.2017.02.003.

82. Wani, I.A., and T. Ahmad. 2013. "Size and shape dependant antifungal activity of gold nanoparticles: A case study of Candida." *Colloids and Surfaces B: Biointerfaces* 101: 162–170 Doi: 10.1016/j.colsurfb.2012.06.005.

83. Zhang, Y., T.P. Shareena Dasari, H. Deng, and H. Yu. 2015. "Antimicrobial activity of gold nanoparticles and ionic gold." *Journal of Environmental Science and Health, Part C* 33 (3): 286–327. Doi: 10.1080/10590501.2015.1055161.

84. López-Miranda, J.L., R. Esparza, G. Rosas, R. Pérez, and M. Estévez-González. 2019. "Catalytic and antibacterial properties of gold nanoparticles synthesized by a green approach for bioremediation applications." *Biotech* 9 (4). Doi: 10.1007/ s13205-019-1666-z.

85. Li, X., S.M. Robinson, A. Gupta, K. Saha, Z. Jiang, D.F. Moyano, A. Sahar, M.A. Riley, and V.M. Rotello. 2014. "Functional gold nanoparticles as potent antimicrobial agents against multi-drug-resistant bacteria." *ACS Nano* 8 (10): 10682–10686. Doi: 10.1021/ nn5042625.

86. Shukla, S.K., R. Kumar, R.K. Mishra, A. Pandey, A. Pathak, M.G.H. Zaidi, S.K. Srivastava, and A. Dikshit. 2015. "Prediction and validation of gold nanoparticles (GNPs) on plant growth promoting rhizobacteria (PGPR): A step toward development of nano-biofertilizers." *Nanotechnology Reviews* 4 (5). Doi: 10.1515/ntrev-2015-0036.

87. Singhal, R.K., B. Gangadhar, H. Basu, V. Manisha, G.R.K. Naidu, and A.V.R. Reddy. 2012. "Remediation of Malathion contaminated soil using zero valent iron nano-particles." *American Journal of Analytical Chemistry* 3 (1): 76–82. Doi: 10.4236/ajac.2012.31011.

88. Pawlett, M., K. Ritz, R.A. Dorey, S. Rocks, J. Ramsden, and J.A. Harris. 2012. "The impact of zero-valent iron nanoparticles upon soil microbial communities is context dependent." *Environmental Science and Pollution Research* 20 (2): 1041–1049. Doi: 10.1007/s11356-012-1196-2.

89. Gohari, G., A. Mohammadi, A. Akbari, S. Panahirad, M.R. Dadpour, V. Fotopoulos, and S. Kimura. 2020. "Titanium dioxide nanoparticles (TiO_2 NPs) promote growth and ameliorate salinity stress effects on essential oil profile and biochemical attributes of Dracocephalum moldavica." *Scientific Reports* 10 (1). Doi: 10.1038/s41598-020-57794-1.

90. Faraji, J., and A. Sepehri. 2019. "Ameliorative effects of TiO_2 nanoparticles and sodium nitroprusside on seed germination and seedling growth of wheat under PEG-stimulated drought stress." *Journal of Seed Science* 41 (3): 309–317. Doi: 10.1590/2317-1545v41n3213139.

91. Gao, F., C. Liu, C. Qu, L. Zheng, F. Yang, M. Su, and F. Hong. 2007. "Was improvement of spinach growth by nano-TiO_2 treatment related to the changes of Rubisco activase?" *BioMetals* 21 (2): 211–217. Doi: 10.1007/s10534-007-9110-y.

92. Das, S., R. Aswani, S.J. Midhun, E.K. Radhakrishnan, and J. Mathew. 2020. "Advantage of zinc oxide nanoparticles over silver nanoparticles for the management of Aeromonas veronii infection in Xiphophorus hellerii." *Microbial Pathogenesis* 147: 104348. Doi: 10.1016/j.micpath.2020.104348.

93. Swain, P., S.K. Nayak, A. Sasmal, T. Behera, S.K. Barik, S.K. Swain, S.S. Mishra, A.K. Sen, J.K. Das, and P. Jayasankar. 2014. "Antimicrobial activity of metal based nanopar-ticles against microbes associated with diseases in aquaculture." *World Journal of Microbiology and Biotechnology* 30 (9): 2491–2502. Doi: 10.1007/s11274-014-1674-4.

94. Lv, J., P. Christie, and S. Zhang. 2019. "Uptake, translocation, and transformation of metal-based nanoparticles in plants: Recent advances and methodological challenges." *Environmental Science: Nano* 6 (1): 41–59. Doi: 10.1039/c8en00645h.

95. Rastogi, A., M. Zivcak, O. Sytar, H.M. Kalaji, X. He, S. Mbarki, and M. Brestic. 2017. "Impact of metal and metal oxide nanoparticles on plant: A critical review." *Frontiers in Chemistry* 5. Doi: 10.3389/fchem.2017.00078.

96. Gholami-Shabani, M., Z. Gholami-Shabani, M. Shams-Ghahfarokhi, F. Jamzivar, and M. Razzaghi-Abyaneh. 2017. "Green nanotechnology: Biomimetic synthesis of metal nanoparticles using plants and their application in agriculture and forestry." *Nanotechnology*, 133–175. Doi: 10.1007/978-981-10-4573-8_8.

97. Kou, J., and R.S. Varma. 2012. "Beet juice utilization: Expeditious green synthesis of noble metal nanoparticles (Ag, Au, Pt, and Pd) using microwaves." *RSC Advances* 2 (27): 10283. Doi: 10.1039/c2ra21908e.

98. Chen, J., L. Sun, Y. Cheng, Z. Lu, K. Shao, T. Li, C. Hu, and H. Han. 2016. "Graphene oxide-silver nanocomposite: Novel agricultural antifungal agent against fusarium gra-minearum for crop disease prevention." *ACS Applied Materials & Interfaces* 8 (36): 24057–24070. Doi: 10.1021/acsami.6b05730.

99. Veisi, M., S.H. Kazemi, and M. Mahmoudi. 2020. "Tunneling-induced optical limiting in quantum dot molecules." *Scientific Reports* 10 (1). Doi: 10.1038/s41598-020-73343-2.

100. Sumanth Kumar, D., B. Jai Kumar, and H.M. Mahesh. 2018. "Quantum nanostruc-tures (QDs): An overview." *Synthesis of Inorganic Nanomaterials*, 59–88. Doi: 10.1016/b978-0-08-101975-7.00003-8.

101. Bakirhan, N.K., and S.A. Ozkan. 2018. "Quantum dots as a new generation nano-materials and their electrochemical applications in pharmaceutical industry." *Handbook of Nanomaterials for Industrial Applications*, 520–529. Doi: 10.1016/b978-0-12-813351-4.00029-8.

102. Efros, A.L. 2019. "Quantum dots realize their potential." *Nature* 575 (7784): 604–605. Doi: 10.1038/d41586-019-03607-z.
103. Lu, H., Z. Huang, M.S. Martinez, J.C. Johnson, J.M. Luther, and M.C. Beard. 2020. "Transforming energy using quantum dots." *Energy & Environmental Science* 13 (5): 1347–1376. Doi: 10.1039/c9ee03930a.
104. Perry, T.S. 2020. "Quantum dots shift sunlight's spectrum to speed plant growth". *IEEE Spectrum: Technology, Engineering, and Science News.* https://spectrum. ieee.org/view-from-the-valley/at-work/start-ups/quantum-dots-shift-sunlights-spectrum-to-speed-plant-growth.
105. Merck Innovation Center. 2020. "Photonic materials for smart farming - Innovation center | Merck KGaA, darmstadt, Germany". https://www.merckgroup.com/en/research/ innovation-center/highlights/light-tuning.html.
106. Liu, Y., Y. Zhao, T. Zhang, Y. Chang, S. Wang, R. Zou, G. Zhu, L. Shen, and Y. Guo. 2019. "Quantum dots-based immunochromatographic strip for rapid and sensitive detection of acetamiprid in agricultural products." *Frontiers in Chemistry* 7. Doi: 10.3389/ fchem.2019.00076.
107. Cotta, M.A. 2020. "Quantum dots and their applications: What lies ahead?" *ACS Applied Nano Materials* 3 (6): 4920–4924. Doi: 10.1021/acsanm.0c01386.
108. Nazemi, P., and M. Razavi. 2017. "Lipid-based nanobiomaterials." *Nanobiomaterials Science, Development and Evaluation*, 125–133. Doi: 10.1016/b978-0-08-100963-5.00006-9.
109. Gopinath, V., and P. Velusamy. 2013. "Extracellular biosynthesis of silver nanoparticles using Bacillus sp. GP-23 and evaluation of their antifungal activity towards Fusarium oxysporum." *Spectrochimica Acta Part A: Molecular and Biomolecular Spectroscopy* 106: 170–174. Doi: 10.1016/j.saa.2012.12.087.
110. Banumathi, B., B. Vaseeharan, P. Rajasekar, N.M. Prabhu, P. Ramasamy, K. Murugan, A. Canale, and G. Benelli. 2017. "Exploitation of chemical, herbal and nanoformulated acaricides to control the cattle tick, Rhipicephalus (Boophilus) microplus – A review." *Veterinary Parasitology* 244: 102–110. Doi: 10.1016/j.vetpar.2017.07.021.
111. Stringer, R.C., S. Schommer, D. Hoehn, and S.A. Grant. 2008. "Development of an optical biosensor using gold nanoparticles and quantum dots for the detection of porcine reproductive and respiratory syndrome virus." *Sensors and Actuators B: Chemical* 134 (2): 427–431. Doi: 10.1016/j.snb.2008.05.018.
112. Avellan, A., M. Simonin, E. McGivney, N. Bossa, E. Spielman-Sun, J.D. Rocca, E.S. Bernhardt, et al. 2018. "Gold nanoparticle biodissolution by a freshwater macrophyte and its associated microbiome." *Nature Nanotechnology* 13 (11): 1072–1077. Doi: 10.1038/s41565-018-0231-y.
113. Al-Rawashdeh, N.A., O. Allabadi, and M.T. Aljarrah. 2020. "Photocatalytic activity of graphene oxide/zinc oxide nanocomposites with embedded metal nanoparticles for the degradation of organic dyes." *ACS Omega* 5 (43): 28046–28055. doi:10.1021/ acsomega.0c03608.
114. Chang, C.-I., C.-C. Wu, T.C. Cheng, J.-M. Tsai, and K.-J. Lin. 2009. "Multiplex nested-polymerase chain reaction for the simultaneous detection of *Aeromonas hydrophila*, *Edwardsiella tarda*, *Photobacterium damselae* and *Streptococcus iniae*, four important fish pathogens in subtropical Asia." *Aquaculture Research* 40 (10): 1182–1190. Doi: 10.1111/j.1365–2109.2009.02214.x.
115. Tombuloglu, H., I. Ercan, T. Alshammari, G. Tombuloglu, Y. Slimani, M. Almessiere, and A. Baykal. 2020. "Incorporation of micro-nutrients (nickel, copper, zinc, and iron) into plant body through nanoparticles." *Journal of Soil Science and Plant Nutrition* 20 (4): 1872–1881. Doi: 10.1007/s42729-020-00258-2.
116. Das, R., Z. Shahnavaz, M.E. Ali, M.M. Islam, and S.B. Abd Hamid. 2016. "Can we optimize arc discharge and laser ablation for well-controlled carbon nanotube synthesis?" *Nanoscale Research Letters* 11 (1). Doi: 10.1186/s11671-016-1730-0.

117. Wang, Y., C.H. Chang, Z. Ji, D.C. Bouchard, R.M. Nisbet, J.P. Schimel, J.L. Gardea-Torresdey, and P.A. Holden. 2017. "Agglomeration determines effects of carbonaceous nanomaterials on soybean nodulation, dinitrogen fixation potential, and growth in soil." *ACS Nano* 11 (6): 5753–5765. Doi: 10.1021/acsnano.7b01337.

118. Kumar, A., A. Singh, M. Panigrahy, P.K. Sahoo, and K.C. Panigrahi. 2018. "Carbon nanoparticles influence photomorphogenesis and flowering time in Arabidopsis thaliana." *Plant Cell Reports* 37 (6): 901–912. Doi: 10.1007/s00299-018-2277-6.

119. Khodakovskaya, M.V., K. De Silva, A.S. Biris, E. Dervishi, and H. Villagarcia. 2012. "Carbon nanotubes induce growth enhancement of tobacco cells." *ACS Nano* 6 (3): 2128–2135. Doi: 10.1021/nn204643g.

120. Hamdi, H., R.D.L. Torre-Roche, J. Hawthorne, and J.C. White. 2014. "Impact of non-functionalized and amino-functionalized multiwall carbon nanotubes on pesticide uptake by lettuce (Lactuca sativaL.)." *Nanotoxicology* 9 (2): 172–180. Doi: 10.3109/17435390.2014.907456.

121. De La Torre-Roche, R., J. Hawthorne, Y. Deng, B. Xing, W. Cai, L.A. Newman, Q. Wang, X. Ma, H. Hamdi, and J.C. White. 2013. "Multiwalled carbon nanotubes and C60 fullerenes differentially impact the accumulation of weathered pesticides in four agricultural plants." *Environmental Science & Technology* 47 (21): 12539–12547. Doi: 10.1021/es4034809.

122. Lahiani, M.H., E. Dervishi, J. Chen, Z. Nima, A. Gaume, A.S. Biris, and M.V. Khodakovskaya. 2013. "Impact of carbon nanotube exposure to seeds of valuable crops." *ACS Applied Materials & Interfaces* 5 (16): 7965–7973. Doi: 10.1021/am402052x.

123. Mukherjee, A., S. Majumdar, A.D. Servin, L. Pagano, O.P. Dhankher, and J.C. White. 2016. "Carbon nanomaterials in agriculture: A critical review." *Frontiers in Plant Science* 7. Doi: 10.3389/fpls.2016.00172.

124. Schäffel, F. 2013. "The atomic structure of graphene and its few-layer counterparts." *Graphene*, 5–59. Doi: 10.1016/b978-0-12-394593-8.00002-3.

125. Husen, A., and K. Siddiqi. 2014. "Carbon and fullerene nanomaterials in plant system." *Journal of Nanobiotechnology* 12 (1): 16. Doi: 10.1186/1477-3155-12-16.

126. Jiang, Y., Z. Hua, Y. Zhao, Q. Liu, F. Wang, and Q. Zhang. 2013. "The effect of carbon nanotubes on rice seed germination and root growth." *Proceedings of the 2012 International Conference on Applied Biotechnology (ICAB 2012)*, 1207–1212. Doi: 10.1007/978-3-642-37922-2_129.

127. Liu, Q., Y. Zhao, Y. Wan, J. Zheng, X. Zhang, C. Wang, X. Fang, and J. Lin. 2010. "Study of the inhibitory effect of water-soluble fullerenes on plant growth at the cellular level." *ACS Nano* 4 (10): 5743–5748. Doi: 10.1021/nn101430g.

128. Shrestha, B., V. Acosta-Martinez, S.B. Cox, M.J. Green, S. Li, and J.E. Cañas-Carrell. 2013. "An evaluation of the impact of multiwalled carbon nanotubes on soil microbial community structure and functioning." *Journal of Hazardous Materials* 261: 188–197. Doi: 10.1016/j.jhazmat.2013.07.031.

129. Brandelli, A. 2015. "Nanobiotechnology strategies for delivery of antimicrobials in agriculture and food." *Nanotechnologies in Food and Agriculture*, 119–139. Doi: 10.1007/978-3-319-14024-7_6.

130. Kabiri, S., F. Degryse, D.N. Tran, R.C. Da Silva, M.J. McLaughlin, and D. Losic. 2017. "Graphene oxide: A new carrier for slow release of plant micronutrients." *ACS Applied Materials & Interfaces* 9 (49): 43325–43335. Doi: 10.1021/acsami.7b07890.

131. Zhang, M., B. Gao, J. Chen, Y. Li, A.E. Creamer, and H. Chen. 2014. "Slow-release fertilizer encapsulated by graphene oxide films." *Chemical Engineering Journal* 255: 107–113. Doi: 10.1016/j.cej.2014.06.023.

132. Gao, B., M. Zhang and Y. Li. 2014. Slow-release fertilizer compositions with graphene oxide films and methods of making slow-release fertilizer compositions. WO 2015066691A1, issued 2014.

133. Zhang, M., B. Gao, J. Chen, & Y. Li. 2015. "Effects of graphene on seed germination and seedling growth. " *Journal of Nanoparticle Research*, 17(2). Doi: 10.1007/s11051-015-2885-9.

134. Park, S., K.S. Choi, S. Kim, Y. Gwon, and J. Kim. 2020. "Graphene oxide-assisted promotion of plant growth and stability." *Nanomaterials* 10 (4): 758. Doi: 10.3390/nano10040758.

135. Bobrinetskiy, I.I., and N.Z. Knezevic. 2018. "Graphene-based biosensors for on-site detection of contaminants in food." *Analytical Methods* 10 (42): 5061–5070. Doi: 10.1039/c8ay01913d.

136. Palaparthy, V.S., H. Kalita, S.G. Surya, M.S. Baghini, and M. Aslam. 2018. "Graphene oxide based soil moisture microsensor for *in situ* agriculture applications." *Sensors and Actuators B: Chemical* 273: 1660–1669. Doi: 10.1016/j.snb.2018.07.077.

137. Akhavan, O., and E. Ghaderi. 2010. "Toxicity of graphene and graphene oxide nanowalls against bacteria." *ACS Nano* 4 (10): 5731–5736. Doi: 10.1021/nn101390x.

138. Mission, E.G., A.T. Quitain, M. Sasaki, and T. Kida. 2017. "Synergizing graphene oxide with microwave irradiation for efficient cellulose depolymerization into glucose." *Green Chemistry* 19 (16): 3831–3843. Doi: 10.1039/c7gc01691c.

139. Zhu, S., J. Wang, and W. Fan. 2015. "Graphene-based catalysis for biomass conversion." *Catalysis Science & Technology* 5 (8): 3845–3858. Doi: 10.1039/c5cy00339c.

140. Denis, P.A., and F. Iribarne. 2013. "Comparative study of defect reactivity in graphene." *The Journal of Physical Chemistry C* 117 (37): 19048–19055. Doi: 10.1021/jp4061945.

141. Lahiani, M.H., J. Chen, F. Irin, A.A. Puretzky, M.J. Green, and M.V. Khodakovskaya. 2015. "Interaction of carbon nanohorns with plants: Uptake and biological effects." *Carbon* 81: 607–619. Doi: 10.1016/j.carbon.2014.09.095.

142. Composite Nanoparticles. 2008. *Encyclopedia of Microfluidics and Nanofluidics* 274–274. Doi: 10.1007/978-0-387-48998-8_243.

143. Soares, D.F.C., S.C. Domingues, D.B. Viana, and M.L. Tebaldi. 2020. "Polymer-hybrid nanoparticles: Current advances in biomedical applications." *Biomedicine & Pharmacotherapy* 131: 110695. Doi: 10.1016/j.biopha.2020.110695.

144. Janczak, C.M., and C.A. Aspinwall. 2011. "Composite nanoparticles: The best of two worlds." *Analytical and Bioanalytical Chemistry* 402 (1): 83–89. Doi: 10.1007/s00216-011-5482-5.

145. Martins, N.C., A. Avellan, S. Rodrigues, D. Salvador, S.M. Rodrigues, and T. Trindade. 2020. "Composites of biopolymers and ZnO NPs for controlled release of zinc in agricultural soils and timed delivery for maize." *ACS Applied Nano Materials* 3 (3): 2134–2148. Doi: 10.1021/acsanm.9b01492.

146. Andelkovic, I.B., S. Kabiri, E. Tavakkoli, J.K. Kirby, M.J. McLaughlin, and D. Losic. 2018. "Graphene oxide-Fe(III) composite containing phosphate – A novel slow release fertilizer for improved agriculture management." *Journal of Cleaner Production* 185: 97–104. Doi: 10.1016/j.jclepro.2018.03.050.

147. Nsibande, S.A., and P.B. Forbes. 2019. "Development of a quantum dot molecularly imprinted polymer sensor for fluorescence detection of atrazine." *Luminescence* 34 (5): 480–488. Doi: 10.1002/bio.3620.

148. Liang, Y., D. Yang, and J. Cui. 2017. "A graphene oxide/silver nanoparticle composite as a novel agricultural antibacterial agent against Xanthomonas oryzae pv. oryzae for crop disease management." *New Journal of Chemistry* 41 (22): 13692–13699. Doi: 10.1039/c7nj02942j.

149. Kumar, R., M. Ashfaq, and N. Verma. 2018. "Synthesis of novel PVA–starch formulation-supported Cu–Zn nanoparticle carrying carbon nanofibers as a nanofertilizer: Controlled release of micronutrients." *Journal of Materials Science* 53 (10): 7150–7164. Doi: 10.1007/s10853-018-2107-9.

150. Guo, D., N. Muhammad, C. Lou, D. Shou, and Y. Zhu. "Synthesis of dendrimer functionalized adsorbents for rapid removal of glyphosate from aqueous solution." *New Journal of Chemistry* 43 (1): 121–129. Doi: 10.1039/c8nj04433c.
151. Madusanka, N., C. Sandaruwan, N. Kottegoda, D. Sirisena, I. Munaweera, A. De Alwis, Veranja Karunaratne, and G.A. Amaratunga. 2017. "Urea–hydroxyapatite-montmorillonite nanohybrid composites as slow release nitrogen compositions." *Applied Clay Science* 150: 303–308. Doi: 10.1016/j.clay.2017.09.039.
152. Ocsoy, I., M.L. Paret, M.A. Ocsoy, S. Kunwar, T. Chen, M. You, and W. Tan. 2013. "Nanotechnology in plant disease management: DNA-directed silver nanoparticles on graphene oxide as an antibacterial against *Xanthomonas perforans*." *ACS Nano* 7 (10): 8972–8980. Doi: 10.1021/nn4034794.
153. Xie, J., Y. Yang, B. Gao, Y. Wan, Y.C. Li, D. Cheng, T. Xiao, et al. 2019. "Magnetic-sensitive nanoparticle self-assembled superhydrophobic biopolymer-coated slow-release fertilizer: Fabrication, enhanced performance, and mechanism." *ACS Nano* 13 (3): 3320–3333. Doi: 10.1021/acsnano.8b09197.
154. Huang, D., and D. Wu. 2018. "Biodegradable dendrimers for drug delivery." *Materials Science and Engineering: C* 90: 713–727. Doi: 10.1016/j.msec.2018.03.002.
155. Govindarajan, M., and G. Benelli. 2016. "A facile one-pot synthesis of eco-friendly nanoparticles using Carissa carandas: Ovicidal and larvicidal potential on malaria, dengue and filariasis mosquito vectors." *Journal of Cluster Science* 28 (1): 15–36. Doi: 10.1007/s10876-016-1035-6.
156. LewisOscar, F., D. MubarakAli, C. Nithya, R. Priyanka, V. Gopinath, N.S. Alharbi, and N. Thajuddin. 2015. "One pot synthesis and anti-biofilm potential of copper nanoparticles (CuNPs) against clinical strains of *Pseudomonas aeruginosa*." *Biofouling* 31 (4): 379–391. Doi: 10.1080/08927014.2015.1048686.
157. Zamani, A., A.P. Marjani, and Z. Mousavi. 2019. "Agricultural waste biomass-assisted nanostructures: Synthesis and application." *Green Processing and Synthesis* 8 (1): 421–429. Doi: 10.1515/gps-2019-0010.
158. Boxall, A.B., K. Tiede, and Q. Chaudhry. 2007. "Engineered nanomaterials in soils and water: How do they behave and could they pose a risk to human health?" *Nanomedicine* 2 (6): Doi: 10.2217/17435889.2.6.919.
159. Ghosh, M., I. Ghosh, L. Godderis, P. Hoet, and A. Mukherjee. 2019. "Genotoxicity of engineered nanoparticles in higher plants." *Mutation Research/Genetic Toxicology and Environmental Mutagenesis* 842: 132–145. Doi: 10.1016/j.mrgentox.2019.01.002.
160. Keller, A.A., S. McFerran, A. Lazareva, and S. Suh. "Global life cycle releases of engineered nanomaterials." *Journal of Nanoparticle Research* 15 (6). Doi: 10.1007/s11051-013-1692-4.
161. Javed, Z., K. Dashora, M. Mishra, V.D. Fasake, and A. Srivastva. 2019. "Effect of accumulation of nanoparticles in soil health- a concern on future." *Frontiers in Nanoscience and Nanotechnology* 5 (2). Doi: 10.15761/fnn.1000181.
162. Jarvis, S.L. and N. Richmond. 2011. "Regulation and governance of nanotechnology in china: Regulatory challenges and effectiveness". *European Journal of Law and Technology* 2 (3). https://www.darryljarvis.com/uploads/2/2/6/9/22690064/nano technology_jarvis__richmond.pdf.
163. Karim M.E., A.B. Munir, and S.H.M. Yasin. 2016. "Update: Nanotechnology and international law research guide". https://www.nyulawglobal.org/globalex/ Nanotechnology_International_Law1.html.
164. ASTM. 2005. "Committee E56 on nanotechnology". https://www.astm.org/COMMITTEE/E56.htm.

9 Utility of Nanomaterials in Food Processing and Packaging

G.A. Lanza, J.A. Perez-Taborda, and A. Avila
Universidad de los Andes

CONTENTS

9.1 INTRODUCTION

The right to food is a fundamental part of the broader framework of the Sustainable Development Goals proposed by the United Nations [1]. This imposes challenges related to food management in order to guarantee quality food for all people around the world. According to the Food and Agriculture Organization (FAO) of the United Nations, studies presented since 2011, each year, approximately 1.3 billion tons of foods are spoiled due to poor management practices in the food industry [2]. In addition, in April 2020, 135 million people suffered from acute hunger in the world, and with the present COVID-19 pandemic, this number doubled at the end of the same year [3]. Because of these facts, changes in the food system are necessary in order to improve interoperability with new technologies such as smart sensing, IoT, blockchain, big data analytics, robotics, among others [4], which could impact the food supply chain. Here, we focus on nanotechnology and how it is transforming the food sector, specifically in food processing and packaging.

The transformation of food sector requires its integration with technologies that generate innovative tools that improve and preserve the food that reaches consumers. It is in this scenario that nanotechnology can make a significant contribution. Nanotechnology is an emerging interdisciplinary technology with a broad range of applications and can offer pioneering strategies in this sector. The manipulation of matter at the nanoscale is not something new in the food sector. One of the first revolutionary steps in food processing is the pasteurization process introduced by Pasteur to eliminate spoilage bacteria with sizes between 400 and 1000 nm [5]. Nowadays, nanotechnology advances allow the manipulation of matter at the nanoscale (nanomaterials (NMs)) with high precision, which enables the manufacture of new devices and solutions in the food industry [6]. The impact of nanotechnology in this sector is due to the unique and novel characteristics of NMs. NMs have unique properties such as high mechanical [7] and thermal stability [8], high surface area [9], and exceptional optical [10], electrical [11], and magnetic properties [12] that can be exploited in applications for food processing and packaging [13].

Food processing consists of transformation of raw ingredients into food and its derivatives [14]. Their main goals are to preserve the characteristics of the raw ingredients, increase its shelf life, and improve its characteristics such as appearance, texture, flavor, etc. These improvements have an impact on the marketing of food. On the other hand, packaging impacts the distribution of food and has two specific functions: to protect and to communicate. The packaging must be a system that resists handling and environmental conditions that may affect the food. Besides, the packaging must have a label that provides information about the food such as its expiration date, statement of identity and nutritional content [15].

In this chapter, recent advances in the use of NMs in food processing and packaging are discussed. In processing, the nanoencapsulation of elements that define the food characteristics such as flavor and nutrients is used to improve the food and its shelf life. In food packaging, NMs can be used in two types of approaches: active and smart packaging. In active packaging, NMs can be used as additives in the polymeric matrix from which the packaging is made to improve its mechanical, permeability, and microbial properties [16]. In smart packaging, NMs are used for the manufacture of sensors and indicators that monitor and communicate parameters that can define food quality. Finally, this chapter intends to provide useful information on the current state of projects standards and regulations applied to food and NMs.

9.2   NANOMATERIALS IN FOOD SECTOR: KEEPING FOOD QUALITY

In the food industry, food processing and packaging focus on guaranteeing and monitoring the quality of food throughout its supply chain [17]. Quality refers to the combination of characteristics that establish the acceptability of a food product [18].

For consumers, the quality of foods is based on attributes (qualitative) that can easily be determined by the senses such as color, size, texture, and flavor [19]. The characteristics given by consumers can be related to physicochemical properties such as the amount of strength-generating components, the wavelength emitted, the structure and adsorbent properties of the tissue, cell size, maturity, and concentration of sugar, acid, and aromatic compounds [20]. These properties must be maintained in the food supply chain in order to increase the food quality and consequently to make better use of the resources and to avoid its loss and waste.

The food supply chain is composed of the following components: (1) agricultural production and harvest/slaughter/catch; (2) post-harvest/slaughter/catch operations; (3) storage; (4) transportation; (5) processing; (6) wholesale and retail; and (7) consumption by households and food services [2,21]. In all supply chains, food is continually exposed to external factors that affect its shelf life. As shown in Figure 9.1, food has two types of critical factors that decrease its quality: environmental and manipulation factors. Environmental factors are related to the influence of temperature, humidity, pressure, lighting, odor, and contaminants to which food is exposed in the supply chain [22]. Manipulation factors are associated with the physical impacts that the food can receive from the consumer or the producer, as well as the impacts received by the technological equipment arranged in the supply chain. Mechanical damage during food manipulation affects its shell or skin, which facilitates its decomposition process [23].

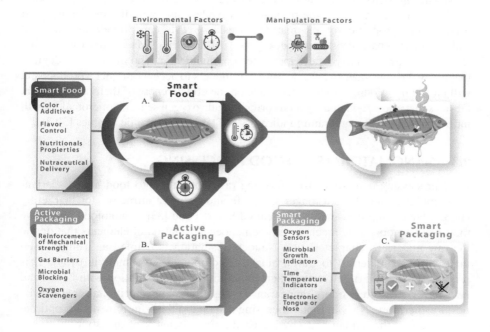

FIGURE 9.1 Applications of nanotechnology in food processing and packaging in the supply chain affected by environmental and manipulation factors. Applications: (A) Smart food to improve the characteristics of the food, (B) Active packaging to protect the food, and (C) Smart packaging to monitor the state of the food.

NMs can be used to avoid problems related to environmental and manipulation factors and thereby enhance the quality and increase the food shelf life [24]. Considering the specific role of NMs in food, NMs applications can be classified into three groups [25]:

   i. Enhancer of food characteristics (smart foods).
   ii. Additives in the packaging (active packaging).
   iii. Indicators for food quality control (smart packaging).

In smart foods (Figure 9.1a), NMs such as micelles, liposomes, dendrimers, carbon nanotubes and, metal oxide nanoparticles, among others, are used directly in foods to preserve and improve the characteristics such as color, flavor, and improvement in the delivery of nutrients and nutraceuticals [26]. There are two ways to carry out the improvement of food characteristics: adding NMs to food or the nanoencapsulation process. For example, TiO_2 nanoparticles are the most widely used food color additive, and their use is approved by regulations such as those of the U.S. Food and Drug Administration (FDA) [27]. The nanoencapsulation process consists of the encapsulation of the elements (as molecules) using nanocomposites, nanoemulsification, and nanostructuration techniques that allow to engineer the food characteristics and extend food preservation [28].

Regarding active packaging (Figure 9.1b), NMs are incorporated into the polymeric material from which the packaging is made, improving the mechanical and permeability properties of the material. Besides, NMs can provide antibacterial and scavenger characteristics to packaging [29]. Finally, NMs in smart packaging (Figure 9.1c) are used to produce integrated communication and information devices that allow the consumer to have decision parameters to accept or reject a product and will enable the producer to have feedback on the effectiveness of their supply chain. Devices that allow food quality monitoring include oxygen sensors, microbial growth indicators, time and temperature indicators (TTIs), and electronic tongues [30,31].

9.3 NANOMATERIALS IN FOOD PROCESSING

Food processing consists of transforming raw materials into food and its derivatives, making them have a longer shelf life and be more attractive for marketing [14,32]. Foods improved with NMs, called Smart Food [33] or nanofood [34], have their primary operating mechanism, the nanoencapsulation of elements that define the raw material's characteristics. Nanoencapsulation is defined as using a shell material (wall material) to trap a bioactive compound (core material) [35]. This process for the smart delivery of elements allows food production with better and new flavors, textures, odors, appearance, sensations, and nutritional characteristics. Furthermore, NMs in food processing respond to the requirements for optimal absorption by the human body by increasing the efficiency of digestion, bioavailability, and metabolism [36].

Among the main characteristics that can define food are its color, flavor, nutritional content, and the nutraceutical components it contains (Figure 9.2). NMs used to encapsulate elements (such as molecules) that determine these characteristics are

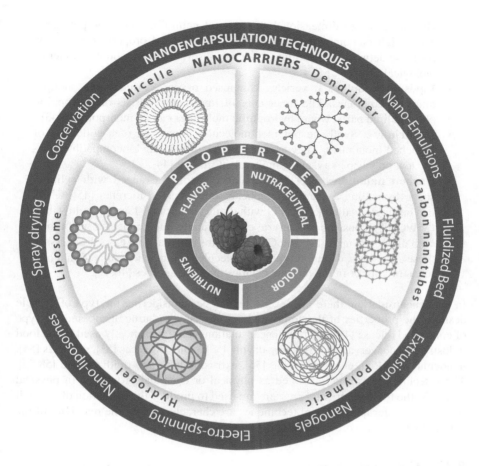

FIGURE 9.2 Nanoencapsulation for food processing: techniques (outer circle) related to the nanocarriers involved (middle circle) and the food characteristics impacted (the inner circle).

called nanocarriers [28,37]. Nanocarriers operate as molecule deliverers, generating systems with high chemical stability and the possibility of releasing the desired elements in a controlled manner [38]. Commonly used nanocarriers are organic based and have the following characteristics:

- **Micelles (amphiphilic colloidal structures)**: They consist of molecules that contain two regions with opposite affinities against water. The core is made up of hydrophobic parts of amphiphilic molecules, and the shell is made up of hydrophilic parts. The diameters of these particles are between 5 and 100 nm [39].
- **Polymeric (colloidal particles)**: The polymers used to synthesize these nanocarriers can be natural (such as carbohydrates and proteins) or synthetic. Polymers are biocompatible and generally hydrophilic [40]. The diameters of these particles are between 10 nm and 1 μm [41].

- **Hydrogel**: Three-dimensional networks of natural or synthetic polymers. Their high biocompatibility and physicomechanical properties characterize these networks. In addition, they can absorb a large amount of water or biological fluids [42].
- **Liposomes**: Spherical vesicles composed of one or more phospholipid bilayer membranes. The inner core of liposomes consists of hydrophilic phospholipid parts, where hydrophilic molecules can be incorporated [43].
- **Dendrimers**: Synthetic polymer molecules composed of a collection of branched monomers. They are characterized by being biocompatible, having low polydispersity, and a multivalent surface [44].
- **Carbon nanotubes (CNTs)**: Tubular structures arranged with cylinder shape by rolling a sheet of graphene (buckminsterfullerene) [45]. CNTs can be single-walled (SWCNT) with diameters around 1 nm or multiwalled (MWCNT), consisting of several concentrically interconnected nanotubes, with diameters up to 100 nm; CNTs' lengths can reach up to millimeters. In these structures, different types of molecules can be encapsulated in a hollow structure or in the internal cavities of the nanotubes [46,47].

In nanoencapsulation, there are several techniques considering the type of capsule desired, the properties of the compound and the encapsulating agent, and the type of controlled release [48–50]. Nanoencapsulation techniques mostly used in the food industry are: coacervation [51], spray-drying [52], nanoliposomes [53], nanogels [54], nanoemulsions [55], electrospinning [56], extrusion [57], and fluidized bed [58,59]. These techniques can be classified as physical or chemical processes. In physical methods, the temperature and pressure are used to produce the formation of the wall in the core material. Chemical techniques are characterized by the reactivity of the external wall in the core material [60].

9.3.1 COLOR ENHANCEMENT

The color of the food has an impact on the consumers' willingness to buy it. Consumers generally consider the food color an indicator of quality, which, in the first instance, results in the acceptance or rejection of the product [61]. For this reason, food producers must allocate a part of production costs in the implementation of natural and synthetic products that improve the color of the food [35]. Most natural colorants are hydrophobic, which determines the proper nanoencapsulation technique to ensure the stability of these water-insoluble pigments. Two of the pigments frequently used in improving the color of the food are carotenoids and flavonoids [62]. These pigments are nanoencapsulated by the nanoliposome technique [63]. The use of digestible lipids favors the absorption of bioactive through the enterocytes since they increase the content of mixed micelles that influence the transport of pigments [64].

9.3.2 FLAVOR CONTROL

The principal aim of food processing is the conservation of flavor to achieve acceptance of the product in the market [65]. Controlling the flavor of food is generally associated with its quality, which predetermines the expectations of consumers and the desire to purchase the product. Flavor systems are complex mixtures of hydrophobic and volatile compound with specific properties, which can be extracted from plants or animals or chemically. Natural flavors are extracted by physical, enzymatic, or microbiological techniques [66]. Chemically, it is also possible to generate synthetic flavors [67, 68].

Nanoencapsulation processes can be used to protect and release the molecules that define flavor in a controlled manner. For this, the choice of the type of nanocarrier depends on the physical and chemical flavor properties (such as solubility and porosity) and the compatibility between the flavor molecule and the nanocarrier [69]. Compatibility ensures that the nanocarrier composition does not react with the flavor molecule, as well as that the nanocarrier is insoluble. The most commonly used nanocarriers to encapsulate flavors are polymers such as carbohydrates, gums, and proteins [70,71]. For the nanoencapsulation of flavor, physical and chemical techniques are used [72]. Extrusion is the most used physical technique for this purpose in industrial processes. This technique is distinguished by the use of temperature and pressure to generate the formation of the wall (nanocarrier) that covers the core material (flavor molecule) [60, 72].

9.3.3 NUTRIENT DELIVERY

For the proper functioning of the human body, biological processes require nutrients such as vitamins and minerals that are absorbed from food. The above indicates that the acquisition of vitamins and minerals not only depends on the amount available in the food but also on their absorption capacity [73,74]. Food nutrients are exposed to degradation by factors such as changes in pH, temperature, presence of oxidants, and storage times and conditions [75]. In this aspect, nanoencapsulation offers the possibility of protecting the nutrient and controlling its release and increasing nutrient availability and solubility [74]. The solubility of vitamins and minerals determines which nanoencapsulation techniques are more efficient to deliver and release them in a controlled manner in the intestine [76]. For example, for vitamin B and C (ascorbic acid), the nanocarriers that are mostly used are polymers and liposomes through the nanospray-drying technique [77]. This technique is based on dissolving or dispersing the bioactive compounds in a biopolymer solution, at that moment atomizing the dispersion that quickly removes solvent and synthesizes a particle [78].

9.3.4 NUTRACEUTICALS DELIVERY

The word nutraceutical comes from the combination of the words "nutrition" and "pharmaceutical" [79]. They are components of oral diet found in food with notable health benefits [80]. Nutraceuticals are used to improve health, prevent chronic diseases, increase life expectancy, and support body function [81]. In the same way, as for nutrients, the main objective of the nanoencapsulation of nutraceutical molecules is to protect and avoid degradation processes, as well as to prolong their release and improve their availability [82].

For nutraceuticals, characteristics such as composition, surface area, size, and surface charge determine the nanoencapsulation technique to use. Techniques such as nanoliposomes and nanogels are here commonly used [83,84]. Nanoliposomes consist of a colloidal-based system that has a vesicular structure. The size range of these systems is between nanometers and micrometers [85]. These systems have an internal aqueous phase within one or more lipid bilayers for the administration of nutraceuticals. On the other hand, nanogels are 3D hydrogels with a diameter between 10 and 100 nm [86]. The hydrogel has a large surface area and an internal network for the incorporation of bioactive molecules. Hydrogels have high water content, biocompatibility, and degradability, and their solubility in water means that they can contain a large number of nutraceuticals in aqueous media [87].

9.4 NANOMATERIALS IN FOOD PACKAGING

Food packaging is a crucial aspect of food quality management because it preserves the food throughout the distribution chain, and the foods avoid direct contact with physical, chemical, and biological contaminants [88]. For this reason, food packaging is being manufactured with new and innovative features (biodegradable and reusable, with dynamic labels, among others) depending on the needs of industry and consumers [89]. According to the European Commission, the materials used in food packaging are classified into two groups depending on their functionality: active and intelligent materials [90]. Active materials are used to extend the shelf life and maintain the condition of the food that has been packaged [91]. Packages with this type of material are called active packaging and are designed to incorporate components that protect the food from internal or external substances to the packaging [90]. Intelligent materials are used to monitor the state of packaged food or the environment surrounding the food [92]. Packages with devices based on these materials are called smart packaging and are a system that monitors the state of food throughout the supply chain [93]. These devices can be adhered to the inside or outside of the package and, according to their operating principle, can offer information about the food quality based on the measures of temperature, the amount of oxygen, microbial growth, and the presence of pathogens [94].

Applications of NMs in packaging are used to offer the consumer a quality product by maintaining the flavor, color, appearance, freshness, and texture of food, as presented in Figure 9.3. To protect the food, NMs in packaging can have functionalities such as reinforcement of mechanical strength [95], gas barriers [96], microbial blocking [97], and oxygen scavenger [98]. For sensing parameters that can define the food quality, there are devices such as oxygen sensors [99], microbial growth indicators [100], TTIs [101], and electronic tongues or noses [102].

The operating mechanisms for these devices are mainly based on physical changes (such as size, shape, concentration, interaction between NMs and the matrix, and the distribution) when nanoparticles are exposed to changes in temperature or the presence of specific elements such as oxygen [31,103]. Mechanisms for applications related to microbial behaviors are characterized by the electrostatic interaction between nanoparticles and the cell membranes of microbes [104]. Another important mechanism that can be exploited in packaging applications consists of distributing nanoparticles in a polymeric matrix of the packaging to avoid gas diffusion and improve mechanical properties. All these mechanisms inherent to nanoparticles

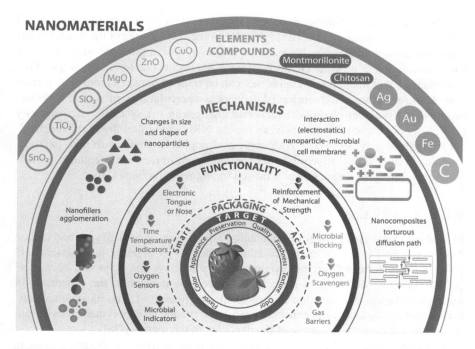

FIGURE 9.3 Use of nanomaterials in food packaging. From outside to inside, information is organized into the composition—elements or compounds, operation mechanisms for the desired functionalities, functionalities according to packaging classification—active or smart, and food properties preserved with the packaging.

make it possible to manufacture devices to protect and sense the properties of food [105]. Finally, regarding NMs, these are used depending on the function or improvement that is given to the packaging. For example, CNTs, polymers, and silicates are usually used for active packaging. For smart packaging, metal and oxides nanoparticles are used [106].

9.4.1 REINFORCEMENT OF MECHANICAL STRENGTH

In the manufacture of polymeric packaging, it is common for additives to be used to improve their mechanical properties, such as strength, stiffness, hardness, and toughness [107]. Traditionally, additives are added in powder form with spherical particles ranging from 500 to 1500 μm [108]. However, the additive characteristics may negatively influence other packaging properties such as transparency and flexibility. The use of NMs (between 5 and 80 nm) as additives in polymers (nanocomposite) can enhance mechanical performance without diminishing other polymer properties [109,110]. A critical aspect that influences the reinforcing effect of nanocomposites is the chemical affinity of the surface of the nanoparticle with the polymer [111].

Mechanisms to explain the reinforcement of nanocomposites can be three: direct interfacial interaction, intercalation of polymer chains between NMs, and establishment of a three-dimensional reinforced network [112]. The direct interfacial interaction is a consequence of the high value of the surface area of the nanoplatelets and nanotubes, which causes the polymer chains to have a hindered mobility [113]. The intercalation of polymer chains between NMs occurs due to the flat shape of the nanoplatelets and the polymer chains that can be trapped between them. The formation of a three-dimensional reinforced network is a percolating structure of NMs and immobilized polymer chains that generate changes in the system's viscoelasticity [114]. Nanoclays and carbon nanotubes are the NMs used to reinforce the mechanical strength of polymers [115]. Nanoclays in nanoplatelet shape with high isotropic modulus in their plane, allowing better reinforcement in random orientations, while nanotubes enhance reinforcement arranged in aligned directions [116].

9.4.2 GAS BARRIERS

In food packaging, reducing the permeability rate of gases such as oxygen and carbon dioxide is a goal to extend the products' shelf life [117]. Currently, gas barrier packaging is manufactured from nanocomposites of nanoclays and polymers. Nanoclays do not allow the diffusion of gases through the polymer due to the generation of tortuous pathways that function as a molecular diffuser [96]. The nanocomposites' performance is based on the nanoclay properties, the intrinsic polymer matrix properties, and the dispersion degree of the nanoclay in the matrix [118].

Montmorillonite (MMT) is the clay mostly used to improve the gas barrier in packaging. MMT nanoclays have a structure consisting of a shared edge octahedral sheet of aluminum oxide (AlO_6) between two layers of tetrahedral silicon oxide (SiO_4) to form a nanoplatelet [119]. There are three possible dispersions in polymer and nanoclay nanocomposite systems: phase-separated, intercalated, and exfoliated [120]. One of the factors that determine the gas barrier properties in a polymer is the nanoplate dispersion. The synthesis of these nanocomposites is usually carried out by applying methods such as *in-situ* polymerization, solution blending, and melt intercalation [118,121].

9.4.3 MICROBIAL BLOCKING

Inhibiting the growth of microbes, such as specific bacteria in food, increases its shelf life considerably [122]. For this, it is possible to use packaging manufactured with nanocomposites that have antibacterial properties (biocidal activity). Typical antimicrobial NMs include nanoparticles of TiO_2, Ag, ZnO, Cu, MgO, and Au in a polymer matrix [123]. There are several mechanisms by which metal nanoparticles can have biocidal activity. For example, ZnO nanoparticles act as reactive oxygen species on microbial cells, which makes them behave as a robust oxidizing agent [124]. Gold nanoparticles modify the charge of the microbial cell membrane, thus suppressing the activities of adenosine triphosphate synthetase to reduce its concentration, causing the metabolic process of the bacteria to slow down [125]. Another gold nanoparticle mechanism consists of producing a biological dysfunction avoiding the assembly of the ribosomal subunit for the tRNA [126]. Cytotoxicity and genotoxicity in microbial cells can also be induced by the release of Ag nanoparticle ions through the Trojan-horse mechanism [127]. In this mechanism, silver nanoparticles enter the cell and subsequently release silver ions, resulting in apoptosis.

9.4.4 Oxygen Scavenger

Oxygen is responsible for food spoilage basically for two reasons: oxidation reactions produced and generation of aerobic decomposition microorganisms [128]. Vacuum food packaging and modified atmosphere packaging are used to avoid food spoilage [129]. These alternatives barely control oxygen from the external environment entering the package through the material of packaging or residual oxygen from modified atmospheres [130]. Currently, active packaging with oxygen scavenging properties is being applied. Iron-based and Titanium Oxide nanoparticles are used as oxygen scavengers [29,131]. Iron oxide nanoparticles with synthetic kaolinite in polymers for food packaging have two functions to eliminate oxygen: react and trap oxygen molecules (active behavior) and the imposition of a tortuous diffusion pathway barrier that prevents oxygen permeation (passive behavior) [132]. On the other hand, the TiO_2 nanoparticles in polymers capture oxygen employing ultraviolet (UV) illumination. Oxygen removal rates of $0.017\,cm^3\,O_2h^{-1}/cm^2$ have been obtained for 24 h [133].

9.4.5 Oxygen Sensors

The variation of electrical and optical properties of NMs when they interact with oxygen molecules allows monitoring the presence of this gas in food packaging [134]. The most widely used colorimetric oxygen sensor is based on oxygen-induced redox reactions [135]. To take a reading from this sensor, it is necessary to activate it by UV radiation. Typically, a UV-activated oxygen sensor is composed of a redox dye, a sacrificial electron donor, and a UV-absorbing photocatalyst [136]. TiO_2 and ZnO nanoparticles are used as photocatalysts, which are suspended in a solution of glycerol and methylene blue. When the TiO_2 and ZnO nanoparticles are irradiated with energy more significant than their bandgap (UV radiation), then, electrons in their valence band are promoted to the conduction band. Resulting excited electrons can be harnessed for redox reactions such as activating redox dyes for oxygen detection. In this mechanism, a mild sacrificial electron donor is also necessary to avoid recombination of the semiconductor electron–hole pairs, ensuring that the excited electrons are available to reduce the redox dye. This reaction causes the dye to become activated, changing from the colored oxidized state to the colorless reduced state [137].

9.4.6 Microbial Growth Indicators

The detection of microbes as pathogens in food is vital for quality control and disease prevention. One of the most successful ways of sensing microbial growth is colorimetric sensors. These are based on the plasmonic response of suspended metallic nanoparticles (such as Gold and Silver) [31,138]. Au nanoparticles are functionalized according to the microbial organism to be recognized. The functionalization seeks to generate electrostatic adhesion between the nanoparticles and the membrane of the microbes. For example, nanoparticles of Au are functionalized with aptamers to recognize *Salmonella typhimurium*. In the presence of the bacterium *S. typhimurium*, aptamers on the surface of the nanoparticles bind to the bacteria, resulting in aggregation of the nanoparticles. This aggregation of nanoparticles produces a change in the color of the suspension [139].

Nanoparticles can also be used to improve the response of other bacterial detection techniques. Ag nanoparticles are used in the Raman-based detection of bacteria. In this approach, Ag nanoparticles are synthesized *in-situ* on the surface of bacteria, which causes the surface charge of the cell wall to change [140]. The change in charge causes the Raman scattering to be significantly amplified, improving the technique's sensitivity.

9.4.7 TIME–TEMPERATURE INDICATORS

TTIs are one of the main components of smart packaging. TTIs are external devices that could easily be attached outside the package and are used to monitor temperature changes that can affect the food in the supply chain. These indicators are classified as irreversible devices that change a physical characteristic, i.e., a visible color, directly correlated to the temperature and time history of the food [141]. Metal nanoparticles are currently being used for the manufacturing of TTIs, owing to their interesting optical properties, such as strong absorption bands in the visible region and coherent oscillation of the conduction electrons [142]. TTIs manufactured with metal nanoparticles base their operation on variations in the morphological and distribution features (particle aggregation or agglomeration) of nanoparticles suspended in a medium due to heat absorption and their capacity to display localized surface plasmon resonances (LSPRs) [101,143].

One of the approaches studied in TTIs is changes in surface morphology and aggregation processes in suspended nanoparticles [144–146]. The macroscopic effect of these variations is an irreversible change in the emission wavelength of the suspension, which the consumer can see as a color change in the indicator. An example of such applications has been the use of triangular Ag nanoplates suspended in an aqueous medium used as a suspension for the manufacture of TTIs [144]. According to the synthesis method, nanoplates can undergo shape or assembly changes when the suspension increases in temperature. Silver nanoplates synthesized by seed-mediated protocol involve reducing $AgNO_3$ by L-ascorbic acid in the presence of Ag seeds, polyvinyl pyrrolidone (PVP), and sodium citrate. The resulting aqueous suspension shows a specific color according to the time and temperature of exposition. With the change in the temperature, the corners of triangular Ag nanoparticles become round. This produces a blue shift for dipole resonance peak position. Peak shift of up to 320 nm for temperature values between 4°C and 47°C has been reported [144]. Another behavior of Ag nanoplates face-to-face assembled by citrate is that they are disassembled quickly by heating; in this process, the LSPR peak shift could be observed [147].

Au nanoparticles are also used for the manufacture of TTIs. In this kind of TTI, nanoparticles are usually suspended in a gelatin matrix. Au nanoparticles can be synthesized by mixing gelatin solution and gold precursor (1 mM of $HAuCl_4$). The concentration of reducing and stabilization establishes the number and morphology of gold nanoparticles synthesized, which determines the color suspension. The operating mechanism for this indicator consists of an LSPR peak red-shifted due to the agglomeration of the Au nanoparticles by temperature variations [145,146].

In recent years, with the number of mobile applications available to facilitate the reading of these types of indicators, smart devices such as smartphones, tablets, and laptops can be used as analyzers of digitized colorimetry information of the samples. Smart device–based sensing technology promises to be applied in next-generation point-of-care testing for food and health safety monitoring [148].

9.5 STANDARDS AND REGULATIONS APPLIED TO NANOMATERIALS IN FOOD INDUSTRY

Food security is a significant part of the fundamental human right to food-related access to sufficient and nutritional food [149]. For an adequate nanotechnology application in food security, it is necessary to implement standards and regulations that guarantee quality and safe products and consumer protection. According to FAO and the World Health Organization, nanotechnology's applications in the food sector will occur in three areas: (1) in advance of nanostructured food products, (2) in the use of nanosized or nanoencapsulated food additives, and (3) in food packaging [150]. In these areas, several authorities, committees, organizations, and standardization bodies have worked on integrating their conventional standards and regulations concerning food and nanotechnology to generate specific guidelines for applying NMs in food.

Table 9.1 provides an overview of the primary standards and up-to-date regulatory documents used in the food industry as well as in nanotechnology. Furthermore, it shows how different standards and regulations, either implicitly or explicitly, cover the NM application in the food sector. International Organization for Standardization (ISO) supports industrial development, balancing the promotion of scientific progress and the development of better regulation [151]. ISO also has standards for food safety management and nanotechnology. The requirements for organizations in the food chain and bodies providing audit and certification of food safety management systems are established in the ISO 22000:2018 and ISO/TS 22003:2013 standards. On the other hand, nanotechnology has standards such as ISO/TS 80004-(1–8):2015, ISO/TR 11360:2010, ISO/TR 13121:2011, and ISO/TR 13014:2012, which describe definitions, classifications, characterization, and risk management of NMs.

Moreover, ISO in its standard ISO/TS 21975:2020 links applications of NMs in the food sector, specifically in food packaging and polymeric nanocomposite films with barrier properties; this is the only standard that involves NMs and food. Similarly, the American Society of Testing Materials (ASTM) has standards related to food management and nanotechnology; however, it does not have any standard that directly describes NMs and food. The ASTM standards related to food are more geared toward the characteristics of the packaging and the indicators that provide information on the product's status, which are ASTM E460-12 and ASTM F1416-96, respectively. ASTM standards for nanotechnology have the same aims as the ISO standards.

Organizations such as the European Union (EU), the Food and Drug Administration (FDA) in the United States, and Food Standards Australia New Zealand (FSANZ) provide legislation intending to have transparency, traceability,

TABLE 9.1

List of the Most Relevant Standards and Regulations for Food, Nanomaterials, and Food in Contact with Nanomaterials, Organized by Authority, Application Sector (Food or Nano or Both), Reference Document, and Description

Authority	Application Sector	Standard (S) or Regulation (R)	Reference document[a]	Description
International Organization for Standardization (ISO)	Food industry/ science	S	ISO 22000:2018	Requirements of a food safety management system for any organization related to the food chain
		S	ISO/TS 22003:2013	Requirements for bodies providing audit and certification of food safety management systems
	Nanotechnology/ nanomaterials	S	ISO/TS 80004-(1–8): 2015	List of technical terms and definitions in the field of nanotechnologies related to core terms, nanoobjects, carbon nanoobjects, nanostructured materials, nano/bio-interface, nanoobject characterization, diagnostics and therapeutics for healthcare, and nanomanufacturing processes
		S	ISO/TR 11360:2010	Methodology for the classification and categorization of nanomaterials
		S	ISO/TS 12805:2011	Guidance on specifying nanoobjects
		S	ISO/TR 13121:2011	Guidance on management of nanomaterial and risk evaluation
		S	ISO/TS 12901-(1–2):2012	Guidance on occupational risk management applied to engineered nanomaterials
		S	ISO/TR 13014:2012	Guidance on physicochemical characterization of engineered nanoscale materials for toxicologic assessment
	Nanomaterials in food package	S	ISO/TS 21975:2020	Specification of characteristics and measurement methods: polymeric nanocomposite films for food packaging with barrier properties

(Continued)

TABLE 9.1 (Continued)

List of the Most Relevant Standards and Regulations for Food, Nanomaterials, and Food in Contact with Nanomaterials, Organized by Authority, Application Sector (Food or Nano or Both), Reference Document, and Description

Authority	Application Sector	Standard (S) or Regulation (R)	Reference document[a]	Description
American Society of Testing Materials (ASTM)	Food industry/science	S	ASTM E460-12	Practice for determining the effect of packaging on food and beverage products during storage
		S	ASTM F1416-96	Guide for selection of TTIs
	Nanotechnology/nanomaterials	S	ASTM E2456-06(2012)	Standard terminology relating to nanotechnology
		S	ASTM E2909-13	Guide for investigation/study/assay tab-delimited format for nanotechnologies (ISA-TAB-Nano): standard file format for the submission and exchange of data on nanomaterials and characterizations
		S	ASTM E2859-11(2017)	Guide for size measurement of nanoparticles using atomic force microscopy
		S	ASTM E2490-09(2015)	Guide for measurement of the particle size distribution of nanomaterials in suspension by photon correlation spectroscopy
		S	ASTM E2524-08(2013)	Standard test method for analysis of hemolytic properties of nanoparticles
European Union	Food industry/science	R	(EC) No 1935/2004	Materials and articles intended to come into contact with food and repealing
		R	(EC) No 1333/2008	List of regulated food additives
		R	(EC) No 450/2009	Active and intelligent materials and articles intended to come into contact with food
	Nanotechnology/nanomaterials	R	(EC) No 1907/2006	Registration, evaluation, authorization, and restriction of chemicals
		R	(EU) No 528/2012	The making available on the market and use of biocidal products
	Nanomaterials in food	R	(EU) No 1169/2011	Provision of food information to consumers

(Continued)

TABLE 9.1 (*Continued*)

List of the Most Relevant Standards and Regulations for Food, Nanomaterials, and Food in Contact with Nanomaterials, Organized by Authority, Application Sector (Food or Nano or Both), Reference Document, and Description

Authority	Application Sector	Standard (S) or Regulation (R)	Reference document[a]	Description
US Food and Drug Administration (FDA)	Food industry/ science	R	FDA-2011-D-0490	Guidance for industry: assessing the effects of significant manufacturing process changes, including emerging technologies, on the safety and regulatory status of food ingredients and food contact substances, including food ingredients that are color additives
	Nanomaterials in food	R	FDA-2010-D-0530	They are considering whether an FDA-regulated product involves the application of nanotechnology
		R	FDA-2013-D-1009	Use of nanomaterials in food for animals
Food Standards Australia New Zealand (FSANZ)	Nanomaterials in food	R	C2018C00243	Composition, labeling, safe handling, and primary production of foods sold in Australia and New Zealand

[a] ISO, International Organization for Standardization; ASTM, American Society of Testing Materials; EU, European Union; FDA, USA, Food and Drug Administration; FSANZ, Food Standards Australia New Zealand.

and information on the use and possible exposure to NMs in food products. EU legislation includes the regulation (EC) No 1333/2008 on food additives, (EC) No 450/2009 on active and intelligent materials, and articles intended to come into contact with food, and (EU) No 1169/2011 for the provision of food information to consumers. In the United States, the FDA guarantees the safety of food additives/food contact materials/feed additives placed on the market and products involved in the application of nanotechnology in standard FDA-2010-D-0530. Australia and New Zealand adopted the Food Standards Code (FSANZ) and risk assessment framework to manage human health risks posed by nanotechnologies. This code includes standards related to foods specifically in additives, vitamins and minerals, processing aids, contaminants and natural toxins, articles, and materials in contact with food and novel foods. Finally, another critical aspect of the legislation is labeling the packaging in which the consumer has information about the NMs applied directly or indirectly in food products. On its labeling, EU legislation makes it mandatory that all ingredients present in foods and biocides in the form of NMs must be indicated in the list of ingredients with these ingredients' names followed by the word "nano" in parentheses [152].

9.6 DISCUSSION

In the food sector, NMs could play a critical role in leveraging food production and reducing the amount of loss and waste that currently exists around the world. Of the three approaches described—smart food, active packaging, and smart packing, only smart food does not have a significant influence on reducing food waste. Smart foods are used to improve the characteristics and food properties, making the products more interesting for the consumers and easier to market.

On the other hand, the use of NMs to improve the internal and external characteristics of packaging does have a high impact on reducing food waste and loss. Improved packaging prevents damage to food products due to environmental and manipulation factors present in supply chains. Currently, the supply chains are limited due to crises such as the one generated by COVID-19, which makes it necessary to produce packages that guarantee the physical protection of food and offer the consumer and producer traceability of the product [153].

A fundamental aspect of the consolidation of technology is the consumer's perception and acceptance. This is one of the main challenges nanotechnology faces in the food sector because due to a lack of information about the effects and risks of the use of NMs, consumers are resistant to accepting this technology. The ultimate success of such applications, and specifically of food packaging applications, will depend on consumer acceptance, which is determined by an understanding of the safety and undesired effects of the NMs, the disposition and location of the NMs on the packaging, and if the NMs are in direct contact with food [154].

In active packaging, NMs are found in the polymer from which the packaging is made, which is in direct contact with food. For these applications such as strength reinforcement, gas barriers, microbial blocking, and oxygen scavenging, consumers can be exposed to hazardous concerns due to the migration of NMs from the polymer to the food. To assess this risk, the key questions are whether nanoparticles can be released from polymers in contact with food, under what circumstances is this possible, and what is their toxicology [155].

In smart packaging, NMs are found in devices externally adhered to the primary packaging; this means that the NMs are never in direct contact with the food. As there are no drawbacks of possible migration of NMs to food, applications such as oxygen sensors, microbial growth indicators, and TTIs would not have problems with consumer acceptance. These applications give the consumers an essential role by providing feedback to the producer on the state of food in the final stage of the supply chain. This is an added value in the interaction between consumer and producer due to the easy access to information on the state of the food and the traceability that can be given.

Another major challenge in the use of NMs in the manufacturing of food packaging is the management of waste from NM-based products and how this could be quantified. There is currently no standard definition of what constitutes waste-containing NMs (WCNMs). WCNMs [156]: (1) in the synthesis of NMs and (2) in the manufacture of products based on NMs. The final fate of the WCNM will depend on the physicochemical properties of the NMs and the matrix in which they are integrated, as well as the capacity of the applied technologies to retain and destroy the NMs [157].

Quantification of manufacturing residues and waste materials containing NMs should be performed by the manufacturers of NMs and NM-based products. However, due to the confidentiality and protection of industrial processes, manufacturers avoid reporting this type of information. This lack of data generates approximate quantities on the use of NMs with significant uncertainties. Problems such as those described above reflect the need to establish unified documentation that guides the correct use of NMs in the different phases of the food sector. Specifically, in the case of sectors such as the food industry, regulations, standards, initiatives, or projects related to the use of NMs are essential because their applications directly influence human health.

As NMs are becoming a crucial value for novel approaches in the food industry, efforts aim to improve consumer and producer safety, and food quality has emerged separately. The quality concept designates the desirable characteristics that add value to the food. Traditionally, this has included forms of production (organic farming, environmental consideration, synthetic, and animal welfare), production areas (designation of origin), and recently the introduction of NMs [158]. Safety is based on ensuring the protection of human health related to food consumption. For this reason, the consumer must have information on how the food they eat is produced, processed, packaged, labeled, sold, and if NMs were used in any process [159]. Below, we list some of the current projects and initiatives that could be the seed of synergies on the two fronts: NMs and food.

On the side of NMs:

- **NanoHarmony**: a project funded by the EU's Horizon 2020 that aims to develop test guidelines and guidance documents (GD) for engineered NMs based on scientifically reliable test methods and acceptable practice documents [160]. These guides are expected to be accepted by the 37 OECD Member countries.

On the side of the food industry:

- **AgriMax**: a project funded by a partnership between the EU and the Bio-based Industries Consortium (BIC). It is a project that establishes the technical and economic viability through the process of biorefining crop residues and food processing to deliver new biocomposites for the chemical, bioplastic, food, fertilizer, packaging, and agriculture sectors [161].
- **BioBarr**: a project funded by a partnership between the EU and the BIC. It is a project to manufacture new biodegradable and bio-based food packaging materials with improvements in its barrier properties and mechanical resistance. These materials are expected to replace conventional polymers [162].

Generating synergies between initiatives related to NMs and food is a challenge that government authorities, committees, private organizations, and standardization bodies must face together. The objective of these synergies should be focused on the generation of standards and regulations that give guidelines on the use of NMs in the progress of technologies that favor the food sector.

9.7 CONCLUSION AND PROSPECT

In summary, this chapter has shown detailed information about recent advances in the use of NMs in the food sector, specifically in food processing and packaging. Smart foods use NMs such as micelles, polymers, hydrogels, liposomes, dendrimers, and carbon nanotubes to nanoencapsulation molecules that determine the characteristics of foods (color, flavor, nutrients, among others). Nanoencapsulation processes improve the characteristics of the food. NMs in packaging are used in two different ways in order to preserve food. Firstly, NMs are used as additives in the polymeric matrices from which the packaging material is made. The inclusion of NMs improves the mechanical, permeability, and antimicrobial characteristics of the package. Secondly, sensors and indicators based on NMs are added to the main food packaging to monitor parameters that define the quality of the food. Usually, metal and metal oxide nanoparticles are used to improve packaging.

Recognition of the usefulness of NMs in the food sector is increasing in the field of research and arouses great interest in the industrial sector. Although commercial uses of NMs in the food sector are currently scarce, the rapid development and understanding of the mechanisms that describe the characteristics of NMs make this field of research promising and soon to be linked to the market. The success of the application of nanotechnology in the food market will depend on the acceptance that this technology has in the consumer. This constitutes a challenge for the scientific community, government organizations, and stakeholders. The scientific community, in addition to generating nanotechnological solutions for the food sector, must produce the scientific basis for the risk assessment of the use of NMs related to toxicity, migration, detection and quantification of NMs and their residues in the environment, final disposal of products, and recycling potential. Government organizations and stakeholders should consolidate regulations and standards for the use of NMs in food, as well as seek the integration of the study of nanotechnology in educational systems since the knowledge of this technology will allow its acceptance by the consumers.

ACKNOWLEDGMENTS

The authors acknowledge the support of the Electrical and Electronics Department at Universidad de los Andes, Colombia. Gustavo Lanza was supported by Agreement Cundinamarca, Center for Interdisciplinary Studies in Basic and Applied Complexity, CEIBA, Doctoral Fellowship program. The authors thank Kelley Crites for her insightful comments in reviewing the manuscript.

REFERENCES

1. DU, COMITE RÉGIONAL. 2016. *Sustainable Development Goals*. World Health Organization.
2. FAO, UN. 2019. *The State of Food and Agriculture: Moving Forward on Food Loss and Waste Reduction*. Food Agriculture Organization.
3. UN. 2020. *The sustainable development goals report 2020*. United Nations.
4. Rabah, K. 2018. "Convergence of AI, IoT, big data and blockchain: A review." *The Lake Institute Journal* 1 (1):1–18.
5. Ewaschuk, J. B, and S. Unger. 2015. "Human milk pasteurization: Benefits and risks." *Current Opinion in Clinical Nutrition and Metabolic Care* 18 (3):269–275.
6. He, X., H. Deng, and H.-m. Hwang. 2019. "The current application of nanotechnology in food and agriculture." *Journal of Food and Drug Analysis* 27 (1):1–21.
7. Wu, Q., W.-s. Miao, H.-j. Gao, and D.d Hui. 2020. "Mechanical properties of nanomaterials: A review." *Nanotechnology Reviews* 9 (1):259–273.
8. Andrievski, R. A. 2014. "Review of thermal stability of nanomaterials." *Journal of Materials Science* 49 (4):1449–1460.
9. Bushell, M., S. Beauchemin, F. Kunc, D. Gardner, J. Ovens, F. Toll, D. Kennedy, K. Nguyen, D. Vladisavljevic, and P. E Rasmussen. 2020. "Characterization of commercial metal oxide nanomaterials: Crystalline phase, particle size and specific surface area." *Nanomaterials* 10 (9):1812.
10. Zhang, J. Z. 2009. *Optical Properties and Spectroscopy of Nanomaterials*. World Scientific.
11. Schmid, G., and U. Simon. 2005. "Gold nanoparticles: Assembly and electrical properties in 1–3 dimensions." *Chemical Communications* (6):697–710.
12. Seip, C. T., E. E Carpenter, C. J. O'Connor, V. T John, and S. Li. 1998. "Magnetic properties of a series of ferrite nanoparticles synthesized in reverse micelles." *IEEE Transactions on Magnetics* 34 (4):1111–1113.
13. Prasad, R. 2019. *Microbial Nanobionics*. Springer.
14. Chellaram, C., G. Murugaboopathi, A. A. John, R. Sivakumar, S. Ganesan, S. Krithika, and G. Priya. 2014. "Significance of nanotechnology in food industry." *APCBEE procedia* 8:109–113.
15. Robertson, G. L. 2016. *Food Packaging: Principles and Practice*. CRC Press.
16. Reig, C. S., A. D. Lopez, M. H. Ramos, and V. A. C. Ballester. 2014. "Nanomaterials: A map for their selection in food packaging applications." *Packaging Technology and Science* 27 (11):839–866.
17. Sacharow, S., and R. C Griffin. 1980. *Principles of Food Packaging*. AVI Pub. Co.
18. Carpenter, R. P, D. H Lyon, and T. A Hasdell. 2012. *Guidelines for Sensory Analysis in Food Product Development and Quality Control*. Springer Science & Business Media.
19. Shewfelt, R. L. 2014. "Measuring quality and maturity." In *Postharvest Handling*, 387–410. Elsevier.
20. Tijskens, L. M. M., and R. E. Schouten. 2014. "Modeling quality attributes and quality related product properties." In *Postharvest Handling*, 411–448. Elsevier.

21. Seuring, S., and M. Müller. 2008. "From a literature review to a conceptual framework for sustainable supply chain management." *Journal of Cleaner Production* 16 (15):1699–1710.
22. Wansink, B. 2004. "Environmental factors that increase the food intake and consumption volume of unknowing consumers." *Annual Review of Nutrition* 24:455–479.
23. Martinez-Romero, D., M. Serrano, A. Carbonell, S. Castillo, F. Riquelme, and D. Valero. 2004. "Mechanical damage during fruit post-harvest handling: Technical and physiological implications." In *Production Practices and Quality Assessment of Food Crops*, 233–252. Springer.
24. Singh, R. P. 2019. "Utility of nanomaterials in food safety." In *Food Safety and Human Health*, 285–318. Elsevier.
25. Konur, O. 2019. "Nanotechnology applications in food: A scientometric overview." In *Nanoscience for Sustainable Agriculture*, 683–711. Springer.
26. Ghorbanpour, M., P. Bhargava, A. Varma, and D. K. Choudhary. 2020. *Biogenic Nano-Particles and Their Use in Agro-Ecosystems*. Springer.
27. He, X., and H.-M. Hwang. 2016. "Nanotechnology in food science: Functionality, applicability, and safety assessment." *Journal of Food and Drug Analysis* 24 (4):671–681.
28. Assadpour, E., and S. M. Jafari. 2019. "A systematic review on nanoencapsulation of food bioactive ingredients and nutraceuticals by various nanocarriers." *Critical Reviews in Food Science and Nutrition* 59 (19):3129–3151.
29. Yildirim, S., B. Röcker, M. K. Pettersen, J. Nilsen-Nygaard, Z. Ayhan, R. Rutkaite, T. Radusin, P. Suminska, B. Marcos, and V. Coma. 2018. "Active packaging applications for food." *Comprehensive Reviews in Food Science and Food Safety* 17 (1):165–199.
30. Yousefi, H., H.-M. Su, S. M Imani, K. Alkhaldi, C. D. M. Filipe, and T. F Didar. 2019. "Intelligent food packaging: A review of smart sensing technologies for monitoring food quality." *ACS Sensors* 4 (4):808–821.
31. Zhang, C., A.-X. Yin, R. Jiang, J. Rong, L. Dong, T. Zhao, L.-D. Sun, J. Wang, X. Chen, and C.-H. Yan. 2013. "Time–Temperature indicator for perishable products based on kinetically programmable Ag overgrowth on Au nanorods." *ACS Nano* 7 (5):4561–4568.
32. Knorr, D., and H. Watzke. 2019. "Food processing at a crossroad." *Frontiers in Nutrition* 6:85.
33. Ameta, S. K., A. K. Rai, D. Hiran, R. Ameta, and S. C Ameta. 2020. "Use of nanomaterials in food science." In *Biogenic Nano-Particles and Their Use in Agro-Ecosystems*, 457–488. Springer.
34. Chaudhry, Q., M. Scotter, J. Blackburn, B. Ross, A. Boxall, L. Castle, R. Aitken, and R. Watkins. 2008. "Applications and implications of nanotechnologies for the food sector." *Food Additives and Contaminants* 25 (3):241–258.
35. Akhavan, S., and S. M. Jafari. 2017. "Chapter 6-Nanoencapsulation of natural food colorants." *Nanoencapsulation of Food Bioactive Ingredients*:223–260.
36. Shafiq, M., S. Anjum, C. Hano, I. Anjum, and B. H. Abbasi. 2020. "An overview of the applications of nanomaterials and nanodevices in the food industry." *Foods* 9 (2):148.
37. Tarhini, M., H. Greige-Gerges, and A. Elaissari. 2017. "Protein-based nanoparticles: From preparation to encapsulation of active molecules." *International Journal of Pharmaceutics* 522 (1–2):172–197.
38. Katouzian, I., and S. M. Jafari. 2016. "Nano-encapsulation as a promising approach for targeted delivery and controlled release of vitamins." *Trends in Food Science & Technology* 53:34–48.
39. Milovanovic, M., A. Arsenijevic, J. Milovanovic, T. Kanjevac, and N. Arsenijevic. 2017. "Nanoparticles in antiviral therapy." In *Antimicrobial Nanoarchitectonics*, 383–410. Elsevier.

40. Hezi-Yamit, A., C. Sullivan, J. Wong, L. David, M. Chen, P. Cheng, D. Shumaker, J. N Wilcox, and K. Udipi. 2009. "Impact of polymer hydrophilicity on biocompatibility: Implication for DES polymer design." *Journal of Biomedical Materials Research Part A: An Official Journal of The Society for Biomaterials, The Japanese Society for Biomaterials, and The Australian Society for Biomaterials and the Korean Society for Biomaterials* 90 (1):133–141.

41. Sharma, M. 2019. "Transdermal and intravenous nano drug delivery systems: Present and future." In *Applications of Targeted Nano Drugs and Delivery Systems*, 499–550. Elsevier.

42. Choudhary, B., S. R Paul, S. K Nayak, D. Qureshi, and K. Pal. 2018. "Synthesis and biomedical applications of filled hydrogels." In *Polymeric Gels*, 283–302. Elsevier.

43. Tekade, R. K. 2018. *Basic Fundamentals of Drug Delivery*. Academic Press.

44. Carvalho, A., A. R Fernandes, and P. V Baptista. 2019. "Nanoparticles as delivery systems in cancer therapy: Focus on gold nanoparticles and drugs." In *Applications of Targeted Nano Drugs and Delivery Systems*, 257–295. Elsevier.

45. ISO. 2020. Nanotechnologies — Characterization of carbon nanotube samples using thermogravimetric analysis. International Organization for Standardization.

46. Inagaki, M., F. Kang, M. Toyoda, and H. Konno. 2013. *Advanced Materials Science and Engineering of Carbon*. Butterworth-Heinemann.

47. Kumari, A., R. Singla, A. Guliani, and S. K. Yadav. 2014. "Nanoencapsulation for drug delivery." *EXCLI Journal* 13:265.

48. Yu, H., J.-Y. Park, C. W. Kwon, S.-C. Hong, K.-M. Park, and P.-S. Chang. 2018. "An overview of nanotechnology in food science: Preparative methods, practical applications, and safety." *Journal of Chemistry* 2018 (1):1–10.

49. Chen, J., and L. Hu. 2020. "Nanoscale delivery system for nutraceuticals: Preparation, application, characterization, safety, and future trends." *Food Engineering Reviews* 12 (1):14–31.

50. Bratovic, A., and J. Suljagic. 2019. "Micro-and nano-encapsulation in food industry." *Croatian Journal of Food Science and Technology* 11 (1):113–121.

51. Battaglia, L., M. Gallarate, R. Cavalli, and M. Trotta. 2010. "Solid lipid nanoparticles produced through a coacervation method." *Journal of Microencapsulation* 27 (1):78–85.

52. Lee, S. H., D. Heng, W. K. Ng, H.-K. Chan, and R. B. H. Tan. 2011. "Nano spray drying: A novel method for preparing protein nanoparticles for protein therapy." *International Journal of Pharmaceutics* 403 (1–2):192–200.

53. Ghorbanzade, T., S. M. Jafari, S. Akhavan, and R. Hadavi. 2017. "Nano-encapsulation of fish oil in nano-liposomes and its application in fortification of yogurt." *Food Chemistry* 216:146–152.

54. Sawada, S.-i., and K. Akiyoshi. 2010. "Nano-encapsulation of lipase by self-assembled nanogels: Induction of high enzyme activity and thermal stabilization." *Macromolecular Bioscience* 10 (4):353–358.

55. Souguir, H., F. Salaün, P. Douillet, I. Vroman, and S. Chatterjee. 2013. "Nanoencapsulation of curcumin in polyurethane and polyurea shells by an emulsion diffusion method." *Chemical Engineering Journal* 221:133–145.

56. Horuz, T. İ., and K. B. Belibağlı. 2018. "Nanoencapsulation by electrospinning to improve stability and water solubility of carotenoids extracted from tomato peels." *Food Chemistry* 268:86–93.

57. Fangmeier, M., D. N. Lehn, M. J. Maciel, and C. F. Volken de Souza. 2019. "Encapsulation of bioactive ingredients by extrusion with vibrating technology: Advantages and challenges." *Food and Bioprocess Technology* 12 (9):1–15.

58. Ezhilarasi, P. N., P. Karthik, N. Chhanwal, and C. Anandharamakrishnan. 2013. "Nanoencapsulation techniques for food bioactive components: A review." *Food and Bioprocess Technology* 6 (3):628–647.

59. Ushak, S., M. Judith Cruz, L. F. Cabeza, and M. Grágeda. 2016. "Preparation and characterization of inorganic PCM microcapsules by fluidized bed method." *Materials* 9 (1):24.
60. Estevinho, B. N., and F. Rocha. 2017. "A key for the future of the flavors in food industry: Nanoencapsulation and microencapsulation." In *Nanotechnology Applications in Food*, 1–19. Academic Press.
61. Spence, C. 2018. "Background colour & its impact on food perception & behaviour." *Food Quality and Preference* 68:156–166.
62. Mortensen, A. 2006. "Carotenoids and other pigments as natural colorants." *Pure and Applied Chemistry* 78 (8):1477–1491.
63. Rostamabadi, H., S. R. Falsafi, and S. M. Jafari. 2019. "Nanoencapsulation of carotenoids within lipid-based nanocarriers." *Journal of Controlled Release* 298:38–67.
64. German, J. B., & Dillard, C. J. 2006. "Composition, structure and absorption of milk lipids: A source of energy, fat-soluble nutrients and bioactive molecules." *Critical Reviews in Food Science and Nutrition*, 46(1):57–92.
65. Etiévant, P., E. Guichard, C. Salles, and A. Voilley. 2016. *Flavor: From Food to Behaviors, Wellbeing and Health*. Woodhead Publishing.
66. Akacha, N. B., and M. Gargouri. 2015. "Microbial and enzymatic technologies used for the production of natural aroma compounds: Synthesis, recovery modeling, and bioprocesses." *Food and Bioproducts Processing* 94:675–706.
67. Burgos, N., A. C. Mellinas, E. García-Serna, and A. Jiménez. 2017. "Nanoencapsulation of flavor and aromas in food packaging." In *Food Packaging*, 567–601. Elsevier.
68. Saffarionpour, S. 2019. "Nanoencapsulation of hydrophobic food flavor ingredients and their cyclodextrin inclusion complexes." *Food and Bioprocess Technology* 12 (7):1157–1173.
69. Saifullah, M., M. R. I. Shishir, R. Ferdowsi, M. R. T. Rahman, and Q. V. Vuong. 2019. "Micro and nano encapsulation, retention and controlled release of flavor and aroma compounds: A critical review." *Trends in Food Science & Technology* 86:230–251.
70. Taheri, A., and S. M. Jafari. 2019. "Gum-based nanocarriers for the protection and delivery of food bioactive compounds." *Advances in Colloid and Interface Science* 269:277–295.
71. Fathi, M., A. Martin, and D. J. McClements. 2014. "Nanoencapsulation of food ingredients using carbohydrate based delivery systems." *Trends in Food Science & Technology* 39 (1):18–39.
72. Jafari, S. M. 2017. "An overview of nanoencapsulation techniques and their classification." In *Nanoencapsulation Technologies for the Food and Nutraceutical Industries*, 1–34. Elsevier.
73. Srinivas, P. R, M. Philbert, T. Q. Vu, Q. Huang, J. L. Kokini, E. Saos, H. Chen, C. M. Peterson, K. E. Friedl, and C. McDade-Ngutter. 2010. "Nanotechnology research: Applications in nutritional sciences." *The Journal of Nutrition* 140 (1):119–124.
74. Karunaratne, D. N., D. A. S. Siriwardhana, I. R. Ariyarathna, R. M. P. I. Rajakaruna, F. T. Banu, and V. Karunaratne. 2017. "Nutrient delivery through nanoencapsulation." In *Nutrient Delivery*, 653–680. Elsevier.
75. Odriozola-Serrano, I., R. Soliva-Fortuny, and O. Martín-Belloso. 2009. "Influence of storage temperature on the kinetics of the changes in anthocyanins, vitamin C, and antioxidant capacity in fresh-cut strawberries stored under high-oxygen atmospheres." *Journal of Food Science* 74 (2):C184–C191.
76. Janjarasskul, T., and P. Suppakul. 2018. "Active and intelligent packaging: The indication of quality and safety." *Critical Reviews in Food Science and Nutrition* 58 (5):808–831.
77. Jafarizadeh-Malmiri, H., Z. Sayyar, N. Anarjan, and A. Berenjian. 2019. "Nanoencapsulation for nutrition delivery." In *Nanobiotechnology in Food: Concepts, Applications and Perspectives*, 95–114. Springer.

78. Piñón-Balderrama, C. I., C. Leyva-Porras, Y. Terán-Figueroa, V. Espinosa-Solís, C. Álvarez-Salas, and M. Z Saavedra-Leos. 2020. "Encapsulation of active ingredients in food industry by spray-drying and nano spray-drying technologies." *Processes* 8 (8):889.
79. Kalra, E. K. 2003. "Nutraceutical-definition and introduction." *Aaps Pharmsci* 5 (3):27–28.
80. Souyoul, S. A, K. P Saussy, and M. P Lupo. 2018. "Nutraceuticals: A review." *Dermatology and Therapy* 8 (1):5–16.
81. Chaudhari, S. P., P. V. Powar, and M. N. Pratapwar. 2017. "Nutraceuticals: A review." *The World Journal of Pharmaceutical Sciences* 6:681–739.
82. Torres-Giner, S., A. Martinez-Abad, M. J. Ocio, and J. M. Lagaron. 2010. "Stabilization of a nutraceutical omega-3 fatty acid by encapsulation in ultrathin electrosprayed zein prolamine." *Journal of Food Science* 75 (6):N69–N79.
83. Akhavan, S., E. Assadpour, I. Katouzian, and S. M. Jafari. 2018. "Lipid nano scale cargos for the protection and delivery of food bioactive ingredients and nutraceuticals." *Trends in Food Science & Technology*, 74:132–146.
84. Zhou, M., T. Wang, Q. Hu, and Y. Luo. 2016. "Low density lipoprotein/pectin complex nanogels as potential oral delivery vehicles for curcumin." *Food Hydrocolloids* 57:20–29.
85. Reza Mozafari, M, C. Johnson, S. Hatziantoniou, and C. Demetzos. 2008. "Nanoliposomes and their applications in food nanotechnology." *Journal of Liposome Research* 18 (4):309–327.
86. Chenthamara, D., S. Subramaniam, S. G. Ramakrishnan, S. Krishnaswamy, M. M. Essa, F.-H. Lin, and M. W. Qoronfleh. 2019. "Therapeutic efficacy of nanoparticles and routes of administration." *Biomaterials Research* 23 (1):1–29.
87. Leena, M. M., L. Mahalakshmi, J. A Moses, and C. Anandharamakrishnan. 2020. "Nanoencapsulation of nutraceutical ingredients." In *Biopolymer-Based Formulations*, 311–352. Elsevier.
88. Badia-Melis, R., P. Mishra, and L. Ruiz-García. 2015. "Food traceability: New trends and recent advances. A review." *Food Control* 57:393–401.
89. Dobrucka, R., and R. Przekop. 2019. "New perspectives in active and intelligent food packaging." *Journal of Food Processing and Preservation* 43 (11):e14194.
90. European Commission. 2009. "Commission Regulation (EC) No 450/2009 of 29 May 2009 on active and intelligent materials and articles intended to come into contact with food." *The Official Journal of the European Union* 135:3–11.
91. Wyrwa, J., and A. Barska. 2017. "Innovations in the food packaging market: Active packaging." *European Food Research and Technology* 243 (10):1681–1692.
92. Ghaani, M., C. A Cozzolino, G. Castelli, and S. Farris. 2016. "An overview of the intelligent packaging technologies in the food sector." *Trends in Food Science & Technology* 51:1–11.
93. Ahvenainen, R. 2003. *Novel Food Packaging Techniques*. Elsevier.
94. Mustafa, F., and S. Andreescu. 2020. "Nanotechnology-based approaches for food sensing and packaging applications." *RSC Advances* 10 (33):19309–19336.
95. Saboori, A., S. K. Moheimani, M. Pavese, C. Badini, and P. Fino. 2017. "New nanocomposite materials with improved mechanical strength and tailored coefficient of thermal expansion for electro-packaging applications." *Metals* 7 (12):536.
96. Adame, D., and G. W. Beall. 2009. "Direct measurement of the constrained polymer region in polyamide/clay nanocomposites and the implications for gas diffusion." *Applied Clay Science* 42 (3–4):545–552.
97. Azeredo, H. M. C. 2013. "Antimicrobial nanostructures in food packaging." *Trends in Food Science & Technology* 30 (1):56–69.
98. Kuswandi, B. 2017. "Environmental friendly food nano-packaging." *Environmental Chemistry Letters* 15 (2):205–221.

99. Chu, C.-S., T.-W. Sung, and Y.-L. Lo. 2013. "Enhanced optical oxygen sensing property based on Pt (II) complex and metal-coated silica nanoparticles embedded in sol–gel matrix." *Sensors and Actuators B: Chemical* 185:287–292.
100. Sutarlie, L., S. Y. Ow, and X. Su. 2017. "Nanomaterials-based biosensors for detection of microorganisms and microbial toxins." *Biotechnology Journal* 12 (4):1–25.
101. Lanza, G., J. Perez-Taborda, and A. Avila. 2019. "Time temperature indicators (TTIs) based on silver nanoparticles for monitoring of perishables products." *Journal of Physics: Conference Series* 1447 (1):012055.
102. Ramgir, N. S. 2013. "Electronic nose based on nanomaterials: Issues, challenges, and prospects." *International Scholarly Research Notices* 2013:1–25.
103. Dhara, K., and R. M. Debiprosad. 2019. "Review on nanomaterials-enabled electrochemical sensors for ascorbic acid detection." *Analytical Biochemistry* 586:113415.
104. Dhas, S. P., P. J. Shiny, S. Khan, A. Mukherjee, and N. Chandrasekaran. 2014. "Toxic behavior of silver and zinc oxide nanoparticles on environmental microorganisms." *Journal of Basic Microbiology* 54 (9):916–927.
105. Emamhadi, M. A., M. Sarafraz, M. Akbari, Y. Fakhri, N. T. T. Linh, and A. M. Khaneghah. 2020. "Nanomaterials for food packaging applications: A systematic review." *Food and Chemical Toxicology* 2020:111825.
106. Huang, Y., L. Mei, X. Chen, and Q. Wang. 2018. "Recent developments in food packaging based on nanomaterials." *Nanomaterials* 8 (10):830.
107. Farah, S., D. G. Anderson, and R. Langer. 2016. "Physical and mechanical properties of PLA, and their functions in widespread applications—A comprehensive review." *Advanced Drug Delivery Reviews* 107:367–392.
108. Ambrogi, V., C. Carfagna, P. Cerruti, and V. Marturano. 2017. "Additives in polymers." In *Modification of Polymer Properties*, 87–108. Elsevier.
109. Bumbudsanpharoke, N., and S. Ko. 2019. "Nanoclays in food and beverage packaging." *Journal of Nanomaterials* 2019:1–13.
110. Martelli-Tosi, M., B. S. Esposto, N. Cristina da Silva, D. R. Tapia-Blácido, and S. M. Jafari. 2020. "Reinforced nanocomposites for food packaging." In *Handbook of Food Nanotechnology*, edited by S. M. Jafari, 533–574. Academic Press.
111. Peng, R. D., H. W. Zhou, H. W. Wang, and L. Mishnaevsky Jr. 2012. "Modeling of nano-reinforced polymer composites: Microstructure effect on Young's modulus." *Computational Materials Science* 60:19–31.
112. Haider, S., A. Kausar, and B. Muhammad. 2016. "Overview of various sorts of polymer nanocomposite reinforced with layered silicate." *Polymer-Plastics Technology and Engineering* 55 (7):723–743.
113. Chan, M.-l., K.-t. Lau, T.-t. Wong, M.-p. Ho, and D. Hui. 2011. "Mechanism of reinforcement in a nanoclay/polymer composite." *Composites Part B: Engineering* 42 (6):1708–1712.
114. Cabedo, L., and J. Gamez-Pérez. 2018. "Inorganic-based nanostructures and their use in food packaging." In *Nanomaterials for Food Packaging*, 13–45. Elsevier.
115. Hosur, M., T. H Mahdi, M. E Islam, and S. Jeelani. 2017. "Mechanical and viscoelastic properties of epoxy nanocomposites reinforced with carbon nanotubes, nanoclay, and binary nanoparticles." *Journal of Reinforced Plastics and Composites* 36 (9):667–684.
116. Liu, H., and L. Catherine Brinson. 2008. "Reinforcing efficiency of nanoparticles: A simple comparison for polymer nanocomposites." *Composites Science and Technology* 68 (6):1502–1512.
117. Arrieta, M. P., L. Peponi, D. López, J. López, and J. M. Kenny. 2017. "An overview of nanoparticles role in the improvement of barrier properties of bioplastics for food packaging applications." In *Food Packaging*, 391–424. Elsevier.
118. Cui, Y., S. Kumar, B. R. Kona, and D. van Houcke. 2015. "Gas barrier properties of polymer/clay nanocomposites." *Rsc Advances* 5 (78):63669–63690.

119. Uddin, F. 2018. "Montmorillonite: An introduction to properties and utilization." *Current Topics in the Utilization of Clay in Industrial and Medical Applications* 1:3–23.
120. Ammar, A., A. Elzatahry, M. Al-Maadeed, A. M. Alenizi, A. F. Huq, and A. Karim. 2017. "Nanoclay compatibilization of phase separated polysulfone/polyimide films for oxygen barrier." *Applied Clay Science* 137:123–134.
121. Saleh, T. A., N. P. Shetti, M. M. Shanbhag, K. R. Reddy, and T. M. Aminabhavi. 2020. "Recent trends in functionalized nanoparticles loaded polymeric composites: An energy application." *Materials Science for Energy Technologies* 3:515–525.
122. Olatunde, O. O., and S. Benjakul. 2018. "Natural preservatives for extending the shelf-life of seafood: A revisit." *Comprehensive Reviews in Food Science and Food Safety* 17 (6):1595–1612.
123. Hoseinnejad, M., S. M. Jafari, and I. Katouzian. 2018. "Inorganic and metal nanoparticles and their antimicrobial activity in food packaging applications." *Critical Reviews in Microbiology* 44 (2):161–181.
124. Sirelkhatim, A., S. Mahmud, A. Seeni, N. H. M. Kaus, L. C. Ann, S. K. M. Bakhori, H. Hasan, and D. Mohamad. 2015. "Review on zinc oxide nanoparticles: Antibacterial activity and toxicity mechanism." *Nano-micro Letters* 7 (3):219–242.
125. El Kurdi, R., and D. Patra. 2018. "Nanosensing of ATP by fluorescence recovery after surface energy transfer between rhodamine B and curcubit [7] uril-capped gold nanoparticles." *Microchimica Acta* 185 (7):349.
126. Folorunso, A., S. Akintelu, A. K. Oyebamiji, S. Ajayi, B. Abiola, I. Abdusalam, and A. Morakinyo. 2019. "Biosynthesis, characterization and antimicrobial activity of gold nanoparticles from leaf extracts of Annona muricata." *Journal of Nanostructure in Chemistry* 9 (2):111–117.
127. El-Shennawy, G. A., R. S. A. Ellatif, S. G. Badran, and R. H. El-Sokkary. 2020. "Silver nanoparticles: A potential antibacterial and antibiofilm agent against biofilm forming multidrug resistant bacteria." *Microbes and Infectious Diseases* 1 (2):77–85.
128. Bradley, D. G, and D. B Min. 1992. "Singlet oxygen oxidation of foods." *Critical Reviews in Food Science & Nutrition* 31 (3):211–236.
129. Djordjevic, J., M. Boskovic, M. Dokmanovic, I. B. Lazic, T. Ledina, B. Suvajdzic, and M. Z. Baltic. 2017. "Vacuum and modified atmosphere packaging effect on Enterobacteriaceae behaviour in minced meat." *Journal of Food Processing and Preservation* 41 (2):e12837.
130. Cruz, R. S., G. P. Camilloto, and A. C. d. S. Pires. 2012. "Oxygen scavengers: An approach on food preservation." *Structure and Function of Food Engineering* 2:21–42.
131. Han, G., J. Y. Kim, K.-J. Kim, H. Lee, and Y.-M. Kim. 2020. "Controlling surface oxygen vacancies in Fe-doped TiO2 anatase nanoparticles for superior photocatalytic activities." *Applied Surface Science* 507:144916.
132. Busolo, M. A, and J. M. Lagaron. 2012. "Oxygen scavenging polyolefin nanocomposite films containing an iron modified kaolinite of interest in active food packaging applications." *Innovative Food Science & Emerging Technologies* 16:211–217.
133. Mills, A., G. Doyle, A. M. Peiro, and J. Durrant. 2006. "Demonstration of a novel, flexible, photocatalytic oxygen-scavenging polymer film." *Journal of Photochemistry and Photobiology A: Chemistry* 177 (2–3):328–331.
134. Carpenter, M. A, S. Mathur, and A. Kolmakov. 2012. *Metal Oxide Nanomaterials for Chemical Sensors.* Springer Science & Business Media.
135. Ward, J. P. T. 2008. "Oxygen sensors in context." *Biochimica et Biophysica Acta (BBA)-Bioenergetics* 1777 (1):1–14.
136. Wen, J., S. Huang, L. Jia, F. Ding, H. Li, L. Chen, and X. Liu. 2019. "Visible colorimetric oxygen indicator based on Ag-loaded TiO$_2$ nanotubes for quick response and real-time monitoring of the integrity of modified atmosphere packaging." *Advanced Materials Technologies* 4 (9):1900121.

137. Mihindukulasuriya, S. D. F., and L.-T. Lim. 2013. "Oxygen detection using UV-activated electrospun poly (ethylene oxide) fibers encapsulated with TiO$_2$ nanoparticles." *Journal of Materials Science* 48 (16):5489–5498.

138. Loiseau, A., V. Asila, G. Boitel-Aullen, M. Lam, M. Salmain, and S. Boujday. 2019. "Silver-based plasmonic nanoparticles for and their use in biosensing." *Biosensors* 9 (2):78.

139. Mocan, T., C. T Matea, T. Pop, O. Mosteanu, A. D. Buzoianu, C. Puia, C. Iancu, and L. Mocan. 2017. "Development of nanoparticle-based optical sensors for pathogenic bacterial detection." *Journal of Nanobiotechnology* 15 (1):25.

140. Zhou, H., D. Yang, N. P. Ivleva, N. E. Mircescu, R. Niessner, and C. Haisch. 2014. "SERS detection of bacteria in water by *in situ* coating with Ag nanoparticles." *Analytical Chemistry* 86 (3):1525–1533.

141. Mercier, S., S. Villeneuve, M. Mondor, and I. Uysal. 2017. "Time–temperature management along the food cold chain: A review of recent developments." *Comprehensive Reviews in Food Science and Food Safety* 16 (4):647–667.

142. Langer, J., S. M Novikov, and L. M Liz-Marzán. 2015. "Sensing using plasmonic nano-structures and nanoparticles." *Nanotechnology* 26 (32):322001.

143. Ghosh, S. K., and T. Pal. 2007. "Interparticle coupling effect on the surface plasmon resonance of gold nanoparticles: From theory to applications." *Chemical Reviews* 107 (11):4797–4862.

144. Zeng, J., S. Roberts, and Y. Xia. 2010. "Nanocrystal-based time–temperature indicators." *Chemistry–A European Journal* 16 (42):12559–12563.

145. Lim, S., S. Gunasekaran, and J.-Y. Imm. 2012. "Gelatin-templated gold nanoparticles as novel time–temperature indicator." *Journal of Food Science* 77 (9):N45–N49.

146. Wang, Y.-C., L. Lu, and S. Gunasekaran. 2015. "Gold nanoparticle-based thermal history indicator for monitoring low-temperature storage." *Microchimica Acta* 182 (7–8):1305–1311.

147. Zhang, X.-Q., J. Ling, C.-J. Liu, Y.-H. Tan, L.-Q. Chen, and Q.-E. Cao. 2018. "An irreversible temperature indicator fabricated by citrate induced face-to-face assembly of silver triangular nanoplates." *Materials Science and Engineering: C* 92:657–662.

148. Wang, Y., P. Zhang, W. Fu, and Y. Zhao. 2018. "Morphological control of nanoprobe for colorimetric antioxidant detection." *Biosensors and Bioelectronics* 122:183–188.

149. Assembly, U. G. 1948. "Universal declaration of human rights." UN General Assembly 302 (2):14–25.

150. WHO. 2010. *FAO/WHO Expert Meeting on the Application of Nanotechnologies in the Food and Agriculture Sectors: Potential Food Safety Implications: Meeting Report.* World Health Organization.

151. Murphy, C. N., and J. Yates. 2009. *The International Organization for Standardization (ISO): Global Governance through Voluntary Consensus.* Routledge.

152. Amenta, V., K. Aschberger, M. Arena, H. Bouwmeester, F. Botelho Moniz, P. Brandhoff, S. Gottardo, H. J. P. Marvin, A. Mech, and L. Quiros Pesudo. 2015. "Regulatory aspects of nanotechnology in the agri/feed/food sector in EU and non-EU countries." *Regulatory Toxicology and Pharmacology* 73 (1):463–476.

153. Cullen, M. 2020. "COVID-19 and the risk to food supply chains: How to respond." *FAO.*

154. Pastrana, H. F., A. X. Cartagena-Rivera, A. Raman, and A. Ávila. 2019. "Evaluation of the elastic Young's modulus and cytotoxicity variations in fibroblasts exposed to carbon-based nanomaterials." *Journal of Nanobiotechnology* 17 (1):32.

155. Störmer, A., J. Bott, D. Kemmer, and R. Franz. 2017. "Critical review of the migration potential of nanoparticles in food contact plastics." *Trends in Food Science & Technology* 63:39–50.

156. Zamengo, L., M. Nasello, B. Branchi, G. Bracalente, W. Vergari, C. Bertocco, and A. Costernaro. 2020. "Risks and implications for health and the environment associated with products and waste containing nanomaterials: Regulatory and management issues in the European framework." *Giornale Italiano di Medicina del Lavoro ed Ergonomia* 42 (1):5–10.

157. UNEP. 2018. "Report on issues related to waste containing nanomaterials and options for further work under the basel convention." 240818.

158. Bojnec, Š., and I. Fertő. 2017. "Quality upgrades of eu agri-food exports." *Journal of Agricultural Economics* 68 (1):269–279.

159. Silano, M., and V. Silano. 2020. *Ensuring Food Safety in the European Union.* CRC Press.

160. NanoHarmony. 2020. "NanoHarmony Project." accessed 25/20/2020. https://nano harmony.eu/.

161. AGRIMAX. 2020. "AgriMax Project." accessed 25/20/2020. http://www.agrimax-project.eu.

162. BIOBARR. 2020. "BioBarr Project." http://www.biobarr.eu.

10 Role of Nanomaterials in Improving the Bioavailability of Functional Components

Shweta Rathee
NIFTEM

Eneyew Tadesse Melaku
Addis Ababa Science and Technology University

Anurag Singh and Ankur Ojha
NIFTEM

CONTENTS

10.1 INTRODUCTION

Bionanotechnology is an emerging technology of the 21st century and deals with the interaction of nutrition, food science, and nanotechnology. It is rapidly entering into various fields, including the food processing industry [1]. It helps in the fabrication of various types of nanomaterials with different properties comprising improvement of physicochemical, biological, antimicrobial, nutritional, and sensory properties and healthfulness in food products. There are different types of nanomaterials including nanoparticles, nanorods, nanofibrils, nanotubes, nanocrystals, nanoemulsions, nanoliposomes, nanocomposites, and nanostructured complexes [2]. These nanomaterials enhanced the physicochemical and thermal stability of functional components (vitamins, minerals, and bioactive compounds). They also mask the unpleasant taste, improve the bioavailability, and give protection against pH, oxygen, light, moisture, and gastric digestion of the functional components during food processing, storage, and inside the food matrix.

Different functional components play different roles like they not only provide us protection from noncommunicable diseases but also offer us nutrients to sustain life. Vitamins and minerals are required in small amounts for the proper functioning of biochemical and physiological processes. Bioactive compounds are physiologically active components present in foods with extra nutritional benefits. In today's world, people are getting more and more health-conscious. Hence, food processing industries invest a lot in making the modern consumer satisfied with nutritional and functional foods. Functional components are added to produce "functional foods" for improving the nutritional status of foods, thereby improving human health [3,4]. Examples of various functional components added into food matrix are essential fatty acids [5], carotenoids [6], vitamins (and minerals) [7], antioxidants [8], phytosterols [9], bioactive peptides [10], and plant essential oils [11]. However, some problems are associated with these because of their poor solubility, instability, undesirable flavor, and low bioavailability. Thus, using bionanotechnology can potentially provide a solution by encapsulating different functional components, thereby creating nanofunctional foods. Researches have shown the superiority of nanomaterials in comparison to microsystems [12,13]. So, the creation of nanofunctional foods will provide healthy and nutritional foods with improved bioavailability of functional components.

This chapter focuses on utilizing different food-grade nanomaterials for improving the bioavailability, protection, stability, and shelf life of different functional components inside the food matrix. Fabrication methods for the development of food-grade nanomaterials are discussed with examples. Safety issues and prospects are also discussed here.

10.2 BIOAVAILABILITY OF FUNCTIONAL COMPONENTS

According to the Food and Drug Administration, the definition of bioavailability is the "amount of component which is absorbed and becomes available."[14] The bioavailability described in Equation 10.1 is the product of FB, the amount of component accessible for intestinal absorption, FT, the amount which can transport across the small intestinal epithelium, and FM, the amount of absorbed component which reaches the systemic circulation [6].

$$F = FB \times FT \times FM \tag{10.1}$$

The factors affecting bioavailability are bioaccessibility and bioactivity explained diagrammatically in (Figure 10.1). Bioaccessibility is the release of a food component in the gastrointestinal tract (GIT) in an absorbable form [15,16]. Bioactivity is the specific effect resulting from exposure to a functional component. It includes tissue uptake and the resulting response. Different functional components are used in the food processing industry as previously described, and their bioavailability played a significant role in their development. The bioavailability of macronutrients is usually very high compared to vitamins, minerals, and bioactive compounds. The efficacy of functional components in providing functional benefits much depends on preserving their bioavailability [17,18]. Bioavailability depends on several factors such as droplet composition, size, interfacial properties, gastric residence time, low permeability, and solubility in the GIT. In the past few years, there were various methods employed for increasing the bioavailability of various functional components. In general, functional components get best absorbed by fully solubilizing in the GIT fluids. The essential steps necessary for the effective absorption of ingredients are explained by diagrammatic representation (Figure 10.2). Different functional components are getting attention due to their bioactivities and ability to reduce the risk of various diseases [19]. However, there are challenges in their bioavailability or sensibility to decomposition during processing and storage [20,21]. So, to overcome such barriers, they need to be delivered using nanomaterials for enhancing their bioavailability. Targeted delivery is one of the most critical issues in the fields of nanoencapsulation in food science. In addition to targeted delivery, they also protect during processing and storage. These nanomaterial-based delivery systems offer certain advantages over other types of

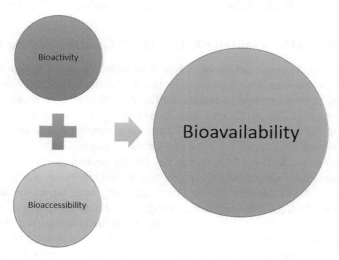

FIGURE 10.1   Bioavailability as a sum of bioaccessibility and bioactivity.

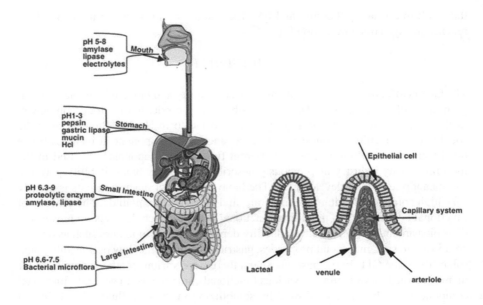

FIGURE 10.2 Diagrammatic representation of functional components' bioavailability pathway.

delivery systems, including clarity, more excellent stability and gravitational separation, and enhanced bioavailability [22]. Their smaller particle size also helps in faster penetration, direct absorption, improved water dispersibility, and optically transparent products [23]. Further enhancement in the bioavailability of encapsulated components may occur because of rapid hydrolysis than larger ones by digestive enzymes in the GIT. Nanoencapsulated components provide improved protection against degradation and interactions with other food ingredients, masking undesirable flavor profiles [24].

10.3 FABRICATION METHODS OF FOOD-GRADE NANOMATERIALS

Food-grade nanomaterials can be fabricated from different food components, including protein, polysaccharides, nucleic acids, lipids, phospholipids, and biosurfactants. There is a continuous increase in demand for the utilization of food-grade nanomaterials as they offer various advantages such as nontoxicity, biocompatibility, and biodegradable and can be easily modified [25]. Information on the nature of nanomaterial components used determines the type of fabrication method used. It is essential to identify the correct food-grade nanomaterial for the desired application. The fabrication methods used should be easy, reproducible, and cheap for industrial scale-up operation [22]. Nanomaterials can be fabricated using both top-down and bottom-up approaches. In top-down approaches, mechanical forces are applied, whereas, in bottom-up approaches, physicochemical and biological methods are used. Both systems differ in the amount of mechanical energy applied, or they can be used in combination according to the need [2]. However, high-energy methods are more suitable for several reasons as low emulsifiers need easy large-scale production and the already available equipment [26]. Both the methods are summarized diagrammatically in Figure 10.3.

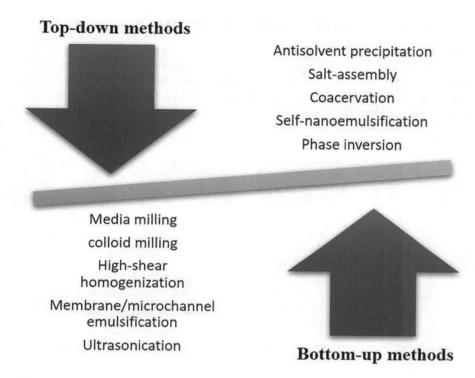

Top-down methods

Antisolvent precipitation

Salt-assembly

Coacervation

Self-nanoemulsification

Phase inversion

Media milling

colloid milling

High-shear
homogenization

Membrane/microchannel
emulsification

Ultrasonication

Bottom-up methods

FIGURE 10.3 An overview of some top-down and bottom-up methods used for fabrication of nanomaterials.

10.3.1 TOP-DOWN METHODS

High-energy methods, including high-pressure homogenizers, milling (media and colloidal), microchannel emulsification, and ultrasonicators, generate intense, disruptive forces and intermix the aqueous lipid phases, which results in the formation of fine lipid particles. Milling includes both media and colloidal used for various processes. High-pressure homogenization is widely used in the food industry to form nanoemulsions because of its possibility of scaling up and versatility. There are several high-pressure homogenizers, with high-pressure valve homogenizers and microfluidizers most widely used [26,27]. There are various edible coatings, edible films, starch nanoparticles, nanoemulsions, and lipid-based nanoparticles.

10.3.2 BOTTOM-UP METHODS

The second "bottom-up" approach is currently rapidly developing and widely used to fabricate nanomaterials because no specialized equipment is required. Spontaneous emulsification, emulsion inversion point, ligand binding, co-assembly, and antisolvent precipitation—such approaches are well developed today with reasonable control of particle size, size distribution, and in some cases, shape.

10.4 CLASSIFICATION OF FOOD-GRADE NANOMATERIALS

There are different types of nanomaterials used for the encapsulation of functional components such as lipid-based nanomaterials, nature-inspired nanomaterials, special equipment–based nanomaterials, and biopolymer nanomaterials. All are discussed below with schematic explanation in Figure 10.4 and diagrammatic explanation in Figure 10.5.

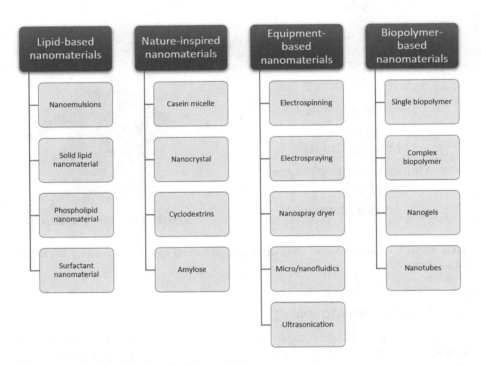

FIGURE 10.4   Classification of nanomaterial systems.

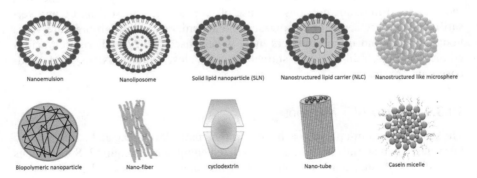

FIGURE 10.5   Nanomaterial types for functional component encapsulation.

10.4.1 Lipid-Based Nanomaterials

Lipid-based nanomaterials are the most promising technologies used in the food processing industry for encapsulation [3]. Despite the good functional performance of different carbohydrate and protein-based nanomaterials, they are challenging to scale up on an industrial scale. These lipophilic bioactive compounds such as triglycerides, carotenoids, tocopherols, flavonoids, polyphenols, and fat-soluble vitamins have various health benefits such as antioxidation, anti-inflammation, wound healing, and anticarcinogenic effects [28,29]. There are different categories of lipid-based delivery systems: nanoemulsion, solid lipid nanomaterials, phospholipid, and surfactant-based nanomaterials.

10.4.1.1 Nanoemulsions

Nanoemulsions are emulsions with droplet size on the order of 100 nm. They consist of a hydrophobic core comprising lipid molecules and a hydrophilic shell composed of surface-active molecules [30]. Different methods are used for preparations such as microfluidizers, ultrasonication, colloid milling, and low-energy methods such as phase inversion temperature or composition, emulsion inversion point, and spontaneous emulsification [31]. They are thermodynamically unstable but kinetically stable, making their formation a nonspontaneous process. Different forms of nanoemulsions can be used to encapsulate components: single, double, and Pickering emulsion. The choice between a simple nanoemulsion depends on, particularly the hydrophilicity or lipophilicity, and on the food matrix. Pickering emulsions are most stable, and their stability arises from the particle's adsorption energy at the interfacial layer. It is one of the main features that conventional surfactant-stabilized emulsions do not possess.

10.4.1.2 Solid Lipid Nanomaterials

Solid lipid nanoparticles (SLNs) and nanostructured lipid carriers (NLCs) are composed of lipids as the inner phase, emulsifiers, and water. Congealing is involved in their formation. They are classified into three categories: SLNs, NLCs, and nanoorganogels [3,32]. There are various top-down technologies like high-pressure homogenization, microfluidization, and membrane contact methods used. Among the bottom-up approaches, emulsification solvent evaporation, emulsification solvent diffusion, antisolvent precipitation, and supercritical techniques are mostly used. SLNs, despite many advantages, have many limitations too such as low-loading capacity and particle aggregation and expel few bioactive compounds. These limitations gave birth to the development of NLCs, the second generation of SLNs. In NLCs, the matrix combines solid and liquid lipids that accommodate more bioactive compounds due to the increased space between solid-state lipids. They provide better loading capacity, better stability, higher efficiency, and more bioavailability. Nanooleogels, colloidal dispersion of gelled lipid nanoparticles, have been used for different applications in recent years because of their exclusive thermal, rheological, and nutritional properties, especially in the food processing and pharmaceutical industries [33].

10.4.1.3 Phospholipid Nanomaterials

Among lipid-based delivery systems, nanomaterials based on phospholipids are among the most commonly used for encapsulation, preservation, and sustained functional component release [34].

10.4.1.3.1 Nanoliposomes

Nanoliposomes are produced from surfactants with average hydrophilic–lipophilic balance ratios in the food processing industry. The most commonly used are prepared from sources such as eggs, soybeans, sunflower, and milk. Thin-film hydration, reverse-phase evaporation, microfluidization, solvent/surfactant displacement, homogenization/sonication, and extrusion are used for their formation.

10.4.1.3.2 Nanophytosomes

Phytosomes are made of phospholipids and phytoactive components by making H-bonds at their polar parts. Their molar ratios are either 1:1 or 2:1 (phospholipid:bioactive), and these differences make them better in absorption and bioavailability than liposomes [35].

10.4.1.4 Surfactant Structures

Surfactants are amphiphilic molecules having two distinct regions of attraction. Self-assembly in aqueous media results in different nanosystems such as niosomes, cubosomes, hexosomes, etc.

10.4.1.4.1 Niosomes

Niosomes are made up of nonionic surfactants because of self-assembly in aqueous media, with structural similarity to liposomes [33]. They are a result of unfavorable interactions between surfactants and aqueous media, resulting in entrapment [36].

10.4.1.4.2 Cubosomes and Hexosome

Cubosomes and hexosomes are present in nonlamellar liquid crystalline nanoparticles resulting from self-assembly, which can be used for encapsulation. These possess unique structural properties and high interfacial area. All the nanobased lipid nanomaterials are summarized in Table 10.1.

10.4.2 NATURE-INSPIRED NANOMATERIALS

Natural biopolymers in the food processing industry have increased in the past few years [33]. These are nontoxic, biodegradable, metabolizable, sustainable, and generally recognized as safe status. Caseins are most studied as vehicles for functional component encapsulation. All these biopolymers are summarized in Table 10.2.

10.4.2.1 Casein Micelle

Caseins consist of about 80% of proteins. There are four types ($\alpha 1$, $\alpha 2$, β, and k) of caseins present in the form of casein micelles with about 104 casein molecules and a diameter ranging from 40 to 300 nm [50]. It is amphiphilic and surface-active and acts as a stabilizer [51]. Due to self-assembly, it forms micelles and aggregates by lowering the pH. It is inexpensive, readily available, biodegradable, highly stable, and safe and having good physicochemical properties. Casein is used in conjugation with other polymers, such as carbohydrates and silica, to formulate nanomaterials for the delivery of folic acid [52], curcumin [53], vitamin D3 [54], and doxorubicin [55].

TABLE 10.1

Examples of Encapsulation Using Lipid-Based Nanomaterials

Type of Lipid Nanomaterials	Functional Components	Log P	Molecular Weight (g/mol)	Water Solubility 20°C (mg/L)	Melting Point (°C)	Ref
Nanoemulsion	Thymol	3.3	150.22	900	49–51	[37]
	D-limonene	3.2	136.23	Insoluble	−74.35	[38]
	Ellagic acid	2.32	302.19	820	>360	[39]
Solid lipid nanoparticles	EGCG	2.38	458.372	73	140–142	[40]
Nanostructured lipid carriers	β-Carotene	14.764	536.87	Insoluble	180	[41]
	Curcumin	3.0	368.38	3.12	183	[42]
	Lycopene	11.93	536.88	Insoluble	175	[43]
	β-Sitosterol	7.27–7.84	414.71	Insoluble	136	[44]
Nanoliposomes	Vitamin C	−1.85	176.12	330g/L	190	[45]
	Vitamin K_1	9.3	450	Insoluble	−20	[46]
Nanophytosomes	Garlic essential oil	1.99	488.9	Insoluble	−85	[47]
	Cinnamon essential oil	1.9	132.16	Insoluble	−7.5	[48]
Niosomes	α-Tocopherol	10.42	430.71	Insoluble	2.5–3.5	[49]

10.4.2.2 Nanocrystal

Nanocrystals are crystalline materials with a size <100 nm. They improve the solubility, thereby increasing the functional component's dissolution rate *in vivo* [9]. They may be utilized as carriers or maybe the bioactive agent itself [3]. The advantage of nanocrystals is that they have 100% loading. Another advantage is that they can be prepared without organic solvents. They have shown improved solubility and improved cellular interaction of the bioactive in nanocrystal form [9]. They have potential applications for the nanoencapsulation of various nutraceuticals, such as flavonoids, silybin, and breviscapine [39]. They are already commercially available as aqueous-based suspensions of multiple drugs such as rifampicin [56], fenofibrate [57], and nimodipine [58].

10.4.2.3 Cyclodextrin

β-Cyclodextrins are cyclic oligosaccharides and have a hydrophilic surface with a hydrophobic cavity. They form inclusion complexes with lipophilic compounds due to self-assembly. It is used in the food processing industry in various applications like encapsulation of quercetin [52] and resveratrol [59], Pickering emulsions [60], volatile essential oils [61], cinnamon essential oils [62], and active packaging of foods [63]. They stabilize drugs, reduce irritation, and enhance drug solubility and permeability by limiting the free drug concentration [6].

10.4.2.4 Amylose

Amylose, a linear component of starch, an inexpensive, abundant, and biocompatible material, has various food processing industry applications as carriers of different bioactive compounds [3]. They can be formed from multiple fabrication methods, including conventional methods, nanoprecipitation, microfluidization, ultrasound-assisted, enzymatic, homogenization, and acidification of alkaline solutions. Various bioactive compounds like curcumin [64], limonene [65], p-coumaric acid [66], ascorbyl palmitate [67], and ß-carotene [68] can be trapped inside the amylose helix.

10.4.3 SPECIAL EQUIPMENT–BASED NANOMATERIAL FORMULATION

Many of the techniques used for nanoencapsulation of food ingredients are based on the knowledge of some well-known equipment such as high-pressure homogenizers, ultrasonication devices, etc. The formation of nanomaterial using top-down methods with the help of specialized types of equipment includes electrospinning, electrospraying, nanospray drying, micro/nanofluidics, high-pressure homogenization, microfluidization, and ultrasonication [33]. Electrospinning is a simple process that passes the functional component solution from a syringe needle toward the collector resulting in the generation of electrospun fibers. It is used for various functional component encapsulation such as essential oil [77], dextran [78], gallic acid [79], and ferulic acid [80]. Electrospraying is a new technique similar to electrospinning, but it generates the nanoparticles instead of nanofibers with enhanced encapsulation used for olive leaf phenolics [81], d-limonene [82], fish oil [83], β-carotene [84], tamoxifen [85], curcumin [86], eugenol [21], hydroxytyrosol [87], saffron extract [88], and coffee bean oil [89]. Nanospray dryer is a relatively new technique and highly efficient in generating fine nanoparticles as compared with conventional dryers.

10.4.4 BIOPOLYMER NANOMATERIALS

Biopolymer nanomaterials are submicron particles, which can be used for nanoencapsulation of functional components. They form different nanomaterials, including nanoparticles, hollow particles, nanogels, nanotubular structures, and nanocomplexes. They can be fabricated from a single biopolymer of proteins (and carbohydrates) or from complex biopolymers using desolvation and precipitation methods. There are different forms like single or complex biopolymers, nanogels, and nanotubes. They can be produced from polysaccharides like chitosan, sodium alginate, and soyabean polysaccharide [2]. Proteins such as zein [90], milk protein concentrate [91], and whey protein isolate [92] are commonly used. Complex polymers are used for thymol [93] and curcumin [12] encapsulation. They provide protection, encapsulation, and delivery of various functional components. These are biocompatible, biodegradable, and safe, and, in some cases, show antibacterial properties [94]. Nanogels are open networks of physiochemically cross-linked biopolymers that trap appreciable amounts of solvent. They exhibit good physicochemical and mechanical properties, just like hydrogels [95]. They can be prepared by both proteins and polysaccharides [96]. They provide promising encapsulation and controlled release for curcumin [97], levodopa [98], and antimicrobial peptides [99], with the

TABLE 10.2

Summary of Functional Component Encapsulation Using Natural Biopolymers

Natural Biopolymers	Functional Components	Encapsulation Method	Wall Material Ingredients	Outcome	Ref
Casein micelles	Curcumin	Ultrasound	Casein amorphous calcium phosphate nanoparticles	Excellent platform for curcumin delivery	[53]
	Doxorubicin	Low cost and facile	Casein nanoparticles	Nontoxic drug delivery	[55]
	Vitamin D3	Hydrophobic interaction	Reassembled casein micelle	Stabilization and uniform dispersion within aqueous food and drinks	[54]
	Quercetin, curcumin	Self-assembly	Reassembled casein micelle and casein nanoparticles	Hydrophobic components in fat-free clear beverages	[69]
Nanocrystal	Quercetin	HPP	Quercetin nanocrystals	Enhancement of oral bioavailability	[70]
	Silybin	HPP	Silybin nanocrystals	Enhancement of oral bioavailability	[71]
Cyclodextrin	Piperine	Freeze-drying	β-Cyclodextrin complexes	Bioaccessibility and antioxidant activity improved	[72]
	Danube common nase (Chondrostoma nasus L.) oil	Kneading method	B-Cyclodextrin and 2-hydroxypropyl-β-cyclodextrin	ω-3-based complexes for functional foods	[73]
Amylose	Garlic bioactive components	Milling method	Amylose complexes	Stable natural flavor compound systems	[74]
	α-Lipoic acid	Simple blending	Octenylsuccinic anhydride-modified high-amylose starch	Efficient carrier of oral delivery	[75]
	Chia seed oil	Alkaline method with modifications	Amylose inclusion complex	Better incorporation of ω-3 and ω-6 fatty acids	[76]

addition of multistimuli-responsive functional property. In comparison, nanotubes have a tubular shape similar to carbon tubes [100]. Their solid and flexible structure makes them suitable as a nanomaterial. The formation of protein nanotubes from milk α-lactoglobulin and whey proteins, such as β-lactoglobulin and bovine serum albumin, has been reported for better delivery of curcumin [95,101]. Halloysite and cyclodextrin are used for the nanocomposite formation for antimicrobial clotrimazole delivery [102]. Researchers developed halloysite (with polylactic acid)-based nanotubes for extending the shelf life of cherry tomatoes [103]. Similarly, halloysite (with essential peppermint oil or *Origanum vulgare* essential oil) is used for developing film with thermosensitive, antioxidant, and antimicrobial activity [104,105].

10.5 SAFETY OF NANOMATERIALS APPLIED IN A FOOD MATRIX

Bionanotechnology has offered various novel opportunities in the food industry. Selected nanomaterials help us to enhance the properties and the potency of components added during food processing, storage, and delivery. Nanomaterial safety is determined by its biocompatibility, biodegradability, and its absorption, distribution, metabolism, and excretion (ADME) profile. Scientists believe that nanomaterials can be metabolized, excreted, or accumulated within specific tissues because of their small size. The risks of nanomaterials can be checked using *in vitro* and *in vivo* studies on cell lines and animal models. However, clinical trials have not yet been carried on the human body, which needs to be investigated [13]. There is little knowledge of the effect of nanomaterials on human health and the environment [106]. Nanomaterials provide an advantage of encapsulation, but they can cause some problems when present at high levels. So, there is a need for interdisciplinary studies for their optimum amount to be present in foods for improving the bioavailability of functional components. These safety studies will be a boom for food processing studies, making the "nanofunctional foods" scale up with improved bioavailability. This will end a gap between the demand and supply of functional foods for the customers' growing needs.

10.6 CONCLUSION AND PROSPECTS

Nanomaterials offer various physicochemical advantages for improved bioavailability and stability of components. Functional component properties, along with delivery purpose, determine the type of nanomaterial to be used. The matrix properties will further define the nanobiological interactions and ADME profile. The functional components' fate dramatically depends on where it is released and its chemical, physical, and morphological properties. Functional components' release can be customized using nanomaterials with specific properties. However, commercialization of nanofunctional foods is still in the nascent stage. Regulations on the use of nanomaterials in the food processing industry are not structured or framed. International and national bodies should increase initiatives to control, regulate, and promote proper development and utilization of nanomaterials in the food processing industry. Meanwhile, more studies should be carried out to analyze these nanomaterials' toxicological and safety aspects. In a nutshell, bionanotechnology has revolutionized the food processing industry and will open new horizons.

ACKNOWLEDGMENTS

The authors would like to thank the National Institute of Food Technology Entrepreneurship and Management, Sonipat, Haryana, India, for providing all the required facilities during the preparation of this manuscript.

REFERENCES

1. Shafiq, M., S. Anjum, C. Hano, I. Anjum, and B. H. Abbasi. 2020. "An overview of the applications of nanomaterials and nanodevices in the food industry." *Foods* 9 (2): 148. Doi:10.3390/foods9020148.
2. Pan, K., and Q. Zhong. 2016. "Organic nanoparticles in foods: Fabrication, characterization, and utilization." *Annual Review of Food Science and Technology* 7(1): 245–66. Doi: 10.1146/annurev-food-041715-033215.
3. Jafari, S. M., and D. J. McClements. 2017. "Nanotechnology approaches for increasing nutrient bioavailability." In *Advances in Food and Nutrition Research* 81:1–30 Academic Press Inc. Doi: 10.1016/bs.afnr.2016.12.008.
4. Oehlke, K., M. Adamiuk, D. Behsnilian, V. Gräf, E. Mayer-Miebach, E. Walz, and R. Greiner. 2014. "Potential bioavailability enhancement of bioactive compounds using food-grade engineered nanomaterials: A review of the existing evidence." *Food and Function* 5(7). Doi: 10.1039/c3fo60067j.
5. Anez-Bustillos, L., D. T. Dao, G. L. Fell, M. A. Baker, K. M. Gura, B. R. Bistrian, and M. Puder. 2018. "Redefining essential fatty acids in the era of novel intravenous lipid emulsions." *Clinical Nutrition* 37(3): 784–9. Doi: 10.1016/j.clnu.2017.07.004.
6. Fernández-García, E., I. Carvajal-Lérida, M. Jarén-Galán, J. Garrido-Fernández, A. Pérez-Gálvez, and D. Hornero-Méndez. 2012. "Carotenoids bioavailability from foods: From plant pigments to efficient biological activities." *Food Research International* 46 (2): 0–450. Doi: 10.1016/j.foodres.2011.06.007.
7. Sanchez, B. A. O., S. M. C. Celestino, M. B. de Abreu Gloria, I. C. Celestino, M. I. O. Lozada, S. D. A. Júnior, E. R. de Alencar, and L. d. L. de Oliveira. 2020. "Pasteurization of passion fruit passiflora setacea pulp to optimize bioactive compounds retention." *Food Chemistry X* 6: 100084. Doi: 10.1016/j.fochx.2020.100084.
8. Gligor, O., A. Mocan, C. Moldovan, M. Locatelli, G. Cri an, and I. C. F. R. Ferreira. 2019. "Enzyme-assisted extractions of polyphenols – A comprehensive review." *Trends in Food Science and Technology* 88: 302–15. Doi: 10.1016/j.tifs.2019.03.029.
9. Borel, T., and C. M. Sabliov. 2014. "Nanodelivery of bioactive components for food applications: Types of delivery systems, properties, and their effect on ADME profiles and toxicity of nanoparticles." *Annual Review of Food Science and Technology* 5 (1): 197–213. Doi: 10.1146/annurev-food-030713-092354.
10. Tonolo, F., A. Folda, L. Cesaro, V. Scalcon, O. Marin, S. Ferro, A. Bindoli, and M. P. Rigobello. 2020. "Milk-derived bioactive peptides exhibit antioxidant activity through the Keap1-Nrf2 signaling pathway." *Journal of Functional Foods* 64: 103696. Doi: 10.1016/j.jff.2019.103696.
11. Jugreet, B. S., S. Suroowan, R. R. Kannan Rengasamy, and M. F. Mahomoodally. 2020. "Chemistry, bioactivities, mode of action and industrial applications of essential oils." *Trends in Food Science and Technology*. Doi: 10.1016/j.tifs.2020.04.025.
12. Esfanjani, F. A., and S. M. Jafari. 2016. "Biopolymer nano-particles and natural nano-carriers for nano-encapsulation of phenolic compounds." *Colloids and Surfaces B: Biointerfaces* 146: 532–543. Doi: 10.1016/j.colsurfb.2016.06.053.
13. Katouzian, I., and S. M. Jafari. 2016. "Nano-encapsulation as a promising approach for targeted delivery and controlled release of vitamins." *Trends in Food Science and Technology* 53: 34–48. Doi: 10.1016/j.tifs.2016.05.002.

14. FDA. 2002. *Bioavailability and Bioequivalence Studies for Orally Administered Drug Products d General Considerations*. Rockville, MD: F. A. D. Administration, U.S. Department of Health and Human Services, Center for Drug Evaluation and Research, p. 27.
15. Marze, S. 2013. "Bioaccessibility of nutrients and micronutrients from dispersed food systems: Impact of the multiscale bulk and interfacial structures." *Critical Reviews in Food Science and Nutrition* 53 (1): 76–198. Doi: 10.1080/10408398.2010.525331.
16. Carbonell-Capella, J. M., M. Buniowska, F. J. Barba, M. J. Esteve, and A. Frígola. 2014. "Analytical methods for determining bioavailability and bioaccessibility of bioactive compounds from fruits and vegetables: A review." *Comprehensive Reviews in Food Science and Food Safety* 13 (2): 155–71. Doi: 10.1111/1541-4337.12049.
17. Gonçalves, R. F. S., J. T. Martins, C. M. M. Duarte, A. A. Vicente, and A. C. Pinheiro. 2018. "Advances in nutraceutical delivery systems: From formulation design for bioavailability enhancement to efficacy and safety evaluation." *Trends in Food Science and Technology* 78: 270–291. Doi: 10.1016/j.tifs.2018.06.011.
18. Ting, Y., Y. Jiang, C. T. Ho, and Q. Huang. 2014. "Common delivery systems for enhancing bioavailability and biological efficacy of nutraceuticals." *Journal of Functional Foods* 7: 112–28. Doi: 10.1016/j.jff.2013.12.010.
19. Bao, C., P. Jiang, J. Chai, Y. Jiang, D. Li, W. Bao, B. Liu, B. Liu, W. Norde, and Y. Li. 2019. "The delivery of sensitive food bioactive ingredients: Absorption mechanisms, influencing factors, encapsulation techniques and evaluation models." *Food Research International* 120: 130–40. Doi: 10.1016/j.foodres.2019.02.024.
20. Joye, I. J. 2019. "Cereal biopolymers for nano- and microtechnology: A myriad of opportunities for novel (functional) food applications." *Trends in Food Science and Technology* 83: 1–11. Doi: 10.1016/j.tifs.2018.10.009.
21. Veneranda, M., Q. Hu, T. Wang, Y. Luo, K. Castro, and J. M. Madariaga. 2018. "Formation and characterization of zein-caseinate-pectin complex nanoparticles for encapsulation of Eugenol." *LWT - Food Science and Technology* 89: 596–603. Doi: 10.1016/j.lwt.2017.11.040.
22. Zou, L., B. Zheng, R. Zhang, Z. Zhang, W. Liu, C. Liu, H. Xiao, and D. J. McClements. 2016. "Food-grade nanoparticles for encapsulation, protection and delivery of curcumin: Comparison of lipid, protein, and phospholipid nanoparticles under simulated gastrointestinal conditions." *RSC Advances* 6 (4): 3126–36. Doi: 10.1039/c5ra22834d.
23. Artiga-Artigas, M., I. Odriozola-Serrano, G. Oms-Oliu, and O. Martín-Belloso. 2018. "Nanostructured systems to increase bioavailability of food ingredients." In *Nanomaterials for Food Applications*, 13–33. Elsevier. Doi: 10.1016/B978-0-12-814130-4.00002-6.
24. Bourbon, A. I., M. A. Cerqueira, and A. A. Vicente. 2016. "Encapsulation and controlled release of bioactive compounds in lactoferrin-glycomacropeptide nanohydrogels: Curcumin and caffeine as model compounds." *Journal of Food Engineering* 180: 110–19. Doi: 10.1016/j.jfoodeng.2016.02.016.
25. Yang, J., S. Han, H. Zheng, H. Dong, and J. Liu. 2015. "Preparation and application of micro/nanoparticles based on natural polysaccharides." *Carbohydrate Polymers* 123: 53–66. Doi: 10.1016/j.carbpol.2015.01.029.
26. Salvia-Trujillo, L., R. Soliva-Fortuny, M. A. Rojas-Graü, D. Julian McClements, and O. Martín-Belloso. 2017. "Edible nanoemulsions as carriers of active ingredients: A review." *Annual Review of Food Science and Technology* 8: 439–66. Doi: 10.1146/annurev-food-030216-025908.
27. Schuh, R. S., É. Poletto, F. N. S. Fachel, U. Matte, G. Baldo, and H. F. Teixeira. 2018. "Physicochemical properties of cationic nanoemulsions and liposomes obtained by microfluidization complexed with a single plasmid or along with an oligonucleotide: Implications for CRISPR/Cas technology." *Journal of Colloid and Interface Science* 530: 243–55. Doi: 10.1016/j.jcis.2018.06.058.

28. Shin, G. H., J. T. Kim, and H. J. Park. 2015. "Recent developments in nanoformulations of lipophilic functional foods." *Trends in Food Science and Technology* 46 (1): 144–57. Doi: 10.1016/j.tifs.2015.07.005.

29. Assadpour, E., and S. M. Jafari. 2018. "Nanoencapsulation: Techniques and developments for food applications." In *Nanomaterials for Food Applications*, 35–61. Elsevier. Doi: 10.1016/B978-0-12-814130-4.00003-8.

30. Yao, M., H. Xiao, and D. J. McClements. 2014. "Delivery of lipophilic bioactives: Assembly, disassembly, and reassembly of lipid nanoparticles." *Annual Review of Food Science and Technology* 5 (1): 53–81. Doi: 10.1146/annurev-food-072913-100350.

31. Assadpour, E., and S. M. Jafari. 2019. "An overview of specialized equipment for nanoencapsulation of food ingredients." In *Nanoencapsulation of Food Ingredients by Specialized Equipment*, 1–30. Academic Press. Doi: 10.1016/b978-0-12-815671-1.00001-9.

32. Aditya, N. P., and S. Ko. 2015. "Solid lipid nanoparticles (SLNs): Delivery vehicles for food bioactives." *RSC Advances* 5 (39): 30902–11. Doi: 10.1039/c4ra17127f.

33. Assadpour, E., and S. M. Jafari. 2019. "Nanoencapsulation. Nanomaterials for food applications." In *Nanomaterials for Food Applications*, 35–61. Elsevier. Doi: 10.1016/b978-0-12-814130-4.00003-8.

34. Ghorbanzade, T., S. M. Jafari, S. Akhavan, and R. Hadavi. 2017. "Nano-encapsulation of fish oil in nano-liposomes and its application in fortification of yogurt." *Food Chemistry* 216: 146–152. Doi: 10.1016/j.foodchem.2016.08.022.

35. Babazadeh, A., M. Zeinali, and H. Hamishehkar. 2017. "Nano-phytosome: A developing platform for herbal anti-cancer agents in cancer therapy." *Current Drug Targets* 18 (999): 1–1. Doi: 10.2174/1389450118666170508095250.

36. Moghassemi, S., and A. Hadjizadeh. 2014. "Nano-niosomes as nanoscale drug delivery systems: An illustrated review." *Journal of Controlled Release* 185: 22–36. Doi: 10.1016/j.jconrel.2014.04.015.

37. Li, X., X. Yang, H. Deng, Y. Guo, and J. Xue. 2020. "Gelatin films incorporated with thymol nanoemulsions: Physical properties and antimicrobial activities." *International Journal of Biological Macromolecules* 150: 161–68. Doi: 10.1016/j.ijbiomac.2020.02.066.

38. Feng, J., Rong Wang, Z. Chen, S. Zhang, S. Yuan, H. Cao, S. M. Jafari, and W. Yang. 2020. "Formulation optimization of D-limonene-loaded nanoemulsions as a natural and efficient biopesticide." *Colloids and Surfaces A: Physicochemical and Engineering Aspects* 596: 124746. Doi: 10.1016/j.colsurfa.2020.124746.

39. Wang, S. T., C. T. Chou, and N. W. Su. 2017. "A food-grade self-nanoemulsifying delivery system for enhancing oral bioavailability of ellagic acid." *Journal of Functional Foods* 34: 207–15. Doi: 10.1016/j.jff.2017.04.033.

40. Shtay, R., J. K. Keppler, K. Schrader, and K. Schwarz. 2019. "Encapsulation of (−)-epigallocatechin-3-gallate (EGCG) in solid lipid nanoparticles for food applications." *Journal of Food Engineering* 244: 91–100. Doi: 10.1016/j.jfoodeng.2018.09.008.

41. Pezeshki, A., H. Hamishehkar, B. Ghanbarzadeh, I. Fathollahy, F. K. Nahr, M. K. Heshmati, and M. Mohammadi. 2019. "Nanostructured lipid carriers as a favorable delivery system for β-carotene." *Food Bioscience* 27: 11–17. Doi: 10.1016/j.fbio.2018.11.004.

42. Chuacharoen, T., and C. M. Sabliov. 2019. "Comparative effects of curcumin when delivered in a nanoemulsion or nanoparticle form for food applications: Study on stability and lipid oxidation inhibition." *LWT* 113: 108319. Doi: 10.1016/j.lwt.2019.108319.

43. Liang, X., C. Ma, X. Yan, X. Liu, and F. Liu. 2019. "Advances in research on bioactivity, metabolism, stability and delivery systems of lycopene." *Trends in Food Science and Technology*. 93: 185–196. Doi: 10.1016/j.tifs.2019.08.019.

44. Soleimanian, Y., S. A. H. Goli, J. Varshosaz, and F. Maestrelli. 2018. "Propolis wax nanostructured lipid carrier for delivery of β sitosterol: Effect of formulation variables on physicochemical properties." *Food Chemistry* 260: 97–105. Doi: 10.1016/j.foodchem.2018.03.145.

45. Hamadou, A. H., W. C. Huang, C. Xue, and X. Mao. 2020. "Formulation of vitamin C encapsulation in marine phospholipids nanoliposomes: Characterization and stability evaluation during long term storage." *LWT* 127: 109439. Doi: 10.1016/j.lwt.2020.109439.
46. Samadi, N., P. A. Azar, S. W. Husain, H. I. Maibach, and S. Nafisi. 2020. "Experimental design in formulation optimization of vitamin K1 oxide-loaded nanoliposomes for skin delivery." *International Journal of Pharmaceutics* 579: 119136. Doi:10.1016/j.ijpharm.2020.119136.
47. Nazari M., B. Ghanbarzadeh, H. Samadi Kafil, M. Zeinali, H. Hamishehkar. (2019a). "Garlic essential oil nanophytosomes as a natural food preservative: Its application in yogurt as food model." *Colloid and Interface Science Communications* 30: 100176. https://doi.org/10.1016/j.colcom.2019.100176
48. Nazari, M., H. Majdi, M. Milani, S. Abbaspour-Ravasjani, H. Hamishehkar, and L. T. Lim. 2019. "Cinnamon nanophytosomes embedded electrospun nanofiber: Its effects on microbial quality and shelf-life of shrimp as a novel packaging." *Food Packaging and Shelf Life* 21: 100349. Doi: 10.1016/j.fpsl.2019.100349.
49. Basiri, L., G. Rajabzadeh, and A. Bostan. 2017. "α-tocopherol-loaded niosome prepared by heating method and its release behavior." *Food Chemistry* 221: 620–28. Doi: 10.1016/j.foodchem.2016.11.129.
50. Xu, G., L. Li, X. Bao, and P. Yao. 2020. "Curcumin, casein and soy polysaccharide ternary complex nanoparticles for enhanced dispersibility, stability and oral bioavailability of curcumin." *Food Bioscience* 35. Doi: 100569doi:10.1016/j.fbio.2020.100569.
51. Chen, H., H. Wooten, L. Thompson, and K. Pan. 2019. "Nanoparticles of casein micelles for encapsulation of food ingredients." In *Biopolymer Nanostructures for Food Encapsulation Purposes*, 39–68. Doi: 10.1016/B978-0-12-815663-6.00002-1.
52. Penalva, R., I. Esparza, M. Agüeros, C.J. Gonzalez-Navarro, C. Gonzalez-Ferrero, and J. M. Irache. 2015. "Casein nanoparticles as carriers for the oral delivery of folic acid." *Food Hydrocolloids* 44: 399–406. Doi: 10.1016/j.foodhyd.2014.10.004.
53. Niu, B., J. Guo, X. Guo, X. Sun, C. Rao, C. Liu, J. Zhang, C. Zhang, Y. Y. Fan, and W. Li. 2020. "(NaPO3)6-assisted formation of dispersive casein-amorphous calcium phosphate nanoparticles: An excellent platform for curcumin delivery." *Journal of Drug Delivery Science and Technology* 55: 101412. Doi: 10.1016/j.jddst.2019.101412.
54. Cohen, Y., S. Ish-Shalom, E. Segal, O. Nudelman, A. Shpigelman, and Y. D. Livney. 2017. "The bioavailability of vitamin D3, a model hydrophobic nutraceutical, in casein micelles, as model protein nanoparticles: Human clinical trial results." *Journal of Functional Foods* 30: 321–25. Doi: 10.1016/j.jff.2017.01.019.
55. Gandhi, S., and I. Roy. 2019. "Doxorubicin-loaded casein nanoparticles for drug delivery: Preparation, characterization and in vitro evaluation." *International Journal of Biological Macromolecules* 121: 6–12. Doi: 10.1016/j.ijbiomac.2018.10.005.
56. Melo, K. J. C., M. A. B. Henostroza, R. Löbenberg, and N. A. Bou-Chacra. 2020. "Rifampicin nanocrystals: Towards an innovative approach to treat tuberculosis." *Materials Science and Engineering C* 112: 110895. Doi: 10.1016/j.msec.2020.110895.
57. Ige, P. P., R. K. Baria, and S. G. Gattani. 2013. "Fabrication of fenofibrate nanocrystals by probe sonication method for enhancement of dissolution rate and oral bioavailability." *Colloids and Surfaces B: Biointerfaces* 108: 366–73. Doi: 10.1016/j.colsurfb.2013.02.043.
58. Fu, Q., J. Sun, X. Ai, P. Zhang, M. Li, Y. Wang, L. Xiaohong, et al. 2013. "Nimodipine nanocrystals for oral bioavailability improvement: Role of mesenteric lymph transport in the oral absorption." *International Journal of Pharmaceutics* 448 (1): 290–97. Doi: 10.1016/j.ijpharm.2013.01.065.
59. Qiu, C., D. J. McClements, Z. Jin, Y. Qin, Y. Hu, X. Xu, and J. Wang. 2020. "Resveratrol-loaded core-shell nanostructured delivery systems: Cyclodextrin-based metal-organic nanocapsules prepared by ionic gelation." *Food Chemistry* 317: 126328. Doi: 10.1016/j.foodchem.2020.126328.

60. Hu, Y., C. Qiu, Z. Jin, Y. Qin, C. Zhan, X. Xu, and J. Wang. 2020. "Pickering emulsions with enhanced storage stabilities by using hybrid β-cyclodextrin/short linear glucan nanoparticles as stabilizers." *Carbohydrate Polymers* 229: 115418. Doi: 10.1016/j.carbpol.2019.115418.

61. Pires, F. Q., J. K. R. da Silva, L. L. Sa-Barreto, T. Gratieri, G. M. Gelfuso, and M. Cunha-Filho. 2019. "Lipid nanoparticles as carriers of cyclodextrin inclusion complexes: A promising approach for cutaneous delivery of a volatile essential oil." *Colloids and Surfaces B: Biointerfaces* 182: 110382. Doi: 10.1016/j.colsurfb.2019.110382.

62. Matshetshe, K. I., S. Parani, S. M. Manki, and O. S. Oluwafemi. 2018. "Preparation, Characterization and in vitro release study of β-cyclodextrin/chitosan nanoparticles loaded cinnamomum zeylanicum essential oil." *International Journal of Biological Macromolecules* 118: 676–82. Doi: 10.1016/j.ijbiomac.2018.06.125.

63. Andrade-Del Olmo, J., L. Pérez-Álvarez, E. Hernáez, L. Ruiz-Rubio, and J. L. Vilas-Vilela. 2019. "Antibacterial multilayer of chitosan and (2-Carboxyethyl)- β-cyclodextrin onto polylactic acid (PLLA)." *Food Hydrocolloids* 88: 228–36. Doi: 10.1016/j.foodhyd.2018.10.014.

64. Liu, C. H., G. W. Lee, W. C. Wu, and C. C. Wang. 2020. "Encapsulating curcumin in ethylene diamine-β-cyclodextrin nanoparticle improves topical cornea delivery." *Colloids and Surfaces B: Biointerfaces* 186: 110726. Doi: 10.1016/j.colsurfb.2019.110726.

65. Ganje, M., S. M. Jafari, A.M. Tamadon, M. Niakosari, and Y. Maghsoudlou. 2019. "Mathematical and fuzzy modeling of limonene release from amylose nanostructures and evaluation of its release kinetics." *Food Hydrocolloids* 95: 186–94. Doi: 10.1016/j.foodhyd.2019.04.045.

66. Wang, S., L. Kong, Y. Zhao, L. Tan, J. Zhang, Z. Du, and H. Zhang. 2019. "Lipophilization and molecular encapsulation of P-coumaric acid by amylose inclusion complex." *Food Hydrocolloids* 93: 270–75. Doi: 10.1016/j.foodhyd.2019.02.044.

67. Bamidele, O. P., and M. N. Emmambux. 2019. "Storage stability of encapsulated ascorbyl palmitate in normal and high amylose maize starches during pasting and spray dryin." *Carbohydrate Polymers* 216: 217–23. Doi: 10.1016/j.carbpol.2019.04.022.

68. Kong, L., R. Bhosale, and G. R. Ziegler. 2018. "Encapsulation and stabilization of β-carotene by amylose inclusion complexes." *Food Research International* 105: 446–52. Doi: 10.1016/j.foodres.2017.11.058.

69. Ghayour, N., S. M. H. Hosseini, M. H. Eskandari, S. Esteghlal, A. R. Nekoei, H. H. Gahruie, M. Tatar, and F. Naghibalhossaini. 2019. "Nanoencapsulation of quercetin and curcumin in casein-based delivery systems." *Food Hydrocolloids* 87: 394–403. Doi: 10.1016/j.foodhyd.2018.08.031.

70. Karadag, A., B. Ozcelik, and Q. Huang. 2014. "Quercetin nanosuspensions produced by high-pressure homogenization." *Journal of Agricultural and Food Chemistry* 62 (8): 1852–59. Doi: 10.1021/jf404065p.

71. Yi, T., C. Liu, J. Zhang, F. Wang, J. Wang, and J. Zhang. 2017. "A new drug nanocrystal self-stabilized pickering emulsion for oral delivery of silybin." *European Journal of Pharmaceutical Sciences* 96: 420–27. Doi: 10.1016/j.ejps.2016.08.047.

72. Quilaqueo, M., S. Millao, I. Luzardo-Ocampo, R. Campos-Vega, F. Acevedo, C. Shene, and M. Rubilar. 2019. "Inclusion of piperine in β-cyclodextrin complexes improves their bioaccessibility and in vitro antioxidant capacity." *Food Hydrocolloids* 91: 143–52. Doi: 10.1016/j.foodhyd.2019.01.011.

73. Hădărugă, N. G., R. N. Szakal, C. A. Chirilă, A. T. Lukinich-Gruia, V. Păunescu, C. Muntean, G. Rusu, G. Bujancă, and D. I. Hădărugă. 2020. "Complexation of danube common nase (Chondrostoma Nasus L.) oil by β-cyclodextrin and 2-hydroxypropyl-β-cyclodextrin." *Food Chemistry* 303: 125419. Doi: 10.1016/j.foodchem.2019.125419.

74. Zhang, L., P. Guan, Z. Zhang, Y. Dai, and L. Hao. 2018. "Physicochemical characteristics of complexes between amylose and garlic bioactive components generated by milling activating method." *Food Research International* 105: 499–506. Doi: 10.1016/j.foodres.2017.11.068.

75. Li, Y. X., Y. J. Kim, C. Koteswara Reddy, S. J. Lee, and S. T. Lim. 2019. "Enhanced bioavailability of alpha-lipoic acid by complex formation with octenylsuccinylated high-amylose starch." *Carbohydrate Polymers* 219: 39–45. Doi: 10.1016/j.carbpol.2019.04.082.

76. Di Marco, A. E., V. Y. Ixtaina, and M. C. Tomás. (2020). Inclusion complexes of high amylose corn starch with essential fatty acids from chia seed oil as potential delivery systems in food. *Food Hydrocolloids* 108: 106030. Doi: 10.1016/j.foodhyd.2020.106030.

77. Göksen, G., M. J. Fabra, H. I. Ekiz, and A. López-Rubio. 2020. "Phytochemical-loaded electrospun nanofibers as novel active edible films: Characterization and antibacterial efficiency in cheese slices." *Food Control* 112: 107133. Doi: 10.1016/j.foodcont.2020.107133.

78. Moydeen, A. M., M. Syed Ali Padusha, E. F. Aboelfetoh, S. S. Al-Deyab, and M. H. El-Newehy. 2018. "Fabrication of electrospun poly(vinyl alcohol)/dextran nanofibers via emulsion process as drug delivery system: Kinetics and in vitro release study." *International Journal of Biological Macromolecules* 116: 1250–59. Doi: 10.1016/j.ijbiomac.2018.05.130.

79. Chuysinuan, P., T. Thanyacharoen, S. Techasakul, and S. Ummartyotin. 2018. "Electrospun Characteristics of gallic acid-loaded poly vinyl alcohol fibers: Release characteristics and antioxidant properties." *Journal of Science: Advanced Materials and Devices* 3 (2): 175–80. Doi: 10.1016/j.jsamd.2018.04.005.

80. Sharif N., M.T. Golmakani, M. Niakousari, S. Hosseini, B. Ghorani, and A. Lopez-Rubio. (2018). Active food packaging coatings based on hybrid electrospun gliadin nanofibers containing ferulic acid/hydroxypropyl-beta-cyclodextrin inclusion complexes. *Nanomaterials,* 8(11): 919. Doi: 10.3390/nano8110919.

81. Soleimanifar, M., S. M. Jafari, and E. Assadpour. 2020. "Encapsulation of olive leaf phenolics within electrosprayed whey protein nanoparticles; production and characterization." *Food Hydrocolloids* 101: 105572. Doi: 10.1016/j.foodhyd.2019.105572.

82. Khoshakhlagh, K., M. Mohebbi, A. Koocheki, and A. Allafchian. 2018. "Encapsulation of D-limonene in alyssum homolocarpum seed gum nanocapsules by emulsion electrospraying: Morphology characterization and stability assessment." *Bioactive Carbohydrates and Dietary Fibre* 16: 43–52. Doi: 10.1016/j.bcdf.2018.03.001.

83. Miguel, G. A., C. Jacobsen, C. Prieto, P. J. Kempen, J. M. Lagaron, I. S. Chronakis, and P. J. García-Moreno. 2019. "Oxidative stability and physical properties of mayonnaise fortified with zein electrosprayed capsules loaded with fish oil." *Journal of Food Engineering* 263: 348–58. Doi: 10.1016/j.jfoodeng.2019.07.019.

84. Rodrigues, R. M., P. E. Ramos, M. F. Cerqueira, J. A. Teixeira, A. A. Vicente, L. M. Pastrana, R. N. Pereira, and M. A. Cerqueira. 2020. "Electrosprayed whey protein-based nanocapsules for β-carotene encapsulation." *Food Chemistry* 314: 126157. Doi: 10.1016/j.foodchem.2019.126157.

85. Liu, Z. P., Y. Y. Zhang, D. G. Yu, D. Wu, and H. L. Li. 2018. "Fabrication of sustained-release zein nanoparticles via modified coaxial electrospraying." *Chemical Engineering Journal* 334: 807–16. Doi: 10.1016/j.cej.2017.10.098.

86. Xue, J., T. Wang, Q. Hu, M. Zhou, and Y. Luo. 2018. "Insight into natural biopolymer-emulsified solid lipid nanoparticles for encapsulation of curcumin: Effect of loading methods." *Food Hydrocolloids* 79: 110–16. Doi: 10.1016/j.foodhyd.2017.12.018.

87. Malapert, A., E. Reboul, M. Tourbin, O. Dangles, A. Thiéry, F. Ziarelli, and V. Tomao. 2019. "Characterization of hydroxytyrosol-β-cyclodextrin complexes in solution and in the solid state, a potential bioactive ingredient." *LWT* 102: 317–23. Doi: 10.1016/j.lwt.2018.12.052.

88. Kyriakoudi, A., and M. Z. Tsimidou. 2018. "Properties of encapsulated saffron extracts in maltodextrin using the Büchi B-90 nano spray-dryer." *Food Chemistry* 266: 458–65. Doi: 10.1016/j.foodchem.2018.06.038.

89. Prasad Reddy, M. N., S. Padma Ishwarya, and C. Anandharamakrishnan. 2019. "Nanoencapsulation of roasted coffee bean oil in whey protein wall system through nanospray drying." *Journal of Food Processing and Preservation* 43 (3): 13893. Doi: 10.1111/jfpp.13893.

90. Yao, K., W. Chen, F. Song, D. J. McClements, and K. Hu. 2018. "Tailoring zein nanoparticle functionality using biopolymer coatings: Impact on curcumin bioaccessibility and antioxidant capacity under simulated gastrointestinal conditions." *Food Hydrocolloids* 79 (June): 262–72. Doi: 10.1016/j.foodhyd.2017.12.029.

91. Luo, Y., Y. Zhang, K. Pan, F. Critzer, P. M. Davidson, and Q. Zhong. 2014. "Self-emulsification of alkaline-dissolved clove bud oil by whey protein, gum Arabic, lecithin, and their combinations." *Journal of Agricultural and Food Chemistry* 62 (19): 4417–24. Doi: 10.1021/jf500698k.

92. Yi, J., T. I. Lam, W. Yokoyama, L. W. Cheng, and F. Zhong. 2015. "Beta-carotene encapsulated in food protein nanoparticles reduces peroxyl radical oxidation in Caco-2 cells." *Food Hydrocolloids* 43: 31–40. Doi: 10.1016/j.foodhyd.2014.04.028.

93. Robledo, N., P. Vera, L. López, M. Yazdani-Pedram, C. Tapia, and L. Abugoch. 2018. "Thymol nanoemulsions incorporated in quinoa protein/chitosan edible films; antifungal effect in cherry tomatoes." *Food Chemistry* 246: 211–19. Doi: 10.1016/j.foodchem.2017.11.032.

94. Rostami, M. R., M. Yousefi, A. Khezerlou, M. A. Mohammadi, and S. M. Jafari. 2019. "Application of different biopolymers for nanoencapsulation of antioxidants via electrohydrodynamic processes." *Food Hydrocolloids*. Doi: 10.1016/j.foodhyd.2019.06.015.

95. Salgado, P. R., L. D. Giorgio, Y. S. Musso, and A. N. Mauri. 2018. "Bioactive packaging: Combining nanotechnologies with packaging for improved food functionality." In *Nanomaterials for Food Applications*, 233–70. Doi: 10.1016/B978-0-12-814130-4.00009-9.

96. Xue, Y., X. Xia, B. Yu, X. Luo, N. Cai, S. Long, and F. Yu. 2015. "A green and facile method for the preparation of a PH-responsive alginate nanogel for subcellular delivery of doxorubicin." *RSC Advances* 5 (90): 73416–23. Doi: 10.1039/c5ra13313k.

97. Hosseini, S. M. H., Z. Emam-Djomeh, P. Sabatino, and P. Van der Meeren. 2015. "Nanocomplexes arising from protein-polysaccharide electrostatic interaction as a promising carrier for nutraceutical compounds." *Food Hydrocolloids* 50: 16–26. Doi: 10.1016/j.foodhyd.2015.04.006.

98. Zhou, M., T. Wang, Q. Hu, and Y. Luo. 2016. "Low density lipoprotein/pectin complex nanogels as potential oral delivery vehicles for curcumin." *Food Hydrocolloids* 57: 20–29. Doi: 10.1016/j.foodhyd.2016.01.010.

99. Bardajee, G. R., N. Khamooshi, S. Nasri, and C. Vancaeyzeele. 2020. "Multi-stimuli responsive nanogel/hydrogel nanocomposites based on κ-carrageenan for prolonged release of levodopa as model drug." *International Journal of Biological Macromolecules* 153: 180–89. Doi: 10.1016/j.ijbiomac.2020.02.329.

100. Jafari, S. M. 2017. "An introduction to nanoencapsulation techniques for the food bioactive ingredients." In *Nanoencapsulation of Food Bioactive Ingredients*, 1–62. Elsevier. Doi: 10.1016/b978-0-12-809740-3.00001-5.

101. Maldonado, L., S. Chough, J. Bonilla, K. H. Kim, and J. Kokini. 2019. "Mechanism of fabrication and nano-mechanical properties of α-lactalbumin/chitosan and BSA/κ-carrageenan nanotubes through layer-by-layer assembly for curcumin encapsulation and determination of in vitro cytotoxicity." *Food Hydrocolloids* 93: 293–307. Doi: 10.1016/j.foodhyd.2019.02.040.

102. Ghaeini-Hesaroeiye, S., S. Boddohi, and E. Vasheghani-Farahani. 2020. "Dual responsive chondroitin sulfate based nanogel for antimicrobial peptide delivery." *International Journal of Biological Macromolecules* 143: 297–304. Doi: 10.1016/j.ijbiomac.2019.12.026.

103. Massaro, M., and S. Riela. 2018. "Organo-clay nanomaterials based on halloysite and cyclodextrin as carriers for polyphenolic compounds." *Journal of Functional Biomaterials* 9 (4): 61. Doi: 10.3390/jfb9040061.
104. Risyon, N. P., S. H. Othman, R. K. Basha, and R. A. Talib. 2020. "Characterization of polylactic acid/halloysite nanotubes bionanocomposite films for food packaging." *Food Packaging and Shelf Life* 23: 100450. Doi: 10.1016/j.fpsl.2019.100450.
105. Biddeci, G., G. Cavallaro, F. Di Blasi, G. Lazzara, M. Massaro, S. Milioto, F. Parisi, S. Riela, and G. Spinelli. 2016. "Halloysite nanotubes loaded with peppermint essential oil as filler for functional biopolymer film." *Carbohydrate Polymers* 152: 548–57. Doi: 10.1016/j.carbpol.2016.07.041.
106. Yousefi, P., S. Hamedi, E. R. Garmaroody, and M. Koosha. 2020. "Antibacterial Nanobiocomposite based on halloysite nanotubes and extracted xylan from bagasse pith." *International Journal of Biological Macromolecules* 160: 276–87. Doi: 10.1016/j.ijbiomac.2020.05.209.
107. (2009). The potential risks arising from nanoscience and nanotechnologies on food and feed safety. *EFSA Journal* 7(3). Doi: 10.2903/j.efsa.2009.958.

11 Advancement of Nanomaterials in the Biomedical Field for Disease Diagnosis

Najla Bentrad
University of Sciences and Technology
Houari Boumediene (USTHB)

Asma Hamida-Ferhat
University of Algiers 1

CONTENTS

11.1 INTRODUCTION

Nanotechnology is now an emerging area of science and technology because there is an interest in the design and development of nanoscale materials that stimulates concern and significant health problems. The use of nanoparticles (NPs) in diagnostics and medical imaging will lead to real change, as NPs play the role of markers or contrast agents, where nanomedicine can make it possible to improve early detection and treatment of multiple pathologies. The application of multifunctional nanoparticles (MNPs) in medical diagnostics needs sufficient control of their biodistribution within an organism or in situ [1]. Most MNPs are, in particular, hybrid structures comprising a surface layer of an organic composition and a functional part of medical imaging contrast agents to allow the visualization of tissues, cells, and metabolic processes (Figure 11.1) [2]. NPs may be designed with various properties of contrast. Furthermore, it is important to establish alternative methods for the isolation and purification of NPs and characterize their biomedical application using physicochemical measurements [3,4]. Intensive research is ongoing to design and develop novel nanohybrids for a range of nanomedicine applications with exceptionally sensitive anatomical localization [5]. It is possible to change the surface of NPs through further studies using highly sensitive metabolic assays [6]. For example, an in vivo fluorescence imaging analysis showed the activity of PEGylated nanoparticles, which is based on polyethylene glycol (PEG) chain type and length [7].

Various nanotechnologies are becoming highly popular [8–11], for rapid identification of in medical diagnosis [12]. NPs are used for several purposes such as identifying lesions at a very early stage of disease development and testing the efficacy of drugs targeting cell diseases and pathologic tissues [8]. Using nanomaterials contrast agents in structural imagery provides information on the anatomy of organs (volume, location, potential existence of lesions, etc.). Nanotechnologies provide theoretically enticing methods for an advanced bioimaging approaches and clinical measures as part of the continuing search for means to reduce mortality due to multiple pathologies related to cancer and inflammatory diseases.

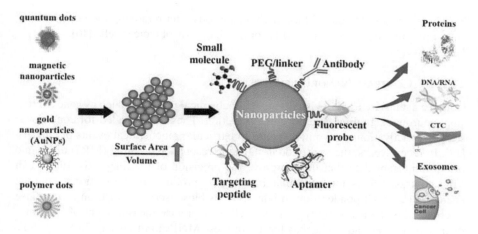

FIGURE 11.1 Nanotechnology involved in detection and diagnosis processes. (Adapted with permission from ref [1].)

However, functional imaging allows noninvasive visualization of biological processes at the molecular or genetic level [9]. In brief, this book chapter examines the latest generation of fluorescent tagging and advanced NPs used as biosensors or biomarkers and explores the application of these nanotechnologies in the diagnosis of various human diseases.

11.2 NANOPARTICLES IN ANATOMICAL AND FUNCTIONAL IMAGING

11.2.1 GOLD NANOPARTICLES

Advanced bioimaging methods that record the energy of the photon incident are more accurate by calculating their standard X-ray attenuation profiles and bio-diffusion of contrast agents [13] to close the distinction between cell distribution and patient outcomes [14,15]. Spectral photon-counting computed tomography (SPCCT) imaging-related nanoparticles as contrast agents consist of elements such as gold, bismuth, and ytterbium, which can be measured in relation to the inductive plasma spectrometry of optical emissions in organs of interest [14,15]. The SPCCT used for monitoring and quantifying therapeutic cells pre-labeled with gold NPs, i.e., polarized M2 macrophages is derived from bone marrow transplanted in brain-damaged rats [15]. The effectiveness of therapeutic cell surveillance has been successfully demonstrated in rats with brain damage caused by SPCCT imaging. The imaging method made it possible to monitor cells in vivo until a couple of weeks with a detection limit of only 5,000 cells with voxel dimensions of $250 \times 250 \times 250$ μm [15]. Contrast imaging can quantify and specify the distribution of gold and iodine nanoparticles in target tissues leading to their predicted pharmacokinetics [13]. Overall, the prepared nanoprobe has an enhanced effect of radiotherapy (RT) and phototherapy (PTT) in both X/CT modes. Multifunctional nanoprobes are expected to perform. Currently, the silica used to replace cetyltrimethylammonium bromide particles and the folic poster have been anchored to the surface of gold nanorods after removing their cytotoxicity and improving their biocompatibility for the diagnosis of cancer cells [16].

11.2.2 Contrast Nanoparticles

Contrast agents are used in visualizing of biochemical processes such as gene expression, neuronal signaling, and hormone secretion. They are intended for conditional storage or activation in vivo in response to particular biochemical events of concern [17]. In recent years, the multimodal magnetic resonance imaging (MRI) technique offers important data that may improve the precision of the diagnosis. Interest in MRI contrast agents for molecular imaging by production of functional contrast agents (T1 and T2 ponderation) might serve as biosensors and turn on in response to complex biological activity [18]. As well as, a precise measurement of their biodistribution is very advantageous for use of these MNPs in in vivo applications. NPs that contain a gadolinium oxide nucleus will either combine fluorescence or resonant imaging with radioactive therapy [7]. For example, post-functionalization of gadolinium oxide by four different PEGylated nanoparticles after intravenous administration of HEK-beta3 in tumor-bearing mice is evaluated for their efficiency and their biodistribution by fluorescence imaging [7]. Many solutions to analytical and therapeutic radiological problems will be accomplished by bioreactive agents mainly focused on gadolinium compounds, small molecule agents, which are dispersed in all extracellular tissue spaces offering no detailed biological data [17]. Some recent methods have been introduced to increase the sensitivity of these gadolinium compounds, target particular tissues, and develop new methods for measuring complex biological processes [19]. The gadolinium used in MRI is not metabolized and then excreted by the kidneys. Because of high intestinal toxicity, it should be delivered as gadolinium chelates (GC) to support clinical detection [20]. The kinetic stability of macrocyclic GC in incorporated ligand cavity is much higher than that of linear GC open-chain form. This class of drugs has an overall safe reputation according a recent study, which reported severe disorder, nephrogenic systemic fibrosis [20]. A rapid use of a molecular probe to detect zinc, achieve improvement in the MRI signal at T3 ponderation, while the same agent does not detectable respond using conventional MRI and this is for the loss of performance of large field reactive probes, with MRI with cyclic field changes [21]. The MRI study reported that gadolinium contrast was weakly bound to human serum albumin (HSA), unlike zinc, which binds strongly to HSA [22,23]. However, in vitro, the gadolinium detects about $30\,\mu M$ of zinc in the presence of HSA in the blood [21]. The use relaxation rate data measured in the absence and presence of a biomarker in medical imaging [20] can be generalized to any biomarker as long as the detection is based on the variation of fluorescence of the probe. On the other hand, simple methods of contrast agent synthesis are therefore required to establish multimodal MRI. The experimental results showed that the successfully constructed and synthesized ferrosoferric oxide/gadolinium oxide nanotubes are important instruments to improve the effectiveness of the MRI imaging in dual-mode (T1 and T2). Also, it allow accurate clinical diagnosis, improve the hydrophilicity and biocompatibility of nanotubes coated with 3,4-dihydroxyhydrocinnamic acid, a nontoxic element [24].

11.2.3 MAGNETIC NANOPARTICLES (MNP)

Magnetic iron oxide nanoparticles have supermagnetism, high saturation area, stability, and biocompatibility properties and are commonly used in biomedical applications such as controlled metabolic processes, pathogen detection, cancer therapy, and other therapies [25,26]. The octopod iron oxide nanoparticles (edge length 30 nm) show a very high transverse relaxation value suggesting that they are effective for precise and early cancer detection compared with conventional iron oxide nanoparticles [27]. Iron oxide nanoparticles have been the first contrast agent used due to their chemical stability, low toxicity, and biodegradability. These oxides are part of NPs called positive contrast agents that are superparamagnetic. By being transported throughout the body by substances considered to be alien to the body, they enter the tumor. The use of magnetic nanoparticles for the diagnosis of cancer seems, therefore, to be a very useful option. Currently, its magnetic properties make it completely aligned with some methods of scanning. High magnetization, length just under 20 nm, specific transfer of particle sizes, and special surface coating are required for the biological application of these nanomaterials, which can limit toxicity and can be combined with biomolecules [28]. The core of iron oxide is coated with oleic acid (OA) and then coated with OA-PEG to form a formulation of MNP soluble in water. To prolong the drug delivery time, the hydrophobic doxorubicin may be delivered into the OA substrate. The OA-PEG N-hydroxysuccinimide group was used to conjugate the antibody's amine feature to selectively target breast cancer cell lines. The formulation increases T2 MRI contrast and has a longer circulating time in the body with a 30% relative concentration 50 min after injection [29]. As well as, polyethyleneimine-coated magnetic iron oxide nanoparticles susceptible to oxidation-reduction are prepared for gene delivery and MRI. It has a high T2 relaxivity of 81 L mmol^{-1}/s in MRI detection [30]. Next, under high electric field intensity, we influenced the length of the acyl chain in the lipid bilayer and its unsaturation, charge, and cholesterol presence. The complexity depends on the magnetic particles' formation. By reducing the MRI signal in the distribution area, resulting in dark images or reduced contrast, samples containing cholesterol display a significant effect on T2 spin-spin relaxation time. In comparison, the insaturation of the acylated lipid chains and the presence of negatively charged lipids in the bilayer appear to indicate a negative rise, making these magnetic liposomes hepatic contrast agents [31,32].

11.3 FLUORESCENCE IMAGING

Imaging methods may promote the detection of subclinical lesions as well as more focused care. Scientists have come a long way to change the structures and properties of these dyes to satisfy several biological needs, synthesizing methods of these dyes are now established, and their spectral properties can be modified, including the precise targeting of biomolecules and coordination of molecular processes through fluorescence activation [33].

11.3.1 RADIOACTIVE PROBES

Magnetic resonance imaging (MRI) using fluorinated probes such as ^{19}F provides a specific high-contrast signal for bioimaging with optimal properties for many in vivo biomedical applications. However, fluorinated probes with a high content of fluorine atoms are required to ensure adequate sensitivity in MRI. Most fluorinated probe agents are perfluorocarbon emulsions that have a wide variety of applications in molecular imaging, but the fluorine content of these molecules is restricted [34].

The divalent ^{99m}Tc-hydroxyamide complex is the radioactive element that can be used as a sensitive imaging tool to assess early thrombus aggregation in LVAD at high operating speeds and high shear rates [35]. A ^{99m}Tc-hydroxyamide amide complex with a divalent amyloid ligand has been identified, which shows strong binding affinity to β-amyloid Aβ peptide aggregates that are often present in the form of AD. In vitro cerebral angiopathy amyloid (CAA) inhibition assays and brain absorption experiments using A aggregates, ^{99m}Tc-Ham complexes were mainly used. Autoradiography of human brain sections and mice cells were subjected to the divalent ^{99m}Tc-Ham complex, which information obtained in connection labeling of β-amyloid (Aβ) deposits in human brain sections and affinity for CAA. And transgenic selective mice. All ^{99m}Tc-Ham complexes with Aβ ligands have shown a very low cerebral cortex absorption. These results provided new possibilities for the production of ^{99m}Tc-Ham probes that complex for CAA-specific imaging [36]. Advanced cancer imaging using ^{64}Cu radiopharmaceuticals was used for hypoxia imaging and targeting smaller molecules (tumor receptor peptides, antigen monoclonal antibodies, and larger and slower proteins and NPs).

A new investigational drug application for ^{64}Cu-ATSM, which paved the way for a multicenter trial to confirm the potency of these promising outcomes, was approved by the FDA [37] for normal therapeutic use. By integrating anatomical and functional data with hybrid approaches, nuclear medicine may play a major role in prostate cancer assessment. In order to target particular prostate cancer biomarkers, various PET radiopharmaceuticals have been used. Ideally, research should concentrate on producing radiopharmaceuticals that address overexpressed antigens in prostate cancer, as opposed to normal prostate tissue. In this respect, outstanding candidates are gastrin-releasing peptide receptors (GRPR). Because of many benefits, such as abundance, ease of radiochemistry, half-life, and cost, Gal-68 (^{68}Ga) is an attractive PET radioisotope. This research focuses on ^{68}Ga named bombesin analogs in prostate cancer [38,39]. Then, the resulting NODAGA-SCH1 was radiolabeled with Ga and assessed for PCa PET imaging. Ga-NODAGA-SCH1 demonstrated outstanding PET/CT imaging properties in nude mice carrying PC-3 tumors, such as high tumor absorption and tumor contrast/high muscle, compared with the Ga-NODAGA-JMV594 probe. Importantly, biodistribution evidence revealed that a relatively equal concentration of Ga-NODAGA-SCH1 was found in the liver and kidneys, ensuring that both the kidney and the liver preceded by clearance. As an imaging agent, iodine 123 (^{131}I) has superior physical properties and its biochemical function is the same when radiation is distributes to the thyroid [40].

The availability of $_{123}$I-IMP (iodine 133 N-isopropyl-paraiodoamphetamine) technology is minimal relative to technical approaches, as iodine 123 ($_{123}$I) must produce a cyclotron. The rCBF technology has been shown to be accurate. Brain scans can be performed within 1 hour of the injection if the redistribution of the drug in the body is very swift. In cerebral acidosis, which is a hallmark associated with acute cerebral ischemia, the ^{123}I-IMP technique can underestimate the rCBF, and the picture of ^{123}I-IMP is marginally inferior to the technology-based process [41], due to low dosimetry and low photon flux. With low side effects, the ^{123}I gamma emission offers excellent imaging. In cerebral acidosis, which is a hallmark associated with acute cerebral ischemia, the ^{123}I-IMP technology can underestimate rCBF [48]. The image efficiency of ^{123}I-IMP is significantly poorer than that of technological approaches because of the lower dosage process and lower photon flux.

Better resolution than Single-photon emission computed tomography (SPECT) is given by positron emission tomography. Short-lived isotopes with fewer than 2 hours, such as ^{15}O, ^{13}N, ^{11}C, and ^{18}F, are commonly used to avoid toxicity, but often the removal time of labeled substances is longer than radioactive decay [42]. Then, when the disturbance can be severe, relevant residual activity associated with unmetabolized compounds and circulating metabolites should be assessed.

11.3.2 INDOCYANINE GREEN FLUORESCENCE (ICG)

Using more affordable and harmless fluorescence in clinical guided surgery has improved massively to help hepatobiliary and pancreatic surgery [43]. Fluorescence contrast agent such as indocyanine green (ICG) when illuminated by laser sources to aid the surgical procedure in detecting a Subclinical dysplastic lesion by the fluorescence method is based on the presence of a real-time contrast (Figure 11.2) between pathological (fluorescent) and healthy (non-fluorescent) tissue [43,44]. Both clinical diagnoses in dermatology have increased the need for noninvasive diagnosis using reflectance laser scanning microscopy approach, which focused only on variations of reflective properties of epidermal and dermal structures [44]. So, fluorescence has not commonly been used in dermatology and has several limits. Therefore, it was found a fluorophore alternative, which corresponds to a modern skin imaging system to advance this innovative diagnostic strategy [44].

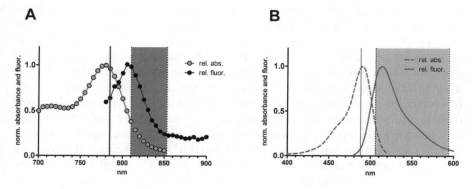

FIGURE 11.2 Absorbance and fluorescence emission spectra of ICG and fluorescein. (Adapted with permission from ref [34].)

11.3.3 CADMIUM SELENIDE FLUORESCENCE IMAGING

Homogeneous and highly fluorescent "core/shell" cadmium selenide (CdSe/ZnS) nanocrystals (NC) have been synthesized. Combined with proteins, DNA, or drugs, NC can be excited at any UV-visible spectral wavelength, offering color spectrum and fluorescence based on their single-diameter routine fluorescence microscopy. These properties offer excellent prospects for high-throughput multiplexing and long-term monitoring of labeled precursors for days or even weeks, both at cellular level and in pathological human surgical samples [45]. CdSe/CdS/ZnS "double shell" nanorods have been synthesized, and the first "point-in-rod" core CdSe/CdS coatings have been prepared. The ZnS coating was then epitaxially formed on these CdSe/ CdS nanotubes for cell labeling experiments. Quantum performance is associated with the design of the nanotube and the thickness of the ZnS coating around the initial CdSe/CdS spins, varying from 1 to 4 monolayers [45]. Cadmium selenide crystals are coated with zinc and other polymers to improve binding ability with targeted molecules. The encapsulation of cadmium selenide (CdSe) nanocrystals in single-wall carbon nanotube cavities of various sizes coated simultaneously with β-D-glucan. This process is supplemented by a complete characterization of single-wall carbon nanotube using spectroscopy, high-resolution electron microscopy, and X-ray analysis, shed light on the composition [46]. The fluorescence of the new hybrid nanomaterial is observed in aqueous dispersions, and spontaneous biocompatibility was shown to be due to the activity of β-D-glucan, a biopolymer isolated from barley [46]. Quantum dot is doping with metal ion transitions such as Cu^+ and Mn^{2+}; it is considered as efficient strategy to improve photoluminescence (Figure 11.3). QD-light-emitting diodes (QLEDs) with a mole fraction of 0.05 Cu showed a higher photoluminescence (peaked at 57.5%) and a maximum external quantum efficiency of 3.5% relative to the molar fraction of 0.1 Cu [47].

11.3.4 NEAR-INFRARED FLUORESCENCE (NIR)

Near-infrared fluorescence (NIR) is widely used in clinical imaging, providing highly specific images of the body or target tissue. Imaging in the near infrared (650–900 nm) is especially useful due to many unique characteristics such as reduced autofluorescence and biomolecular absorption of tissue as well as low light dispersion [48,49]. Compared with visible wavelengths, fluorescent NIR light is transparent, allowing very sensitive real-time imaging in human surgery without altering the surgical environment [48,49]. The advantage of using fluorescent NIR as a clinical imaging technology is its molecular fluorescence as an exogenous contrast agent. In overall, the development of NIR fluorophores such as ICG and methylene blue (MB) is required for early detection. The appropriate combination of fluorescent NIR may provide fluorescence in real time without altering the bioimaging process of specific target tissues during surgery [49]. Certainly, NIR imaging technology has eventually reached clinical use, mainly for the late production of sufficiently innovative contrast agents for existing imaging systems. Both processes require optimization of the structure of the gloves to achieve not only an appropriate NIR window imaging agent but also high molecular visibility, water solubility, biocompatibility,

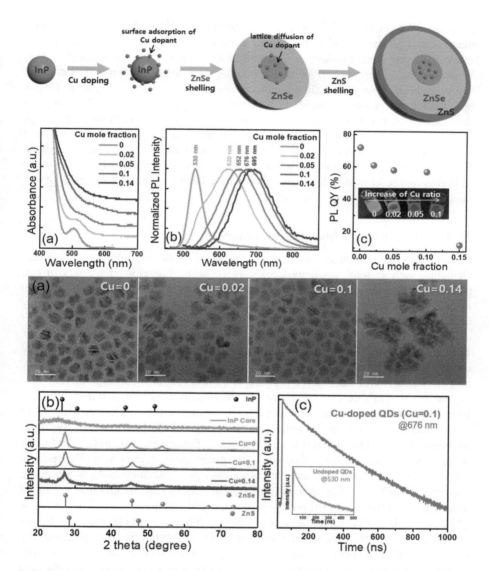

FIGURE 11.3 Schematic of synthesis of InP:Cu/ZnSe/ZnS quantum dots (QDs) by a surface adsorption-lattice diffusion strategy. (Adapted with permission from ref [45].)

and precise targeting using recognized spectrally distinct fluorophore in proper time [48]. Dynamic NIR fluorescence imaging after ICG injection can provide the array of fluorescent light extends in intestine, and various models of intestinal motility, such as peristalsis and segmental phases, have been dynamically visualized. The highest strength indicates ICG transit waves, showing the overall distribution of the wave along the intestine in the mouse [50]. Selective targeting of bone structure has recently been identified as a new technique focused on insertion of specific molecules to chemical structure of near-infrared contrast agents (NIRs). A sequence of imino-diacetic and phosphonated NIR contrast agents (bisphosphonates) for target bone tissue have provided excellent in vivo ability to target metastases bones [51]. This would be a novel approach to focused bone imaging using bifunctional NIR contrast agents and important in real-time fluorescence-guided surgery investigations [51]. A new fluorescent probe permeable to living cells comprises the hydrophilic peptide Asp-Glu-Val-Asp (DEVD) relevant to caspase. Also, the hydrophobic unit tetraphenylethylene pyridinium for cell apoptosis imaging and drug screening has been developed for fluorescence imaging. The probe shows an activated caspase-specific light response with a high signal-to-background ratio. So, because of its good water solubility and biocompatibility, these probes have been shown to be a promising element in diagnosis and cancer therapies [52].

11.3.5 Fluorescence Microscopic Imaging

Cellular process is characterized by precise spatial-time modulation of different elements, such as ions, small molecules, or proteins. Therefore, the study of cell physiology involves an optical record of these systems, in particular the use of fluorescent biosensors. Fluorogenic ratios comprise a protein tag or an RNA that may be a natural or synthetic dye (called a fluorogen) complex. The recent production of diverse fluorogenic systems in the design of these sensors has increased considerably. The advancement of high-speed microscopic imaging technology has allowed the use of automated fluorescence microscopy in high performance for drug discovery assays. The array of fluorescent proteins (FP) provides a wide range of new biological imaging methods [53]. Initially, the green fluorescent protein (GFP) extracted from *Aequorea victoria* (jellyfish) and later the anthozoan fluorescent protein (AFP) were effective components as indicators for living cells [54]. Currently, the light-oxygen-voltage (iLOV) derived from flavin-based fluorescent protein (FbFP) family is considered an alternative component to the green fluorescent protein. The spectral properties of iLOV have been modeled as well as a point mutation in iLOV to achieve a more red-shifted replica. According to these simulations, mutation is predicted to lead to an optimal redshift of about 50 nm, thereby acquiring a new color in the FbFP palette [55]. Certain amino acid residues in flavin protein (Figure 11.4) did not change [56]. The structure of the fluorogenic chromophore and the high tunability of the related chromophore have intriguing properties such as the emission from far red to near infrared. However, the yellow marker that induces and switches the fluorescence absorption of the Y-FAST protein differs from other labeling schemes because the fluorine binding is very complex and reversible, allowing rapid labeling and new development of sequential multiplexing imaging protocols for fluorogen Y-FAST [57]. The aptitude to monitor metabolic processes with precision is therefore important for understanding the functions of living systems. Fluorescence has recently been expanded to include advanced fluorescent compounds that attach to fluorogenic chromophores that are used endogenously or exogenously and trigger their fluorescence, such as current GFP [58]. On the other hand, a new genetically encoded fluorescent marker has been evolved for the microscopy of living cells, a marker consisting of a fluorogen-activating protein, and FAST a shifted absorption

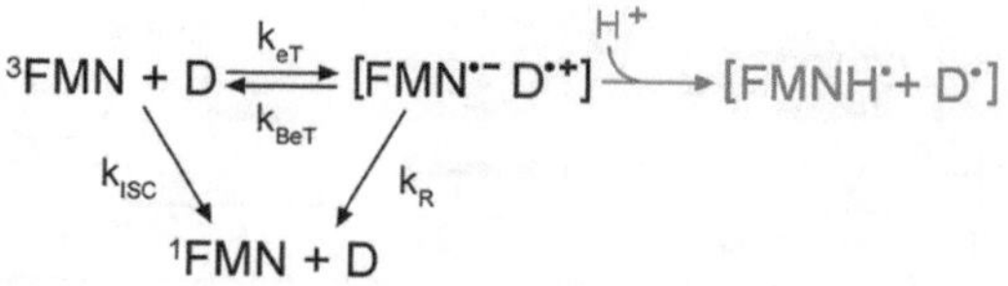

$$^3\text{FMN} + \text{D} \underset{k_{\text{BeT}}}{\overset{k_{\text{eT}}}{\rightleftharpoons}} [\text{FMN}^{\cdot-} \text{D}^{\cdot+}] \overset{\text{H}^+}{\longrightarrow} [\text{FMNH}^{\cdot} + \text{D}^{\cdot}]$$

$$k_{\text{ISC}} \searrow \quad \swarrow k_{\text{R}}$$

$$^1\text{FMN} + \text{D}$$

FIGURE 11.4 Photochemical reactions' triplet-state mechanism of parental iLOV (black) and iLOV-Q489D (red). (Adapted with permission from ref [54].)

marker, and a fluorogenic GFP used for fluorescence activation. Reversible binding of FAST fluorogen is followed by an increase in red fluorescence (580–650 nm) [59]. The proposed fluorogen immediately stains the target cellular proteins bound to the swift rinse within 1 minute and reveals a higher photostability of the fluorescence signal under confocal microscopy [59]. Certain far-red or near-infrared biliverdin proteins have opened up alternative possibilities for in vivo imaging, and the use of infrared excitations requires deeper tissue imaging. Some noticeable innovations include the use of tagging schemes (such as SNAP-tag, PYP-tag, CRABP II, and FAP; see Figure 11.5). These proteins are smaller than GFP and fluoresce regardless of the presence of oxygen [58]. Endogenous or synthetic fluorogens, when reacting with proteins, require a suitable comparison. However, the interaction between the protein site and the fluorogen is non-covalent and reversible to monitor the fluorescence strength when required by integrating (ON) or dismissing the fluoride (OFF) [58]. An exhaustive library of tissue-specific fluorophores is available in the literature, many of which are discussed here, providing clinicians a variety of resources that will improve intraoperative effectiveness and long-term postoperative prognosis [48]. These photoreceptors or protein markers were rapidly identified as promising contrast agents for the detection of biological events [60].

11.3.6 NANOPHOSPHORS

The nanophosphors are ideal for bioimaging and biodetection applications have been assessed, and their colloidal stability in various biologically relevant buffering media [61]. Nanophosphors based on rare earth vanadates (REVO4) were homogeneously synthesized using europium nitrate (EU), bismuth (Bi), and sodium orthovanadate as precursors. The precipitation is done by a combination of ethylene glycol/water at 120°C. The nanoparticle size can also be adjusted by changing polyacrylic acid (PAA) amount or poly(hydrochloride) form [61]. Also, nanophosphorus based on EU-Bi-doped REVO4 exhibits red luminescence at 342 nm [61]. The effectiveness of these contrast agents is also assessed using X-ray tomography, compared with that of the commercial probe, and seems to be more advantageous for medical diagnosis [62].

11.3.7 IMMUNOFLUORESCENCE IMAGING

This section outlines that some determination was carried out in human embryonic stem cell (HESC) immunofluorescence (IF) microscopy and quantitative PCR analysis (qPCR) specifically grown under non-food conditions to explain cellular

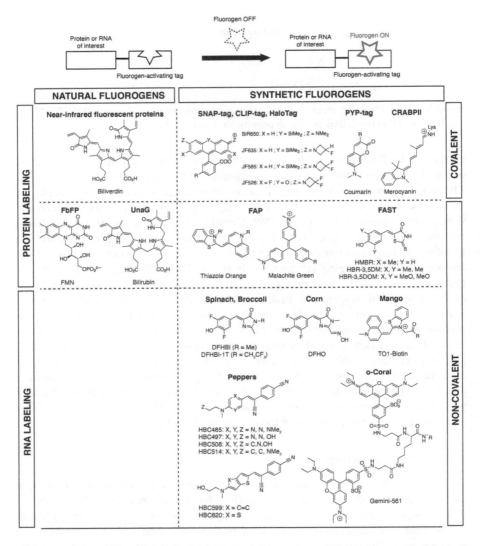

FIGURE 11.5 Some functional biosensor-based fluorogens that bind and activates their fluorescence in endogene or exogene tag. (Adapted with permission from ref [56].)

localization of transcription factors and signaling pathways in stem cell maintenance and differentiation processes [63]. In epidermal carcinoma of the head and neck, antibody treatment for anti-EGFR (epidermal growth factor receptor) approaches has been successfully applied to clarify the changes in EGFR-mediated cell signaling. The therapeutic efficacy, administration, and histological localization of antibody-based therapeutic agents at tumor (peripheral) sites are related to these changes. A clinical study of the commonly used therapeutic antibody (cetuximab-IRDye800CW) reveals that molecular Near-infrared (NIR) imaging (so-called in vivo fluorescent immunohistochemistry) helps one to determine the location of fluorescence labeled with cetuximab. Fluorescence-labeled optical molecular imaging will obviously contrast the morphological aspects of each tumor with the degree of antibody distribution, so that the therapy model can be improved on the basis of an interpretation of the real transmission of tumor antibodies [64]. The immunofluorescence antibody (IFA) test is a highly sensitive and precise technique for the detection and diagnosis of *Enterocytozoon bieneusi* parasite in immunocompromised patients, but it was also afflicted with a brain worm. The diagnostic precision of the IFA test was 99.5% [65].

11.3.8 HOLOGRAPHIC X-RAY IMAGING

Holographic X-ray imaging is a biologically relevant microscopy method, allowing the distribution of the fluorescence-labeled actin cytoskeleton in the cells (heart tissue) to be mapped and overlapped with the phase diagram of the X-ray holography that can be explained by the spatial correlation of the biomolecular shape and all contributing scattering components. The program recorded a series of local far-field diffraction patterns and provided in situ STED recordings to develop diffraction data models based on co-location constraints [66,67] and it was proved that ultrathin films can guide X-ray waves. Combined with model-independent reconstruction algorithms based on a strict dynamic scattering theory, fluorescence holography of thin-film X-ray waveguides has become a unique time-resolved in situ imaging probe that can illuminate the dynamics of nanostructures in real time with unprecedented resolution. Combined with chemically sensitive spectroscopic analysis, the reconstruction can lead to the morphology of specific nanostructural elements for the integration of ultrathin films [67].

11.3.9 FLUORESCENCE RECOVERY AFTER PHOTOBLEACHING

In order to observe the molecular details of clusters related to the activation of the epidermal growth factor receptor, the migration rate and aggregation status of Within the cell membrane of the mammal ovary, EGFR-labeled was evaluated over time. The scattering of larger molecules over longer distances was studied by fluorescence recovery after photobleaching. Fluorescence correlation spectroscopy has been used to detect small molecular movement over a short distance and to report aggregation status [68]. After stimulation with 50 nM EGF, it was observed that the diffusion coefficient of the local membrane receptor decreased significantly from 0.11 to 0.07 µm^2/s and then returned to the original value in about 20 min. The transient attachment of ligand-bound EGFR to fixed or slow structures (such as the cytoskeleton or large aggregates of previously photoblanched receptors) may better explain our observations. On the other hand, the apparent brightness of the diffusion material remains unchanged, and the frap striping experiment leads to a decrease in the direct long-range molecular mobility after stimulation, which can be proven by increasing the recovery time of the slow component, 13–21.9 s [68]. In a mouse model of prediabetes, lacunar junction binding is considered to be impaired. Loss of gap junction coupling contributes to changes in kinetics of insulin secretion and glucose homeostasis disruptions. While methods for measuring gap junction coupling have been developed, they lack cellular accuracy, lack quantification of coupling or sufficient spatial resolution, or are intrusive [69]. The recovery of fluorescence after photobleaching (FRAP) is a means to calculate gap junction coupling in pancreatic islets effectively and safely. These experiments demonstrate that FRAP is a valid tool that can provide cellular resolution to measure Cx36 gap junction coupling distribution and control in particular populations of islet cells. As a result, future research using this technology could explain the role of gap junction coupling in diabetes development and define the gap junction control mechanism for possible therapies.

11.4 NANOPARTICLES FOR CELLULAR AND MOLECULAR DIAGNOSTICS

Fluorescence is a technique since the emission of excitation probes can be visualized with eye or at higher resolution under a microscope as a research instrument widely employed in different biological structures due to technology. This segment focuses on the different insights for the creation of special and creative biosensors, expanding the limits of cell imaging.

11.4.1 POLYDOPAMINE (PDA)

Polydopamine (PDA) is a current nature-inspired biopolymer substance with many intriguing properties, including self-assembly and uniform adhesion [61]. The PDA is used in engineering processes and synthesis of metal nanoparticles, and it has a catalytic hydrogen evolution reaction. It is also capable of forming covalent bonds with different metal ions. NP structure was analyzed through X-ray diffraction in situ and microscopy techniques [61]. The high adhesion of byssus mussels under water enhances PDA materials. Furthermore, at the air/water interface, we found the possibility of making PDA films. Studying this process has allowed self-supporting membranes and reactive stimuli to appear. Dopamine oligomerization in the simple setting causes a PDA coating to be formed on any substance. It is environmentally safe and efficient, in addition to the process's simplicity. We have deposited a dense, superhydrophilic, and biocompatible coating on all substrates by choosing the necessary oxidant [70,71]. PDA consists of indole and dopamine units, a polymer which is readily obtained by dopamine oxidation. Due to the huge number of catecholics supported by amine communities, it adheres to any form of soil, even under water. Two methods are used to acquire analogs: post-modification of PDA and oxidative polymerization of dopamine analogues. The scope and drawbacks of these techniques are clarified, and future study in this field is then intrigued [72]. In order to allow the development of thickness films, the mechanical properties of these PDA films should be improved. The PDA film at the solvent/liquid interface is also seen to be able to stabilize the microemulsion. The chemical properties of catecholamines (mainly dopamine) can be applied to the water/air interface under certain experimental conditions, resulting in a thin film that can be modified and relocated to a solid substrate. The spontaneous oxidation reaction will cover all forms of established brown-black insoluble materials when the catecholamine solution is put in the presence of an oxidant [72].

11.4.2 Diamond Nanoparticles

Among the various drug delivery and imaging systems under development, diamond nanoparticles or nanodiamonds (NDs) have many significant qualities that may be helpful in increasing the effectiveness and safety of clinical diagnosis [73]. The creation of a multifunctional nanoscale sensor (hybrid sensor) is made by adding paramagnetic gadolinium complexes to ND with nitrogen vacancy centers (NV) via surface engineering. The experimental results are replicated by numerical simulation between NV spin and gadolinium complexes covering the ND.

The measurements of pH or redox potential shifts at a submicron length scale in a microfluidic environment enable calculating spin relaxation rates [74]. Diamond nanoparticles have been identified as possible optical markers in medical bioimaging. Fluorescence is produced by a point of defects in particles at a wavelength of 550–800 nm. However, for fluorescent defects to be effective, there are a number of hardware problems that must be overcome; for example, fluorescence can be triggered by high energy of ion beam irradiation and possible thermal correction [75].

Nanodiamonds can be photoluminescent with excellent photostability, and this property allows the control of a single particle over a long period [76]. So, the intracellular localization of NDs using immunofluorescence and transmission electron microscopy has revealed a high interaction between vesicles and ND aggregates [76].

11.4.3 Quantum Dots

A new generation of quantum dots (semiconductor nanocrystals) have the huge ability to research single-molecule intracellular processes, high-resolution cell biochemical pathways, and diagnosis during surgery [77]. The quantum dots may be used in biological fluids or in aqueous media to modulate the surface of the quantum dots to change their solubility and incorporate additional chemical functions. Quantum dots might provide the special optical properties of these nanoscale semiconductor crystals to make an interesting imaging fluorescence method and sensory applications [77].

Various polymeric coatings are used to accomplish this aim by providing water solubility and additional functional groups for attachment [77,78]. Carbon dots (CDs) have received much interest due to their specific characteristics such as weak synthetic methods, costly surface alteration but simple, excellent photoluminescence, water solubility, and low toxicity. Also, CD has been extensively used in numerous research fields [78]. An easy hydrothermal approach was used for dysprosium-doped magneto-fluorescent bimodal carbon (Dy-CD) points, and as a result, the Dy-CDs remain stable in aqueous solution (Figure 11.6) [79].

11.4.4 Carbon-Based Nanoparticle

Researchers often attracted the interest of carbon nanotubes, fullerenes, and graphene in many research areas, including biomedicine. Their nanoscale size and extensive possibilities for surface modification make carbon nanotubes an effective nanostructured substance in the diagnosis and medication [80,81].

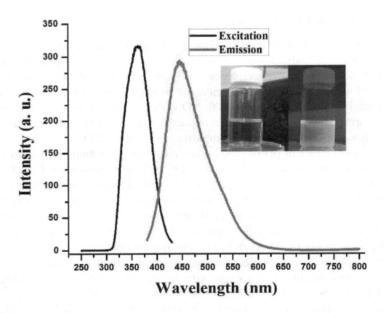

FIGURE 11.6 Photoluminescence excitation and emission spectra of the prepared Dy-CDs. (Adapted with permission from ref [79].)

11.4.4.1 Carbon Nanotubes (CNTs)

Carbon nanotubes (CNTs) are one of the most frequently used nanomaterials and can be quickly functionalized on the surface to bind proteins and nucleic acids and for the delivery of therapeutic molecules in nanoformulations [81]. Due to their nanoscale structure electronic properties and their thermal stability, they are attracting attention as nanoscale materials, encouraging their development in promising for medical applications [81]. The development of the structural activity paradigm, the analysis of the scale of similarity of different types of carbon nanotubes and asbestos fibers in terms of potential toxicity and phagocytosis efficiency, and the size effect and classification of carbon nanotubes are reviewed [82,83]. They also have certain inherent properties that are of great concern for their biological safety particularly on issues related to solubility/dispersibility and safety of CNTs [81,82].

11.4.4.2 Fullerenes

Fullerenes and their derivatives are considered one of the most promising nanomaterials due to their unique properties, which can be used in a variety of medical applications. Fullerene derivatives may achieve the goals of early diagnosis and early treatment of cancer as antitumor drugs [84,85]. In addition, fullerenes is used as scavengers to capture free radicals on cell lines such as tumor and apoptosis cells in the body, and it has a protective effect from various cytotoxic drugs and radiation [84]. Fullerene is also employed against human immunodeficiency virus, it prevents catalysis reaction site of HIV protease [86] and can also be used for serum protein analysis and biomarker discovery and in many other applications [86].

11.4.4.3 Graphene

The emerging science of graphene-based engineering nanomaterials in the form of nanomedicine or dental materials is developing [87]. Biomedical applications of graphene have proven to be widely popular among academic and industrial partners in the creation of next-generation medical systems and therapy; its application is due to graphene's positive properties, including heat, electricity, resistance, and elasticity; and it is one of the fastest two-dimensional nanomaterials, van der Waals monolayers and atomic layers with extremely low atomic thickness used [87–90]. Graphene-based nanomaterials and its derivatives and how they affect the behavior of nerve cells, including adhesion, proliferation, growth, and differentiation of neurites, are increasing. Although their effects on the nervous system in vivo have not yet been explored, encouraging results indicate that graphene-based nanomaterials have great potential as novel treatments for neurodegenerative diseases [90]. The effect of graphene on immune cells, as an immune biosensor, and antibodies targeting graphene-coupled tumors seems realistic in the near future to assume that certain graphene immunoconjugates with defined immune properties can pass preclinical tests and be successfully used in nanomedicine [91].

In addition, graphene and its derivatives can also be functionalized with several biologically active molecules, so that they can be combined and enhanced with different scaffolds used in regenerative dentistry especially on the control of dental stem cells [87].

11.4.5 POLYMERIC NANOPARTICLES (PN)

Polymers such as NPs have been used as drug carriers. Polymeric nanoparticles (PN) (polysaccharide chitosan-polylactic acid, polylactic acid coglycolic acid, polycaprolactone, and chitosan) combined with contrast agents offer significant advantages in molecular imaging and multimodal imaging [92]. Cancer, neurodegenerative diseases, and cardiovascular disease are areas affected by PN technology, pushing the scientific frontiers of revolutionary advances in nanomedicine [93]. The uses of polymer PNs include drug delivery technologies such as drug binding and packaging, prodrugs, stimulus-response systems, imaging methods, and diagnostics revolutionize modern medicine. PN and their application in vivo imaging have made world-renowned achievements, but they are still at the pioneering stage of development as highly sensitive molecular imaging agents [92–94].

In summary, a hybrides-polymériques and multifunctional platform that can be used for image-guided treatment of malignant glioma cancers(remains one of the most notorious incurable cancers)[95]. An MNP-based temozolomide delivery system has improved treatment efficacy in malignant glioma. Compared with the control experiments, the infusion of nanoparticles prolonged the survival duration of intracranial xenografts and do not cause any observable animal toxicity [95].

11.4.6 LIPOSOMES

Paramagnetic liposomes are bifunctional and bimodal paramagnetic nanoparticles comprising an aqueous interior surrounding a lipid bilayer membrane [96]. Liposomes are biocompatible and biodegradable, inert, and non-immunogenic lipids, and liposomes have also been extensively studied and used to treat a variety of diseases. By improving drug absorption, thus eliminating or mitigating accelerated deterioration and side effects, prolonging biological half-life, and reducing toxicity, liposomes enhance treatment effectiveness [97].

In preclinical and clinical trials, there are many new liposome formulations at various stages of development that can improve the therapeutic efficacy of both new and established drugs. In order to better diagnose diseases and monitor treatments, the latest developments in multimodal imaging have begun to use liposomes as diagnostic tools [97,98]. All these characteristics of liposomes and the increased flexibility of targeting groups in surface modification make liposome candidates more suitable for use as drug delivery vehicles [96,97].

11.4.7 SOLID LIPID NANOPARTICLES

These systems require specific contrast agents to obtain appropriate images of the best quality. Nanoparticles are innovative tools in the field of imaging research and diagnosis of various diseases (including cancer). A new form of colloidal drug carrier device sufficient for intravenous administration [99] has also been suggested. The development of novel image sensors that incorporate biomaterials and contrast agents (CA)/imaging probes (IP) strengthens the diagnostic and therapeutic uses of diagnostics that are the subject of several research projects. [100,101]. In the nanometer scale, the device consists of spherical rigid lipid particles, which are dispersed in water or in a surfactant solution. Among NPs, because of their many beneficial properties, lipid-based nanoparticles, such as nanostructured lipid carriers, solid lipid nanoparticles, and liposomes, are the most widely used carriers for imaging [101]. In particular, a lot of attention was provided to classical aliphatic (co) poly (lactide) (PLA), poly (lactide-co-glycolide) (PLGA), and poly (caprolactone) (PCL) poly (co) aliphatic polyesters [100].

11.4.8 MICROARRAYS

In theory, high-quality protein microarrays allow the functional analysis of the human proteome including the discovery of biomarkers, characterization of immune responses, recognition of enzyme substrates, quantification of proteins, and interactions of small molecules, protein-protein, and protein-DNA/RNA. Some of the most widely used full-length recombinant protein microarray formats, including human protein microarrays, programmable nucleic acid arrays, and immune protein arrays, have some limitations [102,103]. Also, microfluidic approach is proposed using components for liposomes processing, and various microfluidic instruments may have considerable potential in the efficient preparation of various lipid-based colloidal structures for biomedical applications. Microfluidic devices with various geometries are easily designed and assembled to allow the creation with closely distributed and highly reproducible monolayer liposomes [104,105]. Microarrays are sensitive miniaturized instruments that have been developed by producing an optical signal to detect the binding of complementary molecules to microarrays. The disadvantage of its usage in laboratories and clinical departments is the expensive implementation due to the specialized hardware needed for platform design, hybridization, and image analysis from DNA chip-based studies. Functional DNA and protein nanoparticles (FNP) are another cheaper technique of similar precision and may be used to identify defined DNA and protein sequences or mutated genes associated with human diseases. They can be used to test for nucleotide polymorphism (SNP) and identify biomarkers of cancer, diseases, and cardiovascular disease as hybridization probes [106]. In addition, peptide nucleic acid (PNA) is a 2-([2-aminoethyl] amino) acetic acid backbone synthetic DNA/RNA analog and has distinct antisense and antigenic properties with affinity and distinctive specificity. They are mainly used in modulation/mutation of polymerase chain reaction, fluorescent in situ hybridization, and microarray as a probe; they are often used in numerous in vitro and in vivo experiments and in micro- and nanometric biosensor/chip/matrix technologies' manufacturing [107].

11.4.9 REDOX SENSOR: PROTON TRANSISTOR

This part highlights the development of nanoparticle "proton transistors," ultra-pH-sensitive nanoparticles similar to electronic transistors that can digitize electronic signals allowing the transmission of data and reactions from molecular processes and biological phenomena [108].

Chemically modified field-effect transistor (FET) is highly sensitive as a detection method. The surface of the nanosensor is covalently modified with a reversible redox group that oxidizes readily in the presence of H_2O_2. So, the reversible redox transformation of molecules confined to the surface is carried out by an electron process, modulating the current source electron via a nanoFET detection system, which can chemically control disruption added to the redox mechanism. In addition, it can be used for the quantification of the related molecular metabolite on the surface of the nanosensor [109].

11.5 NANOBIOSENSORS

Biosensors have seen their number increase dramatically in recent years, especially in research and medical diagnostics. In general, they use a biological component to detect a target and a transducer to translate this detection into a quantifiable signal. There are different types of biosensors, depending on how the detected signal is converted into a quantifiable signal: "optical, electrochemical, mass sensitive, or thermal" sensors. They can be sensitive to metabolites, to antibodies or antigens, to nucleic acids, or to enzymes. The production of nanobiosensors has been made possible thanks to several achievements in nanotechnology and manufacturing processes.

Nanotechnology-enabling electrochemical sensing technologies could allow rapid β-A detection and promote rapid delivery of personalized health care. Recently, an electrochemical immunosensor approach based on β-A antibodies allows the detection of β-A [110]. Antibodies have many important features, high specificity, and structural integrity; they are used to design multiple antibodies against a wide variety of highly specific targets as they reflect a new framework to examine the antibody arsenal and fast identification for diagnosis of susceptible monoclonal antibodies and nanobiosensors [111].

11.5.1 Nanobiosensors Using Tissues and Cells

Nanorod gold has developed a blocked nucleic acid nanosensor for the complex study of single-cell gene expression in living cells and tissues. The nanoparticle enables the distribution of endocytosis and complex gene expression control at the unicellular level in human umbilical endothelial cells, mouse skin tissue, mouse retinal tissue, and mouse corneal tissue [112]. A lot of research has been achieved to build innovative, effective biosensors for the rapid and accurate diagnosis of Alzheimer's disease (AD). The work and association of the tau protein with AD are illustrated, as well as research on the development of nanobiosensors based on optical, electrochemical, and piezoelectric approaches [112].

11.5.2 Nanobiosensor Enzyme

To simulate the coating of an atomic force microscope (AFM) tip with enzymes described as rigid polyhedra, a simulation of adsorption processes was created. The technique was used to assess the frequency of the association between the enzyme acetyl-CoA carboxylase (ACC enzyme) and its substrate intended for molecular dynamics, thereby demonstrating the efficacy of the technique suggested in the experiments about nanobiosensors [113]. A main stage in the development of nanobiosensors for use in the detection process is atomic simulation of the ACC enzyme on the surface of an AFM. The related enzymatic surface orientations have been examined with the active sites of the ACC enzyme available for interaction with mass molecules using electrostatic considerations. For chromatographic applications, antigen-antibody interactions as well as enzymatic immobilization on silica can be tested using the enzymatic surface model with minor changes [114].

Related to the prevalence of diabetes in the world, the identification of glucose in the diagnosis and treatment of diabetes is of considerable significance. The diabetic patients' glucose biosensor controls their sugar levels many times a day, making it an important commercial market. Amperometric devices play a significant role in glucose control because of their wide use. The outline of the various glucose sensor generations is used to explain in detail how amperometric sensors operate and how the application of mediators will enhance sensor efficiency. Due to its ability to monitor changes in resistance and capacitance through binding activities [115], electrochemical impedance spectroscopy is an increasingly convenient method in devices. Various methods, such as electrochemical or optical processes, have accomplished the identification of glucose. The development of advanced glucose sensors with the highest sensitivity and convenience has been facilitated by other nanomaterial transducers using fluorescence techniques. After this, the standard nanomaterials are used in different glucose experiments for fluorescence-emitting/interacting [116].

11.6 CANCER DIAGNOSIS

The use of a wide variety of nanomaterials in the production of electrochemical biosensors has contributed to the development of future nanobiosensors. The study identified the detection limit of electrochemical nanobiosensors, as well as the challenges for the development, design, and evaluation of nanomaterial-based nanobiosensors in the sense of the medicine of the future. Indeed, the design of nanomaterial-based electrodes needs a simple, effective, and inexpensive way to improve sensitivity and reproducibility in the detection, treatment, or screening of cancer [117]. The lack of successful and early detection techniques for the same cancer type may explain the low survival rate. The conversion of nanoparticles doped with gadolinium ions may enhance the efficacy diagnosis of early pancreatic cancer considered as the major cause of death cancer. Gadolinium ion-grafted anti-human CD326 micelles displayed promising successful targeting capabilities by performing dual-mode imaging conducted in the xenograft mouse model of human pancreatic cancer [110]. In in vitro and in vivo toxicity studies, micelles have shown a strong biocompatibility. It indicates a successful active return of these micelles after intravenous injection. This suggests that gadolinium ion-grafted to anti-human CD326 micelles have a great potential for early detection of pancreatic cancer[110]. Hepatocellular carcinoma (HCC) is often metastatic cancer when it is detected [118], *in vivo* bioimaging after intravenous injection of selected micelles MR/UCL exhibit a superior fluorescent/magnetic property, an appropriate biocompatibility, and high specificity for HCC cells [118]. Tumor cells can lead to the parallel development of primary tumors and metastatic tumors in the blood [119]. The present problems and attractive prospects are examined in the assessment of aptamer-based electrochemical cytosensors (EC). The remarkable developments of EC focused on aptamers for the identification of cell type, cell counting, and detection of important proteins. Indeed, aptamers are single-stranded oligonucleotides that can bind with tumor cells in specific structures [12,119]. Aptamer-conjugated nanoparticles combined with particular gold nanoparticles (Apt-AuNPs) allow early detection and drug delivery [12]. So, recent expansion of Apt-AuNPs may efficiently interact with intra- and extracellular biomolecules.

11.7 INFECTIOUS DISEASE DIAGNOSIS

Fluorescence imaging provides easy information on bacterial infection and pathogenicity of microbial agents (e.g., anthrax) by identifying a number of relevant sequences. This technique allows the measures of viral DNA or RNA diffusion in near-infrared laser frequency [120]. The approach offers fast diagnostics (60 s) for detecting and characterizing reproducible spectra-generating viruses without viral manipulation. The spectroscopic analysis based on silver nanotubes amplifies significantly the signal for fast identification of virus [120], allowed enhanced screening, debridement targeting, and improved wound diagnosis, as well as location of low to high bacterial load regions [121]. Fluorescence imaging was carried out in the routine evaluation of diabetic foot ulcers (DFUs) to determine the required degree and location of debridement based on bacterial fluorescence signals. In a clinical trial, the sample showed clear bacterial fluorescence signals [121]. In order to determine the occurrence of bacterial infections, a research was intended to determine that the bacterial fluorescence can be used with evaluation of the signs and symptoms of infection combined with the use of bacterial fluorescence bioimaging, which increased accuracy to recognize low to high bacterial load wounds (≥ 104 CFU/g) [122]. Two of the recent synthesized compounds chosen for their non-cytotoxic existence and tested for bioimaging of bacterial cells using a high-quality screening assay demonstrate that the molecule is appropriate for imaging microbe's diagnosis. The novel substituted pyrroles were synthesized by the reaction of phenyl chalcone and methyl iodide under mild conditions. The pyrene chromophore is a synthesized compound responsible for strong photophysical properties characterized by absorption at 340 nm and emission at 410 nm. The cytotoxicity test has confirmed that these substances are also nontoxic to mammalian cells [123].

Nanobiosensors based on electrochemical luminescent aptamers are biosensors that, compared with conventional methods, offer high sensitivity, rapid response, specificity, and desired portability, as well as simplicity and reduced cost. They are mainly used for the determination of biomarkers of proteins, in particular for cancer detection [124].

Aptamers are small molecules of single-stranded DNA or RNA, which fold into a single form and bind to pathogens. Nanoscale aptasensors, in conjunction with nanomaterials, provide efficient analytical platforms for *Mycobacterium tuberculosis* [125], tuberculosis, mycobacterial proteins, and IFN-γ tuberculosis diagnosis.

11.8 IMPLANTS AND TISSUE ENGINEERING

Researchers are trying to extend bone prostheses by developing nanocomposite materials, and the goal is to prevent the patient's immune system from rejecting the graft. The strength of these composite materials is increasing. A new composition of the prosthesis will reduce the frequency of replacement, thus avoiding major surgical operations by moving toward a nanoporous coating that promotes blood flow through the implant with antibacterial and anti-inflammatory properties.

Radiography and X-ray scanners are used, but these technologies cannot be evaluated accurately. To date, medical imaging technology has not been able to evaluate, visualize, and monitor the integration of biological materials. For example, metal orthopedic implants slowly corrode in the internal environment. Continued tissue exposure to corrosive metal products can limit the effectiveness of implants. However, biological material selenium (Se) in the field of orthopedics showed that the adhesion of osteoblasts to the particle surface of traditional selen metal tablets was improved compared with the non-articulated forged titanium reference plate [126].

The nanoparticle-fibrin-poly (lactide-co-caprolactone) (PLCL) feature was adopted in this analysis. The PLCL scaffold had a porosity of 85% and pores of 300–500 microns. By a solvent diffusion process, heparin-functionalized nanoparticles were prepared. TGF-beta1 was mounted onto the NPs and was injected for 5 days in nude mice. Therefore, the differentiation of chondrogenic lineage HASCs in situ can be caused by hybridization of the three-dimensional organization of fibrin and PLCL scaffold cells and effective delivery of TGF-β1 using heparin-functionalized nanoparticles [127].

Tissue engineering could be an alternative to prosthetics by attempting to regenerate the tissue or organ itself. It involves the cultivation of cells from the tissue to be healed using chemical stimulation of cell growth. The cells will develop on a three-dimensional support, which must be biocompatible. In tissue and cell engineering growth, biomaterials play a key role. For example, gellan xanthan gum comprises fibroblast growth factor (bFGF) and bone morphogenetic protein 7 (BMP7) nanoparticles that are used in a dual growth factor delivery system to support human fetal osteoblasts for tissue engineering applications and to explain that the encapsulation and stabilization of growth factors in nanoparticles and gels are shown by chitosan nanoparticles. Compared with the gel loaded with single growth factor, the gel loaded with double growth factor showed higher alkaline phosphatase and calcium deposition.

It has also been shown that Xanthan gellan gum has antibacterial effects against *Pseudomonas aeruginosa, Staphylococcus aureus, and Staphylococcus epidermidis* (common implant failure pathogens) [128]. Molded hydrogel has broad possibilities for tissue engineering combined with stem cells. For the necessary time, they fix the encapsulated cells while allowing osmotic exchange with the ambient atmosphere of nutrients and substances. We hope to provide external osteogenesis cues in the form of nanoparticles, namely bioactive glass nanoparticles (BGn). The involvement of BGn enhanced the osteogenic differentiation of encapsulated stem cells of rat mesenchymal cells, exhibiting higher levels of activity of alkaline phosphatase and bone-bound genes including collagen and salivary proteins of form I. It may be used for bone tissue engineering to include targeted stem cells. [129]. The development of bioartificial draft organs or tissue models for cosmetology and pharmacology is also being considered.

11.9 ADVANTAGES AND LIMITATIONS OF NANO-BASED MATERIALS IN MEDICAL FIELD

Nanomedicine has become a very promising field and can also be used for radiotherapy, thermotherapy, photodynamic therapy, and diagnosis. In the following chapter, different medical and biomedical imaging technologies and their operating principles, advantages, and disadvantages will be presented. It has a nanometric size ranging from nanometers to hundreds of nanometers. It is biodegradable or biocompatible. It can be associated with active treatment ingredients and/or probes (contrast agents, luminescent elements, radioactive elements) used for the diagnosis or targeting of cells. It is authorized to be associated to be used in the diagnosis of cancer patients. It has good physical and chemical stability and is easy to internalize in the target cells by endocytosis, easy to be ingested, and biodegradable. The absence of drug interactions with healthy cells, thus avoiding side effects, is one of the key advantages.

Quantification of the number of phagocytic particles appears to be the key factor for insight into the mechanisms of toxicity of nanoparticles. Flow cytometry (FCM) and confocal microscopy quantify absorption to distinguish entirely absorbed micro-sized fluorescent particles from those that essentially bind to the cell membrane and compare these effects with in-depth measurements of toxicity. Biological toxicity was assessed in the processing of particles [lactate dehydrogenase release (LDH), tumor necrosis factor-alpha (TNF-α), and reactive oxygen species (ROS)]. These findings indicate that the degree of phagocytosis mainly depends on the size of the chemical classes of particles and surface chemicals. After 24 h of incubation, only the production of TNF-α and the complete ROS had significantly improved [130]. Following the discovery of the various types of nanomaterials, attention is currently being given to the possible hazards related to exposure to these particles, with a focus on the proper evaluation of health threats and the criteria to be established under appropriate regulations. As regards the nanoscale risk evaluation and special characteristics, they should be handled separately because they are very different from typical chemicals [131]. Many concerns remain unanswered regarding the effect of nanoparticles introduced into the living organism. Potential toxic effects are related only to the in vivo use of nanoparticles such as QDs made of fluorescent calcium selenide or zinc sulfide and cadmium and zinc ions [45]. The chronic toxicity of nanoparticles of $ZnSO_4$ and ZnO was investigated in large chips up to the end of their life and analyzed under a microscope. While a very low dosage and a very low dose frequency (0.3 mg Zn/L) were used, ultrastructural analysis of the midgut epithelial cells indicated that the NPs also were internalized around 48 h and 9 days and passed to the other tissues. This raises concerns about their use, which should again be carefully considered. Twenty days later, the areas most impacted by these two compounds were reported. Specifically, samples exposed to ZnO nanoparticles displayed mitochondrial swelling, while a large amount of autophagous vacuoles were found in samples exposed to $ZnSO_4$. In the cytotoxicity of ZnO nanoparticles, the soluble kind plays a key role, and the shape of nanoparticles may increase intracellular Zn content locally, as well as in the ovaries [132].

In other recent studies, carbon nanomaterials, an important part of the circulatory system responsible for blood clotting and hemostasis, have been shown to adversely affect platelets. Significant factors that lead to graphene toxicity may be the shape of graphene sheets or their agglomeration.

The involvement of a core of heavy metal in QDs poses significant questions about their toxicity potential. An outbreak of research on the possible uses of QDs in in situ and in vitro biological imaging has occurred. A lack of sufficient knowledge of the in vitro toxicity of QD in humans has precluded its application. For the reduction of toxicity, detoxifying agents such as glutathione or gelatin and peptide coating have been discovered. An exciting solution to this concern is the use of nontoxic QD formulations. The possible toxicity of carbon nanotubes is characterized by their inhalation-related uniform volume, aspect ratio, and other parameters.

The current assessment of the potential toxicological effects can be consistent with other available studies and previous calibration analyses. Through macrophage-based normalization, the critical value of important dimensionless parameters is characterized. The critical value characterizes the phagocytosis efficiency of different nanotubes. The study concludes that the potential toxicity caused by inhalation of long and short carbon nanotubes is similar to asbestos fibers [133].

Through macrophage-based normalization, the critical value of important dimensionless parameters is characterized, and the critical value characterizes the phagocytosis efficiency of different nanotubes. The present measurement of the toxicological impact of the height/width of high carbon nanotube ratios can be consistent with other available studies and previous calibration analyses characterized by standardized length, aspect ratio, and other parameters related to its inhalation, macrophage phagocytosis, and phagocytosis efficiency [84]. Fullerene is a model of nanoparticles constructed from carbon. In several fields of study, including materials science and biomedical applications, it has generated a lot of interest. Questions regarding their protection and environmental effects have been posed by the use of fullerenes. This chapter presents fullerene or C60 (the most prevalent fullerene) toxicity studies and its derivatives [134].

11.10 CONCLUSION AND PROSPECTS

Nanotechnologies made from nanobiosensor and nanobiomarker materials allow rapid diagnosis and engage in the potential development of these analytical approaches. This suggests that the instruments can quickly activate the separation of chemical sources, providing high-precision point detection.

Another trend has been made so that fluorescent labeling methods using fluorescent nanoparticles allow the development of microscopic sensors capable of detecting chemical patterns in the fluid. However, it is currently difficult to set up the definition of a toxic dose allowing the properties of NPs to be taken into account. In large variety and sizes of nanoparticles, analytical results may differ. NPs also spread differently and can have harmful effects and even toxicity. Noninvasive monitoring of these new diagnostic approaches is necessary. The nanomaterials and the instruments that can be produced are very effective for diagnostics and selective drug delivery, and future advances in nanotechnology diagnostics will continue. The prospects of nano-application seem limitless, for example, the incorporation of nanomaterials into medicines for improving the diagnosis and treatment of all sorts of diseases, cosmetics, textiles, or food contact materials. Moreover, it is expected that emerging technology using nanomaterials can play a key role in waste cleaning and creating effective energy storage systems.

Nanotechnology-based analysis instruments are considered as useful prototypes and can make thousands of measurements rapid and very efficient. The current trend is to either create essential diagnostic devices or to make some changes that constitute functional nanoparticles for fluorescent labeling methods for medical diagnostics, efficient and effective medical delivery, and artificial cell production. Currently, these methods can easily induce in vivo differentiation of a single-cell-sized chemical source from the related chemical concentration, offering high-resolution sensing for both time and space. Future challenges are to design and produce contrast agents and safe polymeric nanomaterials.

ACKNOWLEDGEMENT

We thank Dr. Ravindra Pratap Singh for assistance with Mr. Kshitij RB Singh for comments that greatly improved the manuscript.

REFERENCES

1. Y. Zhang, M. Li, X. Gao, Y. Chen and T. Liu. 2019. "Nanotechnology in cancer diagnosis: progress, challenges and opportunities". *J Hematol Oncol* 12: 137. Doi: 10.1186/s13045-019-0833-3.
2. J. Jose and K. Burgess. 2006. "Syntheses and properties of water-soluble Nile Red derivatives". *J Org Chem* 71(20): 7835–9. Doi: 10.1021/jo061369v.

3. H. L. Wong, R. Bendayan, A. M. Rauth, Y. Li and X. Y. Wu. 2007. "Chemotherapy with anticancer drugs encapsulated in solid lipid nanoparticles". *Adv Drug Deliv Rev* 59(6): 491–504. Doi: 10.1016/j.addr.2007.04.008.

4. A. C. Faure, S. Dufort, V. Josserand, P. Perriat, J. L. Coll, S. Roux and O. Tillement. 2009. "Control of the *in vivo* biodistribution of hybrid nanoparticles with different poly(ethylene glycol) coatings". *Small* 5(22): 2565–2575. Doi: 10.1002/smll.200900563.

5. J. Zhang, B. Liu, H. Liu, X. Zhang and W. Tan. 2013. "Aptamer-conjugated gold nanoparticles for bioanalysis". *Nanomedicine* 8(6): 983–93. Doi: 10.2217/nnm.13.80.

6. S. Si-Mohamed, D. P. Cormode, D. Bar-Ness, M. Sigovan, P. C. Naha, J.-B. Langlois, L. Chalabreysse, P. Coulon, I. Blevis, E. Roessl, K. Erhard, L. Boussel and P. Douek. 2017. "Evaluation of spectral photon counting computed tomography K-edge imaging for determination of gold nanoparticle biodistribution in vivo". *Nanoscale* 9(46): 18246–18257. Doi: 10.1039/c7nr01153a

7. A. Sukhanova, L. Venteo, J.H.M. Cohen, M. Pluot and I. Nabiev. 2006. "Nanobiocapteurs pour la recherche et les diagnostics des maladies inflammatoires et du cancer". *Annales Pharmaceutiques Françaises* 64(2): 125–134. Doi: 10.1016/S0003-4509(06)75305-6.

8. T. F. Massoud and S. S. Gambhir. 2003. "Molecular imaging in living subjects: seeing fundamental biological processes in a new light". *Genes Dev* 17(5): 545–80. Doi: 10.1101/gad.1047403.

9. N. V. S. Vallabani, S. Singh and A. S. Karakoti. 2019. "Magnetic nanoparticles: Current Trends and future aspects in diagnostics and nanomedicine". *Curr Drug Metab* 20(6): 457–472. Doi: 10.2174/1389200220666181122124458.

10. K. K. Jain. 2005. "The role of nanobiotechnology in drug discovery". *Drug Discov Today* 10(21): 1435–1442. Doi: 10.1016/S1359-6446(05)03573-7.

11. Y. Liu, M. K. Shipton, J. Ryan, E. D. Kaufman, S. Franzen and D. L. Feldheim. 2007. "Synthesis, stability, and cellular internalization of gold nanoparticles containing mixed peptide-poly(ethylene glycol) monolayers". *Anal Chem* 79(6): 2221–9. Doi: 10.1021/ac061578f.

12. D. P. Cormode, S. Si-Mohamed, D. Bar-Ness, M. Sigovan, P. C. Naha, J. Balegamire, F. Lavenne, P. Coulon, E. Roessl, M. Bartels, M. Rokni, I. Blevis, L. Boussel and P. Douek. 2017. "Multicolor spectral photon-counting computed tomography: *in vivo* dual contrast imaging with a high count rate scanner". *Sci Rep* 7: 4784. Doi: 10.1038/s41598-017-04659-9.

13. E. Cuccione, P. Chhour, S. Si-Mohamed, C. Dumot, J. Kim, V. Hubert, C. C. Da Silva, M. Vandamme, E. Chereul, J. Balegamire, Y. Chevalier, Y. Berthezène, L. Boussel, P. Douek, D. P. Cormode and M. Wiart. 2020. "Multicolor spectral photon counting CT monitors and quantifies therapeutic cells and their encapsulating scaffold in a model of brain damage". Nanotheranostics, 4(3): 129–141. Doi: 10.7150/ntno.45354.

14. P. Huang, L. Bao, C. Zhang, J. Lin, T. Luo, D. Yang, M. He, Z. Li, G. Gao, B. Gao, S. Fu and D. Cui. 2011. "Folic acid-conjugated silica-modified gold nanorods for X-ray/CT imaging-guided dual-mode radiation and photo-thermal therapy". *Biomaterials* 32(36): 9796–9809. Doi: 10.1016/j.biomaterials.2011.08.086.

15. H. Li and T. J. Meade. 2019. "Molecular magnetic resonance imaging with Gd(III)-based contrast agents: Challenges and key advances". *J Am Chem Soc* 141(43): 17025–17041. Doi: 10.1021/jacs.9b09149.

16. A. Louie. 2013. "MRI biosensors: A short primer". *J Magn Reson Imaging* 38(3): 530–539. Doi: 10.1002/jmri.24298

17. J. Lux and A. D. Sherry, 2018. "Advances in gadolinium-based MRI contrast agent designs for monitoring biological processes *in vivo*". *Curr Opin Chem Biol* 45: 121–130. Doi: 10.1016/j.cbpa.2018.04.006.

18. J. M. Idée, N. Fretellier, M. M. Thurnher, B. Bonnemain and C. Corot. 2015. "Physico-chimie et profil toxicologique d'agents de contraste pour l'imagerie par résonance magnétique, les chélates de gadolinium [Physico-chemical and toxicological profile of gadolinium chelates as contrast agents for magnetic resonance imaging]". *Ann Pharm Fr*, 73(4): 266–276. Doi: 10.1016/j.pharma.2015.01.001.

19. M. Bödenler, K. P. Malikidogo, J.-F. Morfin, C. S. Aigner, É. Tóth, C. S. Bonnet, H. Scharfetter. 2019. "High-field detection of biomarkers with fast field-cycling MRI: The example of zinc sensing". *Chemistry* 25(35): 8236–8239. Doi: 10.1002/chem.201901157.

20. A. C. Esqueda, J. A. López, G. Andreu-de-Riquer, J. C. Alvarado-Monzón, J. Ratnakar, A. J. M. Lubag, A. D. Sherry and L. M. De León-Rodríguez. 2009. "A new gadolinium-based MRI zinc sensor". *J Am Chem Soc* 131(32): 11387–11391. Doi: 10.1021/ja901875v.

21. A. F Martins, V. C. Jordan, F. Bochner, S. Chirayil, N. Paranawithana, S. Zhang, S.-T. Lo, X. Wen, P. Zhao, M. Neeman and A. Dean Sherry. 2018. "Imaging insulin secretion from mouse pancreas by MRI is improved by use of a zinc-responsive MRI sensor with lower affinity for Zn^{2+} ions". *J Am Chem Soc* 140(50): 17456–17464. Doi: 10.1021/jacs.8b07607.

22. M. Qin, Y. Peng, M. Xu, H. Yan, Y. Cheng, X. Zhang, D. H. W. Chen and Y. Meng. 2020. "Uniform Fe_3O_4/Gd_2O_3-DHCA nanocubes for dual-mode magnetic resonance imaging". *Beilstein J Nanotechnol* 11: 1000–1009. Doi: 10.3762/bjnano.11.84.

23. K. Niemirowicz, K. H. Markiewicz, A. Z. Wilczewska and H. Car. 2012. "Magnetic nanoparticles as new diagnostic tools in medicine". *Adv Med Sci* 57(2): 196–207. Doi: 10.2478/v10039-012-0031-9.

24. H. Shao, C. Min, D. Issadore, M. Liong, T.-J. Yoon, R. Weissleder and H. Lee. 2012. "Nanoparticules magnétiques et microNMR pour les applications de diagnostic". *Theranostics* 2(1): 55–65. Doi: 10.7150/thno.3465.

25. Z. Zhao, Z. Zhou, J. Bao, Z. Wang, J. Hu, X. Chi, K. Ni, R. Wang, X. Chen, Z. Chen and J. Gao. 2013. "Octapod iron oxide nanoparticles as high-performance T2 contrast agents for magnetic resonance imaging". *Nat Commun* 4: 2266. Doi: 10.1038/ncomms3266.

26. F. Sabban, P. Collinet, M. Cosson, S. Mordon. 2004. "Technique d'imagerie par fluo-rescence: intérêt diagnostique et thérapeutique en gynécologie". 33(8): 734–738. Doi: 10.1016/S0368-2315(04)96635-5.

27. D. Jirak, A. Galisova, K. Kolouchova, D. Babuka and M. Hruby. 2019. "Fluorine polymer probes for magnetic resonance imaging: quo vadis?" *MAGMA* 32(1): 173–185. Doi: 10.1007/s10334-018-0724-6.

28. P. Ünak. 2008. "Imaging and therapy with radionuclide labeled magnetic nanoparticles". *Braz Arch Biol Technol* 52: 31–37.

29. M. M. Yallapu, S. P. Foy, T. K. Jain, and V. Labhasetwar. 2010. "PEG-functionalized magnetic nanoparticles for drug delivery and magnetic resonance imaging applications." *Pharma Res* 27(11): 2283–2295. Doi: 10.1007/s11095-010-0260-1.

30. S. Peng, Q.-y. Wang, X. Xiao, R. Wang, J. Lin, Q.-h. Zhoua and L.-n. Wu. 2020. "Redox-responsive polyethyleneimine-coatedmagnetic iron oxide nanoparticlesfor controllable gene delivery and magneticresonance imaging". *Polym Int* 69: 206–214. Doi: 10.1002/pi.5943.

31. R. Martínez-González, J. Estelrich and M.A. Busquets. 2016. "Liposomes loaded with hydrophobic iron oxide nanoparticles: Suitable T2 contrast agents for MRI". *Int J Mol Sci* 17(8): 1209. Doi: 10.3390/ijms17081209.

32. N. Kostevšek, C.C.L. Cheung, I. Serša, M.E. Kreft, I. Monaco, M.C. Franchini, J. Vidmar, and W.T. Al-Jamal. 2020. "Magneto-liposomes as MRI contrast agents: A systematic study of different liposomal formulations". *Nanomaterials* 10(5): 889. Doi: 10.3390/nano10050889.

33. G. L. Baiocchi, M. Diana and L. Boni. 2018. "Indocyanine green-based fluorescence imaging in visceral and hepatobiliary and pancreatic surgery: State of the art and future directions". *World J Gastroenterol* 24(27): 2921–2930. Doi: 10.3748/wjg.v24.i27.2921.

34. C. Jonak, H. Skvara, R. Kunstfeld, F. Trautinger and J. A. Schmid. 2011. "Intradermal indocyanine green for *in vivo* fluorescence laser scanning microscopy of human skin: A pilot study". *PLoS One* 6(8): e23972. Doi: 10.1371/journal.pone.0023972.

35. G. Cui, W. J. Akers, M. J. Scott, M. Nassif, J. S. Allen, A. H. Schmieder, K. S. Paranandi, A. Itoh, D. D. Beyder, S. Achilefu, G. A. Ewald, and G. M. Lanza. 2018. "Diagnosis of LVAD thrombus using a high-avidity fibrin-specific[99mTc] probe". *Theranostics* 8(4): 1168–1179. Doi: 10.7150/thno.20271.

36. S. Iikuni, M. Ono, H. Watanabe, K. Matsumura, M. Yoshimura, H. Kimura, H. Ishibashi-Ueda, Y. Okamoto, M. Ihara and H. Saji. 2016. "Imaging of cerebral amyloid angiopathy with bivalent[99mTc]-hydroxamamide complexes". *Sci Rep* 16(6): 25990. Doi: 10.1038/srep25990.

37. C. J. Anderson and R. Ferdani. 2009. "Copper-64 radiopharmaceuticals for PET imaging of cancer: advances in preclinical and clinical research." *Canc Biother Radiopharmac.* 24(4): 379–93. Doi: 10.1089/cbr.2009.0674.

38. I. Sonni, L. Baratto and A. Iagaru. 2017. "Imaging of prostate cancer using gallium-68-labeled bombesin". *PET Clin* 12(2): 159–171. Doi: 10.1016/j.cpet.2016.11.003.

39. Y. Sun, X. Ma, Z. Zhang, Z. Sun, M. Loft, B. Ding, C. Liu, L. Xu, M. Yang, Y. Jiang, J. Liu, Y. Xiao, Z. Cheng and X. Hong.2016. "Preclinical study on GRPR-targeted (68) Ga-probes for PET imaging of prostate cancer". *Bioconjug Chem* 17;27(8): 1857–1864. Doi: 10.1021/acs.bioconjchem.

40. F. A. Mettler, M. J. Guiberteau. 2012. "Thyroid, parathyroid, and salivary glands". *Essentials of Nuclear Medicine Imaging* (Sixth Edition), 99–130. Doi: 10.1016/B978-1-4557-0104-9.00004-4.

41. B. Infeld, S. M. Davis. 2004. "Single-photon emission computed tomography". *Pathophysiology, Diagnosis, and Management, Stroke* (Fourth Edition) Doi: 10.1016/B0-44-306600-0/50027-4.

42. S. M. Evans and C. J. Koch. 2003. "Prognostic significance of tumor oxygenation in humans". *Can Lett* 195: 1–16. Doi: 10.1016/s0304-3835(03)00012-0.

43. S. Deka, A. Quarta, M. G. Lupo, A. Falqui, S. Boninelli, C. Giannini, G. Morello, M. De Giorgi, G. Lanzani, C. Spinella, R. Cingolani, T. Pellegrino and L. Manna. 2009. "CdSe/CdS/ZnS double shell nanorods with high photoluminescence efficiency and their exploitation as biolabeling probes". *J Am Chem Soc* 131(8): 2948–2958. Doi: 10.1021/ja808369e.

44. D. G. Calatayud, H. Ge, N. Kuganathan, V. Mirabello, R. M. J. Jacobs, N. H. Rees, C. T. Stoppiello Andrei, N. Khlobystov, R. M. Tyrrell, E. Da Como and S. I. Pascu. 2018. "Encapsulation of cadmium selenide nanocrystals in biocompatible nanotubes: DFT calculations, X-ray diffraction investigations, and confocal fluorescence imaging". *Chem Open* 7(2): 144–158. Doi: 10.1002/open.201700184.

45. H.-J. Kim, J.-H. Jo, S.-Y. Yoon, D.-Y. Jo, H.-S. Kim, B. Park and H. Yang. 2019. "Emission enhancement of Cu-doped InP quantum dots through double shelling scheme". *Materials* 12(14): 2267. Doi: 10.3390/ma12142267.

46. E. A. Owens, M. Henary, G. E. Fakhri, and H. S. Choi. 2016. "Tissue-specific near-infrared fluorescence imaging". *Acc Chem Res* 49(9): 1731–1740. Doi: 10.1021/acs.accounts.6b00239.

47. D. Jo and H. Hyun. 2017. "Structure-inherent targeting of near-infrared fluorophores for image-guided surgery". *Chonnam Med J* 53(2): 95–102. Doi: 10.4068/cmj.2017.53.2.95.

48. S. Kwon and E. Sevick. 2011. "Non-invasive near-infrared fluorescence imaging of murine intestinal motility". The World Molecular Imaging Congress (WMIC), Scientific Session 10: *In Vivo. Translational Molecular Imaging*, 13:30-13:45. http://www.wmis.org/abstracts/2011/data/papers/T107.html.

49. J. S. Jung, D. Jo, G. Jo and H. Hyun. 2019. "Near-infrared contrast agents for bone-targeted imaging". *Tissue Eng Regen Med* 16(5): 443–450. Doi: 10.1007/s13770-019-00208-9.

50. H. Shi, N. Zhao, D. Ding, J. Liang, B. Z. Tang and B. Liu. "Fluorescent light-up probe with aggregation-induced emission characteristics for *in vivo* imaging of cell apoptosis". *Org Biomol Chem* 11(42): 7289–7296. Doi: 10.1039/c3ob41572d.

51. N. C. Shaner, P. A. Steinbach and R. Y. Tsien. 2005. "Steinbach? Tsien RY. A guide to choosing fluorescent proteins". *Nat Meth* 2(12): 905–909. Doi: 10.1038/nmeth819.

52. M. Wolff, J. Wiedenmann, G. U. Nienhaus, M. Valler and R. Heilker. 2006. "Novel fluorescent proteins for high-content screening". *Drug Discov Today* 11(23–24): 1054–1060. Doi: 10.1016/j.drudis.2006.09.005.

53. M. G. Khrenova, A. V. Nemukhin and T. Domratcheva. 2015. "Theoretical characterization of the flavin-based fluorescent protein iLOV and its Q489K mutant". *J Phys Chem B* 119(16): 5176–5183. Doi: 10.1021/acs.jpcb.5b01299.

54. B. Kopka, K. Magerl, A. Savitsky, M. D. Davari, K. Röllen, M. Bocola, B. Dick, U. Schwaneberg, K.-E. Jaeger and U. Krauss. 2017. "Electron transfer pathways in a light, oxygen, voltage (LOV) protein devoid of the photoactive cysteine". *Sci Rep* 7:13346. Doi: 10.1038/s41598-017-13420-1.

55. M.-A. Plamont, E. Billon-Denis, S. Maurin, C. Gauron, F. M. Pimenta, C. G. Specht, J. Shi, J. Quérard, B. Pan, J. Rossignol, K. Moncoq, N. Morellet, M. Volovitch, E. Lescop, Y. Chen, A. Triller, S. Vriz, T. Le Saux, L. Jullien, and A. Gautier. 2016. "Small fluorescence-activating and absorption-shifting tag for tunable protein imaging *in vivo*". *Proc Natl Acad Sci USA* 113(3): 497–502. Doi: 10.1073/pnas.1513094113.

56. T. Péresse and A. Gautier. 2019. "Next-generation fluorogen-based reporters and biosensors for advanced bioimaging". *Int J Mol Sci* 20(24): 6142. Doi: 10.3390/ijms 20246142μ.

57. N. V. Povarova, S. O. Zaitseva, N. S. Baleeva, A. Y. Smirnov, I. N. Myasnyanko, M. B. Zagudaylova, N. G. Bozhanova, D. A. Gorbachev, K. K. Malyshevskaya, A. S. Gavrikov, A. S. Mishin and M. S. Baranov. 2019. "Red-shifted substrates for FAST fluorogen-activating protein based on the GFP-like chromophores". *Chemistry* 25(41): 9592–9596. Doi: 10.1002/chem.201901151.

58. F. Broch and A. Gautier. 2020. "Illuminating cellular biochemistry: Fluorogenic chemogenetic biosensors for biological imaging". *Chempluschem* 85(7): 1487–1497. Doi: 10.1002/cplu.202000413.

59. A. Escudero, C. Carrillo-Carrión, M. V. Zyuzin, S. Ashraf, R. Hartmann, N. O. Núñez, M. Ocañab and W. J. Paraka. 2016. "Synthesis and functionalization of monodispers near-ultraviolet and visible excitable multifunctional Eu^{3+}, Bi^{3+}:$REVO^4$ nanophosphors for bioimaging and biosensing applications". *Nanoscale* 8: 12221–12236. Doi:10.1039/ c6nr03369e.

60. M. Laguna, N. O. Nuñez, A. I. Becerro, G. Lozano, M. Moros, J. M. de la Fuente, A. Corral, M. Balcerzyk and M. Ocaña. 2019. "Synthesis, functionalization and properties of uniform europium-doped sodium lanthanum tungstate and molybdate (NaLa(XO4)2, X = Mo,W) probes for luminescent and X-ray computed tomography bioimaging". *J Colloid Interface Sci* 554: 520–530. Doi: 10.1016/j.jcis.2019.07.031.

61. H. Li, J. Xi, A. G. Donaghue, J. Keum, Y. Zhao, K. An, E. R. McKenzie and F. Ren. 2020. "Synthesis and catalytic performance of polydopamine supported metal nanoparticles". *Sci Rep* 10: 1–7. Doi: 10.1038/s41598-020-67458-9.

62. D. Ho. 2015. "Nanodiamond-based chemotherapy and imaging". *Cancer Treat Res* 166: 85–102. Doi: 10.1007/978-3-319-16555-4_4.

63. A. Awan, R. S. Oliveri, P. L. Jensen, S. T. Christensen and C. Y. Andersen. 2010. "Analyse immunoflourescence and mRNA analysis of human embryonic stem cells (hESCs) grown under feeder-free conditions. In: Turksen K. (eds) *Human Embryonic Stem Cell Protocols. Methods in Molecular Biology (Methods and Protocols)*. Humana Press 584: 195–210 Doi: 10.1007/978-1-60761-369-5_11.

64. E. De Boer, J. M. Warram, M. D. Tucker, Y. E. Hartman, L. S. Moore, J. S. de Jong, T. K. Chung, M. L. Korb, K. R. Zinn, G. M. van Dam, E. L. Rosenthal and M. S. Brandwein-Gensler. 2015. "In Vivo fluorescence immunohistochemistry: Localization of fluorescently labeled cetuximab in squamous cell carcinomas". *Sci Rep* 29(5): 10169. Doi: 10.1038/srep10169.

65. U. Ghoshal, S. Khanduja, P. Pant and U.C. Ghoshal. 2016. "Evaluation of Immunoflourescence antibody assay for the detection of Enterocytozoon bieneusi and Encephalitozoon intestinalis". *Parasitol Res* 115(10): 3709–3713. Doi: 10.1007/s00436-016-5130-2.

66. M. Bernhardt, J. D. Nicolas, M. Osterhoff, H. Mittelstädt, M. Reuss, B. Harke, A. Wittmeier, M. Sprung, S. Köster and T. Salditt. 2018. "Correlative microscopy approach for biology using X-ray holography, X-ray scanning diffraction and STED microscopy". *Nat Commun* 9(1): 3641. Doi: 10.1038/s41467-018-05885-z.

67. Z. Jiang, J. W. Strzalka, D. A. Walko and J. Wang. 2020. "Reconstruction of evolving nanostructures in ultrathin films with X-ray waveguide fluorescence holography". *Nat Commun* 11(1): 3197. Doi: 10.1038/s41467-020-16980-5.

68. G. Vámosi, E. Friedländer-Brock, S. M. Ibrahim, R. Brock, J. Szöllősi and G. Vereb. 2019. "EGF receptor stalls on activation as evidenced by complementary fluorescence correlation spectroscopy and fluorescence recovery after photobleaching measurements". *Int J Mol Sci* 20(13): 3370. Doi: 10.3390/ijms20133370.

69. N. L. Farnsworth, A. Hemmati, M. Pozzoli and R. K. Benninger. 2014. "Fluorescence recovery after photobleaching reveals regulation and distribution of connexin36 gap junction coupling within mouse islets of Langerhans". *J Physiol* 592(20): 4431–4446. Doi: 10.1113/jphysiol.2014.276733. Epub 2014 Aug 28. Erratum in: J Physiol. 2015 Jul 15;593(14): 3223.

70. F. Ponzio. 2016. "Synthesis at different interfaces of bio-inspired films from mussels' byssus: influence of the oxidant nature at the solid/liquid interface and the addition of polymer at the air/water interface". Theoretical and/or physical chemistry. Université de Strasbourg, English. (NNT: 2016STRAE041). (tel-01673807v2).

71. J. Liebscher. 2019. "Chemistry of polydopamine – scope, variation, and limitation". *Eur J Org Chem* 4976–4994. Doi: 10.1002/ejoc.201900445.

72. F. Ponzio and V. Ball. 2016. "Polydopamine deposition at fluid interfaces". *Polym Int* 65: 1251–1257. Doi: 10.1002/pi.5124.

73. T. Rendler, J. Neburkova, O. Zemek, J. Kotek, A. Zappe, Z. Chu, P. Cigler and J. Wrachtrup. 2017. "Optical imaging of localized chemical events using programmable diamond quantum nanosensors". *Nat Commun* 8: 14701. Doi: 10.1038/ncomms14701.

74. A. S. Barnard. 2009. "Diamond standard in diagnostics: nanodiamond biolabels make their mark". *Analyst* 134(9): 1751–1764. Doi: 10.1039/b908532g.

75. O. Faklaris, V. Joshi, T. Irinopoulou, P. Tauc, M. Sennour, H. Girard, C. Gesset, J.-C. Arnault, A. Thorel, J.-P. Boudou, P. A. Curmi and F. Treussart. 2009. "Photoluminescent diamond nanoparticles for cell labeling: study of the uptake mechanism in mammalian cells". *ACS Nano* 3(12): 3955–3962. Doi: 10.1021/nn901014j.

76. A.F. Hezinger, J. Tessmar and A. Göpferich. 2008. "Polymer coating of quantum dots--a powerful tool toward diagnostics and sensorics". *Eur J Pharm Biopharm* 68(1):138–152. Doi: 10.1016/j.ejpb.2007.05.013.

77. A. Sharma and J. Das. 2019. "Small molecules derived carbon dots: synthesis and applications in sensing, catalysis, imaging, and biomedicine". *J Nanobiotechnology* 17(1): 92. Doi: 10.1186/s12951-019-0525-8.

78. T. S. Atabaev, Z. Piao and A. Molkenova. 2018. "Carbon dots doped with dysprosium: A bimodal nanoprobe for MRI and fluorescence imaging". *J Funct Biomater* 9: 35. Doi: 10.3390/jfb9020035.

79. M. Rothbauer, V. Charwat and P. Ertl. 2016. "Cell microarrays for biomedical applications". *Methods Mol Biol* 1368: 273–291. Doi: 10.1007/978-1-4939-3136-1_19.

80. G. Bogdanović and A. Djordjević. 2016. "Carbon nanomaterials: Biologically active fullerene derivatives". *Srp Arh Celok Lek* 144 (3–4): 222–231.

81. P. Sharma, N. K. Mehra, K. Jain and N. K. Jain. 2016. "Biomedical applications of carbon nanotubes: A critical review". *Curr Drug Deliv* 13(6): 796–817. Doi: 10.2174/1567 201813666160623091814.

82. M. I. Sajid, U. Jamshaid, T. Jamshaid, N. Zafar, H. Fessi and A. Elaissari. 2016. "Carbon nanotubes from synthesis to in vivo biomedical applications". *Int J Pharm* 501(1–2): 278–299. Doi: 10.1016/j.ijpharm.2016.01.064.

83. V. M. Harik. 2017. "Geometry of carbon nanotubes and mechanisms of phagocytosis and toxic effects". *Toxicol Lett* 273: 69–85. Doi: 10.1016/j.toxlet.2017.03.016.

84. A. Djordjević, G. Bogdanović and S. Dobrić. 2006. "Fullerenes in biomedicine". *J BUON* 11(4): 391–404.

85. Z. Chen, R. Mao and Y. Liu. 2012. "Fullerenes for cancer diagnosis and therapy: preparation, biological and clinical perspectives". *Curr Drug Metab* 13(8): 1035–1045. Doi: 10.2174/138920012802850128.

86. R. Bakry, R. M. Vallant, M. Najam-ul-Haq, M. Rainer, Z. Szabo, C. W. Huck and G. K. Bonn 2007. "Medicinal applications of fullerenes". *Int J Nanomedicine* 2(4): 639–649.

87. M. Tahriri, M. Del Monico, A. Moghanian, M. Tavakkoli Yaraki, R. Torres, A. Yadegari and L. Tayebi. 2019. "Graphene and its derivatives: Opportunities and challenges in dentistry". *Mater Sci Eng C Mater Biol Appl* 102: 171–185. Doi: 10.1016/j.msec.2019.04.051.

88. P. Bollella, G. Fusco, C. Tortolini, G. Sanzò, G. Favero, L. Gorton and R. Antiochia. 2017. "Beyond graphene: Electrochemical sensors and biosensors for biomarkers detection". *Biosens Bioelectron* 89: 152–166. Doi: 10.1016/j.bios.2016.03.068.

89. M. E. Foo and S. C. B. Gopinath. 2017. "Feasibility of graphene in biomedical applications". *Biomed Pharmacother* 94: 354–361. Doi: 10.1016/j.biopha.2017.07.122.

90. Q. Wang, Y. H. Li, W. J. Jiang, J. G. Zhao, B.G. Xiao, G. X. Zhang and C. G. Ma. 2018. "Graphene-based nanomaterials: Potential tools for neurorepair". *Curr Pharm Des* 24(1): 56–61. Doi: 10.2174/1381612823666170828130526.

91. M. Orecchioni, C. Ménard-Moyon, L. G. Delogu and A. Bianco 2016. "Graphene and the immune system: Challenges and potentiality". *Adv Drug Deliv Rev* 105: 163–175 Doi: 10.1016/j.addr.2016.05.014.

92. R. Srikar, A. Upendran and R. Kannan. 2014. "Polymeric nanoparticles for molecular imaging". *Wiley Interdiscip Rev Nanomed Nanobiotechnol* 6(3): 245–267. Doi: 10.1002/wnan.1259.

93. B. L. Banik, P. Fattahi and J. L. Brown. 2016. "Polymeric nanoparticles: The future of nanomedicine". *Wiley Interdiscip Rev Nanomed Nanobiotechnol* 8(2): 271–299. Doi: 10.1002/wnan.1364"002/wnan.1364.

94. G. M. Bernal, M. J. LaRiviere, N. Mansour, P. Pytel, K. E. Cahill, D. J. Voce, S. Kang, R. Spretz, U. Welp, S. E. Noriega, L. Nunez, G. F. Larsen, R. R. Weichselbaum and B. Yamini. 2014. "Convection-enhanced delivery and in vivo imaging of polymeric nanoparticles for the treatment of malignant glioma". *Nanomedicine* 10(1): 149–157. https://doi.org/10.1016/j.nano.2013.07.003.

95. D. J. Tng, P. Song, G. Lin, A. M. Soehartono, G. Yang, C. Yang, F. Yin, C. H. Tan and K. T. Yong. 2015. "Synthesis and characterization of multifunctional hybrid-polymeric nanoparticles for drug delivery and multimodal imaging of cancer". *Int J Nanomed* 10: 5771–5786. Doi: 10.2147/IJN.S86468.

96. N. Kamaly and A. D. Miller. 2010. "Paramagnetic liposome nanoparticles for cellular and tumour imaging". *Int J Mol Sci* 11(4): 1759–1776. Doi: 10.3390/ijms11041759.

97. N. Lamichhane, T. S. Udayakumar, W. D. D'Souza, C. B. Simone, S. R. Raghavan, J. Polf and J. Mahmood. 2018. "Liposomes: Clinical applications and potential for image-guided drug delivery". *Molecules* 23(2): 288. Doi: 10.3390/molecules23020288.

98. Y. Xia, C. Xu, X. Zhang, P. Ning, Z. Wang, J. Tian and X. Chen. 2019. "Liposome-based probes for molecular imaging: From basic research to the bedside". *Nanoscale* 11(13): 5822–5838. Doi: 10.1039/c9nr00207c.

99. A. J. Almeida and E. Souto. 2007. "Solid lipid nanoparticles as a drug delivery system for peptides and proteins". *Adv Drug Deliv Rev* 59: 478–490.

100. B. Nottelet, V. Darcos and J. Coudane. 2015. "Aliphatic polyesters for medical imaging and theranostic applications". *Eur J Pharm Biopharm* 97: 350–370. Doi: 10.1016/j.ejpb.2015.06.023.

101. M. Mirahadi, S. Ghanbarzadeh, M. Ghorbani, A. Gholizadeh and H. Hamishehkar. 2018. "A review on the role of lipid-based nanoparticles in medical diagnosis and imaging". *Ther Deliv* 9(8):557–569. Doi: 10.4155/tde-2018-0020.

102. J. G. Duarte and J. M. Blackburn. 2017. "Advances in the development of human protein microarrays". *Expert Rev Proteomics* 14(7): 627–641. Doi: 10.1080/14789450.2017.

103. W. G. Lee, Y.-G. Kim, B. G. Chung, U. Demirci and A. Khademhosseini. 2010. "Nano/Microfluidics for diagnosis of infectious diseases in developing countries". *Adv Drug Deliv Rev*, 62(4–5): 449–457. Doi: 10.1016/j.addr.2009.11.016.

104. E. Bottaro and C. Nastruzzi. 2016. "Off-the-shelf microfluidic devices for the production of liposomes for drug delivery". *Mater Sci Eng C Mater Biol Appl* 64: 29–33. Doi: 10.1016/j.msec.2016.03.056.

105. Q. Feng, J. Wilhelm and J. Gao. 2019. "Transistor-like ultra-pH-sensitive polymeric nanoparticles". *Acc Chem Res* 52(6): 1485–1495. Doi: 10.1021/acs.accounts.9b00080.

106. S. Pedroso and I. A. Guillen. 2006. "Microarray and nanotechnology applications of functional nanoparticles". *Comb Chem High Throughput Screen* 9(5): 389–397. Doi: 10.2174/138620706777452438.

107. K. R. B. Singh, P. Sridevi and R. P. Singh. 2020. "Potential applications of peptide nucleic acid in biomedical domain". *Eng Rep* 2: e12238. Doi: 10.1002/eng2.12238.

108. V. Krivitsky, M. Zverzhinetsky and F. Patolsky. 2020. "Redox-reactive field-effect transistor nanodevices for the direct monitoring of small metabolites in biofluids toward implantable nanosensors arrays". *ACS Nano* 14(3): 3587–3594. Doi: 10.1021/acsnano.9b10090.

109. Y. Han, Y. An, G. Jia, X. Wang, C. He, Y. Dinga and Q. Tang 2018. "Theranostic micelles based on upconversion nanoparticles for dual-modality imaging and photodynamic therapy in hepatocellular carcinoma". *Nanoscale* 10(14): 6511–6523. Doi: 10.1039/C7NR09717D.

110. A. Kaushik, R. D. Jayant, S. Tiwari, A. Vashist and M. Nair. 2016. "Nano-biosensors to detect beta-amyloid for Alzheimer's disease management." *Biosens Bioelectron* 15(80): 273–287. Doi: 10.1016/j.bios.2016.01.065.

111. Y. Hillman, D. Lustiger and Y. Wine. 2019. "Antibody-based nanotechnology". *Nanotechnology*. 30(28): 282001. Doi: 10.1088/1361-6528/ab12f4.

112. S. Wang, R. Riahi, N. Li, D.D. Zhang and P. K. Wong. 2015. "Single cell nanobiosensors for dynamic gene expression profiling in native tissue microenvironments". *Adv Mater* 27(39): 6034–6038. Doi: 10.1002/adma.201502814.

113. A. M. Amarante, G. S. Oliveira, C. C. Bueno, R. A. Cunha, J. C. Ierich, L. C. Freitas, E. F. Franca, O. N. Oliveira and F. L. Leite. 2014. "Modeling the coverage of an AFM tip by enzymes and its application in nanobiosensors". *J Mol Graph Model* 53: 100–104. Doi: 10.1016/j.jmgm.2014.07.009.

114. G. S. Oliveira, F. L. Leite, A. M. Amarante, E. F. Franca, R. A. Cunha, J. M. Briggs and L. C Freitas. 2013. "Molecular modeling of enzyme attachment on AFM probes". *J Mol Graph Model* 45: 128–36. Doi: 10.1016/j.jmgm.2013.08.007.

115. J. L. Hammond, N. Formisano, P. Estrela, S. Carrara and J. Tkac. 2016. "Electrochemical biosensors and nanobiosensors". *Essays Biochem* 60(1): 69–80. Doi: 10.1042/EBC20150008.

116. L. Chen, E. Hwang and J. Zhang. 2018. "Fluorescent nanobiosensors for sensing glucose". *Sensors* 18(5): 1440. Doi: 10.3390/s18051440.

117. M. Sharifi, M. R. Avadi, F. Attar, F. Dashtestani, H. Ghorchian, S. M. Rezayat, A. A. Saboury and M. Falahati. 2019. "Cancer diagnosis using nanomaterials based electrochemical nanobiosensors". *Biosens Bioelectron* 1(126): 773–784. Doi: 10.1016/j.bios.2018.11.026.

118. Y. Han, Y. An, G. Jia, X. Wang, C. He, Y. Ding and Q. Tangl. 2018. "Facile assembly of up conversion nanoparticle-based micelles for active targeted dual-mode imaging in pancreatic cancer". *J Nanobiotechnology* 16(1): 7. Doi: 10.1186/s12951-018-0335-4.

119. D. Sun, J. Lu, L. Zhang and Z. Chen. 2019. "Aptamer-based electrochemical cytosensors for tumor cell detection in cancer diagnosis: A review". *Anal Chim Acta* 1082: 1–17 Doi: 10.1016/j.aca.2019.07.054.

120. J. Cheon and J.-H. Lee. 2008. "Synergistically integrated nanoparticles as multimodal probes for nanobiotechnology". *Acc Chem Res* 41(12): 1630–1640. Doi: 10.1021/ar800045c.

121. R. Raizman, D. Dunham, L. Lindvere-Teene, L. M. Jones, K. Tapang, R. Linden and M. Y. Rennie. 2019. "Use of a bacterial fluorescence imaging device: wound measurement, bacterial detection and targeted debridement". *J Wound Care* 28(12): 824–834. Doi: 10.12968/jowc.2019.28.12.824.

122. T. E. Serena, K. Harrell, L. Serena and R. A. Yaakov. 2019. "Real-time bacterial fluorescence imaging accurately identifies wounds with moderate-to-heavy bacterial burden". *J Wound Care* 28(6): 346–357. Doi: 10.12968/jowc.2019.28.6.346.

123. M. Arun Divakar and S. Shanmugam. 2017. "Live cell imaging of bacterial cells: Pyrenoylpyrrole-based fluorescence labeling". *Chem Biol Drug Des* 90(4): 554–560. Doi: 10.1111/cbdd.12978.

124. A. Hanif, R. Farooq, M. U. Rehman, R. Khan, S. Majid and M. A. Ganaie. 2019. "Aptamer based nanobiosensors: Promising healthcare devices". *Saudi Pharm J* 27(3): 312–319. Doi: 10.1016/j.jsps.2018.11.013.

125. B. Golichenari, R. Nosrati, A. Farokhi-Fard, K. Abnous, F. Vaziri and J. Behravan. 2018. "Nano-biosensing approaches on tuberculosis: Defy of aptamers". *Biosens Bioelectron* 117: 319–331. Doi: 10.1016/j.bios.2018.06.025.

126. V. Perla and T. J. Webster. 2005. "Better osteoblast adhesion on nanoparticulate selenium- A promising orthopedic implant material". *J Biomed Mater Res A* 75(2): 356–364. Doi: 10.1002/jbm.a.30423.

127. Y. Jung, Y. I. Chung, S. H. Kim, G. Tae, Y. H. Kim, J. W. Rhie, S. H. Kim and S. H. Kim. 2009. "In situ chondrogenic differentiation of human adipose tissue-derived stem cells in a TGF-beta1 loaded fibrin-poly(lactide-caprolactone) nanoparticulate complex". *Biomaterials* 30(27): 4657–64. Doi: 10.1016/j.biomaterials.2009.05.034.

128. D. Dyondi, T. J. Webster and R. Banerjee. 2013. "A nanoparticulate injectable hydrogel as a tissue engineering scaffold for multiple growth factor delivery for bone regeneration". *Int J Nanomedicine* 8: 47–59. Doi: 10.2147/IJN.S37953.

129. J. Olmos Buitrago, R. A. Perez, A. El-Fiqi, R. K. Singh, J. H. Kim and H. W. Kim. 2015. "Core-shell fibrous stem cell carriers incorporating osteogenic nanoparticulate cues for bone tissue engineering". *Acta Biomater* 28: 183–192. Doi: 10.1016/j.actbio.2015.09.021.

130. L. Leclerc, D. Boudard, J. Pourchez, V. Forest, O. Sabido, V. Bin, S. Palle, P. Grosseau, D. Bernache, M. Cottier, 2010. "Quantification of microsized fluorescent particles phagocytosis to a better knowledge of toxicity mechanisms". *Inhal Toxicol* 22(13): 1091–100. Doi: 10.3109/08958378.2010.522781.

131. A. Haase, J. Tentschert and A. Luch. 2012. "Nanomaterials: A challenge for toxicological risk assessment?" *Exp Suppl* 101: 219–250. Doi: 10.1007/978-3-7643-8340-48.

132. R. Bacchetta, N. Santo, M. Marelli, G. Nosengo and P. Tremolada. 2017. "Chronic toxicity effects of ZnSO$_4$ and ZnO nanoparticles in Daphnia magna". *Environ Res* 152: 128–140 Doi: 10.1016/j.envres.2016.10.006.
133. A. Rode, S. Sharma and D. K. Mishra. 2018. "Carbon nanotubes: Classification, method of preparation and pharmaceutical application". *Curr Drug Deliv* 15(5): 620–629. Doi: 10.2174/1567201815666171211124711.
134. J. Kolosnjaj, H. Szwarc and F. Moussa. 2007. "Toxicity studies of fullerenes and derivatives". *Adv Exp Med Biol* 620: 168–80. Doi: 10.1007/978-0-387-76713-0_13.

12 Advancement of Metal Nanomaterials in Biosensing Application for Disease Diagnosis

Jin-Ha Choi, Jinho Yoon, Minkyu Shin, Hye Kyu Choi, and Jeong-Woo Choi
Sogang University

CONTENTS

12.1 INTRODUCTION

In the biomedical field, the development of biosensors is important for precise diagnoses and effective therapy of several diseases. Biosensors are the analytical platforms for detecting biological or chemical molecules related to the disease and its harmful effects on the body [1,2]. The most important criteria in the development of biosensors are the accurate and rapid detection of target molecules to prevent the occurrence and spread of the diseases or provide early medical treatment [3]. In particular, early diagnosis can increase the probability of complete recovery and prevent the contagion of infectious diseases, including viral diseases. The recent COVID-19 pandemic has emphasized the importance of biosensors to prevent epidemics and produce treatments [4,5]. To this end, high sensitivity and selectivity are the most important requirements in the development of biosensors. To achieve this, many analytical techniques have been utilized, including electrical, electrochemical, fluorescent, and plasmonic techniques [6–8]. Nanomaterials may help overcome the limitations of biosensors using these techniques due to their advantages, such as high conductivity and superior optical, fluorescent, and plasmonic properties [9–11].

The interest and research on nanomaterials are increasing across various fields [12,13]. Nanomaterials are defined as having a single unit with a size ranging from one to one hundred nanometers. One of the most important characteristics of nanomaterials is that certain properties can be exhibited that were not present at the bulk scale [14,15]. The nanosize of these particles allows for communication with biomolecules that can be decoded to analyze biochemical and physicochemical properties. For instance, gold nanoparticles (AuNPs) exhibit different colorimetric results depending on nanometer-scale differences in the AuNP diameter. These can be utilized to develop AuNP-based colorimetric biosensors that cannot function the same way at the bulk scale [16]. In addition, the small size of nanomaterials produces a greater surface area, which then enhances the electrochemical, electrical, and plasmonic effects [17,18].

Because of their huge potential and unique properties, nanoparticles have been studied and applied in various scientific fields such as energy and biology [19,20]. Among its widespread applications, nanomaterials are useful in the biomedical field due to the suitable size and biocompatibility for administering substances to the human body or specific cells [21,22]. As such, nanomaterials are being used in many studies related to drug delivery, stem cell differentiation, and cell therapy [23,24].

There are some unique characteristics of metal nanomaterials in particular, such as the surface plasmonic effect of AuNPs, which improve the sensitivity, rapidity, and specificity of biosensing systems [25–27]. For example, a gold solution has a golden yellow color, whereas a solution of 20 nm-sized AuNPs has a red ruby color, and a 200 nm-sized AuNP solution has a bluish color [28]. In addition, the high conductivity and redox characteristics of nanomaterials are helpful for electrochemical biosensors, high fluorescence-emitting metal nanomaterials can be used to apply ultrasensitive fluorescent/colorimetric biosensors, and certain nanomaterials can improve Raman scattering for surface-enhanced Raman scattering (SERS)-based biosensors [29–31]. Moreover, metal nanomaterials in biosensors can enable surface modification to immobilize biomolecules or sensing probe materials with excellent biocompatibility [15,32,33]. Numerous biosensors have been developed based on these inherent properties of metal nanomaterials. Furthermore, due to the increasing interest in wearable and portable devices, many studies are being conducted on metal nanomaterial-based point-of-care testing (POCT) biosensors [34,35].

This chapter includes a selective overview of studies on novel biosensors based on the inherent properties of functional metal nanomaterials such as AuNPs (Figure 12.1).

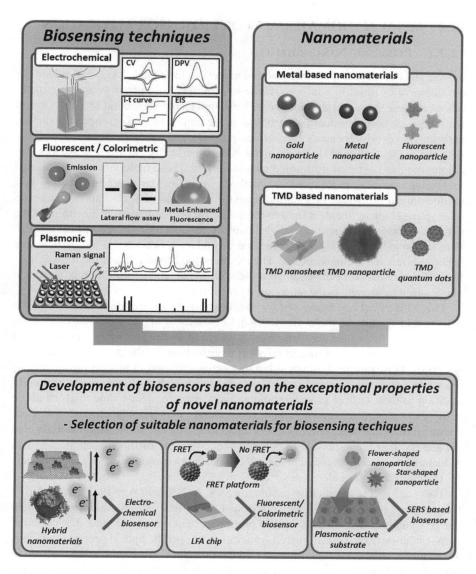

FIGURE 12.1 Biosensors based on the exceptional properties of metal nanomaterials using the applicable techniques.

This review is composed of three sections: types of metal nanomaterials and their properties for biosensing applications, different types of analytical methods using metal nanomaterials (electrochemical, fluorescent, plasmonic, and magnetic detections), and the transition metal dichalcogenide (TMD)-based biosensing platform. This chapter provides an overview of how to utilize functional metal nanomaterials to develop innovative biosensors.

12.2 METAL NANOMATERIALS FOR BIOSENSOR APPLICATION

12.2.1 Plasmonic Nanomaterials

For the last two decades, noble metal nanomaterials have played a crucial role in the development of novel biosensors and the improvement of current biosensing methods to achieve more specific and sensitive biomolecular detection [36,37]. One of the unique physicochemical properties of noble metals is the presence of a surface plasmon wave, which causes the surface plasmon resonance (SPR) in accordance with the incident light [38]. Noble metal NPs, specifically AuNPs and AgNPs, are among the most extensively studied nanomaterials. These materials have been used to develop countless techniques and methods for molecular diagnostics using their unique optical properties originating from the excitation of SPR. This phenomenon of coupling resonant incident light to the collective oscillation of the electrons on the surface of the noble nanomaterials leads to its enhanced or altered optical absorption. It also exhibits strong scattering as a secondary process when the particle size is above a few tens of nanometers. The extent of the scattering and the total extinction efficiency can vary depending on the size, shape, and composition of the noble nanomaterials, as well as the surrounding medium.

This SPR effect has been used for the diagnosis of several diseases through the detection of specific biomolecules [39,40]. In biosensing applications, the changes in the plasmonic resonance wavelength or the range of absorbance values are measured as a change in the chemical and physical environment on the surface of the nanomaterials. For example, biological binding or dissociation events on the surface of the noble metal nanomaterials could be measured as a change in their absorption. In addition to their optical properties, noble metal nanomaterials show high catalytic activity in redox reactions with organic compounds, which can be utilized in electrochemical measurement. The electrochemical properties of noble nanomaterials also depend on their size, shape, and composition. Synthetic methods such as chemical reduction and electrochemical reduction can be used to produce metal nanomaterials for the modification of electrodes.

There are two representative plasmonic NPs: AuNPs and AgNPs. First, an AuNP is a nanometer-sized particle, ranging from 1 to 100 nm, as a colloid form. Colloidal AuNP solutions have different optical properties than bulk-sized gold. The colloidal AuNP solution displayed strong red color due to its specific optical property, which is an exclusive interaction with light [41]. In the incidence of the oscillating electron in the light, the free electrons on the surface of the metal NPs induce an oscillation that causes a change in the ionic lattice. This process produces plasmonic resonance at a certain frequency of the incident light (localized surface plasmon resonance, LSPR). For example, 10 nm-sized AuNPs have a strong absorption peak at 520 nm owing to the LSPR effect of AuNPs. These AuNPs exhibit a red shift of absorbance peak along with an increasing diameter because of the effect of electromagnetic delays in larger particles. In addition, the plasmonic properties of colloidal AuNPs depend on their shape. For example, as shown in Figure 12.2a, Au nanorods have both a transverse and longitudinal absorption peak [42]. Because of these particular optical properties, Au nanomaterials have been used in significant research on applications such as bio-imaging and biosensors for diagnosing several diseases [43,44].

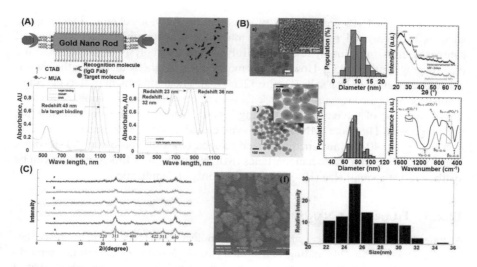

FIGURE 12.2 (a) Schematic diagram for detection of the target molecule by using Au nanorod, and the TEM image of Au nanorod, its target molecule detection measured by UV-vis. (Reprinted with permission from [42] Copyright (2007) ACS publishing). (b) TEM images and characteristic of synthesized MNPs. (Reprinted with permission from [50] Copyright (2020) ACS publishing). (c) XRD pattern and SEM image of MNPs synthesized with various salts (Reprinted with permission from [53] Copyright (2016) Elsevier publishing).

AgNPs are particles of silver or particles with a large percentage of silver oxide. Like AuNPs, there are attempts to integrate AgNPs into several diagnostic systems for sensitive and precise detection [45,46]. The SPR effect of AgNPs is similar to AuNPs. Due to the attractive physicochemical properties of SPR, AgNPs have also received extensive attention in the biomedical imaging and diagnostic fields. In fact, the large scattering cross section of AgNPs makes them ideal candidates for measuring diverse biomolecules. Therefore, many kinds of AgNPs are in development for various applications.

12.2.2 MAGNETIC NANOMATERIALS

Magnetic nanoparticles (MNPs) have demonstrated superior properties, such as a unique magnetic response compared with their larger-sized materials that are more similar in size to biomolecules [47]. There are many magnetic materials such as pure metals (Fe, Co, Ni, etc.), alloys (FeCo, alnico, permalloy, etc.), and oxides (Fe_3O_4, $CoFe_2O_4$, etc.) [48]. Among them, one of the most widely used MNPs is iron oxide (Fe_3O_4), which has high biocompatibility and high chemical and colloidal stability. Iron oxide has a paramagnetic property and has a reddish-brown color. Fe_3O_4, which is one of the representative iron oxides, is superparamagnetic in nature [49]. Owing to its biocompatibility and superior magnetic properties, superparamagnetic iron oxide nanoparticles (SPIONs) have been utilized to diverse biomedical applications such as targeted drug delivery, MRI contrast agent, and biomolecular detection for early diagnosis of cancer, diabetes, atherosclerosis, and other inflammatory diseases (Figure 12.2b) [50,51].

Synthesis of MNPs is a complex process that uses various chemical techniques, such as hydrothermal reactions, microemulsions, sonochemical reactions, sol-gel syntheses, flow injection syntheses, hydrolysis and thermolysis of precursors, and electrospray syntheses [52]. However, the most typical synthetic method is the chemical coprecipitation of iron salts (Figure 12.2c) [53]. The main benefit of the coprecipitation process is the large production scale with a controllable and homogeneous size. This is because of the ease of controlling kinetic factors, which can determine the crystal growth. Therefore, MNPs can be synthesized as small nano-sized particles of iron oxide (10–40 nm) and medium-sized nanoparticles of iron oxide (60–150 nm). Monocrystalline iron oxide nanoparticles (MIONs) with cross-linked dextran can be synthesized at a small size (10–30 nm) for bio-imaging applications.

12.2.3 Fluorescent Nanomaterials

Among the several analytical methods for biomolecular detection, the fluorescence-based method is an outstanding and broadly used method for detecting biomarkers and for the precise diagnosis of several diseases [54]. For fluorescence-based measurement, various types of fluorescent nanoparticles have been used as sensitive probes. Quantum dots (QDs) are one of the potential fluorescent probes, and they are composed of semiconductor materials made from silicon or germanium [55,56]. There are many advantages of using QD instead of organic fluorescence dye. One is the high efficiency of the quantum yield, which results in strong fluorescence intensity. In addition, the emission spectrum has a very sharp peak, which enables multi-analyte detection without an overlayed fluorescent signal. Moreover, with excitation wavelengths around the UV region (around 300 nm), even the QDs each emitted different fluorescence wavelengths. The fluorescence emission of QDs can be explained by the band theory in semiconductors. Band gaps denote the energy difference between the valence band (top) and the conduction band (bottom). Metals, which are characterized as semiconductors, have partially filled bands dissociated from the empty conduction band. When we are dealing with interacting molecular orbitals, the two that interact are generally HOMO (highest energy occupied molecular orbital) and LUMO (lowest energy unoccupied molecular orbital). HOMO and LUMO are closer in energy than any pair of orbitals in the two molecules, which allows them to interact strongly. These orbitals are also called Frontier orbitals because they lie at the outermost boundaries of the molecule's electrons. Therefore, quantum mechanics are applicable in describing the energy of the metal NPs. In addition to the QD, the upconversion nanoparticle (UCNP) has enormous potential in the development of fluorescent biosensors because of its unique optical conversion property by long-wavelength light (near-infrared, NIR) excitation for the emission of the short-wavelength light (visible or ultraviolet) [57]. In addition to this property, the UCNP is considered a next-generation nanoprobe due to its unique optical properties, including a narrow emission peak, a non-autofluorescence background to achieve a high signal-to-noise ratio and high photobleaching resistance. Therefore, various types of UCNPs have been researched for biosensors, such as the lanthanide-doped UCNP, quantum nanostructures or defect-doped nanostructures of UCNP, and transition-metal-doped UCNP.

12.3 ANALYTICAL TECHNIQUES FOR METAL NANOMATERIAL-INTEGRATED BIOSENSORS

In this section, we discuss representative types of biosensing techniques applied to develop advanced biosensors. First, electrochemical biosensors are discussed based on the types of electrochemical techniques such as amperometry, cyclic voltammetry, electrochemical impedance spectroscopy (EIS), and differential pulse voltammetry (DPV). Next, the advantages of fluorescent or colorimetric biosensors are discussed with regard to POCT applications. In addition, the plasmonic-based biosensors and magnetic biosensors are introduced to achieve high sensitivity at the attomolar or femtomolar scales. Lastly, magnetic analytical techniques are discussed that use several magnetic nanomaterials to separate and detect biomarkers. Each technique discussed in this section has advantages that are acquired from incorporating nanomaterials.

12.3.1 ELECTROCHEMICAL BIOSENSORS

Electrochemical biosensors have beneficial characteristics such as fast response, ease of miniaturization and operation, and portability [58,59]. In general, electrochemical biosensors use the changes in electrochemical signals derived from capturing target molecules with sensing probes. These changes could include the impedance change or the generation of electrons. Enzyme-based electrochemical biosensors can also use the redox reactions between sensing probes and target molecules directly, which eliminates the need for an additional tagging process on the sensing probes and reduces the number of steps for detecting target molecules [60,61]. In addition, depending on the target materials and analytical methods, electrochemical biosensors can be developed using various electrochemical techniques such as amperometric, cyclic voltammetric, EIS, and DPV techniques.

The amperometric technique is a quantitative analytical tool that measures a change in current from oxidation or reduction reactions by using the redox biomolecules or redox probes such as methylene blue [62]. The representative example of amperometric biosensors is the enzymatic biosensor. Certain enzymes react with target molecules through redox reactions and generate or remove electrons that can be easily measured by the change in the current [63]. For example, an amperometric nitric oxide (NO) biosensor was reported that used the redox reaction between myoglobin, the redox protein, and NO [64]. In this study, the myoglobin located on the nanomaterial-modified substrate excellently detected the added NO. The amperometric technique was used to measure the increasing electrochemical signal with high sensitivity (detected 3.6 nM NO) and selectivity, as shown in Figure 12.3a. By using redox probes, amperometric biosensors can be easily developed without redox enzymes [65].

In addition to this technique, EIS is a type of impedance spectroscopy that uses sinusoidal and small amplitude signals for electrochemical analysis of target materials. It has been widely used to develop biosensors that conduct an electrochemical analysis of interfacial changes on the electrode [66]. Generally, the acquired data from EIS are expressed in the Nyquist plots through the calculation and conversion of

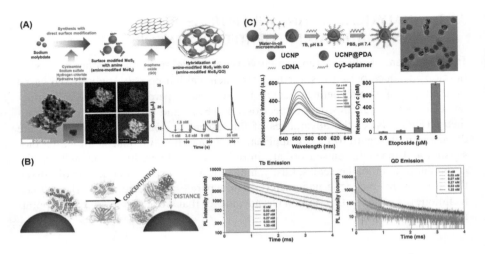

FIGURE 12.3 (a) Schematic diagram of hybridization of amine-modified MoS₂ with GO, and TEM and EDS mapping images of synthesized nanoparticles, and their NO sensing performance measured by amperometric technique. (Reprinted with permission from [64] Copyright (2017) Elsevier publishing). (b) Schematic diagram of ligand binding on the bidomain protein with FRET and photoluminescence intensities of Tb emission and QD emission (Reprinted with permission from [73] Copyright (2020) ACS publishing). (c) Schematic illustration of the fabrication process of Cy3-aptamer modified UCNP@PDA, and TEM image of UCNP@PDA, the fluorescence intensity of Cy3 in the presence of a various concentration of Cty c., and quantitative detection performance of intracellular Cyt c. (Reprinted with permission from [77] Copyright (2017) Elsevier publishing).

components of electrical circuits (e.g., the resistance and capacitance values of electrolytes and materials modified on the electrode). This technique has been utilized to develop electrochemical immunosensors because it can measure the reactions that occur on the surface of the electrodes between substances that do not have redox properties, such as antibody-antigen reactions [67].

The DPV technique measures the response in current changes induced by target molecule detection during the imposition of a voltage in the form of a pulse on the states of potential stairsteps or a linear sweep with a fixed amplitude of the imposed potential pulse. The change in base potential by a constant voltage value is a unique property compared with the other pulse voltammetry techniques and is excellent for detecting target molecules that are mixed with other molecules [68]. Other electrochemical techniques utilized to develop biosensors include cyclic voltammetry (CV) and linear sweep voltammetry (LSV) [69]. In the cases of CV and LSV, these techniques can be utilized not only for detection of current from oxidation or reduction reactions between sensing probes and target molecules, but also for direct detection of redox signals derived from immobilized redox molecules in the absence of redox reaction between sensing probes and target molecules. For example, the redox signals from metalloproteins or redox generators immobilized on the electrode can be measured easily and rapidly, which is hard to be achieved by amperometric measurement or EIS techniques. This detection property can contribute to the expansion of various biosensing applications. Electrochemical techniques require conductivity to analyze the electrical/electrochemical characteristics for target detection, which sometimes restricts the operation conditions of electrochemical biosensors, such as a portable device. As seen here, various electrochemical techniques can be utilized to develop excellent biosensors using nanomaterials that improve the measurement of each electrochemical technique.

12.3.2 FLUORESCENT/COLORIMETRIC BIOSENSORS

12.3.2.1 Fluorescent/FRET-Based Biosensors

In the current situation where the impact of the COVID-19 pandemic and the concern regarding viruses are increasing, rapid quantitative measurement of target molecules is required for POCT applications, and measuring the fluorescence emission of sensing probe materials can satisfy the requirements for this type of biosensor. Since fluorescent probes were chemically synthesized and commercialized, the fluorescent characteristics of sensing probes have been studied extensively [70]. In particular, metal nanomaterials such as QDs and UCNPs have been applied to fluorescent biosensing systems. In general, fluorescent biosensors determine the change in intensity or color of fluorescent nanoprobes or utilize the appearance/disappearance of fluorescent signals through a reaction between target molecules and fluorescent sensing nanoprobes [71]. The following methods are widely used to develop fluorescent biosensors.

The Förster resonance energy transfer (FRET) method is based on FRET pairs consisting of donor and acceptor fluorophores that achieve high fluorescent intensity. A change in emitted wavelength occurs through the energy transfer from donor to acceptor via a nonradiative coupling. This is a type of fluorescent quenching [72]. Léger et al. developed a novel Tb-to-QD FRET nanoprobe that is comprised of C-terminal small artificial proteins (αRep) labeled with a terbium (Tb) complex donor and an N-terminal QD acceptor (Figure 12.3b) [73]. Small scaffold proteins (αRep) provide a specific binding surface and their conformational changes due to target binding results in decreasing FRET signals. Using this phenomenon, sensitive and specific measurement of two different protein targets (blue fluorescence protein, BFP, and enhanced green fluorescence protein, eGFP) was achieved with simple homogeneous (separation-free) binding assays. The sensitivity was improved to as low as the sub-nanomolar level in serum-containing samples.

Among the various fluorescent metal nanomaterials, UCNPs have huge potential in the development of fluorescent biosensors because of their unique optical conversion property of short-wavelength light (NIR) excitation to the emission of long-wavelength light (visible or ultraviolet) [74]. In addition to this property, the UCNP is considered a next-generation nanoprobe due to its unique optical properties, including a narrow emission peak, a non-autofluorescence background to achieve a high signal-to-noise ratio and high photobleaching resistance [75]. Therefore, various types of UCNPs such as lanthanide-doped UCNPs, quantum nanostructures or defect-doped nanostructures of UCNPs, and transition metal-doped UCNPs have been researched for applications in biosensors [76].

By using these advantages of UCNP, a FRET biosensing platform was developed for caspase-9 activity detection using a peptide-functionalized UCNP with silica nanoparticles (UCNP@SiO$_2$@Cy5-pep) (Figure 12.3c) [77]. To fabricate this sensing platform, the surface of the UCNP was coated with silica nanoparticles and modified with a carboxyl group using tetraethylorthosilicate (TEOS) and carboxyethylsilanetriol (CTES) for immobilization of the Cy5-peptide containing the specific peptide sequence (Leu-Glu-His-Asp; LEHD) through the N-(3-Dimethylaminopropyl)-N'-ethylcarbodiimide hydrochloride (EDC)/N-Hydroxysuccinimide (NHS) reaction. Caspase-9 cuts the LEHD motif of the Cy5-peptide, and this unique cleavage activity was measured by a UCNP-based biosensing system through the changes in the fluorescent signals. When the Cy5-peptide was cleaved by caspase-9, the UCNP@SiO$_2$ could emit a strong red fluorescent signal, which was previously quenched by the Cy5-peptide. This biosensing system was tested in in vitro cancer cell conditions. For this study, SW480 (human colon cancer cells) and MG-63 (human osteosarcoma cells) were stained with the UCNP@SiO$_2$@Cy5-pep. Cisplatin, an anticancer drug, was injected to induce increased caspase-9 activity, and the caspase-9 activity was successfully evaluated in the intracellular regions. The developed biosensor exhibited good colloidal stability and low cytotoxicity, and it could monitor caspase-9 activity in vitro within a range of 0.5 to 100 U/mL caspase-9 and a detection limit of 0.068 U/mL caspase-9.

Similar to this study, an aptamer fluorescence biosensor was developed to detect cytochrome c (Cyt c) by using a polydopamine (PDA)-coated UCNP [78]. In this study, the PDA-coated UCNP (UCNP@PDA) was synthesized using the water-in-oil microemulsion method, and the surface of the UCNP@PDA was modified with cDNA for conjugation with Cy3-labeled Cyt c aptamer. The fluorescence emission of Cy3 was quenched by the UCNP@PDA. However, when the Cyt c was added to the UCNP@PDA conjugated with Cy3-labeled Cyt c aptamer, the Cy3-labeled Cyt c aptamer detached from the UCNP@PDA by binding with the Cyt c, and the fluorescence intensity of Cy3 was measured. Using this mechanism, the Cyt c was accurately detected by UCNP@PDA conjugated with Cy3-labeled Cyt c aptamer. This fabricated UCNP-based aptamers' biosensor was also used in HepG-2 cells (human liver carcinoma cell line) for quantitative analysis of the intracellular Cyt c levels. The Cyt c, located in the mitochondrion of the non-apoptotic cells, was released from the mitochondria by causing cell apoptosis when the etoposide was added to the cells. To achieve high sensitivity, the fluorescence intensity of the Cy3 on the Cy3-labeled Cyt c increased depending on the increase in the amount of etoposide. The developed UCNP@PDA-based biosensor showed a highly sensitive detection limit of 20 nM and a wide detection range of 50 nM–10 µM with excellent colloidal and optical stability.

In another study, an upconversion nanosphere composed of UCNP, Au, and CdS was synthesized to detect alpha-fetoprotein (AFP) using the photoelectrochemical immunoassay sensing technique [79]. The proposed upconversion nanosphere could promote the fluorescent property of the UCNP based on the Au in the UCNP nanosphere as the energy conveyor and light concentrator to enhance the interfacial energy in the UCNP system. To synthesize this unique structural upconversion nanosphere, a HAuCl4 aqueous solution was added to the UCNP to precipitate Au (III) ion to the UCNP surface. It was then reacted with an L-cysteine aqueous solution (0.04 M) containing 0.2 mmol cadmium nitrate tetrahydrate. The synthesized UCNP@Au@CdS and the monoclonal mouse antihuman AFP antibody

(Ab1) were sequentially immobilized on the fluorine-doped tin oxide (FTO) electrode (Ab1-coated microplate). GOx and the polyclonal antihuman AFP antibody (Ab2)-labeled AuNP (Ab2-AuNP-GOx) were conjugated to detect the AFP. For AFP detection using this system composed of two different AFP-detecting probes, the AFP was added and specifically captured on the Ab1-coated microplate. Then, the Ab2-AuNP-GOx was injected and immobilized on the AFP captured on the Ab1-coated microplate to form the AFP-mediated sandwich structure. After the formation of the sandwich structure, glucose was injected into the GOx of the immobilized Ab2-AuNP-GOx to generate H_2O_2 by an enzyme immunoreaction. The generated H_2O_2 could enhance the photocurrent of the UCNP@Au@CdS under the irradiation of a 980nm laser. Through this process, AFP levels were successfully measured with high sensitivity. Due to the excellent photocurrent response of the UCNP@Au@CdS to AFP, this system could detect AFP with a dynamic range of 0.01 to 40ng/mL and a highly sensitive detection limit of 5.3 pg/mL.

Yang et al. demonstrated a UCNP (NaYF4: Yb,Er [NaYF]) as a FRET acceptor and rhodamine 6G as a quencher for the detection of coumarin. To specifically bind coumarin, β-cyclodextrin was modified on the surface of the UCNP, and the quencher was bound to the β-cyclodextrin before applying the coumarin. This fluorescent biosensor showed high sensitivity and selectivity, with a detection range of 2.5–32.5 μM and a limit of detection (LOD) as low as 0.74 μM. Wang et al. exhibited a carcinoembryonic antigen (CEA) detection system using a UCNP-based ultrasensitive homogeneous aptasensor [80]. The π-π stacking interaction between the CEA aptamer and graphene oxide (GO), which could be a great quencher, hampered the fluorescent emission of the UCNP due to their proximity. In the presence of CEA, the CEA aptamer changed its conformation and induced the dissociation of UCNP and GO, which resulted in an emitted fluorescence signal. The aptasensor provided a low LOD (7.9 pg/mL) and a wide linear range (0.03 to 6.0ng/mL) in an aqueous solution. Rabie et al. utilized a similar quenching-dequenching effect of a UCNP-aptamer-GO complex to detect intracellular dopamine in differentiating stem cells [81]. To improve the fluorescent emission of UCNP, the authors synthesized a novel sandwich-structured UCNP composed of β-NaYF$_4$: Yb^{3+}@ β-NaYF$_4$: Er^{3+} @ β-NaYF$_4$: Yb^{3+} (Yb@Er@Yb). Even with the use of low power density 980nm NIR excitation, bright emissions were observed to measure dopamine in live stem cells without a heat conversion effect. The authors claimed that dopamine was detected at pM concentrations in stem cell-derived neural interfaces.

12.3.2.2 Metal-Enhanced Fluorescence (MEF)-Based Biosensors

Another widely used method for developing fluorescent biosensors is to use fluorescent sensing probes with noble metal nanomaterials to enhance the fluorescence emission signal. This signal depends on the distance between a quencher and the fluorescent molecule that is altered by the reaction between the fluorescent sensing probe hybrid and target molecules [82]. A distance of less than a few nanometers between the fluorophore and noble nanomaterials could result in the quenching effect, whereas a fluorescent signal could be enhanced at about 7–8 nm of distance [83]. This property could allow for sensitive detection in a simple manner. For example, an exosomal miRNA biosensor was reported that used the quenching and metal-enhanced fluorescent (MEF) effect that occurred at a certain distance (5–10nm) between the bimetallic nanorod and the 5(6)-carboxyfluorescein (FAM) fluorophore to detect target miRNA (miR-124) (Figure 12.4a) [84]. This proposed fluorescent biosensor could detect miR-124 with high sensitivity by the structural change of the sensing probe sequence induced at a certain distance between the nanorod and the FAM fluorophore tagged at the end of the sensing probe sequence. Fluorescent biosensors demonstrate excellent biosensing performance with high sensitivity, but the fluorescent equipment required for highly sensitive analysis poses a limitation for POCT applications.

This MEF effect could also be applied to an electrospun nanofiber-based biosensor [85]. Silica-coated AgNPs have been utilized to enhance the fluorescence signal with a silica spacer between fluorescent dye and AgNPs. These nanoparticles were functionalized on the surface of polycaprolactone (PCL) nanofibers by the photoreducing method. Fluorescent-based immunoassays using IgG and FITC-anti IgG on PCL with silica-coated AgNPs were conducted, and they demonstrated enhanced fluorescence signals and higher sensitivity than assays conducted on a 2D glass slide. Ventura et al. showed a functionalized, superhydrophobic MEF-based detection substrate that achieves rapid detection of human IgG in urine samples (Figure 12.4b) [86]. The system was composed of micro-sized pillars with Au-coated nanostructures that could induce the MEF effect to detect human IgG. The experimental

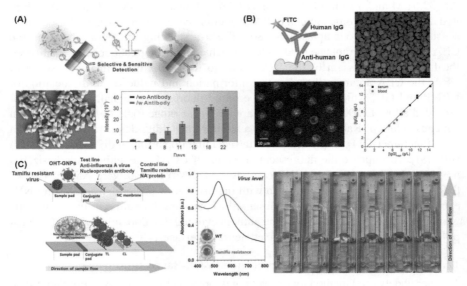

FIGURE 12.4 (a) Scheme of exosomal miRNA detection via MEF, and TEM image of generated magnetic-gold nanorod, fluorescence intensity from time-dependent detection performance of hiPSC-NSC. (Reprinted with permission from [84] Copyright (2019) ACS publishing). (b) Schematic diagram of detection of human IgG by using FITC-tagged secondary antibody, and SEM image of AuNPs modified upper surface of the pillar, the linear curve of IgG levels in serum. (Reprinted with permission from [86] Copyright (2019) ACS publishing). (c) Schematic illustration of oseltamivir hexylthiol (OHT)-based LFA chip for Tamiflu-resistance virus detection, and colorimetric detection of OHT-AuNPs against the real virus, sensitivity analysis of Tamiflu resistance detection. (Reprinted with permission from [91] Copyright (2018) Springer Nature publishing).

results indicated that the device has a LOD of less than 10 ng/mL with a linear range of 10–100 ng/mL. In addition, this superhydrophobic surface could prevent false-negative signals in complex biological mixtures such as urine and blood. Choi et al. exhibited an AuNP-based simple proteolytic enzyme detection system with DNA and peptide functionalization [87]. The distance between the fluorophore and AuNPs (less than 2 nm) was determined by the length of the peptide strand. After caspase-3 degraded to peptide, the single-stranded DNA maintained a distance of around 7 nm between the fluorophore and AuNP, which is an optimal distance for the MEF. Using this MEF-based fluorescent emission, highly sensitive detection of caspase-3 was successfully achieved as low as 78 pg/mL in a cell-free configuration. Moreover, this tiny nanoprobe could be internalized in a live cell to measure intracellular caspase-3 expression levels in pre-apoptotic cells. These studies show that the MEF effect can improve the sensitivity of biosensors with appropriate metal nanomaterials.

12.3.2.3 Colorimetric Biosensors

Because it can be distinguished by the naked eye, colorimetric biosensors may be more suitable for POCT applications. To detect target molecules by colorimetric measurement, metal nanomaterials that exhibit visible colors when aggregated are utilized as the biosensing template for reacting with target molecules through sensing probes modified on the nanomaterials [88]. To develop this type of colorimetric biosensing system, AuNPs or AgNPs that change color depending on the degree of aggregation of nanoparticles are widely used as the template nanomaterials [89]. In addition, certain nanomaterials can change color not only by aggregation but also by specific chemical reactions, which provide a powerful tool for the detection of target molecules with high sensitivity [90]. Furthermore, because of the advantage of being detected rapidly and easily by the naked eye, this colorimetric biosensing system has huge potential in the development of disposable, cheap, and on-site biosensors for POCT applications by combining it with a lateral flow assay (LFA) system capable of rapid quantification (Figure 12.4c) [91]. For example, an LFA biosensor to detect p24 capsid protein, an HIV biomarker, was reported that used the catalytic reaction between platinum nanoparticles modified with sensing probes and hydrogen peroxide (H_2O_2) to enhance the visibility of aggregated platinum nanoparticles on the LFA chip. This biosensor exhibited an ultrabroad detection range at the femtomolar scale (0.8 pg/mL) that could be seen by the naked eye [92]. By introducing novel metal nanomaterials, colorimetric biosensors can contribute to the development of highly sensitive, rapid, and quantitative biosensors for POCT applications.

12.3.3 PLASMONIC-BASED BIOSENSORS

12.3.3.1 Raman-Based Biosensors

In the past, the Raman scattering technique was limited in biosensor applications due to its low signal intensity. However, since the SERS technique with noble metal nanomaterials was discovered, SERS has been one of the most powerful tools for detecting target molecules with extremely high sensitivity [93,94]. Although it requires expensive equipment and expertized technique for operation, the SERS method has shown excellent performance compared with other techniques in terms of sensitivity.

The Raman-active dyes located on the surface of the novel nanometal can exhibit an extremely enhanced Raman signal compared with conventional Raman scattering techniques [95,96]. Enhancement of SERS signals is greatly influenced by the shape and composition of the metal nanomaterial. Therefore, studies on SERS-based biosensors have been conducted to develop excellent nanostructures to maximize the SERS signals. The Raman-active dyes and metallic nanomaterials are the essential components for SERS-based biosensing systems.

Raman-active dyes are employed to develop sensing probes that can exhibit or remove the SERS signals through the detection of target molecules, enabling a "Turn on" and "Turn off" function in biosensors. For example, a SERS-based "Turn off" arginine kinase biosensor was reported that decreased the SERS signal through a reaction between the arginine kinase and the sensing probe nanomaterials composed of AuNP, KRGGGGYIKIIKV (KRV) peptides for detection of arginine kinase, and 4-mercaptobenzoic acid (4-MBA) as the Raman-active dye [97]. After detecting arginine kinase, the phosphorylation of the KRV peptides on the AuNP by arginine kinase could induce the decrease in the SERS signal intensity (peaks at 1,077 and 1,580 cm^{-1}) of 4-MBA, which acted as the sensing probes. This biosensor was highly sensitive with a LOD of 46 pM (Figure 12.5a).

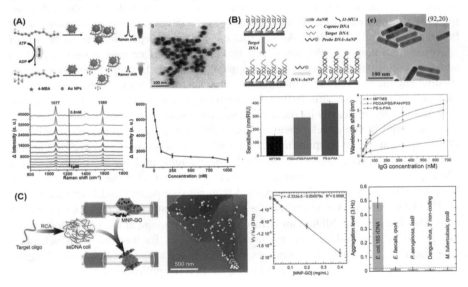

FIGURE 12.5 (a) Scheme of the SERS-based "Turn off" arginine kinase biosensor composed of GNP, KRV, and 4-MBA, and its sensing performance for arginine kinase detection. (Reprinted with permission from [97] Copyright (2018) Elsevier publishing). (b) Schematic diagram of DNA hybridization analysis by AuNP-enhanced plasmonic biosensing platform, TEM image of synthesized AuNR, and sensitivity index of the LSPR peak wavelength and calibration curve measured by AuNR detection with different concentration of IgG. (Reprinted with permission from [104] Copyright (2020) ACS publishing). (c) Schematic diagram of DNA detection by using self-assembled MNP-GO nanocomposites, and TEM image of MNP-GO nanocomposites, optomagnetic peak amplitudes (at 3 Hz) of MNP-GO nanocomposites, and aggregation level of specific and nonspecific DNA sequence. (Reprinted with permission from [109] Copyright (2019) ACS publishing).

To develop excellent SERS-based biosensors, it is crucial that metallic nanostructures are used to improve the SERS signal derived from target detection to enhance the sensitivity of the biosensor. To achieve this, various nanomaterials have been used to develop the SERS-active nanostructures capable of producing a hotspot to enhance the SERS signals [98,99]. A study by Choi et al. demonstrated the sensitive detection of caspase-3 using the hotspot strategy [100]. In this system, the conductive polymer was electrodeposited on an Au working electrode using a stepwise deposition method for the nanohorn structure. On this nanostructure, Ag ions were reduced on the surface of the conductive polymer to provide a large SERS-active substrate. Using the peptide and Raman dye-functionalized AuNPs on the core/double-shell Ag/polymer/Ag nanohorn structures, caspase-3 was measured with a broad detection range of 10 pg/mL–10 μg/mL. In addition, a homogeneous Au nanoarray was developed with GO functionalization to enhance the SERS signal with the synergistic effect of electromagnetic and chemical enhancement [101]. This SERS-sensitive substrate was utilized for in situ monitoring of dopamine released from stem cells to determine neural differentiation levels. For the selective detection of dopamine, dopamine aptamer with Raman dye was immobilized on the surface of GO via π-π interaction. Dopamine was released from the differentiating cells, and the Raman signal from the Raman dye was significantly decreased. In contrast to this "Turn off" system, the "Turn on" system uses the increase in the SERS signal by detection of target molecules [102]. Similar to these examples, SERS-based biosensors are generally developed based on an increase or decrease in the SERS signals through the detection of target molecules.

12.3.3.2 SPR-Based Biosensors

As mentioned in Section 12.2, plasmonic noble metal nanomaterials strongly absorb and scatter light because of their SPR effect. The specific resonance angle depends on the local dielectric environment, which can be changed by biological reactions on the surface of the metal nanomaterials. This phenomenon enables real-time analysis of biomolecules without any specific labeling methods. In fact, the SERS and MEF phenomena also cause the SPR effect on metal nanomaterials. However, in this section, we discuss biosensing methods using only the change in the resonant angle by the localized SPR (LSPR) and biological reactions. Kim et al. developed nanopatterned AuNPs on optical fibers for LSPR-based detection of prostate-specific antigen (PSA) [103]. The fiber-optic platform with metal nanomaterials had advantages in miniaturization, lossless signal delivery, and remote sensing. The optic fiber in this study was fabricated using focused ion beam (FIB) nanopatterning technology, which improved the reproducibility and durability of the sensor. Using this nanofiber substrate, PSA was quantitatively measured within a range of 0.1 pg/mL–1 ng/mL.

Lu et al. used a diblock copolymer-templated Au nanorod monolayer to improve the sensing performances of LSPR biosensors [104]. Poly(styrene)-b-poly(acrylic acid) (PS-b-PAA) diblock copolymer was utilized as a deposition spot for Au nanorods to increase sensitivity. In this system, the aspect ratio, deposition time, and distance between the Au nanorods strongly affected the optical properties of the copolymer-based LSPR biosensors. The highest sensitivity of this LSPR-based sensor was 753 nm/RIU, and the LOD for human IgG and free ssDNA was 0.8 nM and 29 pM, respectively (Figure 12.5b).

Wang et al. reported on a dual Au nanomaterial-based SPR aptasensor to improve the sensitivity of exosome detection [105]. The Au film was functionalized with a capturing aptamer, and the target exosomes were bound by a DNA aptamer on the Au film. Then, aptamer-functionalized AuNPs (T30-AuNPs) were bound to the exosome in a sandwich structure. Thus, target exosomes were detected by a single AuNP-amplified SPR aptasensor. Moreover, more AuNPs (A30-AuNPs) could be captured on the T30-AuNPs through complementary hybridization of DNA, and target exosomes were detected by dual AuNP-amplified LSPR aptasensors. The LOD of the dual AuNP-amplified LSPR aptasensor was as low as 5×10^3 exosomes/mL with high reproducibility. Picciolini et al. developed a system for selective detection of exosomes from different brain plasmas by SPR imaging [106]. An array of different antibodies on the Au substrate was used to separate the subpopulations of exosomes of different neural origins. The relative amount of CD81 and GM1 on the surface of the exosomes was measured, providing evidence of the heterogeneous distribution of exosome subpopulations of neural origin in brain plasma.

12.3.4 Magnetic Biosensors

Magnetic nanomaterial-based biosensors have attracted considerable attention for immunoassay applications because biological samples exhibit very low magnetic susceptibility, which enables very high signal-to-noise ratio magnetic sensing. As a result, magnetoresistive (MR) biosensors have been widely used for biomarker detection [107]. The MR effect is generated from alternating a ferromagnetic and nonmagnetic effect. Therefore, magnetic micro- or nanoparticles can be applied to induce drastic signal changes to measure biomolecules. Several magnetic nanoparticles attached to a polystyrene microbead can enhance the MR signal and improve the sensitivity of the biosensor [108]. On the basis of click chemistry, polystyrene beads and magnetic nanoparticles were bound to each other to analyze three different antibiotics, including chloramphenicol, sulfonamide, and oxytetracycline. This system had high sensitivity and a broad detection range from pg/mL to μg/mL. Moreover, controlling the size of the polystyrene beads could adjust the detection range since the different sizes of polystyrene beads can be functionalized based on different numbers of magnetic nanoparticles.

Tian et al. developed a two-dimensional magnetic nanoparticle-GO nanocomposite using the strong interaction of single-stranded DNA between magnetic nanoparticles and GO (Figure 12.5c) [109]. The rolling circle amplification (RCA) reaction of the target DNA could make the magnetic nanoparticle-GO nanocomposite, and it provided a stronger optomagnetic signal than individual magnetic nanoparticles. Using this method, the *Escherichia coli* 16S rDNA sequence was successfully measured as low as 2 pM within 90 min. In addition, the magnetic nanoparticle-GO nanocomposite could be applied for naked-eye detection for target concentrations higher than 100 pM. In addition to sensing the magnetic properties of magnetic nanoparticles, magnetic separation has also been frequently used for the effective detection of biomolecules in the body fluid complex. The nickel metal in multicomponent nanorods (Au-Ni-Au) was utilized to separate exosomes from the sample into a high concentration of exosomes. This resulted in the sensitive detection of exosomes and exosomal miRNA for neuronal differentiation [84].

Magnetic tunnel junction biosensors are also potential detection methods using magnetic nanoparticles. Sharma et al. developed a sensing system based on magnetic tunnel junction sensors to detect pathogenic DNA [110]. The probe oligonucleotides complementary to the target DNA strands were immobilized on the sensor surface. After hybridization of the target DNA, streptavidin-coated MNPs were added. The real-time signal change of the injection, binding, and washing steps of the MNPs can be read out from the magnetic tunnel junction sensors. The detection limit for target DNA can be as low as 1 nM.

12.4 TRANSITION METAL DICHALCOGENIDE (TMD)-BASED BIOSENSORS

The unique merits of metal-based nanomaterials have huge potential for biosensor applications because of their high conductivity, plasmonic, and fluorescent properties. As mentioned, these properties of nanomaterials are helpful for achieving high sensitivity and selectivity. Furthermore, electrochemical-, fluorescent-, and SERS-based biosensors can be developed depending on the specific properties of the employed nanomaterials that enhance the biosensing techniques. This section discusses highly sensitive biosensors developed using the exceptional properties of transition metal dichalcogenide (TMD) nanomaterials. TMD nanomaterials (MoS_2, WS_2, etc.) have attracted substantial interest, particularly in biochemical applications, because of their exceptional crystal structures, a wide range of chemical compositions, and great material properties such as electro-conductivity and electrocatalytic activity [111]. TMD nanomaterial-based biosensors have enhanced performance due to the inherent advantages of TMD nanomaterials, including high conductivity, high efficiency of ion transport, and huge surface area. In the following studies, TMD-based nanomaterials were used to develop electrochemical, fluorescent, and Raman-based biosensors.

In one example, WS_2 nanosheet-based electrochemical biosensors were developed for H_2O_2 detection using the catalytic property of WS_2 [112]. To develop this biosensor, the WS_2 nanosheet was synthesized using tert-butyllithium for the exfoliation of the WS2 material. As the sensing probe molecule, hemoglobin (Hb), which can detect the H_2O_2 by the enzymatic reaction, was immobilized on the WS_2 nanosheet by physical absorption at room temperature. Glutaraldehyde was introduced as a crosslinking matrix to maintain the structural rigidity of the Hb and to immobilize the WS_2 nanosheet and the Hb on the glassy carbon substrate. The synergistic interaction between the WS_2 nanosheet and Hb induced the electrocatalytic H_2O_2 reduction efficiently through the catalytic property of WS_2 and the H_2O_2 reduction property of Hb. To confirm the electrocatalytic effect of this biosensor, CV was performed in N_2-saturated phosphate-buffered saline (PBS). Because of the synergistic effects of the large catalytic surface area of the WS_2 and the high surface-to-volume ratio and efficient immobilization of the Hb, this biosensor exhibited excellent H_2O_2 detection performance with a wide range of 2.0×10^{-6} to 38×10^{-6} M and a LOD of 0.036×10^{-6}

M. In addition, this WS_2 nanosheet-based biosensor showed the best H_2O_2 sensing performance compared with other TMD-based biosensors prepared with the exact same process but composed of MoS_2, $MoSe_2$, or WSe_2 instead of the WS_2.

TMD nanomaterials can also be applied in a fluorescent detection system with QDs developed with TMD materials. For example, a fluorescent biosensor for measuring copper ions (Cu^{2+}), 2,4,6-trinitrophenol (TNP), and melamine (MA) was developed [113]. To fabricate this biosensor, the authors synthesized surface-modified molybdenum diselenide ($MoSe_2$) QDs with various functional groups such as carboxylic-, amine-, and thiol- groups ($MoSe_2/COOH$, $MoSe_2/NH_2$, and $MoSe_2/SH$). This was completed using thiol-capping agents, including thioglycolic acid (TGA), cysteamine hydrochloride (CSH), and 1,3-propanedithiol (PDT). As shown in Figure 12.6a, the synthesized $MoSe_2$ QD could detect various target molecules through different reaction mechanisms depending on the different surface groups of the $MoSe_2$ QD. For example, the $MoSe_2/COOH$ could react with Cu^{2+}, and the fluorescence signal of the $MoSe_2/COOH$ increased through interaction with metal ions (Cu^{2+}) and showed high selectivity and a sensitivity of 4.6 nM. TNP could be measured by $MoSe_2/NH_2$, and the fluorescence signal of the $MoSe_2/NH_2$ changed due to the fluorescence quenching effect resulting from the binding of TNP and $MoSe_2/NH_2$ (LOD of 45.3 nM for TNP). In addition, the $MoSe_2/SH$ modified with AuNPs was used to detect MA using the FRET method. Before the addition of MA, the synthesized $MoSe_2/SH$ modified with AuNPs exhibited a low fluorescence intensity. After the addition of MA, the fluorescence signal of the $MoSe_2/SH$ modified with AuNPs increased because of the disappearance of the FRET effects through the strong interaction between the AuNP and the MA (LOD of 27.7 nM).

In another study, a fluorescent Alzheimer's disease (AD) biosensor was developed using WS_2 nanosheets with various types of modified dextran polymers (Figure 12.6b) [114]. To fabricate this biosensor, the surface of the WS_2 nanosheet was functionalized with tetramethylammonium (TMA)-modified dextran (TMA-dex) and the fluorescein-labeled DNA probe (FAM-DNA). When it was modified on the WS_2 nanosheet, the FAM-DNA showed no fluorescence signal because of the quenching effect of the WS_2 nanosheet. However, after binding with miR-29a, the AD biomarker, FAM-DNA, could emit a fluorescence signal as the distance between the FAM and the WS_2 nanosheet increased through the double-strand DNA formation. This biosensor exhibited excellent miR-29a sensing capability at 745 pM concentration and showed high selectivity for noncomplementary and single-based mismatched RNA in human serum. Compared with electrochemical or fluorescent biosensors that have been studied at length using TMD, the studies on SERS-based biosensors using TMD have only begun recently. Thus, rather than developing the biosensor by using TMD directly, TMD has mainly been studied to develop SERS-sensitive substrates that can enhance the SERS signal from the Raman-active molecules introduced to biosensing systems. Furthermore, TMD materials are appropriate to develop SERS-sensitive substrates because of unique advantages, such as bonding effectively with Raman-active probe molecules, promoting charge transfer, and exhibiting an excellent chemical enhancement mechanism (CE). Based on these advantages, a SERS-based biosensor for Mb detection was developed using a WS_2 nanosheet and AuNPs [115]. To fabricate this

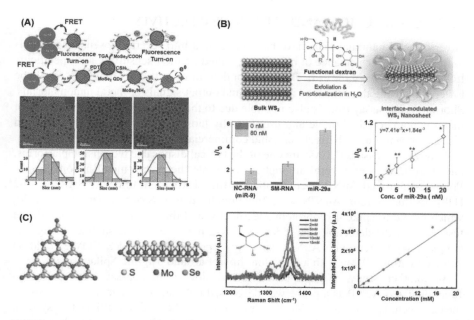

FIGURE 12.6 (a) Scheme of the fluorescence sensing mechanism of the surface-modified MoSe₂ QDs, and FE-TEM images of the MoSe₂/COOH, MoSe₂/NH₂, and the MoSe₂/SH used in this research. (Reprinted with permission from [113] Copyright (2018) ACS publishing). (b) Schematic diagram of WS₂ exfoliation with dextran for the fabrication of interface-modulated WS₂ nanosheet and photoluminescence intensity and a calibration curve of the miR-29a detection. (Reprinted with permission from [114] Copyright (2020) Elsevier publishing). (c) Schematic diagram of MoSSe structure and Raman intensity and linear curve of SERS-based biosensor. (Reprinted with permission from [116] Copyright (2020) RSC publishing).

biosensor, the WS₂ nanosheet was modified with AuNPs using the carboxymethyl cellulose (CMC), and the thiol-labeled anti-Mb DNA aptamer was immobilized on the fabricated AuNP-modified WS₂ nanosheet. The anti-Mb DNA aptamer formed a σ bond on the surface of the AuNP-modified WS₂ nanosheet and could facilitate the charge transfer reaction between the anti-Mb DNA aptamers and the AuNPs. In addition, rhodamine 6G, one of the Raman-active probes, was immobilized on the fabricated biosensor to enhance the SERS performance at 0.2 mW of 532 nm laser. When Mb was added to this biosensor, the Mb interacted with the anti-Mb DNA aptamer to further promote the charge transfer, and the SERS intensity increased. This biosensor showed excellent detection performance with a LOD of 0.5 aM and a wide detection range of 0.5 aM to 5.0 pM.

Many researchers are using TMD to fabricate improved SERS-based biosensors such as the hemispherical WS₂ nanodome modified with AuNP/graphene and MoSSe (Figure 12.6c) [116,117]. By conducting continuous studies to develop TMD with novel structures and optimized compositions of elements, TMD materials are expected to be increasingly utilized in innovative biosensors.

12.5   CONCLUSION AND FUTURE PERSPECTIVES

Diverse metal nanomaterials have been integrated into several biosensing methods to improve the performance of biosensors, including their sensitivity and specificity. LOD, which is one of the critical criteria for determining an early diagnosis, could be lowered by using the enhanced reactivity and output signals of metal nanomaterials. Shortening the assay time is also possible due to the high surface-to-volume ratio of metal nanomaterials. Simple and user-friendly biosensors are needed for increased disease prevention, early treatment, and commercialization of the biosensors.

Lateral flow assays (LFAs) are one of the typical disposable biosensing platforms, and there have been several attempts to integrate them with metal nanomaterials. Polymer-stabilized AuNPs were applied to the LFA system to detect SARS-CoV-2 [118]. In this system, AuNPs were functionalized to α,N-acetyl neuraminic acid, which bound to spike glycoprotein specifically, and they were captured to both the test line and the detection line if SARS-CoV-2 presented in the sample. Using this simple assay system with AuNPs, spike glycoprotein from SARS-COV-2 could be detected within 30 min with a detection limit of around 5 µg/mL spike protein.

In another study, a paper-based lateral flow immunoassay with antibody-functionalized AuNP as a probe was developed for fast, simultaneous, and low-cost measurement of Alzheimer's biomarkers fetuin B and clusterin [119]. In this system, the quantification of fetuin B and clusterin was conducted by analyzing both the color change on the test zone from AuNPs and the oxidation peaks of AuNPs via stripping voltammetry analysis. In addition to the LFA, wearable biosensing systems have been developed for real-time monitoring of pathophysiological signals in biofluids such as sweat, tears, saliva, and interstitial fluid [120,121]. Such real-time monitoring can provide information on well-being, improve the management of chronic diseases such as diabetes, and prevent unexpected situations for the user or medical specialists. For example, Nakata et al. conducted a study on a flexible Schottky-barrier-based charge-coupled device (CCD) that can function as a sensitive pH sensor [122]. The device was integrated with a stable and flexible temperature sensor, which was fabricated using a solution-based process. To measurement pH and temperature, SnO_2 nanoparticles and a single-wall carbon nanotube were utilized. The results indicated that it could be used for real-time monitoring of sweat pH and skin temperature.

On the other hand, there have been several attempts to develop microfluidic systems with biosensing functions. In particular, organ-on-a-chip technology, which consists of three-dimensional cell structure and fluidic flow, can revolutionize in vitro microfluidic cell chips by making it less expensive and introducing a practical drug screening system. Inherent human cell signaling against some drugs could be integrated into the organ-on-a-chip that emulates the human organ and its circulation system under fluidic conditions that mimic the blood and lymphatic vessels. This testing platform would be able to confirm the overall pharmacokinetic and pharmacodynamic properties of nanoparticles and more accurately simulate the human organ system. This organ-on-a-chip technology can be utilized as an effective total analysis system, measuring diverse signals between host-microbe-external stimuli such as drugs, nanotherapeutics, or virulent materials. In fact, microfluidic platforms can provide advantageous conditions to analyze body fluid, including several signaling and marker molecules such as DNA or protein. In microfluidic devices, human samples can be analyzed more accurately, rapidly, and simply because it does not require handling by a person, and it is not affected by external factors that can cause experimental errors. In addition, it is possible to culture cells in microfluidic devices such as organs-on-a-chip, so it is not necessary to collect and inject samples separately when detecting secretome, including cytokine, by external stimuli such as virulent factors or drugs. Microfluidic channels also require smaller sample volumes, reducing the response time and distances for target molecules to diffuse to the capturing ligands. This shortens response time and improves diffusion-limited processes when analyzing drops of blood or any other body fluid.

Metal nanomaterials, which can produce more precise, rapid, and sensitive detection of several biomolecules, can be easily applied in microfluidic channels as capturing and signaling materials since it is small enough to flow liberally in the microchannel. For example, biomarkers released from cells in the microfluidic system can be measured by existing microfluidic sensing systems. This includes a label-free localized SPR microarray with Au nanorods and anti-cytokine for analyzing pro-inflammatory cytokines in a complex serum matrix with high accuracy and sensitivity at concentrations as low as 5–20 pg/mL from a 1 µL sample [123]. Also, it would be possible to test the efficiency of nanotherapeutics more effectively with human organ-on-a-chip systems compared with current in vitro 2D or 3D cell culture systems, as well as in vivo animal systems. In the near future, human organ-on-a-chip technology will provide a unified approach to biology and nanotechnology in highly defined tissue microenvironments, leading to the development of more effective nanotherapeutics with in situ monitoring of biological signals.

In summary, metal nanomaterials can be used in biosensing systems to improve detection properties such as high specificity, high sensitivity, simple and rapid detection, multi-analyte detection, and in situ monitoring. This chapter discusses recent approaches for metal nanomaterial-integrated biomolecular detection systems with several analytical methods, including electrochemical, fluorescent, colorimetric, SPR, and magnetic-based measurement. Different biosensing approaches have their specific inherent properties, and they can be simply exploited to improve sensing performance. Metal nanomaterials are able to facilitate innovative sensing systems because of their high electromagnetic property and plasmonic effect. It is anticipated that advanced sensing platforms using organ-on-a-chip technology or other chip-based systems for simulating human microenvironments with metal nanomaterial-integrated sensing module will offer a robust in vitro drug screening that can be helpful as intermediated testing platform between in vitro and in vivo and provide in vitro personalized analysis in biomedical applications.

AUTHORS' STATEMENT

- **Author contributions**: The book chapter was written through contributions of all authors. J. -H. Choi and J. Yoon wrote about the metal nanomaterials and their application for the biosensors. M. Shin wrote about the TMD materials and their application for the biosensors. H. K. Choi wrote the future perspective for the organ-on-a-chip platform. J. -W. Choi directed the entire book chapter and contributed to this work as the corresponding author. All authors read and approved the submitted manuscript.
- **Conflicts of Interest**: The authors declare no conflict of interest.

REFERENCES

1. Kirsch, J., C. Siltanen, Q. Zhou, A. Revzin, and A. Simonian. 2013. "Biosensor technology: Recent advances in threat agent detection and medicine." *Chemical Society Reviews* 42 (22):8733–8768. Doi: 10.1039/C3CS60141B.
2. Vigneshvar, S., C. C. Sudhakumari, B. Senthilkumaran, and H. Prakash. 2016. "Recent advances in biosensor technology for potential applications – an overview." *Frontiers in Bioengineering and Biotechnology* 4 (11). Doi: 10.3389/fbioe.2016.00011.
3. Roointan, A., T. A. Mir, S. I. Wani, R. Mati ur, K. K. Hussain, B. Ahmed, S. Abrahim, A. Savardashtaki, G. Gandomani, M. Gandomani, R. Chinnappan, and M. H. Akhtar. 2019. "Early detection of lung cancer biomarkers through biosensor technology: A review." *Journal of Pharmaceutical and Biomedical Analysis* 164: 93–103. Doi: 10.1016/j.jpba.2018.10.017.
4. Morales-Narváez, E., and C. Dincer. 2020. "The impact of biosensing in a pandemic outbreak: COVID-19." *Biosensors and Bioelectronics* 163: 112274. Doi: 10.1016/j.bios.2020.112274.
5. Udugama, B., P. Kadhiresan, H. N. Kozlowski, A. Malekjahani, M. Osborne, V. Y. C. Li, H. Chen, S. Mubareka, J. B. Gubbay, and W. C. W. Chan. 2020. "Diagnosing COVID-19: The disease and tools for detection." *ACS Nano* 14 (4): 3822–3835. Doi: 10.1021/acsnano.0c02624.
6. Bansod, B. K., T. Kumar, R. Thakur, S. Rana, and I. Singh. 2017. "A review on various electrochemical techniques for heavy metal ions detection with different sensing platforms." *Biosensors and Bioelectronics* 94: 443–455. Doi: 10.1016/j.bios.2017.03.031.
7. Benito-Peña, E., M. G. Valdés, B. Glahn-Martínez, and M. C. Moreno-Bondi. 2016. "Fluorescence based fiber optic and planar waveguide biosensors. A review." *Analytica Chimica Acta* 943: 17–40. Doi: 10.1016/j.aca.2016.08.049.
8. Li, P., F. Long, W. Chen, Jing Chen, P. K. Chu, and H. Wang. 2020. "Fundamentals and applications of surface-enhanced Raman spectroscopy–based biosensors." *Current Opinion in Biomedical Engineering* 13: 51–59 Doi: 10.1016/j.cobme.2019.08.008.
9. Xie, F., M. Yang, M. Jiang, X.-J. Huang, W.-Q. Liu, and P.-H. Xie. 2019. "Carbon-based nanomaterials – a promising electrochemical sensor toward persistent toxic substance." *TrAC Trends in Analytical Chemistry* 119: 115624. Doi: 10.1016/j.trac.2019.115624.
10. Luo, Y., C. Zhu, D. Du, and Y. Lin. 2019. "A review of optical probes based on nanomaterials for the detection of hydrogen sulfide in biosystems." *Analytica Chimica Acta* 1061:1–12 Doi: 10.1016/j.aca.2019.02.045.
11. Li, X., J. Zhu, and B. Wei. 2016. "Hybrid nanostructures of metal/two-dimensional nanomaterials for plasmon-enhanced applications." *Chemical Society Reviews* 45 (11): 3145–3187. Doi: 10.1039/C6CS00195E.
12. Holzinger, M., A. Le Goff, and S. Cosnier. 2014. "Nanomaterials for biosensing applications: A review." *Frontiers in Chemistry* 2 (63). Doi: 10.3389/fchem.2014.00063.
13. Bruce, P. G, B. Scrosati, and J.-M. Tarascon. 2008. "Nanomaterials for rechargeable lithium batteries." *Angewandte Chemie International Edition* 47 (16): 2930–2946. Doi: 10.1002/anie.200702505.
14. Li, Y., Z. Li, C. Chi, H. Shan, L. Zheng, and Z. Fang. 2017. "Plasmonics of 2D nanomaterials: Properties and applications." *Advanced Science* 4 (8): 1600430. Doi: 10.1002/advs.201600430.
15. Sau, T. K., A. L. Rogach, F. Jäckel, T. A. Klar, and J. Feldmann. 2010. "Properties and applications of colloidal nonspherical noble metal nanoparticles." *Advanced Materials* 22 (16): 1805–1825. Doi: 10.1002/adma.200902557.
16. Guo, L., Y. Xu, A. R. Ferhan, G. Chen, and D.-H. Kim. 2013. "Oriented gold nanoparticle aggregation for colorimetric sensors with surprisingly high analytical figures of merit." *Journal of the American Chemical Society* 135 (33): 12338–12345. Doi: 10.1021/ja405371g.

17. Cao, X., Y. Ye, and S. Liu. 2011. "Gold nanoparticle-based signal amplification for biosensing." *Analytical Biochemistry* 417 (1): 1–16. Doi: 10.1016/j.ab.2011.05.027.
18. Matricardi, C., C. Hanske, J. L. Garcia-Pomar, J. Langer, A. Mihi, and L. M. Liz-Marzán. 2018. "Gold nanoparticle plasmonic superlattices as surface-enhanced Raman spectroscopy substrates." *ACS Nano* 12 (8): 8531–8539. Doi: 10.1021/acsnano.8b04073.
19. Pomerantseva, E., F. Bonaccorso, X. Feng, Y. Cui, and Y. Gogotsi. 2019. "Energy storage: The future enabled by nanomaterials." *Science* 366 (6468): eaan8285. Doi: 10.1126/science.aan8285.
20. Peng, L., Z. Fang, Y. Zhu, C. Yan, and G. Yu. 2018. "Holey 2D nanomaterials for electrochemical energy storage." *Advanced Energy Materials* 8 (9): 1702179. Doi: 10.1002/aenm.201702179.
21. Lee, J. J., L. S. Yazan, and C. A. C. Abdullah. 2017. "A review on current nanomaterials and their drug conjugate for targeted breast cancer treatment." *International Journal of Nanomedicine* 12: 2373–2384. Doi: 10.2147/IJN.S127329.
22. D. Shareena, P. Thabitha, D. McShan, A. K. Dasmahapatra, and P. B. Tchounwou. 2018. "A review on graphene-based nanomaterials in biomedical applications and risks in environment and health." *Nano-Micro Letters* 10 (3): 53. Doi: 10.1007/s40820-018-0206-4.
23. Zhang, B., W. Yan, Y. Zhu, W. Yang, W. Le, B. Chen, R. Zhu, and L. Cheng. 2018. "Nanomaterials in neural-stem-cell-mediated regenerative medicine: Imaging and treatment of neurological diseases." *Advanced Materials* 30 (17): 1705694. Doi: 10.1002/adma.201705694.
24. Mohajeri, M., B. Behnam, and A. Sahebkar. 2019. "Biomedical applications of carbon nanomaterials: Drug and gene delivery potentials." *Journal of Cellular Physiology* 234 (1): 298–319. Doi: 10.1002/jcp.26899.
25. Siavash Moakhar, R., Tamer AbdelFatah, A. Sanati, M. Jalali, S. E. Flynn, S. S. Mahshid, and S. Mahshid. 2020. "A nanostructured gold/graphene microfluidic device for direct and plasmonic-assisted impedimetric detection of bacteria." *ACS Applied Materials & Interfaces* 12 (20): 23298–23310. Doi: 10.1021/acsami.0c02654.
26. Takemura, K., O. Adegoke, N. Takahashi, T. Kato, T.-C. Li, N. Kitamoto, T. Tanaka, T. Suzuki, and E. Y. Park. 2017. "Versatility of a localized surface plasmon resonance-based gold nanoparticle-alloyed quantum dot nanobiosensor for immunofluorescence detection of viruses." *Biosensors and Bioelectronics* 89: 998–1005. Doi: 10.1016/j.bios.2016.10.045.
27. Lerdsri, J., W. Chananchana, J. Upan, T. Sridara, and J. Jakmunee. 2020. "Label-free colorimetric aptasensor for rapid detection of aflatoxin B1 by utilizing cationic perylene probe and localized surface plasmon resonance of gold nanoparticles." *Sensors and Actuators B: Chemical* 320: 128356. Doi: 10.1016/j.snb.2020.128356.
28. Wang, Z., and L. Ma. 2009. "Gold nanoparticle probes." *Coordination Chemistry Reviews* 253 (11): 1607–1618. Doi: 10.1016/j.ccr.2009.01.005.
29. Liu, T., Z. Chu, and W. Jin. 2019. "Electrochemical mercury biosensors based on advanced nanomaterials." *Journal of Materials Chemistry B* 7 (23): 3620–3632. Doi: 10.1039/C9TB00418A.
30. Li, X., Y. Li, Q. Qiu, Q. Wen, Q. Zhang, W. Yang, L. Yuwen, L. Weng, and L. Wang. 2019. "Efficient biofunctionalization of MoS$_2$ nanosheets with peptides as intracellular fluorescent biosensor for sensitive detection of caspase-3 activity." *Journal of Colloid and Interface Science* 543: 96–105 Doi: 10.1016/j.jcis.2019.02.011.
31. Ye, X., X. He, Y. Lei, J. Tang, Y. Yu, H. Shi, and K. Wang. 2019. "One-pot synthesized Cu/Au/Pt trimetallic nanoparticles with enhanced catalytic and plasmonic properties as a universal platform for biosensing and cancer theranostics." *Chemical Communications* 55 (16): 2321–2324. Doi: 10.1039/C8CC10127B.
32. Busseron, E., Y. Ruff, E. Moulin, and N. Giuseppone. 2013. "Supramolecular self-assemblies as functional nanomaterials." *Nanoscale* 5 (16): 7098–7140. Doi: 10.1039/C3NR02176A.

33. Liu, Y., D. Yu, C. Zeng, Z. Miao, and L. Dai. 2010. "Biocompatible graphene oxide-based glucose biosensors." *Langmuir* 26 (9): 6158–6160. Doi: 10.1021/la100886x.

34. Oh, W.-K., O. S. Kwon, and J. Jang. 2013. "Conducting polymer nanomaterials for biomedical applications: Cellular interfacing and biosensing." *Polymer Reviews* 53 (3):407–442. Doi: 10.1080/15583724.2013.805771.

35. You, Z., Q. Qiu, H. Chen, Y. Feng, X. Wang, Y. Wang, and Y. Ying. 2020. "Laser-induced noble metal nanoparticle-graphene composites enabled flexible biosensor for pathogen detection." *Biosensors and Bioelectronics* 150: 111896. Doi: 10.1016/j. bios.2019.111896.

36. Rosati, G., M. Ravarotto, M. Scaramuzza, A. De Toni, and A. Paccagnella. 2019. "Silver nanoparticles inkjet-printed flexible biosensor for rapid label-free antibiotic detection in milk." *Sensors and Actuators B: Chemical* 280: 280–289. Doi: 10.1016/j. snb.2018.09.084.

37. Wang, J. 2012. "Electrochemical biosensing based on noble metal nanoparticles." *Microchimica Acta* 177 (3): 245–270. Doi: 10.1007/s00604-011-0758-1.

38. Jain, P. K., X. Huang, I. H. El-Sayed, and M. A. El-Sayed. 2007. "Review of some interesting surface plasmon resonance-enhanced properties of noble metal nanoparticles and their applications to biosystems." *Plasmonics* 2 (3): 107–118. Doi: 10.1007/s11468-007-9031-1.

39. Choi, J.-H., J.-H. Lee, J. Son, and J.-W. Choi. 2020. "Noble metal-assisted surface plasmon resonance immunosensors." *Sensors* 20 (4). Doi: 10.3390/s20041003.

40. Xi, Z., H. Ye, and X. Xia. 2018. "Engineered noble-metal nanostructures for in vitro diagnostics." *Chemistry of Materials* 30 (23): 8391–8414. Doi: 10.1021/acs. chemmater.8b04152.

41. Underwood, S., and P. Mulvaney. 1994. "Effect of the solution refractive index on the color of gold colloids." *Langmuir* 10 (10):3427–3430.

42. Yu, C., and J. Irudayaraj. 2007. "Multiplex biosensor using gold nanorods." *Analytical Chemistry* 79 (2): 572–579. Doi: 10.1021/ac061730d.

43. Wu, Y., M. R. Ali, K. Chen, N. Fang, and M. A. El-Sayed. 2019. "Gold nanoparticles in biological optical imaging." *Nano Today* 24: 120–140. Doi: 10.1016/j. nantod.2018.12.006.

44. García-Álvarez, R., L. Chen, A. Nedilko, A. Sánchez-Iglesias, A. Rix, W. Lederle, V. Pathak, T. Lammers, G. von Plessen, K. Kostarelos, L. M. Liz-Marzán, A. J. C. Kuehne, and D. N. Chigrin. 2020. "Optimizing the geometry of photoacoustically active gold nanoparticles for biomedical imaging." *ACS Photonics* 7 (3): 646–652. Doi: 10.1021/ acsphotonics.9b01418.

45. Cholula-Díaz, J. L., D. Lomelí-Marroquín, B. Pramanick, A. Nieto-Argüello, L. A. Cantú-Castillo, and H. Hwang. 2018. "Synthesis of colloidal silver nanoparticle clusters and their application in ascorbic acid detection by SERS." *Colloids and Surfaces B: Biointerfaces* 163: 329–335. Doi: 10.1016/j.colsurfb.2017.12.051.

46. Liu, P., Y. Zhou, M. Guo, S. Yang, O. Félix, D. Martel, Y. Qiu, Y. Ma, and G. Decher. 2018. "Fluorescence-enhanced bio-detection platforms obtained through controlled "step-by-step" *Clustering of Silver Nanoparticles. Nanoscale* 10 (2): 848–855. Doi: 10.1039/C7NR07486G.

47. Tansil, N. C., and Z. Gao. 2006. "Nanoparticles in biomolecular detection." *Nano Today* 1 (1): 28–37. Doi: 10.1016/S1748–0132(06)70020-2.

48. Li, J., L. Li, Y. Lv, H. Zou, Y. Wei, F. Nie, W. Duan, M. Sedike, L. Xiao, and M. Wang. 2020. "The construction of the novel magnetic prodrug Fe_3O_4@DOX and its antagonistic effects on hepatocarcinoma with low toxicity." *RSC Advances* 10 (48): 28965–28974. Doi: 10.1039/D0RA01729A.

49. Nakata, K., Y. Hu, O. Uzun, O. Bakr, and F. Stellacci. 2008. "Chains of superparamagnetic nanoparticles." *Advanced Materials* 20 (22): 4294–4299. Doi: 10.1002/ adma.200800022.

50. Kesse, X., A. Adam, S. Begin-Colin, D. Mertz, E. Larquet, T. Gacoin, I. Maurin, C. Vichery, and J.-M. Nedelec. 2020. "Elaboration of superparamagnetic and bioactive multicore–shell nanoparticles (Γ-Fe$_2$O$_3$@SiO$_2$-CaO): A promising material for bone cancer treatment." *ACS Applied Materials & Interfaces* 12 (42):47820–47830. Doi: 10.1021/acsami.0c12769.

51. Spadaro, S., M. Santoro, F. Barreca, A. Scala, S. Grimato, Fortunato Neri, and Enza Fazio. 2017. "PEG-PLGA electrospun nanofibrous membranes loaded with Au@Fe$_2$O$_3$ nanoparticles for drug delivery applications." *Frontiers of Physics* 13 (1): 136201. Doi: 10.1007/s11467-017-0703-9.

52. Amin, P., and M. Patel. 2020. "Magnetic nanoparticles - a promising tool for targeted drug delivery system." *Asian Journal of Nanosciences and Materials* 3 (1): 24–37. Doi: 10.26655/AJNANOMAT.2020.1.3.

53. Yazdani, F., and M. Seddigh. 2016. "Magnetite nanoparticles synthesized by co-precipitation method: The effects of various iron anions on specifications." *Materials Chemistry and Physics* 184: 318–323. Doi: 10.1016/j.matchemphys.2016.09.058.

54. Campbell, R. E. 2009. "Fluorescent-protein-based biosensors: Modulation of energy transfer as a design principle." *Analytical Chemistry* 81 (15): 5972–5979. Doi: 10.1021/ac802613w.

55. Ma, F., C.-c. Li, and C.-y. Zhang. 2018. "Development of quantum dot-based biosensors: Principles and applications." *Journal of Materials Chemistry B* 6 (39): 6173–6190. Doi: 10.1039/C8TB01869C.

56. Lawrie, W. I. L., H. G. J. Eenink, N. W. Hendrickx, J. M. Boter, L. Petit, S. V. Amitonov, M. Lodari, B. Paquelet Wuetz, C. Volk, S. G. J. Philips, G. Droulers, N. Kalhor, F. van Riggelen, D. Brousse, A. Sammak, L. M. K. Vandersypen, G. Scappucci, and M. Veldhorst. 2020. "Quantum dot arrays in silicon and germanium." *Applied Physics Letters* 116 (8): 080501. Doi: 10.1063/5.0002013.

57. Fan, Y., L. Liu, and F. Zhang. 2019. "Exploiting lanthanide-doped upconversion nanoparticles with core/shell structures." *Nano Today* 25: 68–84 Doi: 10.1016/j.nantod.2019.02.009.

58. Zhang, S., G. Wright, and Y. Yang. 2000. "Materials and techniques for electrochemical biosensor design and construction." *Biosensors and Bioelectronics* 15 (5):273–282. Doi: 10.1016/S0956-5663(00)00076-2.

59. Palchetti, I., S. Laschi, and M. Mascini. 2009. "Electrochemical biosensor technology: Application to pesticide detection." In *Biosensors and Biodetection: Methods and Protocols: Electrochemical and Mechanical Detectors, Lateral Flow and Ligands for Biosensors*, edited by A. Rasooly and K. E. Herold, 115–126. Totowa, NJ: Humana Press.

60. Davis, G. 1985. "Electrochemical techniques for the development of amperometric biosensors." *Biosensors* 1 (2): 161–178. Doi: 10.1016/0265-928X(85)80002-X.

61. Pan, C., H. Wei, Z. Han, F. Wu, and L. Mao. 2020. "Enzymatic electrochemical biosensors for in situ neurochemical measurement." *Current Opinion in Electrochemistry* 19:162–167 Doi: 10.1016/j.coelec.2019.12.008.

62. Sadeghi, S. J. 2013. "Amperometric biosensors." In *Encyclopedia of Biophysics*, edited by G. C. K. Roberts, 61–67. Berlin, Heidelberg: Springer Berlin Heidelberg.

63. Yoon, J., T. Lee, B. G, Bharate, J. Jo, B.-K. Oh, and J.-W. Choi. 2017. "Electrochemical H$_2$O$_2$ biosensor composed of myoglobin on MoS$_2$ nanoparticle-graphene oxide hybrid structure." *Biosensors and Bioelectronics* 93: 14–20. Doi: 10.1016/j.bios.2016.11.064.

64. Yoon, J., J.-W. Shin, J. Lim, M. Mohammadniaei, G. B. Bapurao, Taek Lee, and J.-W. Choi. 2017. "Electrochemical nitric oxide biosensor based on amine-modified MoS$_2$/graphene oxide/myoglobin hybrid." *Colloids and Surfaces B: Biointerfaces* 159: 729–736 Doi: 10.1016/j.colsurfb.2017.08.033.

65. Wen, Y., Y. Yuan, L. Li, D. Ma, Q. Liao, and S. Hou. 2017. "Ultrasensitive dnazyme based amperometric determination of uranyl ion using mesoporous silica nanoparticles loaded with methylene blue." *Microchimica Acta* 184 (10): 3909–3917. Doi: 10.1007/s00604-017-2397-7.
66. Bonanni, A., A. H. Loo, and M. Pumera. 2012. "Graphene for impedimetric biosensing." *TrAC Trends in Analytical Chemistry* 37: 12–21. Doi: 10.1016/j.trac.2012.02.011.
67. Ruecha, N., K. Shin, O. Chailapakul, and N. Rodthongkum. 2019. "Label-free paper-based electrochemical impedance immunosensor for human interferon gamma detection." *Sensors and Actuators B: Chemical* 279: 298–304. Doi: 10.1016/j.snb.2018.10.024.
68. Dai, Y., A. Molazemhosseini, and C. C. Liu. 2017. "A single-use, in vitro biosensor for the detection of T-tau protein, a biomarker of neuro-degenerative disorders, in PBS and human serum using differential pulse voltammetry (DPV)." *Biosensors* 7 (1). Doi: 10.3390/bios7010010.
69. Mohammadniaei, M., T. Lee, J. Yoon, D. Lee, and J.-W. Choi. 2017. "Electrochemical nucleic acid detection based on parallel structural DsDNA/recombinant Azurin hybrid." *Biosensors and Bioelectronics* 98: 292–298. Doi: 10.1016/j.bios.2017.07.005.
70. Strianese, M., M. Staiano, G. Ruggiero, T. Labella, C. Pellecchia, and S. D'Auria. 2012. "Fluorescence-based biosensors." In *Spectroscopic Methods of Analysis: Methods and Protocols*, edited by W. M. Bujalowski, 193–216. Totowa, NJ: Humana Press.
71. Tainaka, K., R. Sakaguchi, H. Hayashi, S. Nakano, F. F. Liew, and T. Morii. 2010. "Design strategies of fluorescent biosensors based on biological macromolecular receptors." *Sensors* 10 (2). Doi: 10.3390/s100201355.
72. Watabe, T., K. Terai, K. Sumiyama, and M. Matsuda. 2020. "Booster, a red-shifted genetically encoded Förster resonance energy transfer (FRET) biosensor compatible with cyan fluorescent protein/yellow fluorescent protein-based fret biosensors and blue light-responsive optogenetic tools." *ACS Sensors* 5 (3): 719–730. Doi: 10.1021/acssensors.9b01941.
73. Léger, C., A. Yahia-Ammar, K. Susumu, I. L. Medintz, A. Urvoas, M. Valerio-Lepiniec, P. Minard, and N. Hildebrandt. 2020. "Picomolar biosensing and conformational analysis using artificial bidomain proteins and terbium-to-quantum dot förster resonance energy transfer." *ACS Nano* 14 (5): 5956–5967. Doi: 10.1021/acsnano.0c01410.
74. Tan, G.-R., M. Wang, C.-Y. Hsu, N. Chen, and Y. Zhang. 2016. "Small upconverting fluorescent nanoparticles for biosensing and bioimaging." *Advanced Optical Materials* 4 (7): 984–997. Doi: 10.1002/adom.201600141.
75. Li, H., X. Wang, D. Huang, and G. Chen. 2019. "Recent advances of lanthanide-doped upconversion nanoparticles for biological applications." *Nanotechnology* 31 (7): 072001. Doi: 10.1088/1361-6528/ab4f36.
76. Liu, L., H. Zhang, Z. Wang, and D. Song. 2019. "Peptide-functionalized upconversion nanoparticles-based fret sensing platform for caspase-9 activity detection in vitro and in vivo." *Biosensors and Bioelectronics* 141: 111403. Doi: 10.1016/j.bios.2019.111403.
77. Ma, L., F. Liu, Z. Lei, and Z. Wang. 2017. "A novel Upconversion@Polydopamine Core@Shell nanoparticle based aptameric biosensor for biosensing and imaging of cytochrome C inside living cells." *Biosensors and Bioelectronics* 87: 638–645. Doi: 10.1016/j.bios.2016.09.017.
78. Luo, Z., L. Zhang, R. Zeng, L. Su, and D. Tang. 2018. "Near-infrared light-excited core–core–shell UCNP@Au@CdS upconversion nanospheres for ultrasensitive photoelectrochemical enzyme immunoassay." *Analytical Chemistry* 90 (15): 9568–9575. Doi: 10.1021/acs.analchem.8b02421.
79. Qu, A., X. Wu, L. Xu, L. Liu, W. Ma, H. Kuang, and C. Xu. 2017. "SERS- and luminescence-active Au–Au–UCNP trimers for attomolar detection of two cancer biomarkers." *Nanoscale* 9 (11): 3865–3872. Doi: 10.1039/C6NR09114H.

80. Wang, Y., Z. Wei, X. Luo, Q. Wan, R. Qiu, and S. Wang. 2019. "An ultrasensitive homogeneous aptasensor for carcinoembryonic antigen based on upconversion fluorescence resonance energy transfer." *Talanta* 195: 33–39. Doi: 10.1016/j.talanta.2018.11.011.

81. Rabie, H., Y. Zhang, N. Pasquale, M. J. Lagos, P. E. Batson, and K.-B. Lee. 2019. "NIR biosensing of neurotransmitters in stem cell-derived neural interface using advanced core–shell upconversion nanoparticles." *Advanced Materials* 31 (14): 1806991. Doi: 10.1002/adma.201806991.

82. Ensafi, A. A., P. Nasr-Esfahani, and B. Rezaei. 2017. "Quenching-recovery fluorescent biosensor for DNA detection based on mercaptopropionic acid-capped cadmium telluride quantum dots aggregation." *Sensors and Actuators B: Chemical* 249: 149–155 Doi: 10.1016/j.snb.2017.04.084.

83. Reineck, P., D. Gómez, S. H. Ng, M. Karg, T. Bell, P. Mulvaney, and U. Bach. 2013. "Distance and wavelength dependent quenching of molecular fluorescence by Au@SiO$_2$ core–shell nanoparticles." *ACS Nano* 7 (8): 6636–6648. Doi: 10.1021/nn401775e.

84. Lee, J.-H., J.-H. Choi, S.-T. D. Chueng, T. Pongkulapa, L. Yang, H.-Y. Cho, J.-W. Choi, and K.-B. Lee. 2019. "Nondestructive characterization of stem cell neurogenesis by a magneto-plasmonic nanomaterial-based exosomal mirna detection." *ACS Nano* 13 (8): 8793–8803. Doi: 10.1021/acsnano.9b01875.

85. Yun, B. J., J. E. Kwon, K. Lee, and W.-G. Koh. 2019. "Highly sensitive metal-enhanced fluorescence biosensor prepared on electrospun fibers decorated with silica-coated silver nanoparticles." *Sensors and Actuators B: Chemical* 284: 140–147. Doi: 10.1016/j.snb.2018.12.096.

86. Della Ventura, B., M. Gelzo, E. Battista, A. Alabastri, A. Schirato, G. Castaldo, G. Corso, F. Gentile, and R. Velotta. 2019. "Biosensor for point-of-care analysis of immunoglobulins in urine by metal enhanced fluorescence from gold nanoparticles." *ACS Applied Materials & Interfaces* 11 (4): 3753–3762. Doi: 10.1021/acsami.8b20501.

87. Choi, J.-H., and J.-W. Choi. 2020. "Metal-enhanced fluorescence by bifunctional Au nanoparticles for highly sensitive and simple detection of proteolytic enzyme." *Nano Letters* 20 (10): 7100–7107. Doi: 10.1021/acs.nanolett.0c02343.

88. Sabela, M., S. Balme, M. Bechelany, J.-M. Janot, and K. Bisetty. 2017. "A review of gold and silver nanoparticle-based colorimetric sensing assays." *Advanced Engineering Materials* 19 (12): 1700270. Doi: 10.1002/adem.201700270.

89. Aldewachi, H., T. Chalati, M. N. Woodroofe, N. Bricklebank, B. Sharrack, and P. Gardiner. 2018. "Gold nanoparticle-based colorimetric biosensors." *Nanoscale* 10 (1): 18–33. Doi: 10.1039/C7NR06367A.

90. Xie, Z.-J., M.-R. Shi, L.-Y. Wang, C.-F. Peng, and X.-L. Wei. 2020. "Colorimetric determination of Pb^{2+} ions based on surface leaching of Au@Pt nanoparticles as peroxidase mimic." *Microchimica Acta* 187 (4): 255. Doi: 10.1007/s00604-020-04234-6.

91. Hwang, S. G., K. Ha, K. Guk, D. K. Lee, G. Eom, S. Song, T. Kang, H. Park, J. Jung, and E.-K. Lim. 2018. "Rapid and simple detection of Tamiflu-resistant influenza virus: Development of oseltamivir derivative-based lateral flow biosensor for point-of-care (POC) diagnostics." *Scientific Reports* 8 (1): 12999. Doi: 10.1038/s41598-018-31311-x.

92. Loynachan, C. N., M. R. Thomas, E. R. Gray, D. A. Richards, J. Kim, B. S. Miller, J. C. Brookes, S. Agarwal, V. Chudasama, R. A. McKendry, and M. M. Stevens. 2018. "Platinum nanocatalyst amplification: Redefining the gold standard for lateral flow immunoassays with ultrabroad dynamic range." *ACS Nano* 12 (1): 279–288. Doi: 10.1021/acsnano.7b06229.

93. Chao, J., W. Cao, S. Su, L. Weng, S. Song, C. Fan, and L. Wang. 2016. "Nanostructure-based surface-enhanced Raman scattering biosensors for nucleic acids and proteins." *Journal of Materials Chemistry B* 4 (10): 1757–1769. Doi: 10.1039/C5TB02135A.

94. Hussein, M. A., W. A. El-Said, B. M. Abu-Zied, and J.-W. Choi. 2020. "Nanosheet composed of gold nanoparticle/graphene/epoxy resin based on ultrasonic fabrication for flexible dopamine biosensor using surface-enhanced Raman spectroscopy." *Nano Convergence* 7 (1): 15. Doi: 10.1186/s40580-020-00225-8.

95. Chaney, S. B., S. Shanmukh, R. A. Dluhy, and Y. P. Zhao. 2005. "Aligned silver nanorod arrays produce high sensitivity surface-enhanced Raman spectroscopy substrates." *Applied Physics Letters* 87 (3): 031908. Doi: 10.1063/1.1988980.

96. Vo-Dinh, T. 1998. "Surface-enhanced Raman spectroscopy using metallic nano-structures." *TrAC Trends in Analytical Chemistry* 17 (8): 557–582. Doi: 10.1016/S0165-9936(98)00069-7.

97. Cai, H., B. Huang, R. Lin, P. Xu, Y. Liu, and Y. Zhao. 2018. "A "Turn-Off" SERS assay for kinase detection based on arginine N-phosphorylation process." *Talanta* 189: 353–358. Doi: 10.1016/j.talanta.2018.07.002.

98. Lee, T., S. Kwon, S. Jung, H. Lim, and J.-J. Lee. 2019. "Macroscopic Ag nanostructure array patterns with high-density hotspots for reliable and ultra-sensitive SERS substrates." *Nano Research* 12 (10): 2554–2558. Doi: 10.1007/s12274-019-2484-7.

99. Wang, M., G. Shi, J. Zhu, Y. Zhu, X. Sun, P. Wang, T. Jiao, and R. Li. 2019. "Preparation of a novel SERS platform based on mantis wing with high-density and multi-level "Hot Spots"." *Nanomaterials* 9 (5). Doi: 10.3390/nano9050672.

100. Choi, J. H., W. A. El-Said, and J.-W. Choi. 2020. "Highly sensitive surface-enhanced raman spectroscopy (SERS) platform using core/double shell (Ag/Polymer/Ag) nano-horn for proteolytic biosensor." *Applied Surface Science* 506: 144669. Doi: 10.1016/j.apsusc.2019.144669.

101. Choi, J.-H., T.-H. Kim, W. A. El-said, J.-H. Lee, L. Yang, B. Conley, J.-W. Choi, and K.-B. Lee. 2020. "In situ detection of neurotransmitters from stem cell-derived neural interface at the single-cell level via graphene-hybrid SERS nanobiosensing." *Nano Letters* 20 (10): 7670–7679. Doi: 10.1021/acs.nanolett.0c03205.

102. Shao, H., H. Lin, Z. Guo, J. Lu, Y. Jia, M. Ye, F. Su, L. Niu, W. Kang, S. Wang, Y. Hu, and Y. Huang. 2019. "A multiple signal amplification sandwich-type SERS biosensor for femtomolar detection of miRNA." *Biosensors and Bioelectronics* 143: 111616. Doi: 10.1016/j.bios.2019.111616.

103. Kim, H.-M., M. Uh, D. H. Jeong, H.-Y. Lee, J.-H. Park, and S.-K. Lee. 2019. "Localized surface plasmon resonance biosensor using nanopatterned gold particles on the surface of an optical fiber." *Sensors and Actuators B: Chemical* 280: 183–191. Doi: 10.1016/j.snb.2018.10.059.

104. Lu, M., H. Zhu, L. Hong, J. Zhao, J.-F. Masson, and W. Peng. 2020. "Wavelength-tunable optical fiber localized surface plasmon resonance biosensor via a diblock copolymer-templated nanorod monolayer." *ACS Applied Materials & Interfaces* 12 (45): 50929–50940. Doi: 10.1021/acsami.0c09711.

105. Wang, Q., L. Zou, X. Yang, X. Liu, W. Nie, Y. Zheng, Q. Cheng, and K. Wang. 2019. "Direct quantification of cancerous exosomes via surface plasmon resonance with dual gold nanoparticle-assisted signal amplification." *Biosensors and Bioelectronics* 135: 129–136. Doi: 10.1016/j.bios.2019.04.013.

106. Picciolini, S., A. Gualerzi, R. Vanna, A. Sguassero, F. Gramatica, M. Bedoni, M. Masserini, and C. Morasso. 2018. "Detection and characterization of different brain-derived subpopulations of plasma exosomes by surface plasmon resonance imaging." *Analytical Chemistry* 90 (15): 8873–8880. Doi: 10.1021/acs.analchem.8b00941.

107. Klein, T., W. Wang, L. Yu, K. Wu, K. L. M. Boylan, R. I. Vogel, A. P. N. Skubitz, and J.-P. Wang. 2019. "Development of a multiplexed giant magnetoresistive biosensor array prototype to quantify ovarian cancer biomarkers." *Biosensors and Bioelectronics* 126: 301–307. Doi: 10.1016/j.bios.2018.10.046.

108. Xianyu, Y., Y. Dong, Z. Wang, Z. Xu, R. Huang, and Y. Chen. 2019. "Broad-range magnetic relaxation switching bioassays using click chemistry-mediated assembly of polystyrene beads and magnetic nanoparticles." *ACS Sensors* 4 (7): 1942–1949. Doi: 10.1021/acssensors.9b00900.

109. Tian, B., Y. Han, J. Fock, M. Strömberg, K. Leifer, and M. F. Hansen. 2019. "Self-assembled magnetic nanoparticle–graphene oxide nanotag for optomagnetic detection of DNA." *ACS Applied Nano Materials* 2 (3): 1683–1690. Doi: 10.1021/acsanm.9b00127.

110. Sharma, P. P., E. Albisetti, M. Massetti, M. Scolari, C. La Torre, M. Monticelli, M. Leone, F. Damin, G. Gervasoni, G. Ferrari, F. Salice, E. Cerquaglia, G. Falduti, M. Cretich, E. Marchisio, M. Chiari, M. Sampietro, D. Petti, and R. Bertacco. 2017. "Integrated platform for detecting pathogenic DNA via magnetic tunneling junction-based biosensors." *Sensors and Actuators B: Chemical* 242: 280–287. Doi: 10.1016/j.snb.2016.11.051.

111. Ping, J., Z. Fan, M. Sindoro, Y. Ying, and H. Zhang. 2017. "Recent advances in sensing applications of two-dimensional transition metal dichalcogenide nanosheets and their composites." *Advanced Functional Materials* 27 (19): 1605817. Doi: 10.1002/adfm.201605817.

112. Toh, R. J., C. C. Mayorga-Martinez, Z. Sofer, and M. Pumera. 2017. "1t-Phase WS_2 protein-based biosensor." *Advanced Functional Materials* 27 (5): 1604923. Doi: 10.1002/adfm.201604923.

113. Dhenadhayalan, N., T.-W. Lin, H.-L. Lee, and K.-C. Lin. 2018. "Multisensing capability of $MoSe_2$ quantum dots by tuning surface functional groups." *ACS Applied Nano Materials* 1 (7): 3453–3463. Doi: 10.1021/acsanm.8b00634.

114. Kim, H.-I., D. Yim, S.-J. Jeon, T. W. Kang, I.-J. Hwang, S. Lee, J.-K. Yang, J.-M. Ju, Y. So, and J.-H Kim. 2020. "Modulation of oligonucleotide-binding dynamics on WS_2 nanosheet interfaces for detection of Alzheimer's disease biomarkers." *Biosensors and Bioelectronics* 165: 112401. Doi: 10.1016/j.bios.2020.112401.

115. Shorie, M., V. Kumar, H. Kaur, K. Singh, V. K. Tomer, and P. Sabherwal. 2018. "Plasmonic DNA hotspots made from tungsten disulfide nanosheets and gold nanoparticles for ultrasensitive aptamer-based SERS detection of myoglobin." *Microchimica Acta* 185 (3): 158. Doi: 10.1007/s00604-018-2705-x.

116. Jia, S., A. Bandyopadhyay, H. Kumar, J. Zhang, W. Wang, T. Zhai, V. B. Shenoy, and J. Lou. 2020. "Biomolecular sensing by surface-enhanced Raman scattering of monolayer Janus transition metal dichalcogenide." *Nanoscale* 12 (19): 10723–10729. Doi: 10.1039/D0NR00300J.

117. Gu, B., and Q. Zhang. 2018. "Recent advances on functionalized upconversion nanoparticles for detection of small molecules and ions in biosystems." *Advanced Science* 5 (3): 1700609. Doi: 10.1002/advs.201700609.

118. Baker, A. N., S.-J. Richards, C. S. Guy, T. R. Congdon, M. Hasan, A. J. Zwetsloot, A. Gallo, J. R. Lewandowski, P. J. Stansfeld, A. Straube, M. Walker, S. Chessa, G. Pergolizzi, S. Dedola, R. A. Field, and M. I. Gibson. 2020. "The SARS-COV-2 spike protein binds sialic acids and enables rapid detection in a lateral flow point of care diagnostic device." *ACS Central Science* 6 (11): 2046–2052. Doi: 10.1021/acscentsci.0c00855.

119. Brazaca, L. C., J. R. Moreto, A. Martín, F. Tehrani, J. Wang, and V. Zucolotto. 2019. "Colorimetric paper-based immunosensor for simultaneous determination of Fetuin B and clusterin toward early Alzheimer's diagnosis." *ACS Nano* 13 (11): 13325–13332. Doi: 10.1021/acsnano.9b06571.

120. Padash, M., C. Enz, and S. Carrara. 2020. "Microfluidics by additive manufacturing for wearable biosensors: A review." *Sensors* 20 (15). Doi: 10.3390/s20154236.

121. Kim, J., A. S. Campbell, B. E.-F. de Ávila, and J. Wang. 2019. "Wearable biosensors for healthcare monitoring." *Nature Biotechnology* 37 (4):389–406. Doi: 10.1038/s41587-019-0045-y.

122. Nakata, S., M. Shiomi, Y. Fujita, T. Arie, S. Akita, and K. Takei. 2018. "A wearable pH sensor with high sensitivity based on a flexible charge-coupled device." *Nature Electronics* 1 (11): 596–603. Doi: 10.1038/s41928-018-0162-5.
123. Chen, P., M. T. Chung, W. McHugh, R. Nidetz, Y. Li, J. Fu, T. T. Cornell, T. P. Shanley, and K. Kurabayashi. 2015. "Multiplex serum cytokine immunoassay using nanoplasmonic biosensor microarrays." *ACS Nano* 9 (4): 4173–4181. Doi: 10.1021/acsnano.5b00396.

13 Utility of Nanomaterials in Nanomedicine for Disease Treatment

Rishi Paliwal
Indira Gandhi National Tribal University

Pramod Kumar
Institute of Lung Biology, Helmholtz Zentrum Munich

Shivani Rai Paliwal
Guru Ghasidas Vishwavidyalaya

Rameshroo Kenwat
Indira Gandhi National Tribal University

Otmar Schmid
Institute of Lung Biology, Helmholtz Zentrum Munich

CONTENTS

13.1 INTRODUCTION

Controlled drug delivery with desired objectives has been most demanded after the introduction of the magic bullet concept for drug targeting [1]. Nano-sized materials offer many advantages over micro-sized drug carriers in terms of better penetration in deeper tissues, altered pharmacokinetics, and targetibility [2,3]. Nanomedicine is defined as the drug in nano-form, which is less than 100 nm ideally. A drug or therapeutics once loaded into such nano-sized materials could be delivered for better management of the diseases, which was otherwise difficult to manage with other drug delivery devices [4]. For example, in case of cancer therapy, nanomedicine utilizes the advantages of EPR effect and accumulates passively in the vicinity of the cancer cells to release the entrapped cytotoxic drug [5].

Nanomaterials including nanocomplex or nanohybrid system can be modified over its surface easily for achieving drug targeting particularly in case of diseases like cancer. Nanomaterials are applicable not only in the treatment of the diseases but also in their diagnosis as well [4–8]. Iron-, silica-, and silver-based nanoparticles have been used as nanotheranostics. Lipid materials have been converted into carriers like liposome, niosomes, solid lipid nanoparticles, and lipid nanoemulsion [9–12]. Polymeric materials such as PLGA, chitosan, Eudragit, and dextran have been converted into suitable nanomedicine as they provide several methods of synthesis or preparation [5]. Hybrid lipid-polymeric systems are also being developed to achieve the advantages of both lipid and polymer [13]. Numerous proteins including albumin, gelatin, zein, and casein have been modified as nanoparticles, nanogels, and nanofibers to entrap the desired drug into it for their successful delivery at the site [1].

Nanomaterials are capable in delivery of large variety of molecules including small-sized drug molecules, biomacromolecules (proteins, peptide, or polysaccharides), and genetic materials (mRNA/siRNA/DNA/plasmid) [14]. To achieve high degree of targeting, surface modification with cell surface targeting ligands such as folic acid, fucose, hormones, transferrin, and monoclonal antibodies can be anchored over the surface to make them functionalized and appropriate for improved therapeutics [15–17]. Herbal drug delivery can be also achieved using nanocarriers for better bioavailability, for protection from degradable environment, and hence for improvement of their shelf life [18]. This chapter summarizes the silent remarks and limitations of each class of these nanomaterials with reference to their utilization in nanomedicine development.

13.2 CLASSIFICATION OF NANOMATERIALS AS NANOMEDICINE

The nanoparticles have wider field of applications that include nanoelectronics, tissue engineering, catalytic chemistry, cytocompatibility, nanomedicine, and targeted drug delivery. Various methods of preparation have been proposed depending on the material, purpose, and use but basically they are based on either bottom-up or top-down fashion along with several techniques for nanoparticle modification (Figure 13.1) [19].

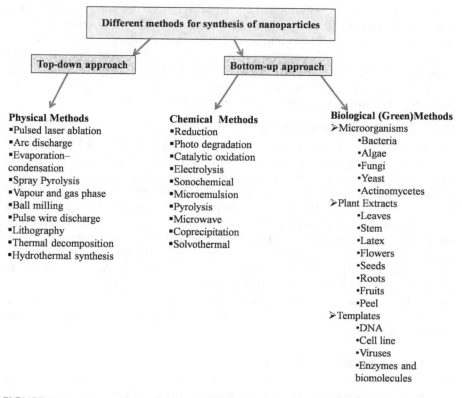

FIGURE 13.1 Physical, chemical, and biological methods for synthesis of nanoparticles.

13.2.1 Polymeric Nanoparticles

The nanoparticles based on organic polymers are generally referred as polymeric nanoparticle (PNP). These are in the size range from 1 to 1,000 nm and are of spherical- or capsular-shaped nanostructures, which can adsorb or entrap active compound within them. The PNP with various physical properties can be prepared having different composition, size, shape, morphology, crystallinity, dispersion state, and surface properties. They are generally characterized by electron microscopy, dynamic light scattering, electrophoresis, various chromatographies, and near-infrared spectroscopy techniques [20].

13.2.1.1 PLGA Nanoparticles

Poly (lactic-co-glycolic acid) (PLGA) is one of the most widely used polymers because of its biodegradability, biocompatibility, and nontoxic nature. These nanoparticles also provide controlled drug release, altered biodistribution, protection of encapsulated compounds from degradation, and opportunity to modify surface with various targeting ligands for site-specific delivery [21]. PLGA NPs are successfully exploited for delivery of several drugs and natural components. Maksimenko et al. (2019) prepared polaxamer-coated PLGA NPs for brain delivery of DOX for the chemotherapy of glioblastoma [22]. These NPs can also be modified so can be used as smart pH-dependent controlled drug release system. Gibbens-Bandala et al. (2019) developed paclitaxel-loaded PLGA NPs for the targeted delivery to breast cancer cells [23]. Gu et al. (2019) explored the use of PLGA NPs as vaccine delivery systems [24]. They studied the effect of antigen loading in NPs and surface charge on immune response to protein antigen. In this sequence, three different formulations were developed using three modified PLGA polymers with different surface charge properties. The results showed positively charged systems were more effective and proved that surface charge significantly affects the immunogenicity of developed system.

13.2.1.2 Eudragit Nanoparticles

Eudragit is a family of copolymers generally used as targeted drug delivery coatings for enteric coating, protective barriers, or sustained release formulations. Eudragit is the trade name of various poly(meth)acrylates polymers, and the chemical composition affects the physiochemical properties of these polymers, which in turn are used for different purposes [25]. One of the copolymer methyl methacrylate has been used in the development of transdermal patches, microsponges, and other film-forming sprays for application on the skin [26]. Katata-Seru et al. (2020) prepared Eudragit-based NP-loaded suppositories for controlled drug delivery and as a alternative route of administration for the treatment of pediatric HIV [27]. Salatin et al. (2020) reported Eudragit RL-100 NP-based Pluronic F-127 hydrogel for nasal administration of rivastigmine with improved therapeutic efficacy [28]. The result showed that the developed formulation demonstrated sustained drug release and higher permeability across the nasal mucosa.

13.2.1.3 Chitosan Nanoparticles

Chitosan is a linear polysaccharide which is composed of β-linked D-glucosamine and N-acetyl-D-glucosamine and is obtained by chemical treatment of chitin shell of shrimp [29]. Due to its pertinent characteristics such as compatibility, biodegradability, and nontoxicity, it plays a promising role in drug delivery. Chitosan and its derivatives can be converted into suitable nanoparticles for diverse applications in pharmaceutical, biomedical, and tissue engineering sciences [30]. It can be suitably modified chemically into a more appropriate derivative for a unique purpose like enhanced mucoadhesiveness, gel characteristics, targeting efficiency, improved solubility, and sustained release either through cross-linking, carboxylation, etherification, etherification, sulfonation, or graft copolymerization. Chitosan can be further combined as part of a nanocomposite along with polymers such as PLGA, carbon nanotubes, and metal oxides for achieving specific features in the nanocarriers for drug delivery. In a report, Kumar et al. (2019) developed chitosan nanoparticle-based mucoadhesive thermoreversible gel of chlorpherinamine maleate for nasal applications [31]. The formulation showed longer duration of action up to 8 h. The authors suggested that chitosan-based in situ gel may be promising formulation for allergic rhinitis treatment. Khan and coworkers (2019) developed lipid-chitosan hybrid nanoparticles loaded with an anticancer drug cisplatin in order to explore the structural advantages of chitosan and biocompatibility of lipids in a single carrier system [32]. The developed nanosystem was tested for cell cytotoxicity and cellular internalization in A2780 ovarian cancer cells. Further, the authors found controlled delivery of the cisplatin in animal studies with improved mean residence time and half-life. Self-gelated insulin-loaded chitosan nanoparticles have been reported with high encapsulation efficiency [33]. Animal studies confirmed the suitability of the developed system for oral insulin delivery with desired hypoglycemic effect and no toxicity.

13.2.2 PROTEIN NPS

Protein as material for nanomedicine development offers features such as safety, effectiveness, and promising results for various routes of delivery including nasal, oral ocular, and pulmonary. Further, these molecules also serve as nanocarriers for non-viral gene delivery, brain targeting, and as immunoadjuvant. Proteins are somewhat superior to synthetic polymers in terms of biocompatibility, biodegradability, and easy surface modifications. Common protein molecules that have been explored for nanomedicine development include albumin, gelatin, zein, soya protein, whey protein, gliadin, elastin, legumin, fibroins, lipoproteins, ferritin, casein, and milk proteins [34]. Some of the protein-based nanoparticles (albumin and gelatin) have reached to the markets as well. The most common techniques that are employed for fabrication of such nanoparticles include complex coacervation, desolvation, emulsification, and electrospray [34]. Batch-to-batch variation of properties of protein-based nanoparticles is one of the limitations in their clinical translation. Table 13.1 summarizes the advantages and limitation of some key protein nanoparticles.

13.2.2.1 Gelatin NPs

Gelatin, a denatured animal protein, is considered as safe in pharmaceutical product as it has GRAS status and therefore has been one of the preferred choices among all protein sources as biomaterials for product development. Due to additional features

TABLE 13.1

Advantage and Limitations of Some Key Protein Nanoparticles

Nanoparticles (NPs)	Advantage	Disadvantage
Gelatin	• Biocompatibility • Biodegradability • Easily modified • Safety • Low cost	• Low mechanical strength • Rapid degradation speed • Batch-to-batch variation
Albumin	• High stability • High solubility in physiological fluids • Biodegradability • Non-immunogenicity • Nontoxic availability and readiness	• Expensive cost • Batch-to-batch variation
Zein	• Degraded into safe by-products • Multiple functional residues • Active/passive targeting features • High availability • High coating capacity • Cytocompatibility and biocompatibility	• High molecular weight distribution • Partial immunogenicity • Batch-to-batch variation

like low cost and easy surface modification, gelatin stands at the top priority in drug delivery development as microparticles and nanoparticles both. Although homogeneity and less polydispersity are the limitation of gelatin, it varies depending upon the source of the protein, process used for its purification, and difference in molecular weight. Recent study reported by Koletti et al. (2020) describes application of gelatin nanoparticles loaded with diclofenac sodium for intravenous administration using two different optimization techniques, i.e., quality by design (QbD) and artificial neuron network (ANN) [35]. The authors observed molecular interaction of the drug with proteins and a biphasic release pattern irrespective of the development method. When compared, multiple linear regression method fitting showed better prediction than ANN. Chuang et al. (2019) developed gelatin-based nanoparticles loaded with kaempferol for topical delivery in eye [36]. This formulation has showed anti-angiogenic activity in human umbilical vein endothelial cells (HUVECs).

13.2.2.2 Albumin NPs

Albumin serves as an endogenous biocompatible, non-immunogenic biomaterial, which is suitable for development of nanomedicine in terms of ease of manufacturing, stability, safety, and half-life. Abraxane® is a commercial nanomedicine product, which is made of albumin and paclitaxel for cancer therapeutics. Recently, albumin-binding prodrugs have gained re-attention after successful commercial outcome of Abraxane®. The flexible surface chemistry of albumin makes it a suitable material for loading of both endogenous and exogenous bioactive [37]. Zhou et al. (2019) developed human serum albumin-based pirarubicin-loaded nanocomplex [38]. The developed NPs have shown better tumor accumulation in animal studies and have also shown antitumor activity by mimicking biomimetic drug delivery pathways. The authors suggested clinical potential of the developed system for tumor-targeted drug delivery.

13.2.2.3 Zein Nanoparticles

Zein is a protein obtained from corn and is amphiphilic in nature. It can be converted into nanoparticles easily and hence could be employed for use as drug carrier system [1]. Being amphiphilic in nature, it has self-assembling properties that can be utilized in developing nanoparticles, nanofibers, microparticles, and films. A sustained release pattern of drugs can be achieved easily as a result of its hydration and developing swellable matrix. Zein has been utilized in controlled drug delivery and tissue engineering. The methods used for developing zein nanoparticles include nanoprecipitation, liquid-liquid dispersion, phase separation, and electrospraying [39]. Some critical factors in developing zein-based nanomedicine are concentration, pH, ionic strength, and its thermal treatment, which are optimized depending upon its application nature of the bioactive to be encapsulated and storage conditions. Being protein, zein nanoparticles may impart some immunogenicity and therefore is subjected for studies in detail. Li et al. (2019) investigated the immunogenicity of zein nanoparticles of different sizes and in different doses after administration through both intramuscular and subcutaneous routes [40]. The authors reported that immunogenic nature of zein nanoparticles was observed; however, the size of zein-based carrier has not impacted much. It was further suggested that low dose and surface modification may reduce their immunogenicity but require further studies. A pH-dependent zein-pectin hydrogel loaded with doxorubicin has been developed for anticancer activity [41]. The authors claimed that the developed system has controlled release with enhanced shelf life and could be used for pH and site-specific responsiveness.

13.2.3 Solid Lipid Nanoparticles (SLNs)

Over the last 2–3 decades, nanoparticles have emerged as a versatile and promising replacement of liposomes as drug carriers. The implementation of nanoparticles for drug delivery depends on their ability to penetrate through several anatomical barriers (mucosa, BBB, epithelium etc.), sustained release of their contents up to molecular level, and stability in the nanometer size range [42]. Introduced in early 1990s as drug carriers [43], SLNs are a newer generation of submicron-sized lipid emulsions where the liquid lipid (oil) has been substituted by a solid lipid. The primary advantage of this aspect is that SLNs are able to encapsulate a high amount of drug bereft their small size but large surface [44].

Most of the lipid ingredients used to develop biodegradable SLNs show excellent biocompatibility, and one can avoid the exposure to organic solvents using water-based preparation techniques. Moreover, the regulatory approval for clinical translation of SLN is faster due to low cost, easy scale-up, simple sterilization strategies, and available validation methods [45]. The key ingredients used to prepare SLN are categorized as solid lipid, emulsifier, and water/solvent. Lipid components in SLN typically include triglycerides (tri-stearin), partial glycerides (Imwitor), fatty acids (stearic acid, palmitic acid), steroids (cholesterol), waxes (cetyl palmitate), and emulsifiers (various surfactants, e.g., macroglycerides, polaxamers, polysorbates, and ethoxylated castor oil). The combination of various emulsifiers (Pluronic F68 and F127) could be used to improve the stability of lipid dispersions by inhibiting agglomeration and subsequent sedimentation of particle [44].

The limitation of SLNs made of conventional lipids is their poor drug loading due to limited solubility of drug in lipid melt and highly ordered crystal structure of lipid. Therefore, the conventional lipids should be replaced with more complex lipids for future SLN formulations. Sometime drug expulsion after polymerization of SLNs or high water content reduces the storage shelf life of such systems [42]. The production methods used for the development of SLNs include primarily based on solidified nanoemulsion technologies. High-pressure homogenization (HPH), high shear homogenization, and ultrasonication for nanoemulsion preparation are the most common methods to develop SLN formulations [46].

13.2.4 NANOSTRUCTURED LIPID CARRIER (NLC)

Nanostructured lipid carriers (NLC) are a modified version of SLNs and are characterized by a solid lipid core consisting of a mixture of solid and liquid lipid. NLC formulations provide enhanced solubility similar to traditional liquid colloidal carrier drug delivery systems, in addition to several supplementary benefits, for example, increased chemical stability of the active pharmaceutical ingredient (API), potential for sustained release, targeted delivery, and lymphatic delivery, which avoids first-pass metabolic degradation and allows for lymphatic targeting [47]. The composition of NLC is similar to SLNs other than the addition of liquid lipids while NLC is a safer drug delivery carrier due to less amount of surfactant [48] and the stability of NLC formulations is also improved due to the lower tendency of crystallization of lipid ingredients [49].

The objective of developing NLC is to provide a precise nanostructure to the lipid matrix to improve the active compound payload and to avoid the expulsion of entrapped therapeutic compounds during storage [50]. For example, the combination of various solid and liquid lipids results in matrix inconsistency, which provides the NLC formulation with a series of advantages over single matrix material SLNs [51]. The liquid lipids also provide higher solvent capacity in the finished NLC due to additional active incorporation space created by crystal imperfections to increase drug load as compared with traditional SLNs [52].

By adjusting the proportions of liquid and solid ratio, NLCs can be tailored to provide a biphasic drug-release pattern with a rapidly released drug fraction from the liquid phase followed by a slower release of the active drug from the solid lipid portion of the NLC [53]. The common techniques for the development of NLC are high-pressure homogenization, solvent emulsification/ evaporation, ultrasonication, spray drying, and microfluidic technology [54].

13.2.5 DENDRIMERS

Dendrimers are well-defined, multivalent molecular synthetic nanoparticles with distinct structural composition, namely inner/central core, well-designed peripheral shell, and abundant terminal groups. These properties of architecture control make the dendrimers one of the advanced drug delivery carriers suitable for several biomedical applications, e.g., drug delivery, gene delivery, or diagnostic applications [55]. The dendrimers used for drug delivery purpose are primarily made up of poly(amidoamine) (PAMAM), poly(propylene imine) (PPI), polyether-copolyester (PEPE), or peptide dendrimer, which can be synthesized by divergent and convergent methods. In divergent methods, the core molecule, e.g., ethylene-diamine (EDA), is modified with four reactive species using Michael addition, and then, EDA is again coupled to these four arms. This reaction is repeated several times, and developed dendrimers are known as G0.5, G1, G2, G3, G4, and G5 generation dendrimers. In convergent methods, the dendrons having terminal groups are synthesized and linked finally to a core molecule to get a complete dendrimer structure [56].

The presence of functionally active chemical structures at dendrimer periphery provides the opportunity for functionalization with various biocompatible improving components, e.g., acetylation, glycosylation, PEGylation, and amino acid. Particularly, the amino acid functionalization of dendrimers has been proved very efficient in dendrimer-based cancer therapy [57]. The ability of dendrimers to be conjugated with several drugs, targeting ligands, and solubilizing agents also provide it the benefit over several linear polymers for biomedical applications [58]. For example, it improves the dendrimer capacity to be used via various potential routes of drug administration, e.g., oral mode of delivery in addition to the transdermal, intravenous, and ocular deliveries [59].

The cytotoxicity of dendrimers is observed in various cell types which is mainly attributed to the high cationic charge density at the periphery, which interacts with the cell membrane and may result in membrane disruption. Cationic dendrimers, e.g., PAMAM, PPI, and poly-l-lysine, show toxicity in a dose-dependent manner. However, negatively charged dendrimers such as sulfonated, carboxylated, phosphonated, or neutral dendrimers, or dendrimers with poly (ethylene oxide), acetyl, carboxyl, mannose, and galactose end groups, show reduced toxicity as compared with cationic dendrimers [60]. To minimize the dendrimer toxicity potential, useful methods are using neutral or anionic biocompatible dendrimers or masking the peripheral charge using chemical modifications [55].

13.2.6 Vesicular Systems

13.2.6.1 Liposomes

Liposomes are the most widely investigated and clinically applied biodegradable, biocompatible, and nonimmunogenic nanocarrier. Liposomes are nanosized vesicles (ca. 100 nm) consisting of an aqueous core surrounded by one or more phospholipid bilayers (2007). Thus, both hydrophobic (in lipid bilayers) [61] and hydrophilic drugs (in aqueous core) [62] can be incorporated effectively in liposomes. Liposomes can be stratified based on their lipid bilayer assemblies (lamellarity of liposomes) into multilammelar ($\geq 0.5\,\mu M$), oligolammelar ($0.1-1.0\,\mu M$), small unilamellar ($25-50\,nM$), large unilamellar ($0.1-1.0\,\mu M$), giant unilamellar ($1-200\,nm$), or multivescicular ($'1\,\mu M$) vesicle liposomes. The basic mechanism of liposome development is the hydrophilic/hydrophobic interaction between lipid-to-lipid and lipid-to-aqueous phase molecules. Moreover, the input of energy (sonication, homogenization, shaking, or heating) results in the arrangement of the lipid molecules to form the bilayered vesicles, which results in thermodynamic equilibrium in the aqueous phase. The lipid membranes of the liposomes efficiently fuse with cell membranes because of their structural similarity as both are bilayered lipid membranes. Fusion of the liposomal and cellular membrane facilitates the delivery of aqueous core materials into the cytosol without causing membrane disruption or cytotoxic effects on the cell membrane [63]. Due to their flexible nature of lipid constituents, liposomes have been investigated for several applications, e.g., anticancer therapy, photodynamic therapy, targeted drug delivery, and triggered drug release strategies for various pathologic conditions. Liposomal nanocarriers are highly useful for cancer-like pathologies as they improve tumor targeting of the cytotoxic agents due to the enhanced permeability and retention (EPR, Figure 13.2) effect for both active and passive targeted liposomes. The EPR effect leverages high extravasation of substances from the blood into cancerous tissue due to its leaky vasculature but also their higher retention in the diseased tissue due to the cancer-induced dysfunctional lymphatic system [63]. The nanometer scale of liposomes prolongs the half-life of encapsulated materials in the body. Moreover, through modification of liposomes, their uptake mechanism or organ/pathologic selectively can be further optimized using PEGylation technology or using the active targeting approach. Coating the surface of liposomes or other nanocarriers with polyethylene glycol (PEG), "PEGylation", is a widely used approach to improve the efficiency of drug and gene delivery to target cells and tissues by avoiding liposome aggregation, opsonization, and phagocytosis and overall prolonging systemic circulation time [64]. However, active ligand binding on the liposome surface facilitates tissue-specific delivery of liposomal drugs by ligand receptor binding [17,61,62]. Figure 13.3 repsents structure of various types of lipid vesicle-based drug delivery systems. Table 13.2 summarizes the list of lipid vesicle-based marketed formulations as nanomedicine.

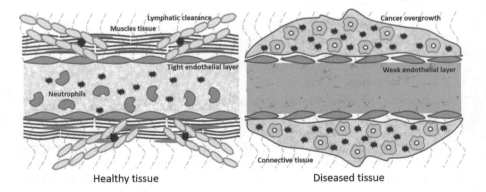

FIGURE 13.2   Schematic enhanced permeability and retention effect in diseased tissue which is useful in drug delivery and their accumulation at target site.

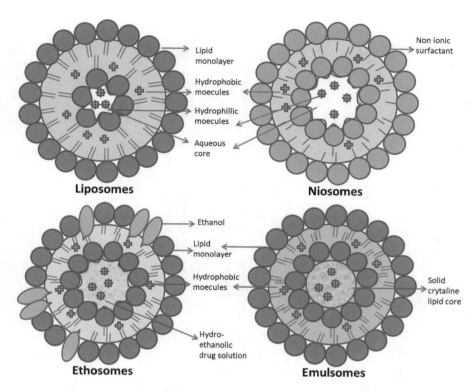

FIGURE 13.3 Schematic presentation of different types of lipid-based nanocarrier drug delivery platform.

13.2.6.2 Niosomes

Niosomes, which have been introduced as a substitute for liposomal carriers, are also a lipid-based vesicular drug delivery system. Similar to liposomes, niosomes are biodegradable, biocompatible, and nonimmunogenic with long shelf life. The high stability enables niosomes for the delivery of drug at target site in a controlled and/or sustained manner [65]. Niosomes are made by self-association of nonionic surfactants and cholesterol in an aqueous phase to provide overall neutral change to the vesicles. In this bilayered structure, hydrophilic and hydrophobic ends of surfactant arrange themselves into an external and internal surface, respectively, which also allows the entrapment of both hydrophobic and hydrophilic drug molecules in the same carrier simultaneously [66]. Recently, the role of niosomes in drug delivery research has significantly improved as the composition, size, lamellarity, surface charge, and stability of niosomes can be controlled by the type of preparation method, the type of surfactant used, the amount of cholesterol, surface charge additives, and formulation concentration [67]. Thus, the most important features of niosomes distinguishing them from liposomes are their relatively stable structure, cost-effective surfactant ingredients, and higher "leakiness" of payload. Occasionally, niosomes show physical instability associated with excessive drug leakage.

TABLE 13.2

List of Marketed Lipid Vesicle-Based Formulation for Nanomedicine Use

S. No.	Type of Vesicles	Drug/or Active API/ Delivery Route	Disease	Regulatory Status	Company	Product Name
1	Liposomes	Tocopheryl Acetate; 3-o-Ethyl Ascorbic Acid; Retinyl Palmitate/ Dermal	Skin nourishes with essential vitamins and protects against free radicals	In market	Rovi cosmetics international GmbH	Rovisome ACE Plus
2	Liposomes	Pro-Retinol A/ Dermal	Antiaging	In market	L'Oreal	Revitalift
3	Liposomes	Amphotericin B/ IV	Systemic fungal infection	In market	Gilead advancing therapeutics	AmBisome®
4	Liposomes	Daunorubicin citrate/ IV	HIV-related KS (IV)	In market	NeXtar, Inc.	DaunoXome®
6	Liposomes	Vincristine sulfate/ IV	Acute lymphoblastic leukemia	In market	Acrotech Biopharma	Marqibo®
7	Liposomes	Doxorubicin hydrochloride / IV	AIDS-related Kaposi's sarcoma, breast, ovarian, and other solid tumors	In market	Baxter Healthcare Corporation	Doxil®
8	Liposomes	Influenza virus antigens/ IM	Influenza vaccine	In market	Crucell Italy S.r.l	Inflexal® V
9	Liposomes	Doxorubicin. HCl/ Infusion	Breast neoplasm	In market	Sopherion Therapeutics, Inc	Myocet®
10	Liposomes	Morphine sulfate/ Epidural	Post-surgical pain	In market	Pacira Pharmaceuticals	DepoDur®
11	Niosomes	-Lipstick/ Dermal application	Lip color and lipstick	In market	Orlane – Lipcolor and Lipstick	Lip Gloss
12	Niosomes	Ketoprofen /NA	Nonsteroidal anti-inflammatory drugs with analgesic/antipyretic effects	Patent awarded	NA	Bayer aktiengesellsschaft

(Continued)

TABLE 13.2 (Continued)
List of Marketed Lipid Vesicle-Based Formulation for Nanomedicine Use

S. No.	Type of Vesicles	Drug/or Active API/ Delivery Route	Disease	Regulatory Status	Company	Product Name
13	Niosomes	Mixed fragrance of Bergamotte, Ananas, Rosmarin, Zitrone, Neroli, Jasmin, Koriander, Veilchen, Eichenmoos/ Spray	Fragrance	In market	Loris Azzaro – Chrome	Chrome Eau De Toilette Spray
14	Niosomes	Louisiana magnolia, golden Anjou pear, lotus flower, tuberose, star jasmine, pink cyclamen, vanilla-infused musk, sandalwood and blonde woods/ Spray	Fragrance	In market	Britney Spears	Curious coffret
15	Niosomes	Composition not disclosed/ Facial cream	Sun soothing moisturizer	In market	Helena Rubinstein - HR - Golden Beauty - Body Care	Golden Beauty After Sun Soothing Moisturizer
16	Niosomes	Vitamin E, panthenol /Nanocapsules/ Dermal	Antiaging formulation	In market	Lancome L' Oreal	Flash Retouch Brush on Concealer
17	Ethosome	Cellulite/ Dermal	Orange peel	In market	Physionics, Nottingham	Skin genuity
18	Ethosome	Acyclovir / Dermal	Herpes virus	In market	Trima Israel	Supravir cream
19	Ethosomes	Minoxidil/ Dermal	Hair growth promoter	In market	Sinere Germany	Nanominox
20	Ethosomes	Decorin cream/ Dermal	Toning, moisturizing antioxidant	In market	Genome cosmetic	Decorin cream
21	Ethosomes	Cellulite buster/ Dermal	Topical anti-cellulite	In market	Novel therapeutic technology	Noicelex
22	Ethosomes	Cellulite/ Dermal	Topical anti-cellulite	In market	Hampden Health	Celltight EF

The surfactant primarily used to develop the niosomes may be of alkyl ether (alkyl glycerol, polyoxyethylene glycol alkyl ethers), alkyl ester (span and tween-based), alkyl amide (glycosides, alkyl polyglucosides), fatty acid or fatty alcohol, or block copolymer Pluronics [67]. The two most relevant parameters for the development of niosome formulations are hydrophilic-lipophilic balance (HLB) and critical packing parameter (CPP). HLB is determined by the measurement of the lipophilic/lipophilicity ratio (range from 1 to 20, 1 means most lipophilic and 20 means most hydrophilic), and the surfactants having a HLB value of 4–8 are ideal to develop niosomes [68]. In a similar way, based on the CPP value the shape of the nanostructure can be anticipated. Bilayered vesicles are formed if $1/2 \leq CPP \leq 1$, which can be calculated from the hydrophilic head group volume (υ), critical hydrophobic group length (lc), and ao is area of hydrophilic group according to

$$CPP = \frac{\upsilon}{lc*a°o}.$$

During the storage of niosomes' dispersion formulations, there is chance of vesicles' aggregation and/ or fusion and hydrolysis of the encapsulated drug due to leakage from the vesicles. They also require specific effort for sterilization as they cannot be sterilized using conventional membrane filtration or heat sterilization methods [69].

13.2.6.3 Ethosomes

Another major phospholipid-based nanovesicular system is ethosomes, which require a high ethanol content (20%–40%) for formation. Thus, the rather soft ethosomal vesicles are composed primarily of phospholipids, ethanol, and an aqueous phase. Occasionally, they may also contain propylene glycol, isopropyl alcohol, transcutol, or other surface modifiers to accelerate the solubility of the drug(s) in the solvents. Similar to other lipid vesicles, they also can carry hydrophobic and hydrophilic drug moieties. Some of the prevailing lipids used to develop ethosomes are phosphatidyl-cholines and phosphatidylethanolamine with compounds such as Phospholipon 90, Soya phosphatidylcholine (S-75), and Lipoid S 100 [70].

Ethanol works as a permeation enhancer and also provides elasticity to the ethosomal vesicles. Ethanol increases the liquid fluidity and cellular membrane fluidity of vesicles by reducing the melting point of the stratum corneal lipid by interacting with the polar head region of the lipid region. Unlike other vesicles, the flexible membrane of the ethosomes allows them to squeeze through pores offering alternative transport routes [71]. Thus, the potential advantages of ethanol-based vesicular system include delivering large amounts of drugs through the stratum corneum into deep regions of the skin. Other key advantages of ethosomes as drug carrier are their ability to carry larger amounts and more diverse types of drug, the well-defined low-risk profile, high patient compliance if used in the form of a gel, and their availability for immediate commercialization. Thus, they are used for various applications in pharmaceutical, veterinary, and cosmetic fields [72]. Albeit the technology on ethosome formulation is growing rapidly, the exact mode and nature of vesicle transport into the skin are not fully understood, yet. Other limitations of using ethosomes as drug carriers include potential allergic reactions to ethanol if patients are sensitive to alcohol, their focused applicability to the transdermal or dermal route of drug delivery, and the higher precautionary measures required during the development process as ethanol is inflammable [73].

13.2.6.4 Emulsomes

Emulsomes are developed as novel bilayered lipoidal vesicular system having an inner solid fat core enclosed into a phospholipid bilayer. It is used for drug candidates with poor bioavailability to provide protection from enzymatically harsh environments such as the gastrointestinal tract. At the same time, the ingredients of emulsome formulations are also cost-effective and hence an economical alternative to current lipid-based formulations. Emulsomes are formed in an aqueous phase, and their core polymeric materials should be solid at room temperature, which are stabilized with cholesterol and soya lecithin in the form of o/w emulsion.

The primarily used lipid core materials are fatty acids and their derivatives or mixtures of lipids. Triglyceride-based lipids are also ideal lipids because these impart the acceptable storage life of o/w emulsion during the preparation. This chemical composition of emulsomes makes them an attractive drug delivery vehicle as gastric harsh environment is unable to digest the triglycerides where drug is incorporated within triglyceride-based lipid core, and thus, oral delivery is facilitated easily [74].

Other ingredients may be antioxidants (butylated hydroxyl toluene (BHT or similar), negatively charged particles (oleic acid, phosphatidic acid) to improve the zeta potential properties (enhanced stability, reduced agglomeration), surfactants (spans, tweens), phosphatidyl choline, and cholesterol. The triglyceride-based lipids are used to form micelles or are ordered as lipid bilayers as their hydrophobic tails line up against each another and the hydrophilic head groups are exposed to the aqueous phase on both sides. This characteristic feature of emulsomes makes them a most suitable vehicle for hydrophobic drugs. Some of the potential drug candidates delivered using emulsome technologies are amphotericin B, methotrexate, and zidovudine [74].

Emulsomes are structurally comparable to natural lipoproteins, and the chylomicrons thus could behave similar to these natural lipoproteins. Thus, emulsomes could facilitate oral delivery of sensitive therapeutic molecules as they are absorbed by enterocytes through endogenous lipid absorption mechanisms of the gastrointestinal route [74]. Emulsome-based genetic drug formulations also (e.g., antisense oligonucleotides and plasmids) have shown clear potential for systemic delivery [75]. The development of emulsome formulations is still largely empirical relying on in vitro models that are predictive of oral bioavailability. Thus, there is a need for more precise in vitro methods for predicting the dynamic changes in the gut in order to assess the solubilization state of the drug in vivo for their better assessment. Similar to other lipid-based formulations, emulsomes are formulated using lipid film formation (hand shaking method), reserve phase evaporation, high-pressure extrusion, or ethanol injection or sonication methods. Currently, no emulsome formulation is approved for clinical or cosmetic applications, yet.

13.2.7 Carbon Nanotubes

The advancement in nanotechnology and their application in the development of medicinal products paves the way for the investigation of various nanomaterials as potential drug delivery carriers such as silica particles, titanium dioxide nanoparticles, and carbon-based nanomaterials. The carbon nanomaterials, including carbon nanotubes (CNTs), fullerenes, and carbon nanohorns (CNHs), have been investigated as drug delivery carriers [76]. These materials hold promise as drug delivery carrier with respect to surface functionalization, high loading capacity, high chemical stability, and feasibility of incorporating both hydrophilic and hydrophobic drug substances [77].

Nanosized carbon-based nanomaterials particularly CNTs have been explored for drug, peptide, and gene delivery. CNTs are cylindrical large structures of hybridized carbon atoms arranged in hexagonal shape, which are formed by rolling up a single sheet of graphene (known as single-walled carbon nanotubes, SWCNTs) or by rolling up multiple sheets of graphene (known as multiwalled carbon nanotubes, MWCNTs). Functionalized CNTs have been used for delivery of drug molecules, which are adsorbed on the surface or inside it. Later, this drug conjugate can be used in the form of oral or intravenous (IV) delivery or even can be actively directed for targeted drug delivery using an external magnetic force by using the magnetic conjugates. Finally, the nanotubes release their contents intracellularly after cell membrane permeation [78]. CNT drug conjugates can be designed for safe and more effective delivery as they have the ability to carry drug molecules across the cytoplasmic membrane as well as nuclear membrane without producing a toxic effect. The cellular permeation ability of CNT can be understood by multiple mechanisms, e.g., their simple hydrophobic and stacking interaction, electrostatic adsorption and covalent bonds in their structure, and also by adsorption of the drug into the hollow cylinder, which increase the permeation capacity via the cell membrane [79].

Nevertheless, one of the most important factors in using CNTs as drug delivery carrier is their potential cytotoxicity. As CNT drug carriers are in an early phase of development, their safety profile has to be examined more deeply before declaring them acceptable as potential drug carrier. Some preliminary tests have shown that CNTs are toxicologically safe, while other studies have shown that CNTs, especially raw materials, are potentially unsafe to use [80]. In light of these uncertainties, CNTs conjugated with therapeutic molecules are not investigated in clinical settings, yet. However, it is predicted that CNTs will reach clinical translation in the near future after successfully passing preclinical studies.

13.2.8 Metallic Nanoparticles

The numbers of metallic nanoparticles of gold, silver, copper, silica, iron, or their oxides have been reported in the literature for various pharmaceutical, chemical, or biotechnological applications. These NPs are generally synthesized by the reduction process using plant extract, yeast, fungus, and bacteria. The reduction of metal ions results in crystallization into metallic NPs [81]. Metallic NPs can also be prepared by chemical synthesis methods. However, these methods involve the use of toxic chemicals, which are not safe for biomedical applications. Metallic NPs have been widely used as drug delivery system, but they also possess inherent cytotoxic and reactive oxygen species (ROS) generation capability. These ROS have multiple effects that include modification of cell signaling pathways involved in cell death, proliferation, and differentiation and are also responsible for toxicity associated with the use of metallic NPs [82]. In this section, we will discuss applications of different metallic NPs.

13.2.8.1   Gold Nanoparticles

Gold nanoparticles (AuNPs) are the most widely used inorganic nanoparticles in diagnostics, therapy, immunoassay, targeted drug delivery, bioimaging, and other medical and biological research due to their optical, electronic, sensing, and biochemical properties. These AuNPs possess high surface area to volume ratio and surface functionalization capability and can be sysnthesized easily. They protect loaded drug from degradation and are biocompatible. Due to all these advantages, AuNPs are the promising delivery system [83]. Chen et al. (2019) developed AuNPs via green synthesis using gelatin/protein nanogel that serves as reducing and stabilizing agents [84]. These AuNPs demonstrated metal enhance luminescence/fluorescence properties and can be used for various diagnosis and therapeutic purposes. Wang et al. (2019) synthesized AuNPs by green synthesis using *Scutellaria barbata*, which is traditionally used as food or medicine and evaluated anticancer activity on pancreatic cancer cell lines. The developed formulation showed strong anticancer activity against pancreatic cancer cells [85]. Sojinrin et al. (2019) developed AuNP-conjugated AFB1 antibody-based diagnostic strips. The colorimetric detection by color change can be monitored by the naked eyes [86].

13.2.8.2   Silver Nanoparticles

Silver nanoparticles (AgNPs) are widely used in various fields, as drug delivery vehicle, diagnosis, as antimicrobial agent and in electronic industry [87]. The AgNPs exhibit higher thermal and electrical conductivities than AuNPs and are more stable. These nanoparticles are also used as antituberculous agent that avoids the problem of drug resistance associated with current TB treatment [88]. The AgNPs are nontoxic, safe inorganic antibacterial agents, and can be successfully used as antibiotic replacement. Feroze et al. (2020) reported a biogenic synthesis method of AgNPs from Penicillium oxalicum metabolites [89]. The synthesized AgNPs were evaluated for antibacterial activity against various strains such as *Staphylococcus aureus, S. dysenteriae, and Salmonella typhi*. The results suggested that these nanoparticles possess antibacterial activity and can be used to prevent infections, as wound healing and anti-inflammatory agent.

13.2.8.3   Silica Nanoparticles

Silica nanoparticles have emerged as promising multifunctional, biocompatible, and biodegradable drug delivery system. These nanoparticles have enormous applications due to higher surface area, pore size adjustability, and surface modification opportunity [90]. Because of such unique properties, higher amount of drug loading is possible for the controlled and targeted drug delivery. Apart from this, silica is FDA approved as an anticaking agent, emulsifier, and defoaming agent [91]. Silica NPs are of two types: nonporous and mesoporous [92]. Li et al. (2019) developed mesoporous silica NPs for sustained drug delivery of an anticancer agent doxorubicin [93]. A multifunctional nanoparticulate system carrying doxorubicin-peptide was prepared as pH-sensitive smart drug delivery vehicle. The drug was released in acidic tumor microenvironment and reached to cell nucleus.

13.2.8.4 Iron Oxide Nanoparticles

Iron oxide nanoparticles (IONPs) are composed of different types of iron oxide and size range from 5 to 380 nm. IONPs are formed by the reduction of ferric/ferrous iron ions which is then crystallized to nanoparticulate form and are stabilized by organic stabilizer coating. The properties depend on the type of iron oxide, and surface charge depends on the nature of material used for coating [94]. These nanoparticles are commercially available for the treatment of cancer and iron deficiency anemia. Some IONPs are developed that can be activated by external magnetic field for targeted drug delivery, thus avoiding side effects to normal cells [95]. Filippi et al. (2019) prepared metronidazole-based dendron-coated IONPs with bioreductive compounds for site-specific delivery to hypoxic tissues [96]. The developed targeted NPs selectively accumulated in hypoxic tissues. Ghosh et al. (2020) reported two types of surface-modified superparamagnetic iron oxide nanoparticles (SPION) for biomedical applications [97]. The surface was modified by poly(lactic-co-glycolic acid), and SPION were synthesized using two different emulsifiers TPGS or DMAB. The coated formulations were less genotoxic and showed reduced ROS generation as compared with uncoated SPION.

13.3 APPLICATIONS OF NANOMATERIALS AS NANOMEDICINE

Nanomaterial either alone or in combination with drug/biomolecules/genetic material may serve as tiny nanomedicine, which may be useful in the treatment of different diseases. Nanomedicines due to their unique small size and modified surface characteristics such as ligand anchoring and charge become specific drug cargos for précise and accurate delivery of drug to the specific locations in the body. The deeper sites such as tumor microenvironment, brain, and posterior ocular sites could be targeted with controlled releases of the drugs. Further, additional characteristics like magnetic and/or optical properties of the some selected nanomaterials (iron, gold, and silver) could further be utilized for therapy and diagnosis both.

13.3.1 Nanomaterials in Drug Delivery

The drug delivery systems could be developed using desired biomaterials depending upon the route of administration, site of application, type of the disease being treated, and physicochemical characteristics of the drugs. The GRAS status of the material favors its higher applications in the drug delivery as fewer complications are received in commercialization of successful products. The small-sized drug molecules can be entrapped in polymeric/protein/lipid-based nanoparticles. The lipid selected may be solid lipid or phospholipid. For example, solid lipid nanoparticles, which are composed of solid lipid and stabilized by surfactants, can accommodate both types of drug molecules, that is, hydrophilic and hydrophobic. Similarly, liposomes have vesicle-like structure and in its core aqueous compartment is available for encapsulating hydrophilic drug. Doxil® is a well-known example of doxorubicin-loaded liposomal formulation for anticancer therapy. Protein molecules like albumin can be associated with drugs like paclitaxel without the use of solvent (Abraxane®) and have been successfully marketed for different cancers of the breast, lung, and pancreas. Further, the surface of these nano-drug carriers can be modified with suitable ligands that make them selective for drug targeting at organ, cellular, and intraceullar levels too. Commonly employed ligands for drug targeting purpose include folic acid, transferrin, monoclonal antibodies, and hormone receptors such as estrogen and progesterone.

13.3.2 Nanomaterials in Gene Delivery

Site-directed intracellular delivery of genetic materials like genes and siRNA has been successfully achieved using nanomaterials designed for this specific purpose. Cationic polymer and cationic lipid/lipid-based systems have been utilized for the delivery of the loaded bioactive cargos. Ligand-directed intracellular delivery via endocytosis is also a common technique. The environment-sensitive drug carrier system like pH-sensitive and redox-sensitive nanomaterials/nano-complexes has been utilized. Inorganic nanomaterials like gold and silver nanoparticles have also been reported for the enhanced delivery of genetic therapeutics.

13.3.3 Nanomaterials in Biomacromolecules' Delivery

Being large in size, biomacromoleucles including proteins and peptides do not cross easily through biological hurdles like skin and GIT epithelium. This leads to low bioavailability of these therapeutics and hence low pharmacological activity in total. Therapeutic proteins such as insulin and peptide like cyclosporine A require drug carrier designed in a customized manner in order to deliver them through tough biological barriers and harsh microenvironments. Nanomaterials such as chitosan nanoparticles, polymeric nanoparticles, and surface-modified lipid carriers encapsulated with either of this class of the therapeutics can be delivered via oral/topical routes.

13.3.4 Nanomaterials in Herbal Drug Delivery

Herbal drugs are regaining interest due to their large safety window. However, most of the isolated active phytochemicals suffer from poor bioavailability upon administration due to their lipophilic characters. Nanomedicine based on various nanomaterials has been developed for delivery of herbal compounds including curcumin, resveratrol, and plant extracts as well. Oral delivery of herbal molecules could be achieved successfully when encapsulated into suitable nanocarriers such as liposomes, niosomes, lipid nanoparticles, polymeric nanoparticles, and lipid-polymer hybrid systems. The encapsulation into nanocarriers offers additional advantages of being protected from oxidation, hydrolysis, and enzymatic degradation during oral administration. The storage shelf life of herbal molecules can be enhanced using nanomaterials successfully.

13.3.5 Nanomaterials in Tissue Engineering

Nanoscale systems have been successfully utilized in tissue engineering and regenerative medicine due to their low toxicity, surface characteristics, suitability for diagnosis, and precisely controlled release and behavior. For tissue engineering, a specialized control over biological happenings and real-time monitoring is a prerequisite. Further, it demands site-specific localized delivery of bioactive such as chemokines, cytokines, genetic material, and growth factors and also for diagnostic agent to detect the sequences of events over time. Nanoparticles serve best carrier for either of these purposes in tissue engineering scaffolds including delivery of growth factors and diagnostic agent simultaneously with desired controlled properties that can be retained during the entire regenerative process. The selection of nanocarriers depends upon its application, and either polymers, proteins, metals, ceramics, lipid, or their composites have been utilized successfully.

13.4 CONCLUSION AND FUTURE PROSPECTS

Biomaterials have been explored for diverse applications in drug delivery for effective management of the diseases. Nano-sized biomaterials loaded with therapeutic moieties are known as nanomedicine. Advanced nanomedicine such as surface-grafted nanocarriers, personalized nanomedicine, theranostic nanomedicine, and genetic nanomedicine could be delivered using nanobiomaterials of various shapes, sizes, and properties. Both organic and inorganic materials have been utilized for nanomedicine development for the treatment of the otherwise incurable diseases like cancer. Hybrid nanobiomaterial with further improved characteristics for more accurate drug delivery kinetics should be tailored following regulatory guidelines.

CONFLICT OF INTEREST

The authors report no conflict of interest.

FUNDING

The support of Department of Biotechnology, Government of India, for the grant number BT/PR26950/NNT/28/1505/2017 to RP as PI and JRF support to one of the authors RK is duly acknowledged.

REFERENCES

1. Paliwal, R., and S. Palakurthi. 2014. "Zein in controlled drug delivery and tissue engineering." *Journal of Controlled Release*. Elsevier B.V. Doi: 10.1016/j.jconrel.2014.06.036.
2. Tabrez, S., N. R. Jabir, V. M. Adhami, M. I. Khan, M. Moulay, M. A. Kamal, and H. Mukhtar. 2020. "Nanoencapsulated dietary polyphenols for cancer prevention and treatment: successes and challenges." *Nanomedicine*. Future Medicine Ltd. Doi: 10.2217/nnm-2019-0398.

3. Paliwal, R., R. J. Babu, and S. Palakurthi. 2014. "Nanomedicine scale-up technologies: Feasibilities and challenges." *Ageing International* 15 (6). Doi: 10.1208/s12249-014-0177-9.
4. Han, H. J., C. Ekweremadu, and N. Patel. 2019. "Advanced drug delivery system with nanomaterials for personalised medicine to treat breast cancer." *Journal of Drug Delivery Science and Technology.* Editions de Sante. Doi: 10.1016/j.jddst.2019.05.024.
5. Lu, B., X. Lv, and Y. Le. 2019. "Chitosan-modified PLGA nanoparticles for control-released drug delivery." *Polymers* 11 (2). MDPI AG. Doi: 10.3390/polym11020304.
6. Paliwal, S. R., R. Kenwat, S. Maiti, and R. Paliwal. 2020. "Nanotheranostics for cancer therapy and detection: State of the art." *Current Pharmaceutical Design* 26 (November). Bentham Science Publishers Ltd. Doi: 10.2174/1381612826666201116120422.
7. Paliwal, S. R., R. Paliwa, G. P. Agrawa, and S. P. Vyas. 2012. "Targeted breast cancer nanotherapeutics: Options and opportunities with estrogen receptors." *Critical Reviews in Therapeutic Drug Carrier Systems* 29 (5): 421–46. Begel House Inc. Doi: 10.1615/CritRevTherDrugCarrierSyst.v29.i5.20.
8. Paliwal, R., S.R. Paliwal, and S.P. Vyas. 2017. "Nanotherapeutics for cancer imaging and therapy." *Mini-Reviews in Medicinal Chemistry* 17 (18): 1686–1687.
9. Vyas, S., S. Rai, R. Paliwal, P. Gupta, K. Khatri, A. Goyal, and Bhuvaneshwar Vaidya. 2008. "Solid lipid nanoparticles (SLNs) as a rising tool in drug delivery science: One step up in nanotechnology." *Current Nanoscience* 4 (1): 30–44. Bentham Science Publishers Ltd. Doi: 10.2174/157341308783591816.
10. Paliwal, R., S. Rai, and S. P. Vyas. 2011. "Lipid Drug Conjugate (LDC) nanoparticles as autolymphotrophs for oral delivery of methotrexate." *Journal of Biomedical Nanotechnology* 7 (1). Doi: 10.1166/jbn.2011.1235.
11. Paliwal, R., S. Rai, B. Vaidya, K. Khatri, A. K. Goyal, N. Mishra, A. Mehta, and S. P. Vyas. 2009. "Effect of lipid core material on characteristics of solid lipid nanoparticles designed for oral lymphatic delivery." *Nanomedicine: Nanotechnology, Biology, and Medicine* 5 (2). Doi: 10.1016/j.nano.2008.08.003.
12. Paliwal, R., S. R. Paliwal, G. P. Agrawal, and S. P. Vyas. 2011. "Biomimetic solid lipid nanoparticles for oral bioavailability enhancement of low molecular weight heparin and its lipid conjugates: In vitro and in vivo evaluation." *Molecular Pharmaceutics* 8 (4). Doi: 10.1021/mp200109m.
13. Bagalkot, V., M. A. Badgeley, T. Kampfrath, J. A. Deiuliis, S. Rajagopalan, and A. Maiseyeu. 2015. "Hybrid nanoparticles improve targeting to inflammatory macrophages through phagocytic signals." *Journal of Controlled Release* 217 (November): 243–55. Elsevier B.V. Doi: 10.1016/j.jconrel.2015.09.027.
14. Rai, S., R. Paliwal, B. Vaidya, K. Khatri, A. Goyal, P. Gupta, and S. P. Vyas. 2008. "Targeted delivery of doxorubicin via estrone-appended liposomes." *Journal of Drug Targeting* 16 (6). Doi: 10.1080/10611860802088481.
15. Paliwal, R., S. Rai, B. Vaidya, S. Mahor, P. N. Gupta, A. Rawat, and S. P. Vyas. 2007. "Cell-selective mitochondrial targeting: Progress in mitochondrial medicine." *Current Drug Delivery* 4 (3). Doi: 10.2174/156720107781023910.
16. Rai, S., R. Paliwal, B. Vaidya, P. N. Gupta, S. Mahor, K. Khatri, A. K. Goyal, A. Rawat, and S. P. Vyas. 2007. "Estrogen(s) and analogs as a non-immunogenic endogenous ligand in targeted drug/DNA delivery." *Current Medicinal Chemistry* 14 (19). Doi: 10.2174/092986707781368432.
17. Paliwal, S. R., R. Paliwal, H. C. Pal, A. K. Saxena, P. R. Sharma, P. N. Gupta, G. P. Agrawal, and S. P. Vyas. 2012. "Estrogen-anchored PH-sensitive liposomes as nano-module designed for site-specific delivery of doxorubicin in breast cancer therapy." *Molecular Pharmaceutics* 9 (1): 176–86. American Chemical Society. Doi: 10.1021/mp200439z.

18. Fabricant, D. S., and N. R. Farnsworth. 2001. "The value of plants used in traditional medicine for drug discovery." *Environmental Health Perspectives* 109 (SUPPL. 1): 69–75. Public Health Services, US Dept of Health and Human Services Doi: 10.1289/ehp.01109s169.

19. Slepička, P., N. SlepičkováKasálková, J. Siegel, Z. Kolská, and V. Švorčík. 2019. "Methods of gold and silver nanoparticles preparation." *Materials* 13 (1): 1. MDPI AG. Doi: 10.3390/ma13010001.

20. Zielińska, A., F. Carreiró, A. M. Oliveira, A. Neves, B. Pires, D. N. Venkatesh, A. Durazzo, et al. 2020. "Polymeric nanoparticles: Production, characterization, toxicology and ecotoxicology." *Molecules* 25 (16): 3731. MDPI AG. Doi: 10.3390/molecules25163731.

21. Renu, S., K. S. Shivashangari, and V. Ravikumar. 2020. "Incorporated plant extract fabricated silver/Poly-D,L-lactide-co-glycolide nanocomposites for antimicrobial based wound healing." *Spectrochimica Acta - Part A: Molecular and Biomolecular Spectroscopy.* Elsevier B.V. Doi: 10.1016/j.saa.2019.117673.

22. Maksimenko, O., J. Malinovskaya, E. Shipulo, N. Osipova, V. Razzhivina, D. Arantseva, O. Yarovaya, et al. 2019. "Doxorubicin-loaded PLGA nanoparticles for the chemotherapy of glioblastoma: Towards the pharmaceutical development." *International Journal of Pharmaceutics* 572 (December). Elsevier B.V. Doi: 10.1016/j.ijpharm.2019.118733.

23. Gibbens-Bandala, B., E. Morales-Avila, G. Ferro-Flores, C. Santos-Cuevas, L. Meléndez-Alafort, M. Trujillo-Nolasco, and B. Ocampo-García. 2019. "177Lu-Bombesin-PLGA (Paclitaxel): A targeted controlled-release nanomedicine for bimodal therapy of breast cancer." *Materials Science and Engineering C* 105 (December): 110043. Elsevier Ltd. Doi: 10.1016/j.msec.2019.110043.

24. Gu, P., A. Wusiman, Y. Zhang, Y. Zhang, Z. Liu, et al. 2019. "Rational design of PLGA nanoparticle vaccine delivery systems to improve immune responses." *Molecular Pharmaceutics* 16 (12): 5000–12. American Chemical Society. Doi: 10.1021/acs.molpharmaceut.9b00860.

25. Iglesias, N., E. Galbis, L. Romero-Azogil, E. Benito, R. Lucas, M. Gracia García-Martín, and M.-Violante de-Paz. 2020. "In-depth study into polymeric materials in low-density gastroretentive formulations." *Pharmaceutics* 12 (7): 636. MDPI AG. Doi: 10.3390/pharmaceutics12070636.

26. Cilurzo, F., F. Selmin, C. G. M. Gennari, L. Montanari, and P. Minghetti. 2014. "Application of methyl methacrylate copolymers to the development of transdermal or loco-regional drug delivery systems." *Expert Opinion on Drug Delivery. Informa Healthcare.* Doi: 10.1517/17425247.2014.912630.

27. Katata-Seru, L., B. Moses Ojo, O. Okubanjo, R. Soremekun, and O. S. Aremu. 2020. "Nanoformulated Eudragit lopinavir and preliminary release of its loaded suppositories." *Heliyon* 6 (5): e03890. Elsevier Ltd. Doi: 10.1016/j.heliyon.2020.e03890.

28. Salatin, S., J. Barar, M. Barzegar-Jalali, K. Adibkia, M. Alami-Milani, and M. Jelvehgari. 2020. "Formulation and evaluation of Eudragit RL-100 nanoparticles loaded in-situ forming gel for intranasal delivery of rivastigmine." *Advanced Pharmaceutical Bulletin* 10 (1): 20–29. Tabriz University of Medical Sciences. Doi: 10.15171/apb.2020.003.

29. Rizeq, B. R., N. N. Younes, K. Rasool, and G. K. Nasrallah. 2019. "Synthesis, bioapplications, and toxicity evaluation of chitosan-based nanoparticles." *International Journal of Molecular Sciences.* MDPI AG. Doi: 10.3390/ijms20225776.

30. Eliyahu, S., A. Almeida, M. H. Macedo, J. das Neves, B. Sarmento, and H. Bianco-Peled. 2020. "The effect of freeze-drying on mucoadhesion and transport of acrylated chitosan nanoparticles." *International Journal of Pharmaceutics* 573 (January). Elsevier B.V. Doi: 10.1016/j.ijpharm.2019.118739.

31. Kumar, M., P. Upadhayay, R. Shankar, M. Joshi, S. Bhatt, and A. Malik. 2019. "Chlorpheniramine maleate containing chitosan-based nanoparticle-loaded thermosensitive in situ gel for management in allergic rhinitis." *Drug Delivery and Translational Research* 9 (6): 1017–26. Springer Verlag. Doi: 10.1007/s13346-019-00639-w.

32. Khan, M. M., A. Madni, V. Torchilin, N. Filipczak, J. Pan, N. Tahir, and H. Shah. 2019. "Lipid-chitosan hybrid nanoparticles for controlled delivery of cisplatin." *Drug Delivery* 26 (1): 765–72. Taylor and Francis Ltd. Doi: 10.1080/10717544.2019.1642420.

33. Mumuni, M. A., F. C. Kenechukwu, K. C. Ofokansi, A. A. Attama, and D. Díaz Díaz. 2020. "Insulin-loaded mucoadhesive nanoparticles based on mucin-chitosan complexes for oral delivery and diabetes treatment." *Carbohydrate Polymers* 229 (February): 115506. Elsevier Ltd. Doi: 10.1016/j.carbpol.2019.115506.

34. Hong, S., D. W. Choi, H. N. Kim, C. G. Park, W. Lee, and H. H. Park. 2020. "Protein-based nanoparticles as drug delivery systems." *Pharmaceutics* 12 (7): 604. MDPI AG. Doi: 10.3390/pharmaceutics12070604.

35. Koletti, A. E., E. Tsarouchi, A. Kapourani, K. N. Kontogiannopoulos, A. N. Assimopoulou, and P. Barmpalexis. 2020. "Gelatin nanoparticles for NSAID systemic administration: Quality by design and artificial neural networks implementation." *International Journal of Pharmaceutics* 578 (March). Elsevier B.V. Doi: 10.1016/j.ijpharm.2020.119118.

36. Chuang, Y. L., H. W. Fang, A. Ajitsaria, K. H. Chen, C. Y. Su, G. S. Liu, and C. L. Tseng. 2019. "Development of kaempferol-loaded gelatin nanoparticles for the treatment of corneal neovascularization in mice." *Pharmaceutics* 11 (12). MDPI AG. Doi: 10.3390/pharmaceutics11120635.

37. Lamichhane, S., and S. Lee. 2020. "Albumin nanoscience: Homing nanotechnology enabling targeted drug delivery and therapy." *Archives of Pharmacal Research* 43 (1): 118–33. Springer. Doi: 10.1007/S12272-020-01204-7.

38. Zhou, C., X. Song, C. Guo, Y. Tan, J. Zhao, Q. Yang, D. Chen, et al. 2019. "Alternative and injectable preformed albumin-bound anticancer drug delivery system for anticancer and antimetastasis treatment." *ACS Applied Materials and Interfaces* 11 (45): 42534–48. American Chemical Society. Doi: 10.1021/acsami.9b11307.

39. Pascoli, M., R. de Lima, and L. F. Fraceto. 2018. "Zein nanoparticles and strategies to improve colloidal stability: A mini-review." *Frontiers in Chemistry*. Frontiers Media S. A. Doi: 10.3389/fchem.2018.00006.

40. Li, F., Y. Chen, S. Liu, X. Pan, Y. Liu, H. Zhao, X. Yin, C. Yu, W. Kong, and Y. Zhang. 2019. "The effect of size, dose, and administration route on zein nanoparticle immunogenicity in BALB/c mice." *International Journal of Nanomedicine* 14 (December): 9917–28. Dove Medical Press Ltd. Doi: 10.2147/IJN.S226466.

41. Kaushik, P., E. Priyadarshini, K. Rawat, P. Rajamani, and H. B. Bohidar. 2020. "PH responsive doxorubicin loaded zein nanoparticle crosslinked pectin hydrogel as effective site-specific anticancer substrates." *International Journal of Biological Macromolecules* 152 (June): 1027–37. Elsevier B.V. Doi: 10.1016/j.ijbiomac.2019.10.190.

42. Mukherjee, S., S. Ray, and R. S. Thakur. 2009. "Solid lipid nanoparticles: A modern formulation approach in drug delivery system." *Indian Journal of Pharmaceutical Sciences*. Wolters Kluwer -- Medknow Publications. Doi: 10.4103/0250-474X.57282.

43. Schwarz, C., W. Mehnert, J. S. Lucks, and R. H. Müller. 1994. "Solid lipid nanoparticles (SLN) for controlled drug delivery. I. production, characterization and sterilization." *Journal of Controlled Release* 30 (1). Elsevier: 83–96. Doi: 10.1016/0168-3659(94)90047-7.

44. Cavalli, R., O. Caputo, and M. R. Gasco. 1993. "Solid lipospheres of doxorubicin and idarubicin." *International Journal of Pharmaceutics* 89 (1). Elsevier: R9–12. Doi: 10.1016/0378-5173(93)90313-5.

45. Jenning, V., M. Schäfer-Korting, and S. Gohla. 2000. "Vitamin A-loaded solid lipid nanoparticles for topical use: Drug release properties." *Journal of Controlled Release* 66 (2–3): 115–26. Doi: 10.1016/S0168–3659(99)00223-0.

46. Lin, C. H., C. H. Chen, Z. C. Lin, and J. Y. Fang. 2017. "Recent advances in oral delivery of drugs and bioactive natural products using solid lipid nanoparticles as the carriers." *Journal of Food and Drug Analysis*. Elsevier Taiwan LLC. Doi: 10.1016/j.jfda.2017.02.001.

47. Khan, A. A., J. Mudassir, N. Mohtar, and Y. Darwis. 2013. "Advanced drug delivery to the lymphatic system: Lipid-based nanoformulations." *International Journal of Nanomedicine*. Doi: 10.2147/IJN.S41521.

48. Čerpnjak, K., A. Zvonar, M. Gašperlin, and F. Vrečer. 2013. "Lipid-based systems as a promising approach for enhancing the bioavailability of poorly water-soluble drugs." *Acta Pharmaceutica* 63 (4): 427–45. Croatian Pharmaceutical Society. Doi: 10.2478/acph-2013-0040.

49. Muchow, M., P. Maincent, and R. H. Müller. 2008. "Lipid nanoparticles with a solid matrix (SLN®, NLC®, LDC®) for oral drug delivery." *Drug Development and Industrial Pharmacy* Taylor & Francis. Doi: 10.1080/03639040802130061.

50. Battaglia, L., M. Gallarate, P. P. Panciani, E. Ugazio, S. Sapino, E. Peira, and D. Chirio. 2014. "Techniques for the preparation of solid lipid nano and microparticles." Doi: 10.5772/58405.

51. Khan, S., S. Baboota, J. Ali, Sana Khan, R. S. Narang, and J. K. Narang. 2015. "Nanostructured lipid carriers: An emerging platform for improving oral bioavailability of lipophilic drugs." *International Journal of Pharmaceutical Investigation*. Wolters Kluwer Medknow Publications. Doi: 10.4103/2230-973X.167661.

52. Zhuang, C. Yang, N. Li, M. Wang, X. N. Zhang, W. S. Pan, J. J. Peng, Y. S. Pan, and X. Tang. 2010. "Preparation and characterization of vinpocetine loaded nanostructured lipid carriers (NLC) for improved oral bioavailability." *International Journal of Pharmaceutics* 394 (1–2): 179–85. Doi: 10.1016/j.ijpharm.2010.05.005.

53. Tiwari, R., and K. Pathak. 2011. "Nanostructured lipid carrier versus solid lipid nanoparticles of simvastatin: Comparative analysis of characteristics, pharmacokinetics and tissue uptake." *International Journal of Pharmaceutics* 415 (1–2). 232–43. Doi: 10.1016/j.ijpharm.2011.05.044.

54. Haider, M., S. M. Abdin, L. Kamal, and G. Orive. 2020. "Nanostructured lipid carriers for delivery of chemotherapeutics: A review." *Pharmaceutics*. MDPI AG. Doi: 10.3390/pharmaceutics12030288.

55. Tsai, H. C., and T. Imae. 2011. "Fabrication of dendrimers toward biological application." In *Progress in Molecular Biology and Translational Science*, 104: 101–40 Elsevier B.V. Doi:10.1016/B978-0-12-416020-0.00003-6.

56. Juthi Singh, B., S. Saini, S. Lohan, and S. Beg. 2017. "Systematic development of nanocarriers employing quality by design paradigms." In *Nanotechnology-Based Approaches for Targeting and Delivery of Drugs and Genes*, 110–48. Elsevier Inc. Doi: 10.1016/B978-0-12-809717-5.00003-8.

57. Cheng, Y., L. Zhao, Y. Li, and T. Xu. 2011. "Design of biocompatible dendrimers for cancer diagnosis and therapy: Current status and future perspectives." *Chemical Society Reviews* 40 (5): 2673–703. The Royal Society of Chemistry. Doi: 10.1039/c0cs00097c.

58. Jain, V., and P. V. Bharatam. 2014. "Pharmacoinformatic approaches to understand complexation of dendrimeric nanoparticles with drugs." *Nanoscale*. The Royal Society of Chemistry. Doi: 10.1039/c3nr05400d.

59. Yellepeddi, V. K., and H. Ghandehari. 2016. "Poly(Amido Amine) dendrimers in oral delivery." *Tissue Barriers*. Taylor and Francis Inc. Doi: 10.1080/21688370.2016.1173773.

60. Madaan, K., S. Kumar, N. Poonia, V. Lather, and D. Pandita. 2014. "Dendrimers in drug delivery and targeting: Drug-dendrimer interactions and toxicity issues." *Journal of Pharmacy and Bioallied Sciences* 6 (3): 139–50. Medknow Publications. Doi: 10.4103/0975-7406.130965.

61. Rai, S., R. Paliwal, B. Vaidya, K. Khatri, A. Goyal, P. Gupta, and S. P. Vyas. 2008. "Targeted delivery of doxorubicin via estrone-appended liposomes." *Journal of Drug Targeting* 16 (6). Taylor & Francis: 455–63. Doi: 10.1080/10611860802088481.

62. Mishra, P. K., A. Gulbake, A. Jain, S. P. Vyas, and S. K. Jain. 2009. "Targeted delivery of an anti-cancer agent via steroid coupled liposomes." *Drug Delivery* 16 (8): 437–47. Doi: 10.3109/10717540903271391.

63. Kumar, P., A. Gulbake, and S. K. Jain. 2012. "Liposomes a vesicular nanocarrier: Potential advancements in cancer chemotherapy." *Critical Reviews in Therapeutic Drug Carrier Systems* 29 (5): 355–419. Doi: 10.1615/CritRevTherDrugCarrierSyst.v29. i5.10.

64. Suk, J. S., Q. Xu, N. Kim, J. Hanes, and L. M. Ensign. 2016. "PEGylation as a strategy for improving nanoparticle-based drug and gene delivery." *Advanced Drug Delivery Reviews*. Elsevier B.V. Doi: 10.1016/j.addr.2015.09.012.

65. Ag Seleci, D., M. Seleci, J. G. Walter, F. Stahl, and T. Scheper. 2016. "Niosomes as nanoparticular drug carriers: Fundamentals and recent applications." *Journal of Nanomaterials*. Hindawi Limited. Doi: 10.1155/2016/7372306.

66. Rahimi, S., S. Khoee, and M. Ghandi. 2019. "Preparation and characterization of rod-like chitosan–quinoline nanoparticles as PH-responsive nanocarriers for quercetin delivery." *International Journal of Biological Macromolecules* 128 (May): 279–89. Elsevier B.V. Doi: 10.1016/j.ijbiomac.2019.01.137.

67. Khoee S., and M. Yaghoobian. 2017, January 1."Niosomes: A novel approach in modern drug delivery systems." *Nanostructures for Drug Delivery*: 207–237. Elsevier.

68. Uchegbu I. F., and A. T. Florence. 1995. Non-ionic surfactant vesicle (niosomes): Physical and pharmaceutical chemistry. *Advances in Colloid and Interface Science*. 58(1): 1–55.

69. Bartelds, R., M. H. Nematollahi, T. Pols, M. C. A. Stuart, A. Pardakhty, G. Asadikaram, and B. Poolman. 2018. *Niosomes, an Alternative for Liposomal Delivery*. Edited by Z. Leonenko. *PLOS One* 13 (4): e0194179. Public Library of Science. Doi: 10.1371/journal. pone.0194179.

70. Sinico C., and A. M. Fadda. 2009 Aug 1. "Vesicular carriers for dermal drug delivery." *Expert Opinion on Drug Delivery* 6(8):813–25.

71. Ainbinder, D., B. Godin, and E. Touitou. 2016. "Ethosomes: Enhanced delivery of drugs to and across the skin." In *Percutaneous Penetration Enhancers Chemical Methods in Penetration Enhancement: Nanocarriers*, 61–75. Springer Berlin Heidelberg. Doi: 10.1007/978-3-662-47862-2_4.

72. Verma, P., and K. Pathak. 2010. "Therapeutic and cosmeceutical potential of ethosomes: An overview." *Journal of Advanced Pharmaceutical Technology and Research*. Doi: 10.4103/0110-5558.72415.

73. Sankar, V., S. Ramesh, and K. Siram. 2018. "Ethosomes: An exciting and promising alcoholic carrier system for treating androgenic alopecia." In *Alopecia*. InTech. Doi: 10.5772/intechopen.79807.

74. Paliwal, R., S. R. Paliwal, N. Mishra, A. Mehta, and S. P. Vyas. 2009. "Engineered chylomicron mimicking carrier emulsome for lymph targeted oral delivery of methotrexate." *International Journal of Pharmaceutics* 380 (1–2). Elsevier: 181–88. Doi: 10.1016/j.ijpharm.2009.06.026.

75. Kumar, R., Nirmala Seth, and S. L. Hari kumar. 2013. "Emulsomes: An emerging vesicular drug delivery system." *Journal of Drug Delivery and Therapeutics* 3 (6). Society of Pharmaceutical Tecnocrats: 133. Doi: 10.22270/jddt.v3i6.665.

76. Yang, S. T., J. Luo, Q. Zhou, and H. Wang. 2012. "Pharmacokinetics, metabolism and toxicity of carbon nanotubes for bio-medical purposes." *Theranostics*. Ivyspring International Publisher. Doi: 10.7150/thno.3618.

77. Yamashita, T., K. Yamashita, H. Nabeshi, T. Yoshikawa, Y. Yoshioka, S. i. Tsunoda, and Y. Tsutsumi. 2012. "Carbon nanomaterials: Efficacy and safety for nanomedicine." *Materials*. Multidisciplinary Digital Publishing Institute (MDPI). Doi: 10.3390/ma5020350.

78. He, H., L. Ai Pham-Huy, P. Dramou, D. Xiao, P. Zuo, and C. Pham-Huy. 2013. "Carbon nanotubes: Applications in pharmacy and medicine." *BioMed Research International* 2013. Doi: 10.1155/2013/578290.

79. Jiang, L., T. Liu, H. He, L. A. Pham-Huy, L. Li, C. Pham-Huy, and D. Xiao. 2012. "Adsorption behavior of pazufloxacin mesilate on amino-functionalized carbon nanotubes." *Journal of Nanoscience and Nanotechnology* 12 (9): 7271–79. Doi: 10.1166/jnn.2012.6562.

80. Ali-Boucetta, H., A. Nunes, R. Sainz, M. A. Herrero, B. Tian, M. Prato, A. Bianco, and K. Kostarelos. 2013. "Asbestos-like pathogenicity of long carbon nanotubes alleviated by chemical functionalization." *AngewandteChemie - International Edition* 52 (8): 2274–78. Doi: 10.1002/anie.201207664.

81. Alphandéry, E. 2020b. "Natural metallic nanoparticles for application in nano-oncology." *International Journal of Molecular Sciences*. MDPI AG. Doi: 10.3390/ijms21124412.

82. AbdalDayem, A., M. Hossain, S. Lee, K. Kim, S. Saha, G.-M. Yang, H. Choi, and S.-G. Cho. 2017. "The role of reactive oxygen species (ROS) in the biological activities of metallic nanoparticles." *International Journal of Molecular Sciences* 18 (1): 120. MDPI AG. Doi: 10.3390/ijms18010120.

83. Singh, P., S. Pandit, V. R.S.S. Mokkapati, A. Garg, V. Ravikumar, and I. Mijakovic. 2018. "Gold nanoparticles in diagnostics and therapeutics for human cancer." *International Journal of Molecular Sciences*. MDPI AG. Doi: 10.3390/ijms19071979.

84. Chen, I. H., Y. F. Chen, J. H. Liou, J. T. Lai, C. C. Hsu, N. Y. Wang, and J. S. Jan. 2019. "Green synthesis of gold nanoparticle/gelatin/protein nanogels with enhanced bioluminescence/biofluorescence." *Materials Science and Engineering C* 105 (December): 110101. Elsevier Ltd. Doi: 10.1016/j.msec.2019.110101.

85. Wang, L., J. Xu, Y. Yan, H. Liu, T. Karunakaran, and F. Li. 2019. "Green synthesis of gold nanoparticles from scutellariabarbata and its anticancer activity in pancreatic cancer cell (PANC-1)." *Artificial Cells, Nanomedicine and Biotechnology* 47 (1): 1617–27. Taylor and Francis Ltd. Doi: 10.1080/21691401.2019.1594862.

86. Sojinrin, T., K. Liu, K. Wang, D. Cui, H. J. Byrne, F. Curtin, and F. Tian. 2019. "Developing gold nanoparticles-conjugated aflatoxin B1 antifungal strips." *International Journal of Molecular Sciences* 20 (24). MDPI AG. Doi: 10.3390/ijms20246260.

87. Tan, P., H. S. Li, J. Wang, and S. C. B. Gopinath. 2020. "Silver nanoparticle in biosensor and bioimaging: Clinical perspectives." *Biotechnology and Applied Biochemistry*, October. Blackwell Publishing Ltd, bab.2045. Doi: 10.1002/bab.2045.

88. Tăbăran, A.-F., C. Tudor Matea, T. Mocan, A. Tăbăran, M. Mihaiu, C. Iancu, and L. Mocan. 2020. "Silver nanoparticles for the therapy of tuberculosis." *International Journal of Nanomedicine* 15 (March): 2231–58. Dove Medical Press Ltd. Doi: 10.2147/IJN.S241183.

89. Feroze, N., B. Arshad, M. Younas, M. I. Afridi, S. Saqib, and A. Ayaz. 2020. "Fungal mediated synthesis of silver nanoparticles and evaluation of antibacterial activity." *Microscopy Research and Technique* 83 (1): 72–80. Wiley-Liss Inc. Doi: 10.1002/jemt.23390.

90. Li, Z., Y. Mu, C. Peng, M. F. Lavin, H. Shao, and Z. Du. 2020. "Understanding the mechanisms of silica nanoparticles for nanomedicine." *Wiley Interdisciplinary Reviews: Nanomedicine and Nanobiotechnology.* Wiley-Blackwell. Doi: 10.1002/wnan.1658.

91. Chen, L., J. Liu, Y. Zhang, G. Zhang, Y. Kang, A. Chen, X. Feng, and L. Shao. 2018. "The toxicity of silica nanoparticles to the immune system." *Nanomedicine.* Future Medicine Ltd. Doi: 10.2217/nnm-2018-0076.

92. Tang, L., and J. Cheng. 2013. "Nonporous silica nanoparticles for nanomedicine application." *Nano Today.* Elsevier B.V. Doi: 10.1016/j.nantod.2013.04.007.

93. Li, X., G. He, H. Jin, J. Tao, X. Li, C. Zhai, Y. Luo, and X. Liu. 2019. "Dual-therapeutics-loaded mesoporous silica nanoparticles applied for breast tumor therapy." *ACS Applied Materials and Interfaces* 11 (50): 46497–503. American Chemical Society. Doi: 10.1021/acsami.9b16270.

94. Alphandéry, E. 2020c. "Bio-synthesized iron oxide nanoparticles for cancer treatment." *International Journal of Pharmaceutics.* Elsevier B.V. Doi: 10.1016/j.ijpharm.2020.119472.

95. Alphandéry, E. 2020a. "Iron oxide nanoparticles for therapeutic applications." *Drug Discovery Today.* Elsevier Ltd. Doi: 10.1016/j.drudis.2019.09.020.

96. Filippi, M., D. V. Nguyen, F. Garello, F. Perton, S. Bégin-Colin, D. Felder-Flesch, L. Power, and A. Scherberich. 2019. "Metronidazole-functionalized iron oxide nanoparticles for molecular detection of hypoxic tissues." *Nanoscale* 11 (46): 22559–74. Royal Society of Chemistry. Doi: 10.1039/c9nr08436c.

97. Ghosh, S., I. Ghosh, M. Chakrabarti, and A. Mukherjee. 2020. "Genotoxicity and biocompatibility of superparamagnetic iron oxide nanoparticles: Influence of surface modification on biodistribution, retention, DNA damage and oxidative stress." *Food and Chemical Toxicology* 136 (February): 110989. Elsevier Ltd. Doi: 10.1016/j.fct.2019.110989.

Index

Note: **Bold** page numbers refer to tables and *italic* page numbers refer to figures.